Inorganic Reactions and Methods

Volume 12A

Inorganic Reactions and Methods

Distribution: VCH Verlagsgesellschaft mbH, P.O. Box 1260/1280, D-6940 Weinheim, Federal Republic of Germany
USA and Canada: VCH Publishers, Inc., 303 N.W. 12th Avenue, Deerfield Beach, FL 33442-1705, USA

Inorganic Reactions and Methods

Volume 12A

The Formation of Bonds to Elements of Group IVB (C, Si, Ge, Sn, Pb) (Part 4)

Founding Editor

J.J. Zuckerman

Editor

A.P. Hagen

VCH

Library of Congress Cataloging-in-Publication Data

Inorganic reactions and methods.

Includes bibliographies and indexes.
Contents: v. 1. The formation of bonds to hydrogen - pt. 2, v. 2. The formation of the bond to hydrogen - v. 15. Electron-transfer and electrochemical reactions; photochemical and other energized reactions.
1. Chemical reaction, Conditions and laws of - Collected works. 2. Chemistry, Inorganic - Synthesis - Collected works. I. Zuckerman, Jerry J.
QD501.I623 1987 541.3'9 85-15627
ISBN 0-89573-250-0 (set)

Printed in the United States of America.

ISBN 0-89573-262-9 VCH Publishers
ISBN 3-527-26270-9 VCH Verlagsgesellschaft

Contents of Volume 12A

How to Use this Book

1. Organization of Subject Matter

1.1. Logic of Subdivision and Add-On Chapters

This volume is part of a series that describes all of inorganic reaction chemistry. The contents are subdivided systematically and so are the contents of the entire series: Using the periodic system as a correlative device, it is shown how bonds between pairs of elements can be made. Treatment begins with hydrogen making a bond to itself in H_2 and proceeds according to the periodic table with the bonds formed by hydrogen to the halogens, the groups headed by oxygen, nitrogen, carbon, boron, beryllium and lithium, to the transition and inner-transition metals and to the members of group zero. Next it is considered how the halogens form bonds among themselves and then to the elements of the main groups VI to I, the transition and inner-transition metals and the zero-group gases. The process repeats itself with descriptions of the members of each successive periodic group making bonds to all the remaining elements not yet treated until group zero is reached. At this point all actual as well as possible combinations have been covered.

The focus is on the primary formation of bonds, not on subsequent reactions of the products to form other bonds. These latter reactions are covered at the places where the formation of those bonds is described. Reactions in which atoms merely change their oxidation states are not included, nor are reactions in which the same pairs of elements come together again in the product (for example, in metatheses or redistributions). Physical and spectroscopic properties or structural details of the products are not covered by the reaction volumes which are concerned with synthetic utility based on yield, economy of ingredients, purity of product, specificity, etc. The preparation of short-lived transient species is not described.

While in principle the systematization described above could suffice to deal with all the relevant material, there are other topics that inorganic chemists customarily identify as being useful in organizing reaction information and that do not fit into the scheme. These topics are the subject of eight additional chapters constituting the last four books of the series. These chapters are systematic only within their own confines. Their inclusion is based on the best judgment of the Editorial Advisory Board as to what would be most useful currently as well as effective in guiding the future of inorganic reaction chemistry.

1.2. Use of Decimal Section Numbers

The organization of the material is readily apparent through the use of numbers and headings. Chapters are broken down into divisions, sections and subsections, which have short descriptive headings and are numbered according to the following scheme:

1. Major Heading
1.1. Chapter Heading
1.1.1. Division Heading
1.1.1.1. Section Heading
1.1.1.1.1. Subsection Heading

Further subdivision of a five-digit "slice" utilizes lower-case Roman numerals in parentheses: (i), (ii), (iii), etc. It is often found that as a consequence of the organization, cognate material is located in different chapters but in similarly numbered pieces, i.e., in parallel sections. Section numbers, rather than page numbers, are the key by which the material is accessed through the various indexes.

1.3. Building of Headings

1.3.1. Headings Forming Part of a Sentence

Most headings are sentence-fragment phrases which constitute sentences when combined. Usually a period signifies the end of a combined sentence. In order to reconstitute the context in which a heading is to be read, superior-rank titles are printed as running heads on each page. When the sentences are put together from their constituent parts, they describe the contents of the piece at hand. For an example, see 2.3 below.

1.3.2. Headings Forming Part of an Enumeration

For some material it is not useful to construct title sentences as described above. In these cases hierarchical lists, in which the topics are enumerated, are more appropriate. To inform the reader fully about the nature of the material being described, the headings of connected sections that are superior in hierarchy always occur as running heads at the top of each page.

2. Access and Reference Tools

2.1. Plan of the Entire Series (Front Endpaper)

Printed on the inside of the front cover is a list, compiled from all 18 reaction volumes, of the major and chapter headings, that is, all headings that are preceded by a one- or two-digit decimal section number. This list shows in which volumes the headings occur and highlights the contents of the volume that is at hand by means of a gray tint.

2.2. Contents of the Volume at Hand

All the headings, down to the title of the smallest decimal-numbered subsection, are listed in the detailed table of contents of each volume. For each heading the table of contents shows the decimal section number by which it is preceded and the number of the page on which it is found. Beside the decimal section numbers, successive indentations reveal the hierarchy of the sections and thereby facilitate the comprehension of the phrase (or of the enumerative sequence) to which the headings of hierarchically successive sections combine. To reconstitute the context in which the heading of a section must be read to become meaningful, relevant headings of sections superior in hierarchy are repeated at the top of every page of the table of contents. The repetitive occurrences of these headings is indicated by the fact that position and page numbers are omitted.

2.3. Running Heads

In order to indicate the hierarchical position of a section, the top of every page of text shows the headings of up to three connected sections that are superior in hierarchy. These running heads provide the context within which the title of the section under discussion becomes meaningful. As an example, the page of Volume 1 on which section 1.4.9.1.3 "in the Production of Methanol" starts, carries the running heads:

1.4. The Formation of Bonds between Hydrogen and O,S,Se,Te,Po
1.4.9. by Industrial Processes
1.4.9.1. Involving Oxygen Compounds

whereby the phrase "in the Production of Methanol" is put into its proper perspective.

2.4. List of Abbreviations

Preceding the indexes there is a list of those abbreviations that are frequently used in the text of the volume at hand or in companion volumes. This list varies somewhat in length from volume to volume; that is, it becomes more comprehensive as new volumes are published.

Abbreviations that are used incidentally or have no general applicability are not included in the list but are explained at the place of occurrence in the text.

2.5. Author Index

The author index is compiled by computer from the lists of references. Thus it tells whose publications are cited and in that respect is comprehensive. It is not a list of authors, beyond those cited in the references, whose results are reported in the text. However, as the references cited are leading ones, consulting them, along with the use of appropriate works of the secondary literature, will rapidly lead to the complete literature related to any particular subject covered.

Each entry in the author index refers the user to the appropriate section number.

2.6. Compound Index

The compound index lists individual, fully specified compositions of matter that are mentioned in the text. It is an index of empirical formulas, ordered according to the following system: the elements within a given formula occur in alphabetical sequence except for C, or C and H if present, which always come first. Thus, the empirical formula

for	$Ti(SO_4)_2$ is	O_8S_2Ti
	$BH_3 \cdot NH_3$	BH_6N
	Be_2CO_3	CBe_2O_3
	$CsHBr_2$	Br_2CsH
	$Al(HCO_3)_3$	$C_3H_3AlO_9$

The formulas themselves are ordered alphanumerically without exception; that is, the formulas listed above follow each other in the sequence BH_6N, Br_2CsH, CBe_2O_3, $C_3H_3AlO_9$, O_8S_2Ti.

A compound index constructed by these principles tells whether a given compound is present. It cannot provide information about compound classes, for example, all aluminum derivatives or all compounds containing phosphorus.

In order to open this route of access as well, the compound index is augmented by successively permuted versions of all empirical formulas. Thus the number of appearances that an empirical formula makes in the compound index is equal to the number of elements it contains. As an example, $C_3H_3AlO_9$, mentioned above, will appear as such and, at the appropriate positions in the alphanumeric sequence, as $H_3AlO_9{*}C_3$, $AlO_9{*}C_3H_3$ and $O_9{*}C_3H_3Al$. The asterisk identifies a permuted formula and allows the original formula to be reconstructed by shifting to the front the elements that follow the asterisk.

Each nonpermuted formula is followed by linerarized structural formulas that indicate how the elements are combined in groups. They reveal the connectivity of the compounds underlying each empirical formula and serve to distinguish substances which are identical in composition but differ in the arrangement of elements (isomers). As an example, the empirical formula $C_4H_{10}O$ might be followed by the linearized structural formulas $(CH_3CH_2)_2O$, $CH_3(CH_2)_2OCH_3$, $(CH_3)_2CHOCH_3$, $CH_3(CH_2)_3OH$, $(CH_3)_2CHCH_2OH$ and $CH_3CH_2(CH_3)CHOH$ to identify the various ethers and alcohols that have the element count $C_4H_{10}O$.

Each linearized structural formula is followed in a third column by keywords describing the context in which it is discussed and by the number(s) of the section(s) in which it occurs.

2.7. Subject Index

The subject index provides access to the text by way of methods, techniques, reaction types, apparatus, effects and other phenomena. Also, it lists compound classes such as organotin compounds or rare-earth hydrides which cannot be expressed by the empirical formulas of the compound index.

For multiple entries, additional keywords indicate contexts and thereby avoid the retrieval of information that is irrelevant to the user's need.

Again, section numbers are used to direct the reader to those positions in the book where substantial information is to be found.

2.8. Periodic Table (Back Endpaper)

Reference to periodic groups avoids cumbersome enumerations. Section headings in the series employ the nomenclature.

Unfortunately, however, there is at the present time no general agreement on group designations. In fact, the scheme that is most widely used (combining a group number with the letters A and B) is accompanied by two mutually contradictory interpretations. Thus, titanium may be a group IVA or group IVB element depending on the school to which one adheres or the part of the world in which one resides.

In order to clarify the situation for the purposes of the series, a suitable labeled periodic table is printed on the inside back cover of each volume. All references to periodic group designations in the series refer to this scheme.

Preface to the Series

Inorganic Reactions and Methods constitutes a closed-end series of books designed to present the state of the art of synthetic inorganic chemistry in an unprecedented manner. So far, access to knowledge in inorganic chemistry has been provided almost exclusively using the elements or classes of compounds as starting points. In the first 18 volumes of **Inorganic Reactions and Methods**, it is bond formation and type of reaction that form the basis of classification.

This new route of access has required new approaches. Rather than sewing together a collection of review articles, a framework has had to be designed that reflects the creative potential of the science and is hoped to stimulate its further development by identifying areas of research that are most likely to be fruitful.

The reaction volumes describe methods by which bonds between the elements can be formed. The work opens with hydrogen making a bond to itself in H_2 and proceeds through the formation of bonds between hydrogen and the halogens, the groups headed by oxygen, nitrogen, carbon, boron, beryllium and lithium to the formation of bonds between hydrogen and the transition and inner-transition metals and elements of group zero. This pattern is repeated across the periodic system until all possible combinations of the elements have been treated. This plan allows most reaction topics to be included in the sequence where appropriate. Reaction types that do not arise from the systematics of the plan are brought together in the concluding chapters on oxidative addition and reductive elimination, insertions and their reverse, electron transfer and electrochemistry, photochemical and other energized reactions, oligomerization and polymerization, inorganic and bioinorganic catalysis and the formation of intercalation compounds and ceramics.

The project has engaged a large number of the most able inorganic chemists as Editorial Advisors creating overall policy, as Editorial Consultants designing detailed plans for the subsections of the work, and as authors whose expertise has been crucial for the quality of the treatment. The conception of the series and the details of its technical realization were the subject of careful planning for several years. The distinguished chemists who form the Editorial Advisory Board have devoted themselves to this exercise, reflecting the great importance of the project.

It was a consequence of the systematics of the overall plan that publication of a volume had to await delivery of its very last contribution. Thus was the defect side of the genius of the system revealed, as the excruciating process of extracting the rate-limiting manuscripts began. Intense editorial effort was required in order to bring forth the work in a timely way. The production process had to be designed so that the insertion of new material was possible up to the very last stage, enabling authors to update their pieces with the latest developments. The publisher supported the cost of a computerized bibliographic search of the literature and a second one for updating.

Each contribution has been subjected to an intensive process of scientific and linguistic editing in order to homogenize the numerous individual pieces, as well as to provide the highest practicable density of information. This had several important consequences. First, virtually all semblances of the authors' individual styles have been excised. Second, it was learned during the editorial process that greater economy of language could be achieved by dropping conventionally employed modifiers (such as very) and eliminating italics used for emphasis, quotation marks around nonquoted words, or parentheses around phrases, the result being a gain in clarity and readability. Because the series focuses on the chemistry rather than the chemical literature, the need to tell who has reported what, how and when can be considered of secondary importance. This has made it possible to bring all sentences describing experiments into the present tense. Information on who published what is still to be found in the reference lists. A further consequence is that authors have been burdened neither with identifying leading practitioners, nor with attributing priority for discovery, a job that taxes even the talents of professional historians of science. The authors' task then devolved to one of describing inorganic chemical reactions, with emphasis on synthetic utility, yield, economy, availability of starting materials, purity of product, specificity, side reactions, etc.

The elimination of the names of people from the text is by far the most controversial feature. Chemistry is plagued by the use of nondescriptive names in place of more expository terms. We have everything from Abegg's rule, Adkin's catalyst, Admiralty brass, Alfven number, the Amadori rearrangement and Adurssov oxidation to the Zdanovskii law, Zeeman effect, Zincke cleavage and Zinin reduction. Even well-practiced chemists cannot define these terms precisely except for their own areas of specialty, and no single source exists to serve as a guide. Despite these arguments, the attempt to replace names of people by more descriptive phrases was met in many cases by a warmly negative reaction by our colleague authors, notwithstanding the obvious improvements wrought in terms of lucidity, freedom from obscurity and obfuscation and, especially, ease of access to information by the outsider or student.

Further steps toward universality are taken by the replacement of element and compound names wherever possible by symbols and formulas, and by adding to data in older units their recalculated SI equivalents. The usefulness of the reference sections has been increased by giving journal-title abbreviations according to the *Chemical Abstracts Service Source Index*, by listing in each reference all of its authors and by accompanying references to patents and journals that may be difficult to access by their *Chemical Abstracts* citations. Mathematical signs and common abbreviations are employed to help condense prose and a glossary of the latter is provided in each volume. Dangerous or potentially dangerous procedures are highlighted in safety notes printed in boldface type.

The organization of the material should become readily apparent from an examination of the headings listed in the table of contents. Combining the words constituting the headings, starting with the major heading (one digit) and continuing through the major chapter heading (two digits), division heading (three digits), section heading (four digits) to the subsection heading (five digits), reveals at once the subject of a "slice" of the plan. Each slice is a self-contained unit. It includes its own list of references and provides definitions of unusual terms that may be used in it. The reader, therefore, through the table of contents alone, can in most instances quickly reach the desired material and derive the information wanted.

In addition there is for each volume an author index (derived from the lists of references) and a subject index that lists compound classes, methods, techniques, apparatus, effects and other phenomena. An index of empirical formulas is also provided. Here in each formula the element symbols are arranged in alphabetical order except that C, or C and H if present, always come first. Moreover, each empirical formula is permuted successively. Each permuted formula is placed in its alphabetical position and cross referenced to the original formula. Therefore, the number of appearances that an empirical formula makes in the index equals the number of its elements. By this procedure all compounds containing a given element come together in one place in the index. Each original empirical formula is followed by a linearized structural formula and keywords describing the context in which the compound is discussed. All indexes refer the user to subsection rather than page number.

Because the choice of designations of groups in the periodic table is currently in a state of flux, it was decided to conform to the practice of several leading inorganic texts. To avoid confusion an appropriately labeled periodic table is printed on the back endpaper.

From the nature of the work it is obvious that probably not more than two persons will ever read it entire: myself and the publisher's copy editor, Dr. Lindsay S. Ardwin. She, as well as Ms. Mary C. Stradner, Production Manager of VCH Publishers, are to be thanked for their unflagging devotion to the highest editorial standards. The original conception for this series was the brainchild of Dr. Hans F. Ebel, Director of the Editorial Department of VCH Verlagsgesellschaft in Weinheim, Federal Republic of Germany, who also played midwife at the birth of the plan of these reaction volumes with my former mentor, Professor Alan G. MacDiarmid of the University of Pennsylvania, and me in attendance, during the Anaheim, California, American Chemical Society Meeting in the Spring of 1978. Much of what has finally emerged is the product of the inventiveness and imagination of Professor Helmut Grünewald, President of VCH Verlagsgesellschaft. It is a pleasure to acknowledge that I have learned much from him during the course of our association. Ms. Nancy L. Burnett is to be thanked for typing everything that had to do with the series from its inception to this time. Directing an operation of this magnitude without her help would have been unimaginable. My wife Rose stood by with good cheer while two rooms of our home filled up with 10,000 manuscript pages, their copies and attendant correspondence.

Finally, and most important, an enormous debt of gratitude toward all our authors is to be recorded. These experts were asked to prepare brief summaries of their knowledge, ordered in logical sequence by our plan. In addition, they often involved themselves in improving the original conception by recommending further refinements and elaborations. The plan of the work as it is being published can truly be said to be the product of the labors of the advisors and consultants on the editorial side as well as the many, many authors who were able to augment more general knowledge with their own detailed information and ideas. Because of the unusually strict requirements of the series, authors had not only to compose their pieces to fit within narrowly constrained limits of space, format and scope, but after delivery to a short deadline were expected to stand by while an intrusive editorial process homogenized their own prose styles out of existence and shrank the length of their expositions. These long-suffering colleagues had then to endure the wait for the very last manuscript scheduled for their volume to be delivered so that their work could be published, often after a further diligent search of the literature to insure that the latest discoveries were being cited and that claims for facts now proved false were eliminated. To these co-workers (270 for the reaction volumes alone), from whom so much was demanded but who continued to place their knowledge and talents unstintingly at the disposal of the project, we dedicate this series.

J. J. ZUCKERMAN
Norman, Oklahoma
July 4, 1985

The scientific community is appreciative of the JJZ vision for a systematic inorganic chemistry. Many of the contributions had been edited prior to his death; therefore, his precise syntax will remain an important part of the series.

A.P. HAGEN
Norman, Oklahoma
July 1, 1990

Editorial Consultants to the Series

Contributors to Volume 12A

Dr. S. Aime
University of Turino
Via Pietro Guiria, 7
10125 Torino, Italy
(*Sections* 5.8.2.7.1–5.8.2.7.3, 5.8.2.7.5)

Dr. B. Akermark
Department of Organic Chemistry
Royal Institute of Technology
Stockholm 70, Sweden
(*Sections* 5.8.2.3.8., 5.8.2.4.4)

Prof. J.-E. Backvall
P. O. Box 532
Uppsala University
S-75121 Uppsala Sweden
(*Section* 5.8.2.3.8)

Mr. R. E. Bachman
Department of Chemistry
Rice University
Houston, TX 77251
(*Section* 5.8.2.2.1)

Prof. D. E. Bergbreiter
Texas A&M University
Department of Chemistry
College Station, TX 77843
(*Section* 5.8.2.3.3)

Prof. W. E. Billups
Department of Chemistry
Rice University
Houston, TX 77251
(*Section* 5.8.2.2.1)

Prof. A. J. Birch
Australian National University
Department of Chemistry
Canberra, Australia ACT 2601
(*Sections* 5.8.2.4.2, 5.8.2.4.3)

Prof. M. Brookhart
Department of Chemistry
University of North Carolina
Chapel Hill, NC 27599
(*Section* 5.8.2.5.2)

Dr. Y. Chauvin
Institut Francais du Petrole
Rueil-Malmaison 92506, France
(*Section* 5.8.2.3.6)

Prof. J. W. Connolly
Department of Chemistry
The University of Missouri at Kansas City
Kansas City, MO 64110
(*Section* 5.8.2.2.6)

Prof. N. J. Cooper
Department of Chemistry
The University of Pittsburgh
Pittsburgh, PA 15260
(*Section* 5.8.2.3.11)

Dr. S. G. Davies
Dyson Perrins Laboratory
The University of Oxford
South Parks Road
Oxford OX1 3QY, England
(*Sections* 5.8.2.3.9–5.8.2.3.10)

Prof. J. Dehand
Lab. Chim. Coord.
Univ. Louis Pasteur
1 rue Blaise Pascal
F-67070 Strasbourg, France
(*Section* 5.8.2.2.3)

Prof. T. C. Flood
Department of Chemisty
University of Southern California
Los Angeles, CA 90089
(*Section* 5.8.2.3.7)

Dr. J. J. Garcia
Department of Chemistry
Princeton University
Princeton, NJ 08544
(*Sections* 5.8.2.6.1–5.8.2.6.2)

Prof. F. R. Hartley
Vice Chancellor
Cranfield Institute of Technology
Central Milton Keynes SN6 8LA, England
(*Sections* 5.8.2.3.1, 5.8.2.5.1)

Prof. R. F. Heck
Department of Chemistry
The University of Delaware
Newark, DE 19711
(*Sections* 5.8.2.3.5, 5.8.2.6.4)

Prof. L. S. Hegedus
Department of Chemistry
Colorado State University
Fort Collins, CO 80523
(*Section* 5.8.2.3.4)

Prof. J. A. Heppert
Department of Chemistry
The University of Kansas
Lawrence, KS 66045
(*Sections* 5.8.2.6.1–5.8.2.6.3)

Dr. A. T. Hutton
Department of Chemistry
University of Cape Town
Rondebosch 7700, South Africa
(*Section* 5.8.2.2.5)

Dr. S. D. Ittel
E.I. duPont de Nemours & Company
Central Research and Development Department
Wilmington, DE 19880-0328
(*Sections* 5.8.2.6.5, 5.8.2.6.6)

Prof. K. J. Klabunde
Department of Chemistry
Kansas State University
Manhattan, KS 66506
(*Sections* 5.8.2.3.2, 5.8.2.4.6, 5.8.2.5.3, 5.8.2.6.7, 5.8.2.7.7)

Prof. C. P. Lillya
Department of Chemistry
The University of Massachusetts
Amherst, MA 01003
(*Section* 5.8.2.4.1)

Mr. M. L. Milligan
Department of Chemistry
The University of Kansas
Lawrence, KS 66045
(*Section* 5.8.2.6.3)

Prof. K. N. Nicholas
The Department of Chemistry and Biochemistry
The University of Oklahoma
Norman, OK 73019
(*Section* 5.8.2.4.5)

Prof. R. Noyori
Department of Chemistry
Nagoya University
Nagoya 464, Japan
(*Section* 5.8.2.2.4)

Ms. R. Pankayatselvan
Department of Chemistry and Biochemistry
The University of Oklahoma
Norman, OK 73019
(*Section* 5.8.2.4.5)

Dr. M. Pfeffer
Lab. Chim. Coord.
Univ. Louis Pasteur
Strasbourg F-67070, France
(*Section* 5.8.2.2.3)

Prof. D. L. Reger
Department of Chemistry
University of South Carolina
Columbia, SC 29208
(*Section* 5.8.2.7.6)

Dr. G. F. Schmidt
Dow Chemical Company
Central Research Bldg. 1776
Midland, MI 48674
(*Section* 5.8.2.5.2)

Prof. M. F. Semmelhack
Department of Chemistry
Princeton University
Princeton, NJ 08544
(*Sections* 5.8.2.6.1, 5.8.2.6.2)

Mr. R. L. Thompson
Department of Chemistry
The University of Pittsburgh
Pittsburgh, PA 15260
(*Section* 5.8.2.3.11)

Prof. K. P. Vollhardt
Department of Chemistry
University of California at Berkeley
Berkeley, CA 94720
(*Section* 5.8.2.7.4)

Dr. S. Wreford
E.I. duPont deNemours, Inc.
Jackson Laboratory
Wilmington, DE 19898
(*Section* 5.8.2.4.5)

Dr. G. B. Young
Department of Chemistry
Imperial College of Science, Technology and Medicine
South Kensington, London SW7 2AY UK
(*Section* 5.8.2.2.2)

Dr. K. Zetterberg
Department of Organic Chemistry
Royal Institute of Technology
Stockholm 70, Sweden
(*Sections* 5.8.2.3.8, 5.8.2.4.4.)

5. The Formation of Bonds to Elements of Group IVB (C, Si, Ge, Sn, Pb) (Part 4)

5.8. Formation of Bonds between Elements of Groups IVB (C, Si, Ge, Sn, Pb) and Transition and Inner-Transition Metals

5.8.1. Introduction

This part begins describing the formation of carbon bonds to transition- and innertransition-metal bonds which continues in Volume 12B.

(A. P. HAGEN, ED.)

5.8.2. Formation of the Carbon-Transition and Inner Transition Metal Bond

5.8.2.1 from the Elements.

This topic is covered in Section 5.10.

5.8.2.2. from Saturated Hydrocarbons.

5.8.2.2.1. η^1 σ-Alkyl Complexes by Direct Hydrocarbon Activation.

The discovery of many examples of intramolecular C—H activation (cyclometalation reactions)[1–11] and the observation[12–15] that $[PtCl_4]^{2-}$ catalyzes H-D exchange between CH_3CO_2D and alkanes has led to the rapid development of routes to σ-alkyl complexes by direct hydrocarbon activation[14,16–23].

The coordinatively unsaturated organometallic fragment required for the activation process can be produced by the extrusion of dihydrogen from the excited state of a dihydride. The resulting reactive fragment may then insert into the C—H bonds of alkanes and cycloalkanes, as exemplified in the reaction of $(\eta^5\text{-}C_5Me_5)Ir(PMe_3)H_2$ with neopentane[24,25]:

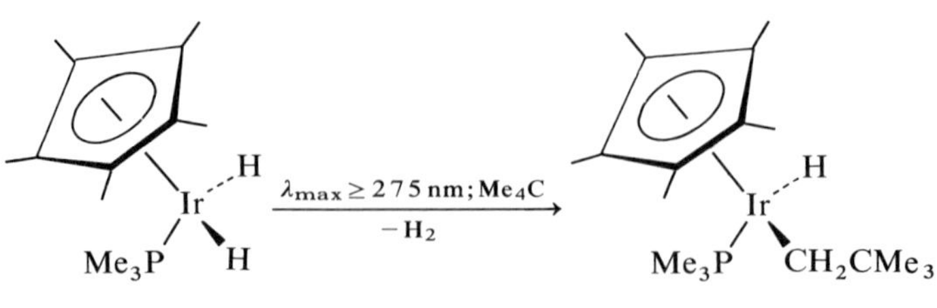

(a)

The complex produced in this reaction, as are those from other alkanes and cycloalkanes, is difficult to obtain pure. However, they can be isolated and purified with some loss of material and characterized fully using conventional spectroscopic techniques.

Photolytic expulsion of CO has been used to accomplish the same type of reaction. The coordinatively unsaturated (η^5-C_5Me_5)IrCO resulting from the photolysis of (η^5-C_5Me_5)Ir$(CO)_2$ has been used to activate neopentane and cyclohexane at RT[26]:

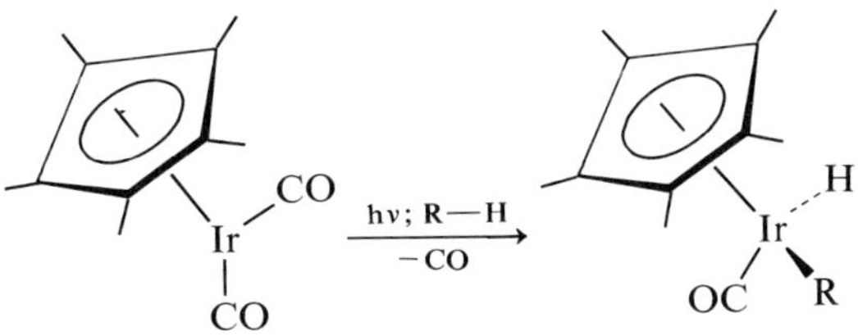

(b)

where R = neopentyl, Chx.

Both systems are capable of methane activation. When a solution of (η^5-C_5Me_5)Ir$(CO)_2$ is irradiated under methane pressure, insertion produces (η^5-C_5Me_5)IrCO(H)(CH_3) in ca. 20% yield[27]. The low kinetic barrier to alkane activation is shown by the reaction of methane even at 12 K with the species generated by photolysis of (η^5-C_5Me_5)Ir$(CO)_2$ to give the hydridomethyl species[28].

Methane activation has been achieved using the dihydride complex by taking advantage of the propensity of the hydridoalkyl complexes to undergo reductive elimination upon heating. In this way the hydridocyclohexyl complex can be converted to the hydridomethyl complex:

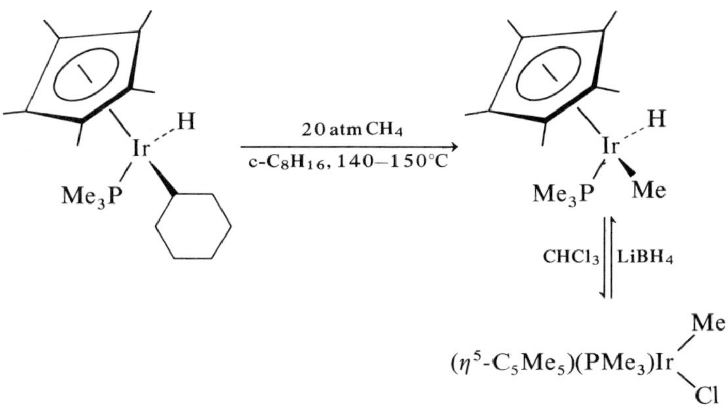

(c)

The latter can be isolated as the chloromethyl complex by treatment with $CHCl_3$. It can be assumed that the hydridocyclooctyl complex (formed from the solvent) is formed reversibly but does not accumulate because of its thermodynamic instability relative to the hydridomethyl complex[29].

σ-Alkyl complexes of Ir can be prepared indirectly by C—H activation from (η^5-C_5Me_5)(η^3-C_3H_5)HIr upon reaction with benzene or cylopropane[30].

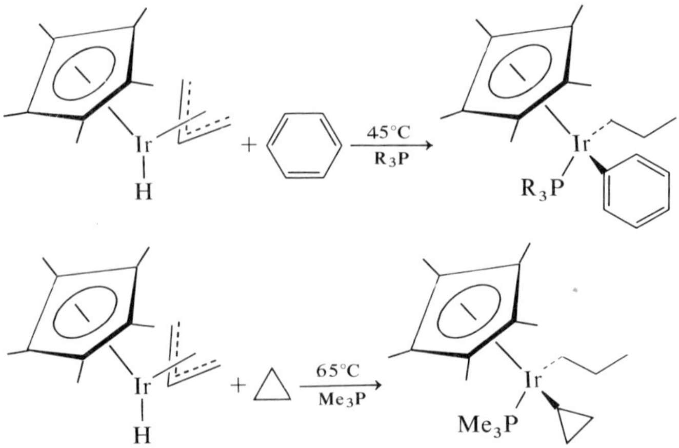

(d)

Oxidative addition of the corresponding Rh complexes to alkane C—H bonds also occurs:

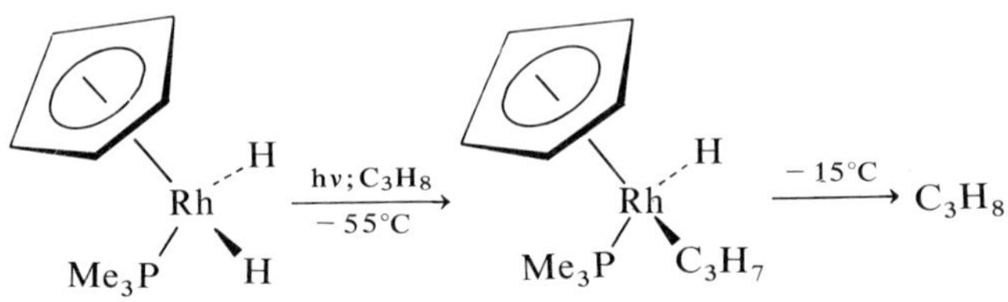

(e)

although the products formed are much less stable and revert to the starting alkane above approximately $-20°C$ [31–33].

The Rh complexes differ markedly in their selectivity toward different C—H bonds within the same molecule. Whereas the Ir complex favors primary insertion over secondary insertion only weakly, it is possible to activate primary C—H bonds selectively in n-alkanes with Rh. No example of tertiary C—H insertion has been observed with either complex. The relative rates of various alkanes and cycloalkanes, versus cyclohexane as 1.0, are presented[29] in Table 1.

Cyclopentadienyl- and pentamethylcyclopentadienyl-Re complexes are capable of intermolecular C—H activation if appropriate electron-donating ligands are attached to the metal center. Primary, cyclopropyl and methane C—H bonds are attacked but the secondary and tertiary C—H bonds are not. The cleanest results are obtained with the $(\eta^5\text{-Cp})Re(PMe_3)_3$ complex, where an adduct of cyclopropane is isolated as a stable complex by recrystallization[34].

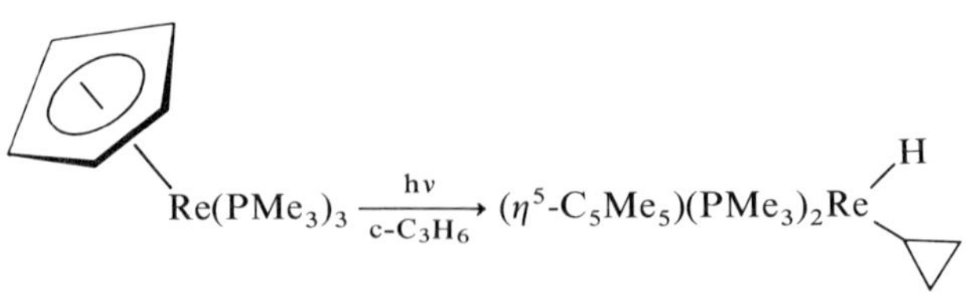

(f)

TABLE 1. Relative Rates

	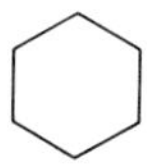	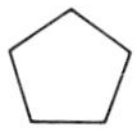			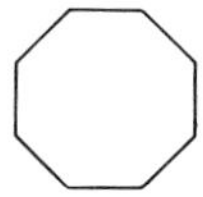	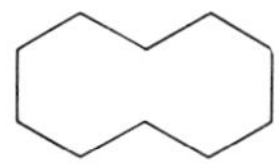
$k_{rel}(Ir)$	1.0	1.1	2.1	—	0.09	0.23
$k_{rel}(Rh)$	1.0	1.8	10.4	0.14	0.06	

Me_4C	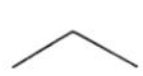	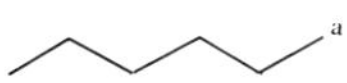 [a]	$H_3C—CH_3$	
1.14	1.5	2.7	—	—
—	2.5(1.9)	5.9(2.4)	2.0	3.6

[a] denotes a primary C—H bond.

The corresponding insertion product of n-hexane is considerably less stable, reductively eliminating n-hexane at 20°C with a halflife of approximately 12 min:

$$(\eta^5\text{-}C_5Me_5)Re(PMe_3)_3 \underset{20°C;\ 20\ min}{\overset{h\nu;\ -30°C}{\rightleftharpoons}} (\eta^5\text{-}C_5Me_5)(PMe_3)_2Re\begin{matrix} H \\ (n\text{-}C_6H_{13}) \end{matrix} \quad (g)$$

Facile C—H activation by bis(pentamethylcyclopentadienyl)-Lu complexes $Lu(\eta^5\text{-}C_5Me_5)R(R = Me, H)$ at 20–50°C has been reported[35]. Methane activation for both the Lu and the Yt complexes can be seen in the exchange:

$$(\eta^5\text{-}C_5Me_5)_2MCH_3 + {}^{13}CH_4 \rightleftharpoons (\eta^5\text{-}C_5Me_5)_2M^{13}CH_3 + CH_4 \quad (h)$$

where M = Lu, Y. These f-block metals, as strong electron-pair acceptor acids but negligible π donors, probably activate alkanes through complexes in which the C—H bond of the alkane is bound to the metal (agostic interaction).

Other multistep reactions may be initiated by intermolecular oxidative additions. For example, Ir complexes react with cyclopentane at 80°C to give the cyclopentadienyl complex in 40% yield[36].

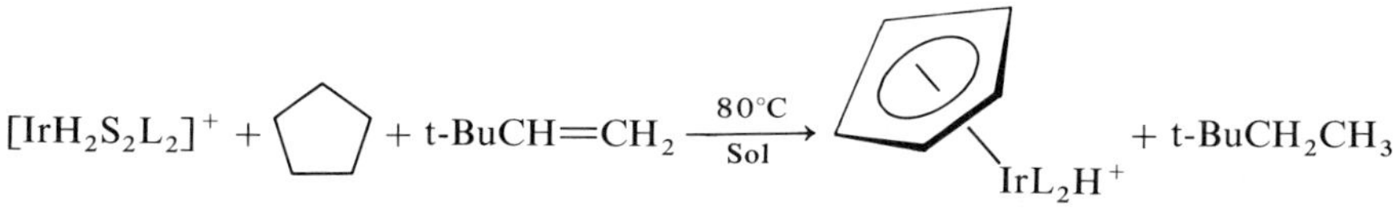

(i)

where Sol = acetone, H_2O. In this case the solvent ligands are so weakly bound to the metal that they are displaced easily by hydrocarbon molecules.

Another system involving Re and 3,3-dimethyl-1-butene as a hydrogen acceptor is used to dehydrogenate cycloalkanes[37].

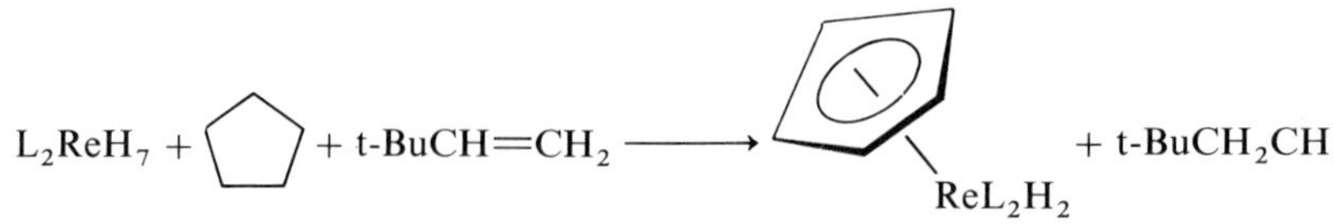

(j)

where L = PPh_3, PEt_2Ph. The initial intermediate in both of the above processes is thought to be a σ-bonded alkylmetal hydride produced when the metal inserts into the C—H bond of the cycloalkane. The thermodynamics of both reactions are favorable, with 3,3-dimethyl-1-butene acting as the hydrogen acceptor.

Oxidative addition reactions of bare metal atoms (Mn, Fe, Co, Ni, Cu, Zn, Ag, Au) with alkanes are observed upon irradiation of the metal in alkane matrices[38–40]. For example, iron atoms in their excited state insert spontaneously into the C–H bonds of methane:

$$CH_4 + Fe \underset{420\,nm}{\overset{300\,nm}{\rightleftharpoons}} CH_3FeH \qquad (k)$$

The process is photoreversible, reverting to starting materials upon photolysis at 420 nm [41]. Higher alkanes yield products resulting from C—H activation only, whereas cyclopropane reacts by attack at the carbon–carbon bond[42]. Insertion adducts of Cu, Ag and Au are susceptible to secondary photolysis, yielding[39] R + MH and RM + H.

(W. E. BILLUPS, R. E. BACHMAN)

1. P. Foley, G. M. Whitesides, *J. Am. Chem. Soc., 101*, 2732 (1979).
2. H. C. Clark, A. B. Goel, S. Goel, *Inorg. Chem., 18*, 2803 (1979).
3. S. J. Simpson, H. W. Turner, R. A. Andersen, *J. Am. Chem. Soc., 101*, 7728 (1979).
4. H. D. Empsall, E. M. Hyde, R. Markam, W. S. MacDonald, M. C. Norton, B. L. Shaw, B. Weeks, *J. Chem. Soc., Chem. Commum.*, 589 (1979).
5. N. A. Al-Salem, W. S. MacDonald, R. Markam, M. C. Norton, B. L. Shaw, *J. Chem. Soc., Dalton Trans., 59* (1980) and refs. therein.
6. R. A. Andersen, R. A. Jones, G. Wilkinson, *J. Chem. Soc., Dalton Trans.*, 446, (1978).
7. S. J. Simpson, H. W. Turner, R. A. Andersen, *Inorg. Chem., 20*, 2991 1981).
8. S. D. Chappel, D. Cole-Hamilton, *J. Chem. Soc., Chem. Commum.*, 238 (1980).
9. T. H. Tulip, D. L. Thorn, *J. Am. Chem. Soc., 103*, 2448 (1981).
10. R. D. Adams, J. P. Selegue, *Inorg. Chem., 19*, 1795 (1980).
11. H. Werner, R. Werner, *J. Organomet. Chem., 209*, C60 (1981).
12. N. F. Gol'dschlenger, M. B. Tyabin, A. E. Shilov, A. A. Shteinman, *Zh. Fiz. Khim., 43*, 2174 (1969): *Russ. J. Phys. Chem. Engl. Transl., 43*, 1222 (1969).
13. A. E. Shilov, *Activation of Saturated Hydrocarbons by Transition Metal Complexes*, Reidel, Dordrecht, 1984.
14. A. E. Shilov, A. A. Shteinman, *Coord. Chem. Rev., 24*, 97 (1977).
15. A. E. Shilov, A. A. Shteinman, *Kinet. Katal., 18*, 1129 (1977); *Chem. Abstr., 88*, 37,147 (1978).
16. R. H. Crabtree, *Chem. Rev., 85*, 245 (1985).
17. G. W. Parshall, *Acc. Chem. Res., 8*, 113 (1975).
18. E. L. Muetterties, *Chem. Soc. Rev., 11*, 283 (1982).
19. J. Halpern, *Inorg. Chim. Acta, 100*, 41 (1985).
20. R. G. Bergman, *Science, 223*, 902 (1984).
21. R. G. Bergman, *Polyhedron, 7*, 1429 (1988).
22. R. H. Crabtree, *The Organometallic Chemistry of the Transition Metals*, Wiley, New York, 1988, Chapter 12, p. 290.
23. G. W. Parshall, in *Catalysis*, Specialist Periodical Reports Vol. 1, C. Kemball ed., The Chemistry Society, London, 1977; p. 335.
24. A. H. Janowicz, R. G. Bergman, *J. Am. Chem. Soc., 104*, 352 (1982).

25. A. H. Janowicz, R. G. Bergman, *J. Am. Chem. Soc., 105*, 3929 (1983).
26. J. K. Hoyano, W. A. G. Graham, *J. Am. Chem. Soc., 104*, 3723 (1982).
27. J. K. Hoyano, A. D. McMaster, W. A. G. Graham, *J. Am. Chem. Soc., 105*, 7190 (1983).
28. A. J. Rest, I. Whitwell, W. A. G. Graham, J. Hoyano, A. D. McMaster, *Chem. Commum*, 624 (1984).
29. A. H. Janowicz, R. A. Periana, J. M. Buchanan, C. A. Kovac, J. M. Stryker, M. J. Wax, R. G. Bergman, *Pure Appl. Chem., 56*, 13 (1984).
30. W. D. McGhee, R. G. Bergman, *J. Am. Chem. Soc., 110*, 4246 (1988).
31. W. D. Jones, F. J. Feher, *J. Am. Chem Soc., 104*, 4240 (1982).
32. W. D. Jones, F. J. Feher, *J. Am. Chem. Soc., 106*, 1650 (1984).
33. W. D. Jones, F. J. Feher, *J. Am. Chem. Soc. 108*, 4814 (1986).
34. R. G. Bergman, P. F. Seidler, T. T. Wenzel, *J. Am. Chem. Soc., 107*, 4358 (1985).
35. P. L. Watson, *J. Am. Chem. Soc., 105*, 6491 (1983).
36. R. H. Crabtree, M. F. Mellea, J. M. Mihelcic, J. M. Quirk, *J. Am. Chem. Soc., 104*, 107 (1982).
37. D. Baudry, M. Ephritikhine, H. Felkin, *Chem. Commum.*,1243 (1980).
38. W. E. Billups, M. M. Konarski, R. H. Hauge, J. L. Margrave, *J. Am. Chem. Soc., 102*, 7393 (1980).
39. G. A. Ozin, D. F. Mclntosh, S. A. Mitchell, *J. Am. Chem. Soc., 103*, 1574 (1981).
40. S.-C. Chang, R. H. Hauge, W. E. Billups, J. L. Margrave, Z. H. Kafafi, *Inorg. Chem., 27*, 205 (1988).
41. G. A. Ozin, J. G. McCaffrey, *J. Am Chem. Soc., 104*, 7351 (1982).
42. Z. H. Kafafi, R. H. Hauge, L. Fredin, W. E. Billups, J. L. Margrave, *Chem. Commun.*, 1230 (1983).

5.8.2.2.2. η^1 σ-Alkyl and -Aryl Complexes by Cyclometallation on γ and More Remote Carbon–Hydrogen Bonds

Cyclometallations in which a carbon–metal bond supplants a carbon–hydrogen bond are frequent in transition-metal chemistry. They usually involve metallation on a hydrocarbyl substituent of a ligating donor other than carbon (see §5.8.2.2.3). The cyclometallation of nitrogen[1–6], phosphorus[1–4,7–11], oxygen[2,12] and sulfur[2,3,13] donor ligands is reviewed. Carbometallation of substituents on coordinated cyclopentadienyl ligands also occurs[2,9,14]. Cyclometallated ligands are also significant in the chemistry of nontransition elements[15,16].

Although this class of reaction is common, and hydrogen easily migrates from β and α positions of hydrocarbyl ligands σ-bound to transition metals, H transfer from more remote carbons has only lately become a significant route to metallacyclic compounds[17]. These reactions feature hydrocarbyl ligands in which the β-hydrogen rearrangement, at least, has been suppressed. The migrating hydrogen is eliminated with a second organic group, where available, as free hydrocarbon. Four-, five- and six-membered rings have been formed in this fashion.

Neopentyl and related derivatives of the group-VIII transition metals in low oxidation states favor such a rearrangement (sometimes called γ elimination), affording 3,3′-disubstituted metallacyclobutanes; the reaction is less available earlier transition-metal analogs[18]. This is an important route to metallacyclobutane, isolable examples of which are still relatively rare, but which are of great interest as reaction intermediates[19–24].

The earliest incidence of this rearrangement is the reaction of $Ru_2(O_2CMe)_4Cl$ with $(Me_3SiCH_2)_2Mg$ in the presence of PMe_3. This does not afford the dialkyl $RuR_2(PMe_3)_4$, as observed for the reaction with Me_2Mg, but instead yields the 1-ruthena-3-silacyclobutane.

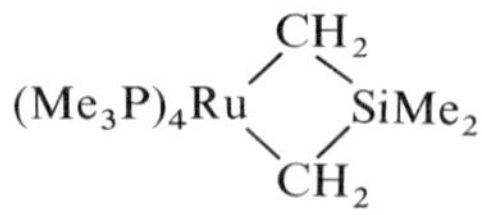

I

More labile ruthenacyclobutane, is obtained similarly using[25] $(Me_3CCH_2)_2Mg$. Broadly parallel chemistry is observed on treating $Os(O_2CMe)_2(PMe_3)_4$ with $(Me_3SiCH_2)_2Mg$ to obtain the silametallacycle fac-$(Me_3P)_4Os(CH_2SiMe_2CH_2)$. Similar reaction with $(Me_3CCH_2)_2Mg$ does not afford the carbometallacycle but leads to metallation on the methylphosphine[26].

Treating $Rh_2(O_2CMe)_4$ with $(Me_3SiCH_2)_2Mg$ gives the 1-rhoda-3-silacyclobitane, $(Me_3P)_3(Me_3SiCH_2)Rh(CH_2SiMe_2CH_2)$. These rearrangements feature γ-hydrogen migration from a coordinated trimethylsilylmethyl or neopentyl $[(CH_3)_3CCH_2]$ ligand, with evolution of 1 equiv of the corresponding alkane, Me_4E. There are no data elucidating the intimate pathway for these events[25,26].

Reaction of (1,5-COD)$Pt[CH_2C(CH_3)_3]_2$ with bulky trialkylposphines, e.g., tri-isopropyl- and tricyclohexylphosphine, leads directly to 1, 1-bis(phosphine)-3,3-dimethyl platinacyclobutane, $(R_3P)_2Pt(CH_2CMe_2CH_2)$, II, plus thermal decomposition and disproportionation products[29]. With Et_3P as substrate, on the other hand, cis-$(Et_3P)_2Pt(CH_2CMe_3)_2$ is isolated; it rearranges thermally at higher T to platinacyclobutane and neopentane. Deuterium labeling and kinetic experiments establish, unequivocally that a methyl C—H bond on the alkyl is activated to produce neopentane. This is envisaged as an oxidative addition after dissociation of one phosphine, to give a hydrido-platinum (IV) species, from which irreversible reductive elimination of neopentane generates **II**. Kinetic parameters are consistent with either initial C—H scission or subsequent C—H formation as rate-limiting step, but the latter is favored[27,28]. Ready neutral ligand dissociation is evidently a critical control on this route to platinacyclobutanes; increasing bulk of the trialkylphosphines favors the rearrangement[29], and relief of steric congestion may be a major thermodynamic motivation. Comparison of the molecular structures of $(Et_3P)_2Pt(CH_2CMe_2CH_2)$ and cis-$(Et_3P)_2Pt(CH_2CMe_3)_2$ supports this contention[30].

In a presumably analogous manner, interaction of $(Ph_3P)_2NiCl_2$ and 2 equiv Me_3CCH_2Li at low T yields a labile dineopentylnickel(II) complex that rearranges at RT to 3,3-dimethyl-l-nickelacylobutane[31] characterized by its subsequent reaction products. dissociation is evidently a critical control on this route to platinacyclobutanes; increasing

The implication of higher valent hydridoalkyl metal intermediates in these transformations is supported by the isolation of $(Me_3P)_3HIr(CH_2CMe_2CH_2)$ as the only organometallic product of the reaction of $[Ir(PMe_3)_4]Cl$ with Me_3CCH_2Li in nonpolar solvents[32]. The presumed precursor to that product, $(Me_3P)_3IrCH_2CMe_3$ cannot be detected, although under the same conditions the analogous neopentylrhodium(I) complex is isolable. In further contrast, interaction of $[Ir(PMe_3)_4]Cl$ with Me_3SiCH_2Li yields the related, less labile monoalkyl $(Me_3P)_3IrCH_2SiMe_3$, which rearranges to fac-$IrH(CH_2SiMe_2CH_2)(PMe_3)_3$ upon heating[32].

Reaction of xs Me_3CCH_2Li or Me_3CCH_2MgX with $(\eta^5\text{-}C_5Me_5)Rh(PPh_3)Cl_2$ produces $(\eta^5\text{-}C_5Me_5)Rh(CH_2CMe_2CH_2)PPh_3$ in modest yield. Again, the alkylmetal precursor to C—H activation is not detected; a hydridorhodium(V) intermediate is invoked. In this instance, activation of an ortho-C—H bond of coordinated PPh_3 is also observed to a comparable extent.

Four-membered ring formation by activation of an aromatic C—H bond is observed under similar conditions. Thus, treatment of $[Ir(PMe_3)_4]Cl$ with $PhCH_2MgCl$ gives only the benzometallacyclobutene at RT. The same skeleton is accessible, in principle, by metallation on the methyl of a 2-methylaryl ligand. It is now clear that the previously reported[34,35] trimethylphosphite analog of benzometallacyclobutene, obtained by treating trans-$(Ph_3P)_2Ir(CO)(Ar)$ with $P(OMe)_3$ (Ar = 2-tolyl, 2-xylyl or 2-mesityl) must be reformulated as the iridaindanones[36]. It is not clear whether migratory insertion of CO occurs before, during or after aliphatic C—H addition to the metal.

When $Os(O_2CMe)_2(PMe_3)_4$ reacts with $PhCH_2MgCl$, a product formulated as the osmacycle is isolated[26]. A related benzomanganacyclobutene is reportedly the thermolysis product from the remarkable inert $(dmpe)Mn(CH_2Ph)_2$ in the presence of xs dmpe. Metallacyclization of this precursor or of $(dmpe)Mn(CH_2SiMe_3)_2$ cannot be photo- or thermochemically induced in the absence of dmpe[37].

The formation of five-membered metallacycles by aliphatic C—H scission is also observed for several of the group-VIII metals. The isolable 2-methylbenzylplatinum(II) complexes cis-$(R_3P)_2Pt(CH_2C_6H_4CH_3\text{-o})_2$ (R = Et, Ph) undergo cyclization in refluxing xylene to the 2-platinaindans, $(R_3P)_2\overline{Pt(CH_2C_6H_4CH_2}\text{-o})$, by attack on the pendant methyl group and elimination of o-xylene[38]. In a related reaction, treating L_3RuCl_2 (L = Ph_2MeP, $PhMe_2P$) with 2 equiv o-$CH_3C_6H_4CH_2Li$ leads to facile elimination of o-xylene, even at RT, and formation of $(Ph_2MeP)_3\overline{Ru(CH_2C_6H_4CH_2}\text{-o})$. The dialkylmetal precursor is not detected[39]. The structure of the ruthenocycle is anomalous, as the metal completes its coordinative and electronic saturation by interacting with the aromatic ring with η^2:η^2 bindings of the xylylene ligand. Thus the molecule resembles the substituted η^4-butadiene rather than the metallaindan favored by $(R_3P)_2\overline{Pt(CH_2C_6H_4CH_2}\text{-o})$.

The 3-oxo-1-ruthenaindan $(Me_3P_3)_3\overline{Rh(C_6H_4O}CH_3\text{-o})$ is a major product of reacting $(2\text{-}MeOC_6H_4)_2Mg$ with $Ru_2(CO_2Me)_4Cl$ in xs PMe_3 [40]. Comparison of longer chain homologs of neopentyl metals, which are capable in principle of forming four-, five- and six-membered rings by C–H activation, indicates the energetics of these routes to metallacycles. The dialkylplatinum $(Et_3P)_2Pt(CH_2CMe_2CH_2CH_3)_2$ forms platinacyclopentane, $(Et_3P)_2\overline{Pt(CH_2CMe_2CH_2CH_2})$, exclusively, while alkane elimination from $(Et_3P)_2Pt(CH_2CMeCH_2CH_2CH_3)_2$ affords a mixture of products, $(Et_3P)_2\overline{Pt(CHMeCH_2CMe_2CH_2})$, $(Et_3P)_2\overline{Pt(CH_2CMeEtCH_2})$ and $(Et_3P)_2)\overline{Pt(CH_2CH_2CH_2CMe_2CH_2})$. The mechanism appears to parallel that forming $(R_3P)_2\overline{Pt(CH_2CMe_2CH_2})$ and production of metallacyclopentane is clearly more favorable than four- or six-membered ring formation, in part because of differences in strain energy. These differences, however, are less acute than those in aliphatic cycloalkanes; the platinacyclobutane, e.g., is estimated to possess intrinsic strain only ca. 1 kJ mol^{-1} greater than the corresponding metallacyclopentane. Preferences between four- and six-membered rings are less easily predicted and depend on the extent and position of substitutions on the metallacyclic product. The interesting complex $(Et_3P)_2Pt(CH_2CMe_2CH_2CMe_3)_2$ yields only[41] six-membered $(Et_3P)_2\overline{Pt(CH_2CMe_2CH_2CMe_3CH_2})$.

Activation of aromatic C—H to form a five-membered metallacycle occurs in $(Ph_3P)_2PtCl(C_{10}H_6CH_3\text{-}1)$. In refluxing toluene platinacycle **III** is formed by elimination of methane[42]:

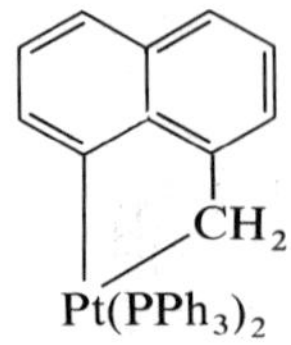

III

A more common instance of aromatic δ-hydrogen abstraction occurs via the neophyl (2-methyl-2-phenylpropyl) ligand. Treating $[Ir(PMe_3)_4]Cl$ with 1 equiv $LiCH_2CMe_2Ph$ leads directly at the hydrido-1-iridaindan; the monodentate alkyl complex cannot be intercepted[32]. Neophylplatinum(II) complexes $L_2Pt(CH_2CMe_2Ph)R$ are isolable, however, and these thermally eliminate RH, affording 1-platinaindans:

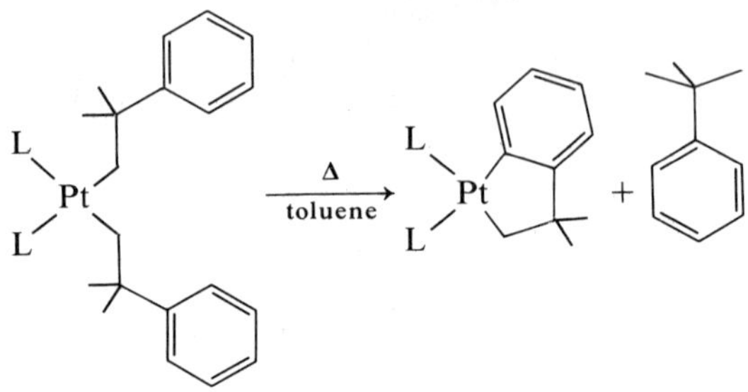

(a)

where L = PEt_3, PPh_3; L_2 = COD, 2,2′-bipy, 2,2′-bipyrimidyl, $Ph_2PCH_2CH_2PPh_2$, $Me_2PC_2H_4PMe_4$; R = neophyl, methl. Lability is controlled by the steric and electronic nature of ancillary ligands L (or bidentate L_2) and, as in the neopentyl analogs, an apparent prerequisite to metallacyclization is the dissociation of one neutral donor group. The methyl derivative is more inert than the dineophylmetal[43]. The corresponding nickel chemistry cannot be studied directly because of the high lability of the organonickel intermediates[44]. Neophypalladium complexes also tend to cyclo metallate[45].

Although rarer than mononuclear metallacyclizations, activation of a C—H bond in a ligand bound to one metal by an adjacent metal center in an agglomerate, forming a bimetallacycle, is known. Treatment of $Mo_2(O_2CMe)_4$ with xs $(CH_2SiMe_3)_2Mg$ and PMe_3 leads to the air-sensitive species:

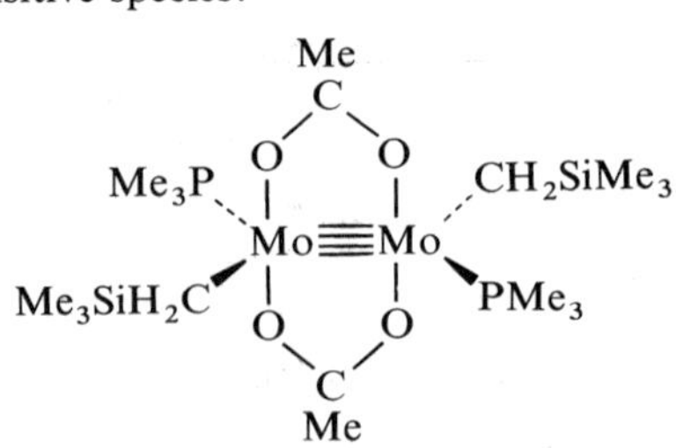

III

which features a quadruple Mo—Mo bond and a formal aysmmetry in oxidation state Mo(I)—Mo(III)[25].

(G. B. YOUNG)

1. G. W. Parshall, *Acc. Chem. Res., 3*, 139, (1970).
2. J. Dehand, M. Pfeffer, *Coord. Chem. Rev., 18*,327, (1976).
3. M. I. Bruce, *Angew. Chem., Int. Ed. Engl., 16*, 73, (1977).
4. C. S. Kraihanzel, *J. Organomet. Chem., 73*, 137, (1974).
5. I. Omae, *Chem. Rev., 79*, 287, (1979).
6. A. J. Deeming, I. P. Rothwell, *Pure Appl. Chem., 52*, 649, (1980).
7. I. S. Kolomnikov, V. P. Kukolev and M. E. Vol'pin, *Russ. Chem. Rev (Engl. Transl.), 43*, 551, (1974),
8. G. W. Parshall, *Acc. Chem. Res., 8*, 113, (1975).
9. D. E. Webster, *Adv. Organomet. Chem., 15*, 147, (1975).
10. I. Omae, *Coord. Chem. Rev., 32*, 235, (1980).
11. P. E. Garrou, *Chem. Rev., 81*, 229, (1981).
12. I. Omae, *Coord. Chem. Rev., 42*, 245, (1982).
13. J. Omae, *Coord. Chem. Rev., 28*, 97 (1979).
14. I. Omae, *Coord. Chem. Rev., 42*, 31, (1982).
15. A. K. Prokof'ev, V. I. Bregadze and O. Yu. Okhlobystin, *Russ. Chem. Rev. (Engl. Transl.), 39*, 196, (1970).
16. A. K. Prokof'ev, *Russ. Chem. Rev. (Engl. Transl.), 45*, 519, (1976).
17. For a review of metallacycles, see S. D. Chappell, D. J. Cole-Hamilton, *Polyhedron, 1*, 739 (1982).
18. See, e.g., P. Diversi, G. Ingrosso, A. Lucherni, W. Porzio, M. Zocchi, *J. Chem. Soc., Dalton Trans.*, 967 (1983); R. R. Schrock, *Acc. Chem. Res., 12*, 98 (1979); D. N. Clark, R. R. Schrock, *J. Am. Chem. Soc., 100*, 6774 (1978).
19. F. J. Brown, *Prog. Inorg. Chem.*, 1980, *27*, 1.
20. N. Calderon, J. P. Lawrence, E. A. Ofstead, *Adv. Organomet. Chem., 17*, 449, (1979).
21. R. H. Grubbs, *Prog. Inorg. Chem., 24*, 1, (1978).
22. T. J. Katz, *Adv. Organomet. Chem., 16*, 283, (1977).
23. S. M. Rocklage, J. D. Fellmann, G. A. Ruprecht, L. Messerle, R. R. Schrock, *J. Am. Chem. Soc., 103*, 1440, (1981).
24. D. A. Straus, R. H. Grubbs, *Organometallics, 1*, 1658, (1982).
25. R. A. Andersen, R. A. Jones, G. Wilkinson, *J. Chem. Soc., Dalton Trans.*, 446 (1978).
26. T. Behling, G. S. Girolami and G. Wilkinson, R. G. Somerville, M. B. Hursthouse, *J. Chem. Soc., Dalton Trans.*, 877 (1984).
27. P. Foley, R. DiCosimo, G. M. Whitesides, *J. Am. Chem. Soc., 102*, 6713 (1980).
28. P. Foley, G. M. Whitesides, *J. Am. Chem. Soc., 101*, 2732 (1979).
29. R. DiCosimo, G. M. Whitesides, *J. Am. Chem. Soc., 104*, 360 (1982); see also D. C. Griffiths, G. B. Young, *Organometallics*, 8, 875 (1989).
30. J. A. Ibers, R. DiCosimo, G. M. Whitesides, *Organometallics, 1*, 13 (1982).
31. R. H. Grubbs, in *Fundamental Research in Homogeneous Catalysis*, Vol. 3, M. Tsutsui, ed. Plenum Press, New York, 1979, p. 151.
32. T. H. Tulip, D. L. Thorn, *J. Am. Chem. Soc., 103*, 2448 (1981).
33. P. Diversi, G. Ingrosso, A. Lucherini, D. Fasce, *J. Chem. Soc., Chem. Commun.*, 945 (1982).
34. L. Dahlenburg, D. Rehder, W. Reith., W. Strauz, D. Thoennes, *Transition Met. Chem., 1*, 206 (1976).
35. L. Dahlenburg, U. Sinwell, D. Thoennes, *Chem. Ber. 111*, 3367 (1978).
36. K. Deuten, L. Dahlenburg, *Transition Met. Chem., 5*, 222 (1980).
37. C. G. Howard, G. S. Girolami, G. Wilkinson, M. Thornton-Pett, M. B. Hursthouse, *J. Chem. Soc., Dalton Trans.*, 2631 (1983).
38. S. D. Chappell, D. J. Cole-Hamilton, *J. Chem. Soc., Dalton Trans.*, 1051, (1983); *J. Chem. Soc., Chem. Commun.*, 238 (1980).
39. S. D. Chappell, D. J. Cole-Hamilton, A. M. R. Galas, M. B. Hursthouse, *J. Chem. Soc., Dalton Trans.*, 1867 (1982).

40. R. A. Jones, G. Wilkinson, *J. Chem. Soc., Dalton Trans.*, 472 (1979).
41. R. DiCosimo, S. S. Moore, A. F. Sowinski, G. M. Whitesides, *J. Am. Chem. Soc.*, *104*, 124 (1982); *103*, 948 (1981).
42. J. A. Duff, B. L. Shaw, B. L. Turtle, *J. Organomet. Chem.*, *66*, C18 (1974).
43. D. C. Griffiths, G. B. Young, *Polyhedron*, 1095 (1983).
44. B. Akermark, A. Ljungqvist, *J. Organomet. Chem.*, *149*, 97 (1978).
45. R. F. Heck, *J. Organomet. Chem.*, 37, 389 (1972).

5.8.2.2.3. η^1 σ-Alkyl Complexes by Cyclometallation of Donor Ligands.

The cyclometallation reaction in which the metallated carbon is aliphatic occurs less frequently[1–3] than does the aromatic carbon reaction (orthometallation) (see §5.8.2.6.4).

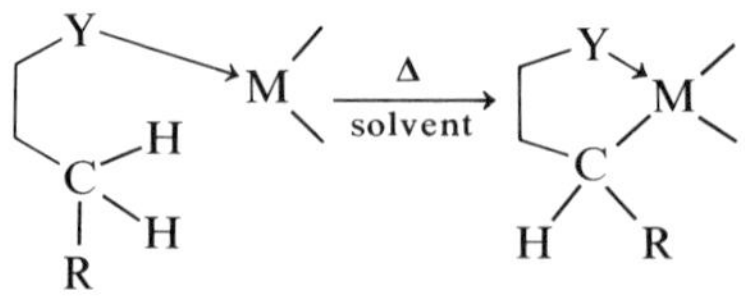

(a)

The carbon–metal σ bond of these heterometallocyclic compounds is obtained by intramolecular activation of a C—H bond by a transition-metal complex. This activation results either from the oxidative addition on a basic metal in a low oxidation state such as Fe(0), Ru(0), Ir(I), or by electrophilic attack of a transition metal in a di- or trivalent states [Ru(II), Pb(II), Pt(II), Ir(III), ...].

(i) Oxidative Addition of C–H Bonds on Electron-Rich Metals. Complexes containing a basic metal lead to a C—H activation reaction providing a coordination site is made available by the loss of a ligand. This is achieved either photochemically or thermally as in the Fe(0) compound:

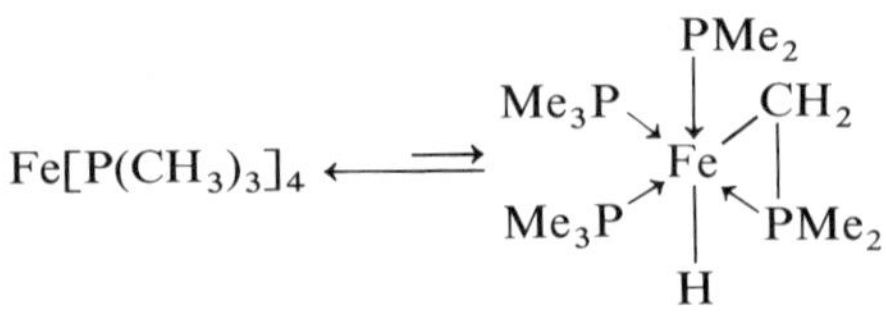

(b)

This fluxional compound is prepared by reducing $FeCl_2(PMe_3)_3$ with Na–Hg in THF at 0°C in the presence of excess trimethylphosphine[4]. The closely related Ru metallocycle $HRuCH_2PMe_2(PMe_3)_3$ [5], obtained in a similar way in benzene, is configurationally stable on the NMR time scale. The compound $Ru(dmpe)_2$ is in fact a Ru(II) hydride dimer[6] resulting from the intermolecular oxidative addition of a C—H bond:

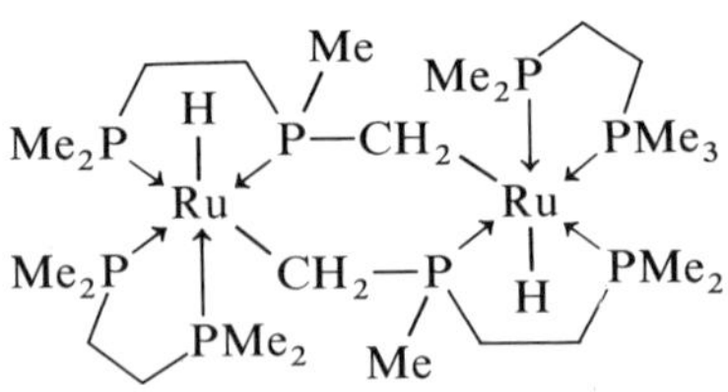

A related oxidative addition is observed on the compound obtained by the reaction of $[Ir(PMe_3)_4]Cl$ with $LiCH_2CMe_3$ in hexane or toluene to yield[7]:

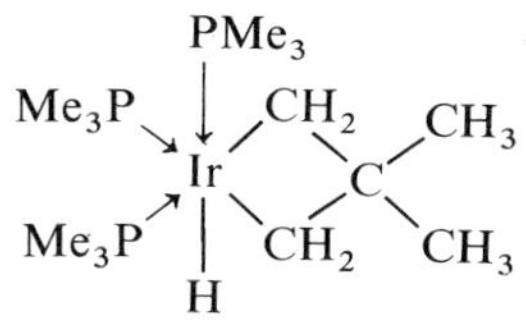

$[IrCl(COT)_2]_2$ (COT = cyclooctene) reacts at 25°C with both $P(t\text{-}Bu)_2CH_2OCH_3$ and $P(t\text{-}Bu)_2CH(CH_3)OCH_3$ to form, in the presence of CO, Ir(III)-metallated complexes by oxidative addition of a C—H bond of the methoxy groups[8].

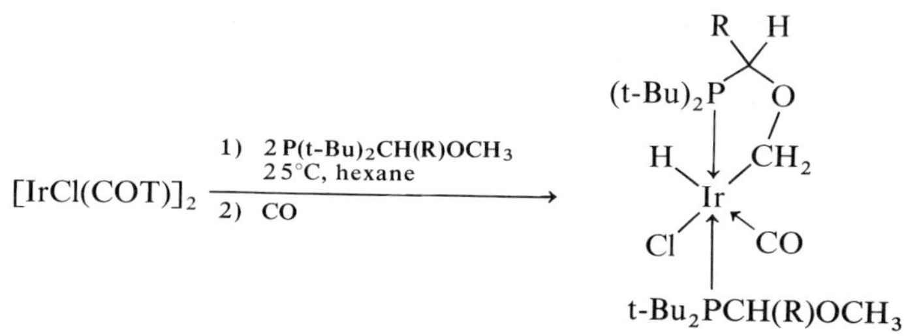

(c)

(ii) Electrophillic Attack of a Transition Metal on an Aliphatic Carbon: (a) Compounds with Y = N. The tetrachloropalladate(II) dianion reacts with 8-methylquinoline (R = H) in methanol at RT to give palladation at the 8-methyl group[9].

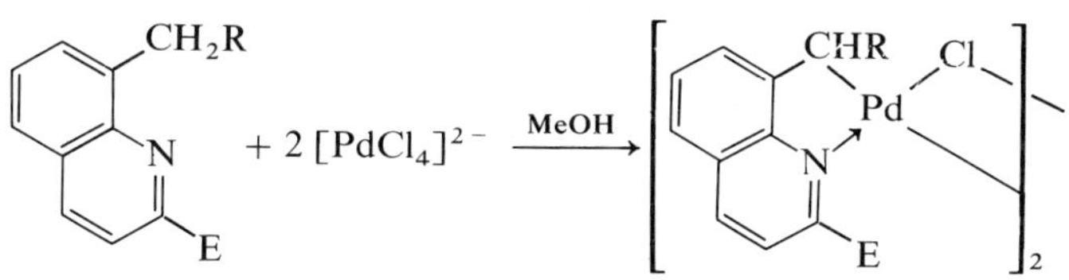

(d)

8-Ethylquinoline (R = CH_3) behaves similarly, but 8-isopropyl quinoline cannot be metallated[10]. In general, better yields are obtained using acetate as leaving group; metallation of the CH_2R unit is assisted when E is CH = NMe, CH_2OH or CO_2H but is prevented with E = Me, Br, or CHO. The key step of the activation of the C—H bond in these reactions should involve the formation of a 3-coordinated Pd moiety. The coordination site thus liberated should be occupied by the C—H bond prior to its cleavage[10].

Methyl groups of other ligands may also be involved in cyclopalladation. Thus N, N-dimethyl-o-toluidine with Pd acetate in acetic acid gives a trinuclear complex $[CH_3COOPd(2\text{-}CH_2C_6H_4)N(CH_3)_2]_2[\mu\text{-}Pd(CH_3COO)_2]$[11], from which by metathetical reaction the μ-dichloro dimer $[Pd(2\text{-}CH_2C_6H_4)NMe_2]_2(\mu\text{-}Cl)_2$ is obtained. The course of this reaction depends dramatically on the starting complex and the solvent used: With Li_2PdCl_4 in MeOH demethylation of the nitrogen atom occurs to afford trans-$PdCl_2[2\text{-}CH_3C_6H_4N(CH_3)(H)]_2$, whereas with $PdCl_2(PhCN)_2$ in CH_2Cl_2 the dimeric complex $Pd_2Cl_{14}[2\text{-}CH_3C_6H_4N(CH_3)_2]_2$, in which the amine is simply coordinated to Pd through the N atom, is obtained[12].

The palladation of the neopenthyl group of dimethylneopenthylamine also leads to a trimetallic compound[13].

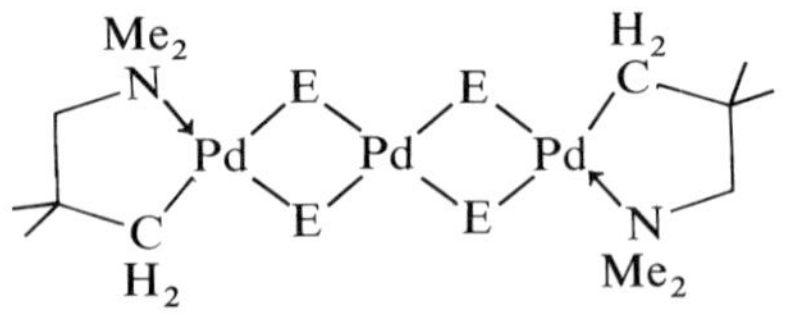

when E = OAc.

t-Butylketoximes (R = Me, Et, Ph) undergo cyclopalladation at t-Bu, giving the chlorobridged dimer $[PdCH_2C(CH_3)_2(C{=}NOH)R]_2(\mu\text{-}Cl)_2$ [14].

A related reaction occurs with one of the methyl of the highly congested 2,2′,6,6′-tetramethylcyclohexanone oxime[15] and even with the nonactivated methyl group of steroids, e.g., (E)-lupanone oxime[15] and lanost-8-en-3-one oxime[16].

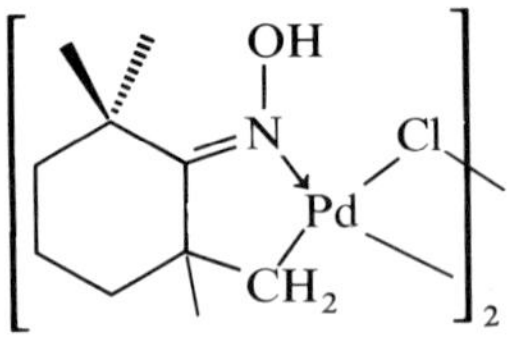

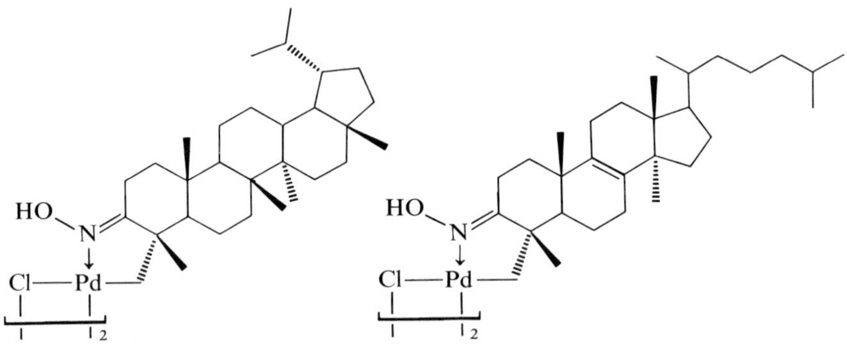

The tendency to form five-membered rings is illustrated in the complex obtained with methyl-t-butyl-N,N-dimethylhydrazone$[Pd{-}CH_2{-}C({=}NNMe_2)Bu\text{-}t]_2(\mu\text{-}Cl)_2$, where t-Bu has not been metallated, which would have given rise to a six-membered ring[17].

However, exceptions to this rule have been found with 2-neopenthylpyridine and 2-pivaloylpyridine for which six-membered palladocycles are easily obtained[18]:

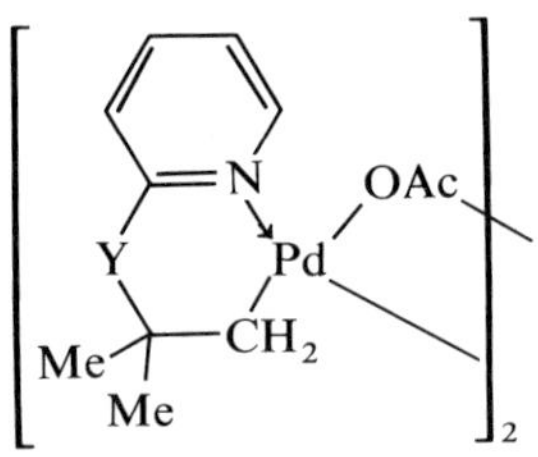

where Y = CO, CH_2.

The electrophilic attack of Pd(II) on aliphatic C—H bonds is surprisingly also observed when the carbon to be metallated is substituted by strong electron-withdrawing group, e.g., with several pyridines orthosubstituted by activated alkyl groups[19–22]:

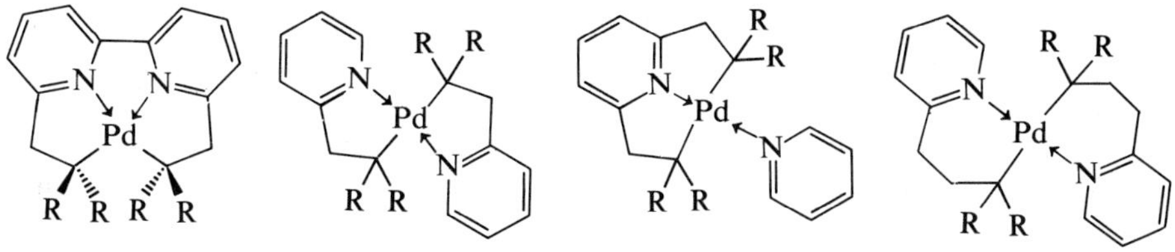

where R = $COCH_3$, CO_2Et, CN.

(b) Compounds with Y = P, As. With alkylphosphines and arsines bulky substituents on the P or As affect the rate and extent of metallation, both by conformational and by internal rotational entropy effects. Internal metallation of tertiary phosphines (or arsines) with bulky substituents is obtained by thermolysis in a polar solvent[23–26]:

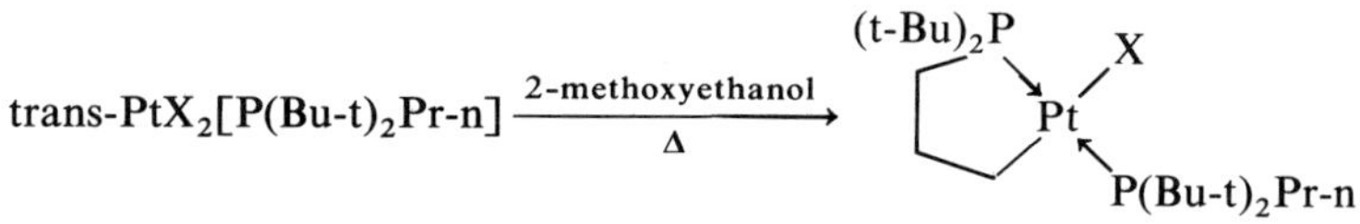

(e)

where X = Cl, Br, I. The reaction also occurs on a methyl group of an o-tolyl unit giving a five-membered ring. With the related $P(Bu\text{-}t)_2CH_2OCH_3$ or $P(Bu\text{-}t)_2OCH_2CH_3$, the formation of the five-membered ring occurs even more readily, emphasizing the activating role of the oxygen atom[8]:

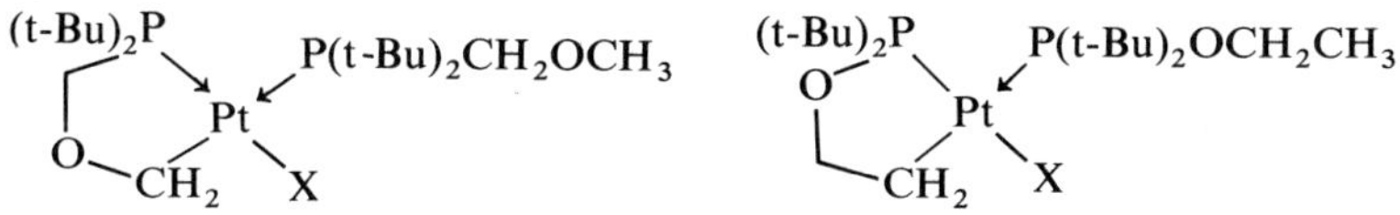

Internal metallation is easier for X = I > Br > Cl and for tertiary phosphines than arsines[27]. The Pd analogs give halide-bridged dimeric compounds[25].

Bis-chelated complexes are obtained by thermolysis of the monomeric compounds[24,25]:

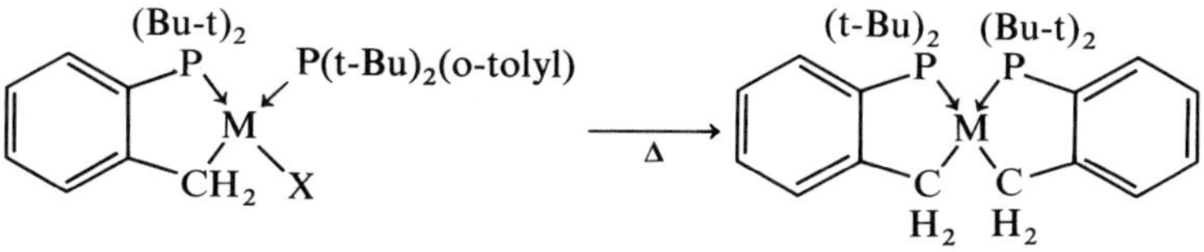

cis or trans

(f)

where X = a halide.

In opposition to what is observed with N-containing ligand, four-membered heterometallocyclic ring is obtained by metallation of tri-t-butylphosphine[28–30].

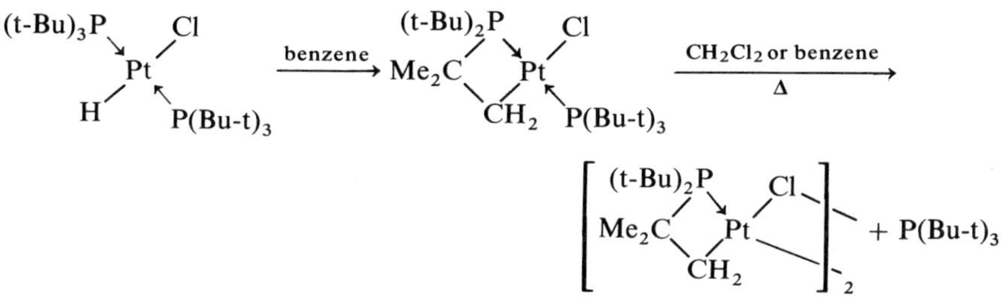

(g)

The compound trans-$PdHCl[As(Bu\text{-}t)_3]_2$ gives, at a slower rate[31], a mixture of cis and trans hydrido-bridged dinuclear Pd(II) complexes, $[PdCH_2C(Me)_2As(Bu\text{-}t)_2]_2(\mu\text{-}H)_2$.

The metallation of a methyl group of PMe_2R coordinated to a metal, leading to three-membered ring, occurs by the reaction with strong bases such as lithium diisopropylamine[32]:

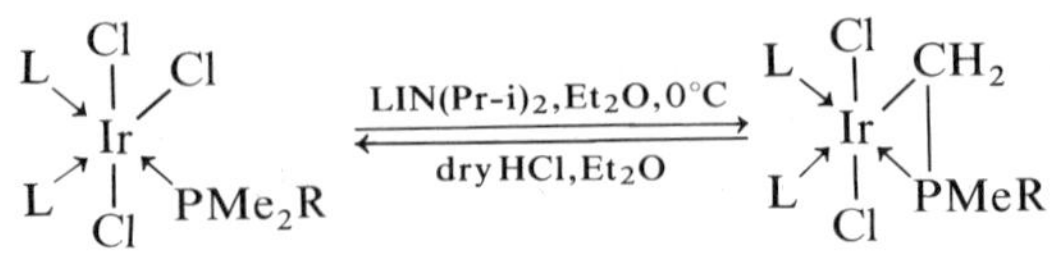

where R = Me, Ph; L = PMe_3, PMe_2Ph.

With alkoxy phenylphosphines, platination may occur on the alkoxy group, leading either a to six- or a five-membered ring by the cleavage of a C—H or a C—O bond, respectively[33].

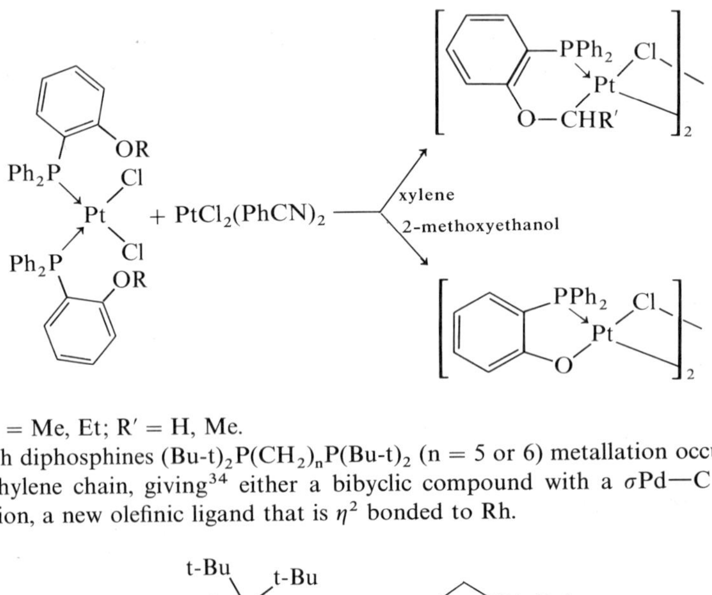

(i)

where R = Me, Et; R′ = H, Me.

With diphosphines $(Bu\text{-}t)_2P(CH_2)_nP(Bu\text{-}t)_2$ (n = 5 or 6) metallation occurs on the polymethylene chain, giving[34] either a bibyclic compound with a σPd—C or, via β elimination, a new olefinic ligand that is η^2 bonded to Rh.

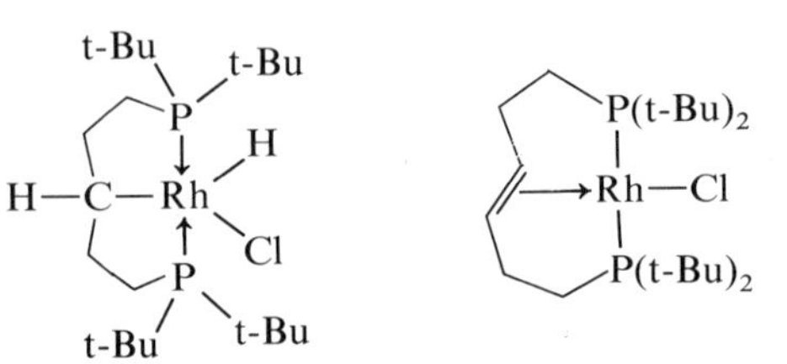

The abstraction of an alkyl proton from 2,2′-bisdiphenylphosphinobibenzyl gives[25] $XMPh_2PC_6H_4CH{-}CH_2C_6H_4PPh_2$(M = Pt, Pd; X = Cl, Br, I).

(c) Miscellaneous. The only examples of S-assisted metallation of a alkyl group are those derived from tertiary thiobenzamines[36] thioalkylamides[37]

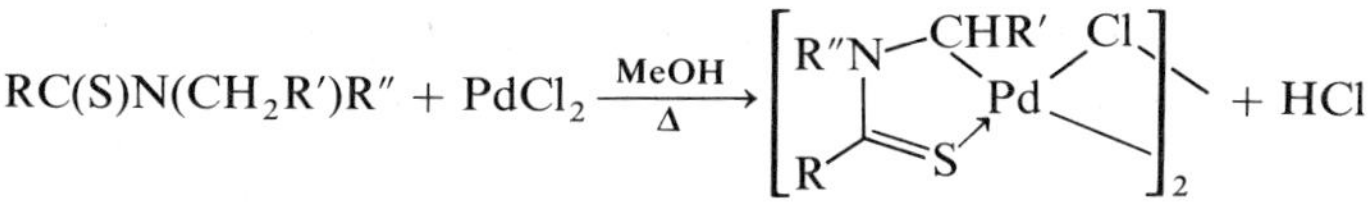

(j)

where R = Ph, t-Bu; R′ = H, CH_3; R″ = CH_3, C_2H_5. The palladation of a NCH_3 group is much easier to achieve than that of the corresponding NC_2H_5 group.

The C—H activation of a alkyl group by early transition metals (M = Ti, Zr, Hf) occurs with aryl oxide in which the aryl rings are orthosubstituted by Bu-t. In this case, however, the oxygen atom is covalently bonded to the metal[38,39].

(iii) Reactivity. Studies of the reactivity of the cyclometallated compounds have two main emphases:

a. By taking advantage of the stabilizing effect of the three-electron donor chelates, new organometallic molecules have been synthesized with unexpected geometries or showing unusual chemical or physical properties.

b. The reactions that can be performed at the metallated carbon atom afford new ways of synthesis of carbon–carbon and/or carbon–heteroatom bonds.

(a) Reactions in which the metallated $\overline{C\quad Y}$ unit is an ancillary ligand. Classical halide bridge cleavage of dimeric complexes occurs with neutral nucleophiles to give monomers in which, in most cases, the nucleophile is coordinated trans to the heteroatom Y:

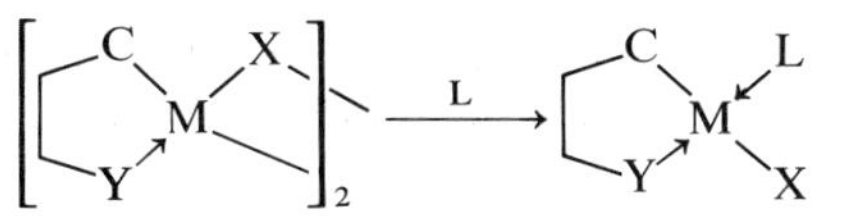

where X = a halogen.

A rare example of a 18-electron palladium(II) compound is synthesized through the reaction of $[\overline{PdCH_2CMe_2P}(Bu\text{-}t)R(\mu\text{-}Cl)]_2$ with TlC_5H_5, affording $(\eta\text{-}^5Cp)\overline{PdCH_2CMe_2P}(Bu\text{-}t)R$ (R = Ph, Bu-t) [30].

The important stabilizing effect of the palladated 8-methylquinoline ligand [hereafter the Pd compounds derived from this ligand are referred to as Pd(8-mq) has been used to build an original chemistry on the positions trans to it. Thus it is possible to isolate phosphoenolates, when functionalized phosphines are bound to palladium, by direct reaction of a lithium phosphoenolate on the palladium compound[40], the nuclearity of the product depending upon the stoichiometry:

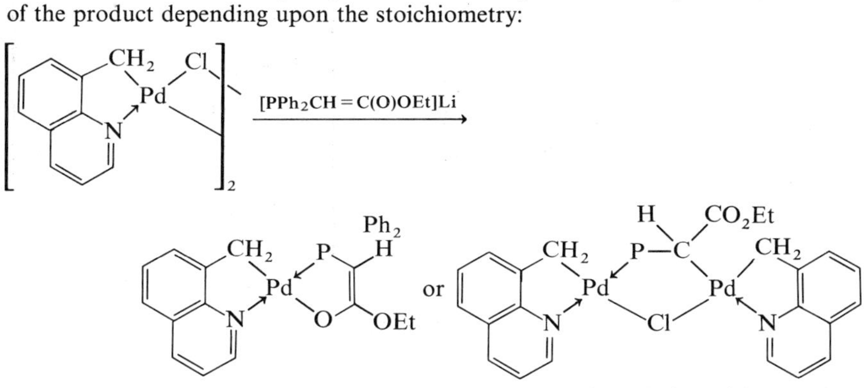

(l)

Carbon–carbon coupling reactions occur at the P–O chelate when these complexes react with heterocumulenes such as CO_2[40] or isocyanates[41]:

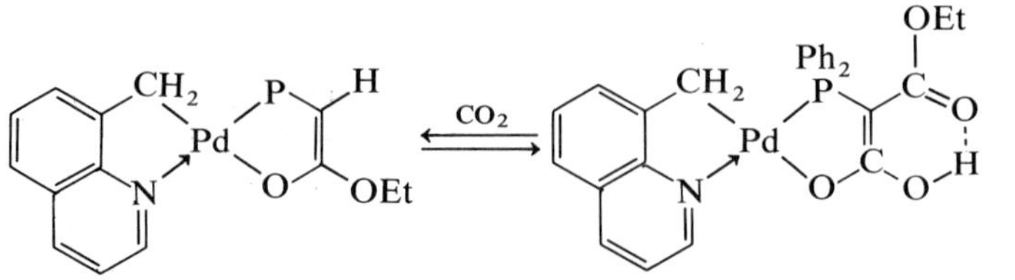

(m)

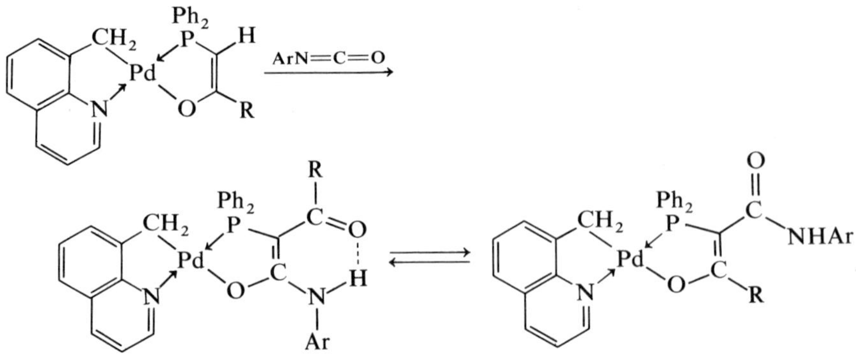

(n)

Polynuclear compounds are obtained by reacting Pd(8-mq) compounds with carbonylmetallate anions. Depending upon the starting material used hetero di-[42] or tri-metallic[43] species are formed.

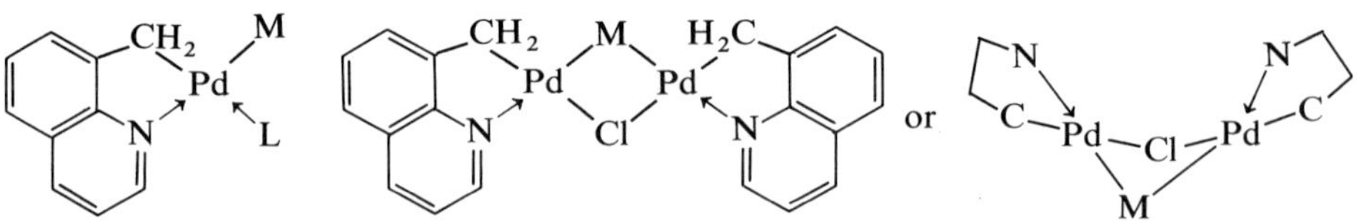

$[(8\text{-mqPd})_2(\mu^2\text{-M})(\mu^2\text{-Cl})]$

when M = $Co(CO)_4$, $Fe(CO)_3NO$, $Mo(CO)_3Cp$-η^5; L = PR_3, py.

The latter trimetallic compounds, moreover, lead to an unusual tetrametallic species via reaction with a cationic Pd derivative[44]:

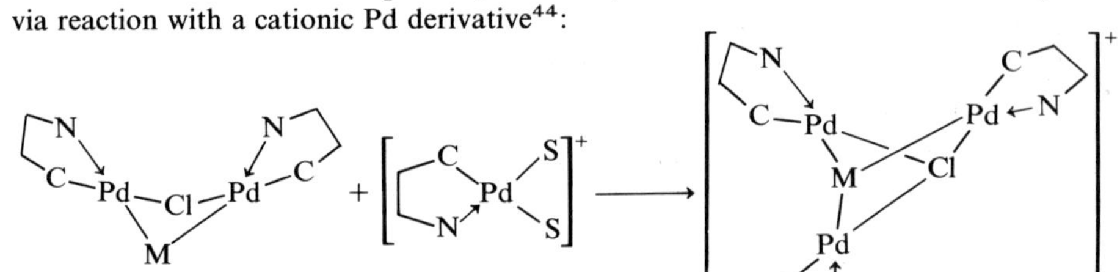

$[(8\text{-mqPd})_2(\mu^2\text{-M})(\mu^2\text{-Cl})]$

$[(8\text{-mqPd})_3(\mu^3\text{-M})(\mu^3\text{-Cl})]^+$

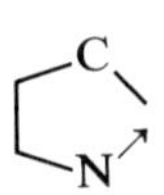 = 8-methylquinolyl (8-mq) (o)

where M = $Mo(CO)_3Cp$-η^5, S = solvent.

(b) C—C and C—Y bond formations. By insertion of nucleophiles into the Pd–C bond of cyclopalladated compounds or through the reaction with electrophiles, new C—C or C—Y bond formation occurs. Chlorination of cyclopalladated 2,2′,6,6′-tetramethylcyclohexanone oxime leads to the substitution of the palladium atom by Cl, whereas oxidation by lead tetraacetate of E-lupanone oxime cyclopalladated at the 23rd position affords the corresponding O-acetoxy derivatives[15].

Phosphorus–carbon coupling is observed when reacting a Pd (8-mq) compound with chlorophosphines[45]:

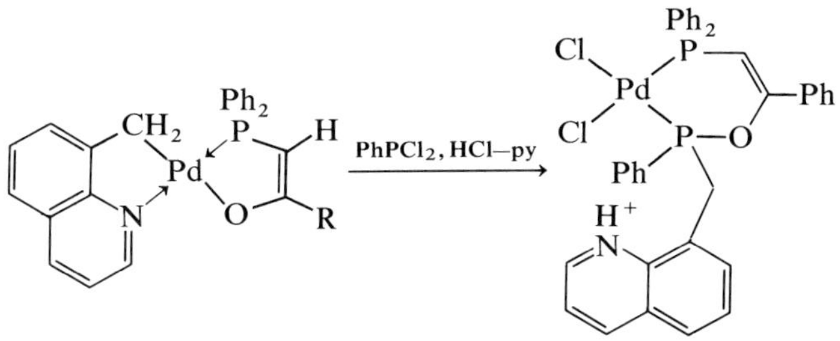

(p)

Insertion of alkynes into the Pd—C bond of cyclopalladated compounds affords reliable ways for making new C—C bonds. Thus, one of three alkynes can be inserted into the Pd—CH_2 bond of Pd(8-mq), leading to new organopalladium compounds [46–49]:

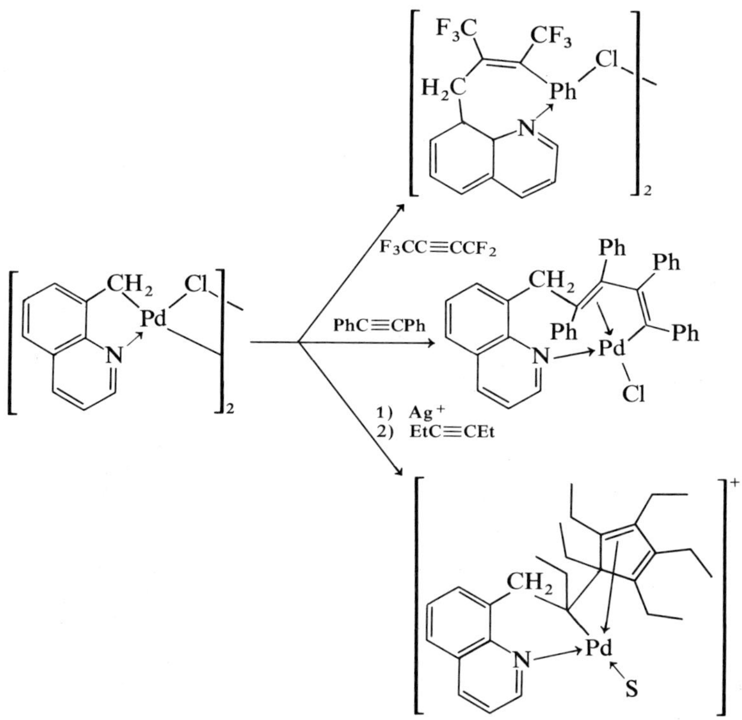

(q)

The used of cationic cyclopalladated species or iodide-containing complexes increases the reactivity of the Pd–C bond and leads to spontaneous depalladation reaction of the new compounds formed after insertion of the alkynes. This afforded in the case of diphenylacetylene an heterocyclic compound through reductive elimination of Pd(0) and formation of Pd(0) and formation of a C—N bond[49]:

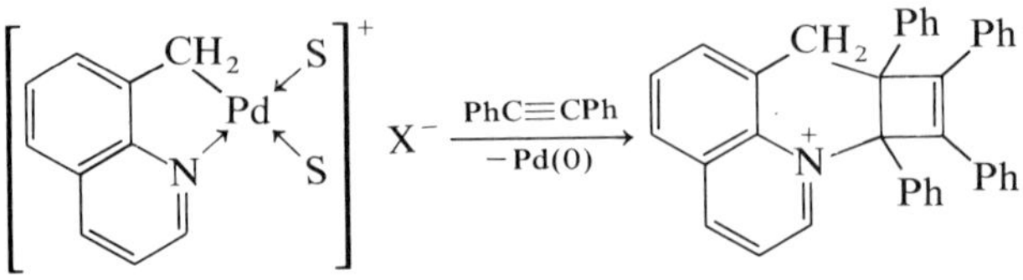

(r)

(J, DEHAND, M. PFEFFER)

1. J. Dehand, M. Pfeffer, *Coord. Chem. Rev., 18*, 327 (1976).
2. M. I. Bruce, *Angew. Chem., Int. Ed. Engl., 16*, 73 (1977).
3. I. Omae, *Coord. Chem. Rev., 79*, 287 (1979).
4. J. W. Rathke, E. L. Muetterties, *J. Am. Chem. Soc., 97* 3272 (1975).
5. H. Werner, R. Werner, *J. Organomet. Chem., 209*, C60 (1981).
6. F. A. Cotton, B. A. Frenz, D. L. Hunter, *J. Chem. Soc., Chem. Commum.*, 755 (1974).
7. T. H. Tulip, D. L. Thorn, *J. Am. Chem. Soc., 103*, 2448 (1981).
8. B. D. Dombek, *J. Organomet. Chem., 169* 315 (1979).
9. G. E. Hartwell, R. V. Lawrence, M. J. Smas, *J. Chem. Soc., Chem. Commun.*, 912 (1970).
10. A. J. Deeming, I. P. Rothwell, *J. Organomet. Chem.*, 105, 117 (1981); A. J. Deeming, I. P. Rothwell, M. B. Hurthouse, K. M. A. Malik, *J. Chem. Soc., Dalton Trans.*, 1899 (1979).
11. M. Pfeffer, E. Wehman, G. van Koten, *J. Organomet. Chem., 282*, 127 (1985).
12. J. Dehand, C. Mutet, M. Pfeffer, *J. Organomet. Chem., 209*, 255 (1981).
13. Y. Fuchita, K. Huraki, Y. Matsumoto, *J. Organomet. Chem., 280*, C51 (1985).
14. A. G. Constable, W. S. McDonald, L. C. Sawkins, B. L. Shaw, *J. Chem. Soc., Dalton Trans.*, 1992 (1980).
15. J. E. Baldwin, R. H. Jones, C. Najera, M. Yus, *Tetrahedron, 41*, 699 (1985).
16. K. Carr, J. K. Sutherland, *J. Chem. Soc., Chem Commun.*,1227 (1984).
17. B. Galli, F. Gasparrini, L. Mareca, G. Natile, G. Palmieri, *J. Chem. Soc., Dalton Trans.*, 1483 (1983).
18. Y. Fuchita, K. Hiraki, T. Uchiyama, *J. Chem. Soc., Dalton Trans.*, 897 (1983).
19. G. R. Newkome, T. Kawato, D. K. Kohli, W. E. Puckett, B. D. Olivier, G. Chiari, F. R. Fronczek, W. A. Deutsch, *J. Am. Chem. Soc., 103*, 3423 (1981).
20. G. R. Newkome, M. Onishi, W. E. Puckett, *J. Am. Chem. Soc., 102* 4551 (1980).
21. G. R. Newkome, V. K. Gupta, F. R. Fronczek, *Organometallics, 1*, 907 (1982).
22. G. R. Newkome, W. E. Puckett, V. K. Gupta, F. R. Fronczek, *Organometallics, 2* 1247 (1983).
23. A. J. Cheney, B. E. Mann, B. L. Shaw, R. M. Slade, *J. Chem. Soc., A*, 3833 (1971).
24. A. J. Cheney, B. L. Shaw, *J. Chem. Soc., Dalton Trans.*, 754 (1972).
25. A. J. Cheney, B. L. Shaw, *J. Chem. Soc., Dalton Trans.*, 860 (1972).
26. D. F. Gill, B. E. Mann, B. L. Shaw, *J. Chem. Soc., Dalton Trans.*, 270 (1973).
27. B. L. Shaw, R. E. Stainbank, *J. Chem. Soc., Dalton Trans.*, 2394, (1973).
28. H. C. Clark, A. B. Goel, R. G. Goel, S. Goel, *Inorg. Chem., 19*, 3220 (1980).
29. H. C. Clark, A. B. Goel, R. G. Goel, S. Goel, *Inorg. Chem., 18*, 2803 (1979).
30. H. Werner, H. J. Kraus, *J. Organomet. Chem., 204*, 415 (1981).
31. R. G. Goel, W. O. Ogini, *Inorg. Chim. Acta, 44*, L165 (1980).
32. S. Aljibori, C. Crocker, W. S. McDonald, B. L. Shaw, *J. Chem. Soc., Dalton Trans.*, 1572 (1981).
33. C. E. Jones, B. L. Shaw, B. L. Turtle, *J. Chem. Soc., Dalton Trans.*, 992 (1974).
34. C. Crocker, R. J. Errington, R. Markham, C. J. Moulton, K. J. Odell, B. L. Shaw, *J. Am. Chem. Soc., 102*, 4373 (1980).
35. M. A. Bennett, P. W. Clark, *J. Organomet. Chem., 110*, 367 (1976).
36. T. J. Grinter, D. Leaver, R. M. O'Neil, *Inorg. Nucl. Chem. Lett., 16*, 145 (1980).
37. Y. Tamaru, M. Kagotami, Z. I. Yoshida, *Angew. Chem., Int. Ed. Engl., 20*, 980 (1981).
38. S. L. Latesky, A. K. McMullen, I. P. Rothwell, J. C. Huffman, *J. Am. Chem. Soc., 107*, 5981 (1985).
39. L. R. Chamberlain, I. P. Rothwell, J. C. Huffman, *J. Am. Chem. Soc., 108*, 1502 (1986).

40. P. Braunstein, D. Matt. Y. Dusausoy, J. Fischer, A. Mitschler, L. Ricard, *J. Am. Chem. Soc., 103*, 5115 (1981).
41. S. E. Bouaoud, P. Braunstein, D. Grandjean, D. Matt, D. Nobel, *Inorg. Chem., 27*, 2279 (1988).
42. M. Pfeffer, D. Grandjean, G. Le Borgne, *Inorg. Chem., 20*, 4426 (1981).
43. M. Pfeffer, J. Fischer, A. Mitschler, *Organometallics, 3*, 1531 (1984).
44. P. Braunstein, J. Fischer, D. Matt, M. Pfeffer, *J. Am. Chem. Soc., 106*, 410 (1984).
45. F. Balegroune, P. Braunstein, D. Grandjean, D. Matt, D. Nobel, *Inorg. Chem., 27*, 3320 (1988).
46. A. Bahsoun, J. Dehand, M. Pfeffer, M. Zinsius, S. E. Bouaoud, G. Le Borgne, *J. Chem. Soc., Dalton Trans.*, 547 (1970).
47. C. Arlen, M. Pfeffer, O. Bars, D. Grandjean, *J. Chem. Soc., Dalton Trans.*, 1535 (1983).
48. G. Wu, A. L. Rheingold, R. F. Heck, *Organometallics, 5*, 1922 (1986).
49. F. Maassarani, M. Pfeffer, G. Le Borgne, *Organometallics, 6*, 2029 (1987).

5.8.2.2.4. Metallacycles by Oxidative Addition of Strained Carbocyclic Rings.

The reaction of transition-metal complexes and strained carbocycles has been examined most extensively with Pt(II) complexes and cyclopropane derivatives. Chloroplatinic acid and cyclopropane react in acetic anhydride at RT to give a stable, polymeric compound, $[PtCl_2(C_3H_6)]_4$(**I**), in 50–70% yield[1]:

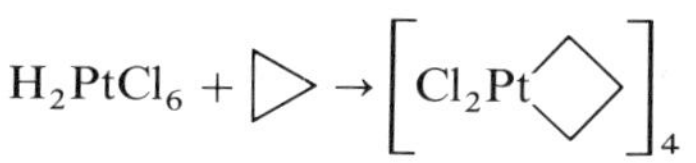

(a)

I

The bromo analog is prepared similarly using bromoplatinic acid[2]. Treatment of these halide-bridged products with py affords compounds of the composition $PtX_2(C_3H_6)(py)_2$ (**II**). The platinacyclopropane structures have been determined by IR[3] and NMR[3,4] analyses and X-ray crystallographic study[5] on the dichloro complex **II** (X = Cl). The geometry of the C_3H_6 ligand rules out the cyclopropane edge structure. Treatment of **I** or **II** with nitrogen such bases as picolines, 2,6-lutidine, 2,2′-bipyridyl or phen produces adducts of type **III** without displacement of the trimethylene moiety[3–7]:

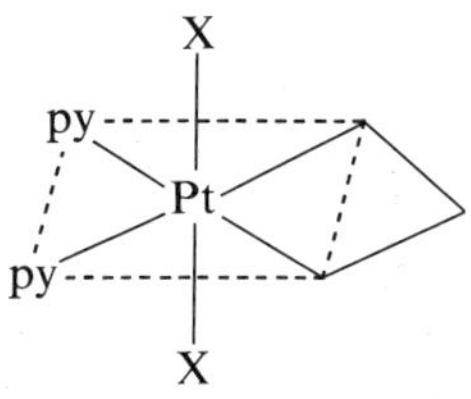

II

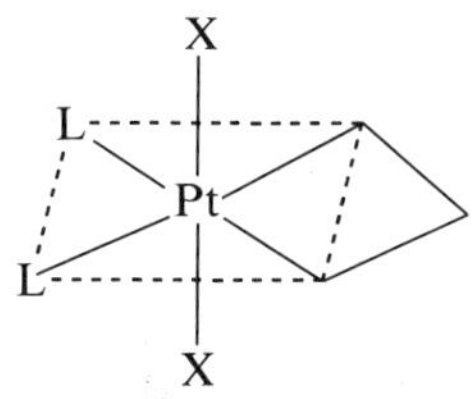

III

The bisammine complexes are prepared by condensing dry ammonia at low temperatures onto **I**[4]. Oxygen ligands such as 1,4-dioxane[2], THF[6] or DMF[8] can be used as well. Simple alkylated cyclopropanes can be converted to the corresponding tetrameric platinacyclobutanes on reaction with chloroplatinic acid in ethyl acetate[9,10], rather than in acetic anhydride[11].

Alkyl- and aryl-substituted cyclopropanes (**IV**) displace ethylene ligand from $[PtCl_2(C_2H_4)]_2$ in dichloromethane or ether, which provides a versatile, general tool for the preparation of substituted platina cyclobutanes[9–13]:

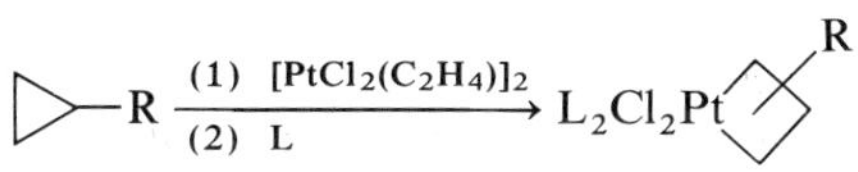

(b)

IV **V**

The yield is moderate to high. The reaction is electrophilic in nature with respect to Pt(II) species and hence a reactive sequence **IV**, R = hexyl > benzyl > phenyl > o-nitrophenyl[12], or R = p-ethoxyphenyl ≫ p-tolyl > > phenyl[14], has been observed. Cyclopropanes with electron-withdrawing cyano, acetayl, and carbomethoxy groups fail to give the adducts. The initially formed tetrameric Pt complexes are difficult to characterize, and the structures have been determined after exposure to a nitrogen or oxygen ligand to give the hexa-coordinated Pt(IV) complex **V**. Products arising from Pt insertion at the least substituted bond are obtained. With cyclopropanes having a long alkyl group, only 3-alkylated platinacyclobutanes have been isolated; with spiro[2.n]alkanes the bond opposite the spiro-carbon is cleaved exclusively[13]. The platinacyclobutane formation is reversible[15–18]; e.g., phenylcyclopropane gives initially a product resulting from insertion at C(1)–C(2), and this compound then undergoes a facile rearrangement to the C(2)–C(3) inserted product. The alkylated platinacyclobutanes isomerize even more rapidly[16]. Therefore, the isolated metallacyclobutanes may represent the outcome of thermodynamic rather than kinetic reaction control. In the kinetic formation of the metallacycle products electronic factors are significant but the position of equilibrium (L = nitrogen or oxygen base) is influenced largely by steric effects:

$$PtCl_2(CHRCH_2CH_2)L_2 \rightleftharpoons PtCl_2(CH_2CHRCH_2)L_2 \qquad (c)$$

Tables 1 and 2 exemplify the formation of stable platinacyclobutanes. Complexes of Pt(0) and Pd(0) of the type ML_n (M = Pt or Pd, L = tertiary phosphines or triphenylarsine, n = 3 or 4) or $Pt(PPh_3)_2(C_2H_4)$ reacts with 1,1,2,2-tetracyanocyclopropane and its 3-substituted derivatives (**VI**) in THF or benzene to give the M(II) adducts (**VII**)[20–22]:

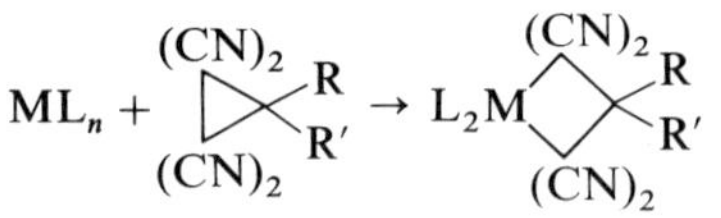

VI **VII** (d)

This metal insertion reaction is nucleophilic with respect to the zerovalent metals and takes place exclusively at the cyclopropane C(1)–C(2) bond (see Table 3).

Reaction of dibenzosemibullvalene (**VIII**) and $Fe_2(CO)_9$ in benzene (55°C, 2 h) gives the ferretane **X** in 25% yield together with some CO insertion product[23,24]. In a similar manner, **VIII** or its dicarbomethoxy derivative **IX** reacts with $[Rh(CO)_2Cl]_2$ in cyclohexane to afford a Rh(III)-containing four-membered ring compound **XI** (RT, 5 h, 85% yield) or **XII** (60–65°C, 36 h, 5%), respectively[25].

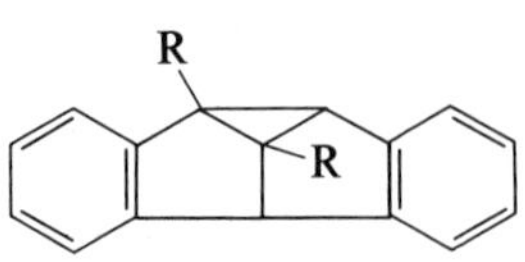

VIII, R = H
IX, R = COOMe

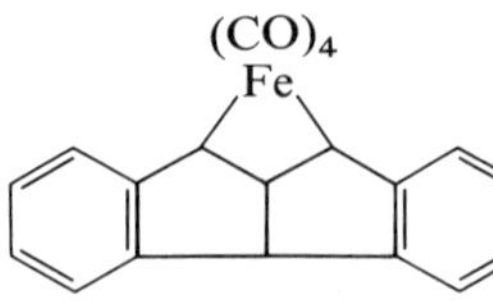

X

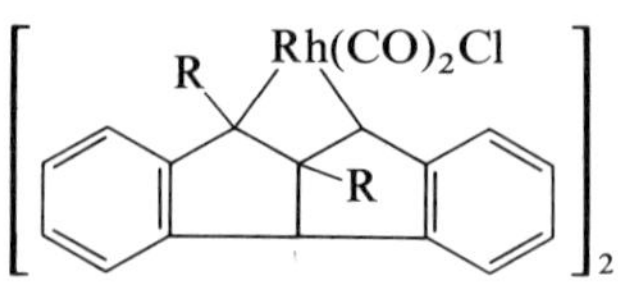

XI, R = H
XII, R = COOMe

Few isolable metallacycles are formed from strained hydrocarbon molecules other than cyclopropane derivatives. Oxidative addition of an electron-deficient σ bond of

TABLE 1. FORMATION OF PtX_2(PROPANE-1,3-DIYL)L_2 COMPLEXES FROM CYCLOPROPANE DERIVATIVES VIA THE TETRAMERIC PtX_2 (PROPANE-1,3-DIYL) COMPOUNDS (X = Cl OR Br)[a]

Cyclopropane	Product	Ref.
Cyclopropane	$PtCl_2(CH_2)_3(py)_2$	1–3, 5
Cyclopropane	$PtBr_2(CH_2)_3(py)_2$	2
Cyclopropane	$PtCl_2(CH_2)_3$(bipy)	2, 3
Cyclopropane	$PtCl_2(CH_2)_3(NH_3)_2$	4
Cyclopropane	$PtCl_2(CH_2)_3$(1,4-dioxane)[b]	2
Hexylcyclopropane	$PtCl_2[CH_2CH(n\text{-}C_6H_{13})CH_2](py)_2$	12
1,2-Dicyclopropylethane	[c]	12
Benzylcyclopropane	$PtCl_2[CH_2CH(CH_2Ph)CH_2](py)_2$	12
Phenylcyclopropane	$PtCl_2(CHPhCH_2CH_2)(py)_2$	15
Phenylcyclopropane	$PtCl_2(CH_2CHPhCH_2)(py)_2$	12, 15
Phenylcyclopropane	$PtCl_2(CHPhCH_2CH_2)$(2-Me-py)$_2$	15
Phenylcyclopropane	$PtCl_2(CHPhCH_2CH_2)$(4-Me-py)$_2$	15
Phenylcyclopropane	$PtCl_2(CHPhCH_2CH_2)$(bipy)	15
Phenylcyclopropane	$PtCl_2(CHPhCH_2CH_2)$(TMED)	15
o-Tolylcyclopropane	$PtCl_2[CH_2CH(o\text{-}CH_3C_6H_4)CH_2](py)_2$	15
o-Nitrophenylcyclopropane	$PtCl_2[CH_2CH(o\text{-}NO_2C_6H_4)CH_2](py)_2$	12
p-Tolylcyclopropane	$PtCl_2[CH(p\text{-}CH_3C_6H_4)CH_2CH_2](py)_2$	12, 15
p-Methoxyphenylcyclopropane	$PtCl_2[CH(p\text{-}MeOC_6H_4)CH_2CH_2](py)_2$	15
1,1-Dimethylcyclopropane	$PtCl_2(CH_2CMe_2CH_2)(py)_2$	9
Spiro[2.2]pentane	$PtCl_2[CH_2C(C_2H_4)CH_2](py)_2$	13
Spiro[2.4]heptane	$PtCl_2[CH_2C(C_4H_8)CH_2](py)_2$	13
Spiro[2.5]octane	$PtCl_2[CH_2C(C_5H_{10})CH_2](py)_2$	13
trans-1,2-Diphenylcyclopropane	$PtCl_2(CHPhCHPhCH_2)(py)_2$[d]	12
1,1,2-Trimethylcyclopropane	$PtCl_2(CHMeCMe_2CH_2)(py)_2$	9

[a] Usually the two-step formation is achieved in a fair to high yield.
[b] Dimeric structure.
[c] Tetrachlorotetra(pyridine)-μ-2,2′-ethylenedipropane-1,3-diplatinum.
[d] For the crystal and molecular structure, see ref. 19.

TABLE 2. FORMATION OF TRANS-DICHLORO(DIAMNE)(PROPANE-1,3-DIYL)PLATINUM(IV) COMPLEXES FROM THE BIS-PYRIDINE COMPLEXES AND DIAMINES

Diamine	Temp. (°C)	Time	Product and yield (%)	Ref.
2,2′-Bipyridyl	73		$PtCl_2[CH_2(CH_2)_3]$(bipy)[a]	2
2,2′-Bipyridyl	RT	7 d	$PtCl_2(CH_2CHPhCH_2)$(bipy), 74	14
1,10-Phenanthroline	RT	2 h	$PtCl_2(CH_2)_3$(phen), 74	14
Ethylenediamine	RT	15 min	$PtCl_2(CH_2)_3$(en)	5
N,N,N′,N′-Tetramethyl-ethylenediamine	RT	7 d	$PtCl_2(CH_2CHPhCH_2)$(TMED), 32	14

[a] The bis-dioxane complex may be used as the starting material.

TABLE 3. REACTION OF ELECTRONEGATIVELY SUBSTITUTED CYCLOPROPANES AND Pt(0) or Pd(0) COMPLEXES

Cyclopropane	Metal complex	Temp.	Time (h)	Product and yield (%)	Ref.
1,1,2,2-Tetracyanocyclopropane	$Pt(PPh_3)_4$	RT	0.5	$Pt[C(CN)_2CH_2C(CN)_2]$-$(PPh_3)_2$, 80[a]	20, 21
1,1,2,2-Tetracyanocyclopropane	$Pt(PMePh_2)_4$	RT		$Pt[C(CN)_2CH_2C(CN)_2]$-$(PMePh_2)_2$, 50	20, 21
1,1,2,2-Tetracyanocyclopropane	$Pt(AsPh_3)_4$	RT	2	$Pt[C(CN)_2CH_2C(CN)_2]$-$(AsPh_3)_2$, 50	20, 21
1,1,2,2-Tetracyanocyclopropane	$Pd(PPh_3)_4$	RT		$Pd[C(CN)_2CH_2C(CN)_2]$-$(PPh_3)_2$, 50	20, 21
1,1,2,2-Tetracyanocyclopropane	$Pd(PMePh_2)_4$	RT	0.17	$Pd[C(CN)_2CH_2C(CN)_2]$-$(PMePh_2)_2$, 80	20, 21
1,1,2,2-Tetracyano-3-phenyl-cyclopropane	$Pt(PPh_3)_2(C_2H_4)$	RT		$Pt[C(CN)_2CHPhC(CN)_2]$-$(PPh_3)_2$[a]	22
1,1,2,2-Tetracyano-3-di-methylcyclopropane	$Pt(PMePh_2)_4$	RT		$Pt[C(CN)_2CMe_2C(CN)_2]$-$(PMePh_2)_2$, 20	20, 21
1,1,2,2-Tetracyano-3,3-di-methylcyclopropane	$Pt(PPh_3)_2(C_2H_4)$	RT	0.5	$Pt[C(CN)_2CMe_2C(CN)_2]$-$(PPh_3)_2$, 80	20, 21
1,1,2,2-Tetracyano-3,3-di-methylcyclopropane	$Pt(AsPh_3)_4$	RT		$Pt[C(CN)_2CMe_2C(CN)_2]$-$(AsPh_3)_2$, 50	20, 21
1-Carbethoxy-1,2,2-tricyano-trans-3-phenylcyclopropane	$Pt(PPh_3)_2(C_2H_4)$	RT		$Pt[CCN(COOEt)CHPh$-$C(CN)_2](PPh_3)_2$[a]	22

[a] The crystal and molecular structure has been determined by X-ray crystallography.

1,1,2,2-tetracyanocyclobutanes, **XIII** and **XIV**, onto Pt(0) species takes place in ether–THF at RT to afford the platinacyclopentane complexes, **XV** and **XVI**, respectively[26]:

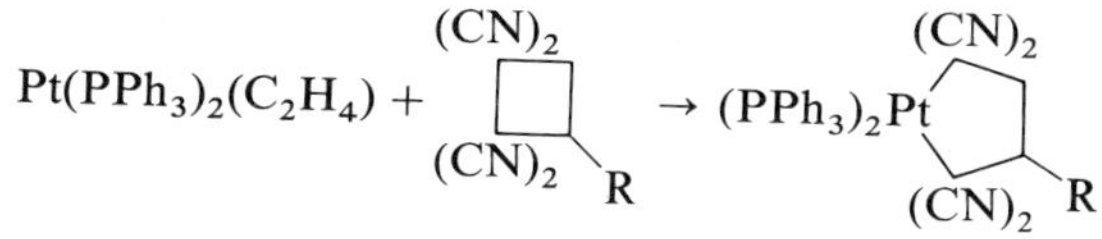

(e)

XIII, R = OEt **XV,** R = OEt
XIV, R = C_6H_4-o-Me **XVI,** R = C_6H_4-o-Me

Reaction of 1,3-bishomocubane (**XVII**) and $[Ir(CO)_3Cl]_2$ in benzene at 120°C gives rise to the dimeric product **XVIII**, possessing an iridacyclopentane structure[24]:

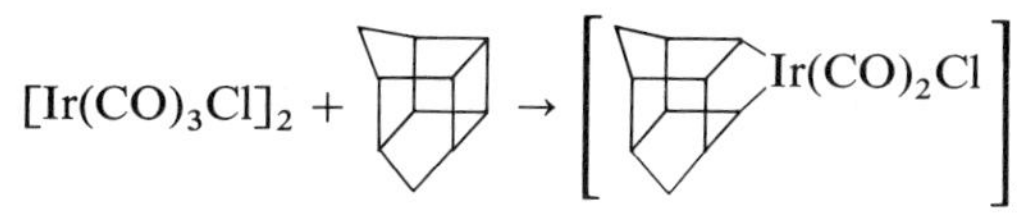

(f)

XVII **XVIII**

Unstable metallacycloalkane intermediates play an important role in a variety of transition-metal-catalyzed isomerizations or rearrangements of strained carbocyclic systems and olefin metathesis reactions[28–34].

(R. NOYORI)

1. C. F. H. Tipper, *J. Chem. Soc.*, 2045 (1955).
2. S. E. Binns, R. H. Cragg, R. D. Gillard, B. T. Heaton, M. F. Pilbrow, *J. Chem. Soc., A*, 1227 (1969).
3. D. M. Adams, J. Chatt, R. G. Guy, N. Sheppard, *J. Chem. Soc.*, 738 (1961).
4. P. W. Hall, R. J. Puddephatt, C. F. H. Tipper, *J. Organomet. Chem.*, *71*, 145 (1974).
5. R. D. Gillard, M. Keeton, R. Mason, M. F. Pilbrow, D. R. Russell, *J. Organomet. Chem.*, *33*, 247 (1971).
6. P. W. Hall, R. J. Puddephatt, K. R. Seddon, C. F. H. Tipper, *J. Organomet. Chem.*, *81*, 423 (1974).
7. F. Iwanciw, M. A. Quyser, R. J. Puddephatt, C. F. H. Tipper, *J. Organomet. Chem.*, *113*, 91 (1976).
8. P. W. Hall, R. J. Puddephatt, C. F. H. Tipper, *J. Organomet. Chem.*, *84* 407 (1975).
9. B. M. Cushman, S. E. Earnest, D. B. Brown, *J. Organomet. Chem.*, *159*, 431 (1978).
10 T. H. Johnsón and S.-S. Cheng, *J. Am. Chem. Soc.*, *101*, 5277 (1979).
11. S. E. Earnest, D. B. Brown, *J. Organomet. Chem.*, *120*, 135 (1976).
12. F. J. McQuillin, K. G. Powell, *J. Chem. Soc., Dalton, Trans.*, 2123 (1972).
13. D. B. Brown, V. A. Viens, *J. Organomet. Chem.*, *142*, 117 (1977).
14. R. J. Al-Essa, R. J. Puddephatt, M. A. Quyser, C. F. H. Tipper, *J. Organomet. Chem.*, *150*, 295 (1978).
15. R. J. Al-Essa, R. J. Puddephatt, M. A. Quyser, C. F. H. Tipper, *J. Am. Chem. Soc.*, *101*, 364 (1979).
16. R. J. Al-Essa, R. J. Puddephatt, C. F. H. Tipper, P. J. Thompson, *J. Organomet. Chem.*, *157*, C40 (1978).
17. C. P. Casey, D. M. Scheck, A. J. Shusterman, *J. Am. Chem. Soc.*, *101*, 4233 (1979).
18. T. H. Johnson, *J. Org. Chem.*, *44*, 1356 (1979).
19. J. A. McGinnety, *J. Organomet. Chem.*, *59*, 429 (1973).
20. M. Lenarda, R. Ros, M. Graziani, U. Belluco, *J. Organomet. Chem.*, *65*, 407 (1974).
21. D. J. Yallow, J. A. Ibers, M. Lenarda, M. Graziani, *J. Organomet. Chem.*, *70*, 133 (1974).
22. J. Rajarman, J. A. Ibers, *J. Am. Chem. Soc.*, *100*, 829 (1978).
23. R. M. Moriarty, K.-N. Chen, C.-L. Yeh, *J. Am. Chem. Soc.*, *94*, 8944 (1972).
24 S. W. Tam, *Tetrahedron Lett.*, 2385 (1974).

25. B. F. G. Johnson, J. Lewis, S. W. Tam, *J. Organomet. Chem.*, *105*, 271 (1976).
26. R. Ros, M. Lenarda, N. B. Pahor, M. Callingaris, P. Delise, L. Randaccio, M. Graziani, *J. Chem. Soc., Dalton Trans*, 1937 (1976).
27. J. Blum and C. Zlotogorski, *Tetrahedron Lett.*, 3501 (1979).
28. K. C. Bishop III. *Chem. Rev.*, *76*, 461 (1976).
29. F. D. Mango, *Coord. Chem. Rev.*, *15*, 109 (1975).
30. R. Noyori, *Inorganic Compounds with Unusual Properties-II*, Advances in Chem. Ser., No. 173, R. B. King, ed., American Chemical Society, Washington D. C., 1979, p 307.
31. G. Wilke, *Pure Appl. Chem.*, *50*, 677 (1978).
32. J. P. Collman, L. S. Hegedus, *Principles and Applications of Organotransition Metal Chemistry*, University Science Books, Mill Valley, CA, 1980, p 506.
33. N. Calderon, E. A. Ofstead, W. A. Judy, *Angew. Chem., Int. Ed. Engl.*, *15*, 401 (1976).
34. T. J. Katz, *Adv. Organomet. Chem.*, *16*, 283 (1977).

5.8.2.2.5. η^2 π-Olefin Complexes by β-Hydride Elimination.

Organometallic σ complexes having hydrogen on a β carbon commonly react with triphenylmethyl (or trityl) cation. While main-group alkyls fragment to alkene, metal cation and triphenylmethane, comparable transition-metal derivatives produce rearranged cationic olefin π complexes and triphenylmethane:

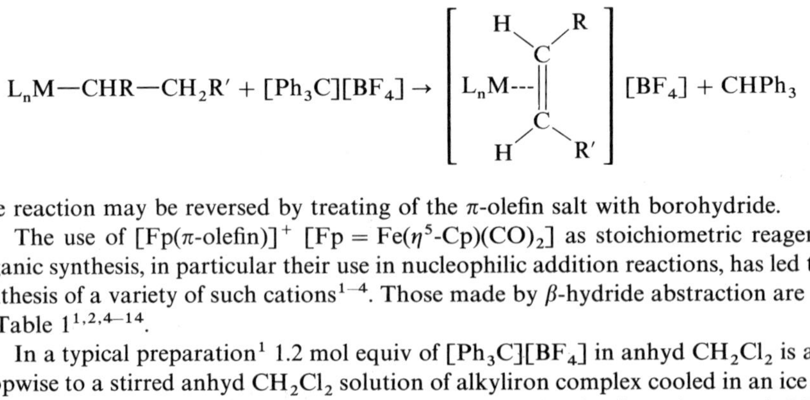
(a)

The reaction may be reversed by treating of the π-olefin salt with borohydride.

The use of $[Fp(\pi\text{-olefin})]^+$ $[Fp = Fe(\eta^5\text{-Cp})(CO)_2]$ as stoichiometric reagents in organic synthesis, in particular their use in nucleophilic addition reactions, has led to the synthesis of a variety of such cations[1–4]. Those made by β-hydride abstraction are listed in Table 1[1,2,4–14].

In a typical preparation[1] 1.2 mol equiv of $[Ph_3C][BF_4]$ in anhyd CH_2Cl_2 is added dropwise to a stirred anhyd CH_2Cl_2 solution of alkyliron complex cooled in an ice bath. After addition of the trityl cation is complete, the reaction is allowed to reach RT and continue to completion as determined by IR monitoring of the ν_{CO} region of the spectrum (ca. 1 h). The reaction solution is then cooled to at least 0°C and treated with diethyl ether to precipitate the product. After collection by vacuum filtration, washing with several small portions of ether and drying in vacuo, the yellow olefin complexes are recrystallized from CH_2Cl_2-ether with ca. 10% acetone at −78°C.

Solvents previously reported to be suitable for use in the abstraction reaction (i.e., THF and MeCN)[5,6] are now considered unsatisfactory because of the reactivity of THF with trityl cation[15] and the coordinating ability of acetonitrile[16]. The commercially available trityl salts, stored under nitrogen at 0°C, are stable. However, even carefully stored but aged samples generally give lower yields of the olefin complexes and the trityl salts are often best freshly prepared[8,17]. Normally the trityl salts $[Ph_3C][X]$ ($X = BF_4^-$, PF_6^-, or less often SbF_6^-, ClO_4^-) are used indiscriminantly, although the potential exists for different reactivities based upon variations in the counterion X[18]. The borate, phosphate or antimonate salts of the metal–olefin complexes have the advantage of being nonexplosive, although the $[BF_4]^-$ salts crystallize less readily.

Table 1 shows that relatively high yields of terminal olefin complexes are obtained, without isomerization, from primary alkyliron substrates. The formation of the complex of 3-phenylpropene rather than that of its more stable, conjugated isomer β-methylstyrene, from $FpCH_2CH_2CH_2Ph$ is noteworthy, as is the formation of the exocyclic olefin complexes $[Fp\{\eta^2\text{-}CH_2{=}\overline{C(CH_2)_nCH_2}\}]^+$ (n = 3, 4, 5). Elimination reactions from

TABLE 1. TYPICAL [Fp(π-OLEFIN)][X] COMPLEXES RESULTING FROM TREATMENT OF Fp(ALKYL) COMPLEXES WITH $[Ph_3C][X]$[a]

Fp(alkyl)	[Fp(π-olefin)][X]	Yield (%)	Ref.
$FpCH_2CH_3$	$[Fp(\eta^2\text{-}C_2H_4)][X]$[b]		
	$X = ClO_4^-$	c	5, 6
	$X = PF_6^-$	77	7
	$X = BF_4^-$	95, 57	8, 9
$FpCH_2CH_2Me$	$[Fp(\eta^2\text{-}CH_2{=}CHMe)][X]$		
	$X = ClO_4^-, BF_4^-$	c	5, 6
	$X = BF_4^-$	99	10
$FpCHMe_2$	$[Fp(\eta^2\text{-}CH_2{=}CHMe)][X]$		
	$X = ClO_4^-, BF_4^-$	c	5, 6
$FpCDMe_2$	$[Fp(\eta^2\text{-}CH_2{=}CDMe)][BF_4]$	86	6
$FpCH_2CH_2Ph$	$[Fp(\eta^2\text{-}CH_2{=}CHPh)][PF_6]$	71	8, 11
$FpCH_2CH_2CH_2Ph$	$[Fp(\eta^2\text{-}CH_2{=}CHCH_2Ph)][BF_4]$	79	1
$FpCH_2CHMe_2$	$[Fp(\eta^2\text{-}CH_2{=}CMe_2)][BF_4]$	70	8
$FpCD_2CHMe_2$	$[Fp(\eta^2\text{-}CD_2{=}CMe_2)][BF_4]$	48	12
$FpCDMePh$	$[Fp(\eta^2\text{-}CH_2{=}CDPh)][PF_6]$	54	13
$Fp\overline{CHCH_2CH_2CH_2}$	$[Fp(\eta^2\text{-}\overline{CH{=}CHCH_2CH_2})][BF_4]$	81	12
$Fp\overline{CHCH_2CH_2CH_2CHR}$	$[Fp(\eta^2\text{-}\overline{CH{=}CHCH_2CH_2CHR})][BF_4]$		
	$R = CH(COOEt)_2$	76	2, 4
	$R = CH(COOMe)_2$	72	4
$FpCH_2CH_2CHMe_2$	$[Fp(\eta^2\text{-}CH_2{=}CHCHMe_2)][BF_4]$	75	10
$FpCH_2CH_2CH_2Me$	$[Fp(\eta^2\text{-}CH_2{=}CHCH_2Me)][BF_4]$	56	1
$FpCHMeCH_2Me$	$[Fp(\eta^2\text{-}MeCH{=}CHMe)][BF_4]$	54	
	and $[Fp(\eta^2\text{-}CH_2{=}CHCH_2Me)][BF_4]$	27	1
$FpCHMe(CH_2)_nMe$	$[Fp\{(\eta^2\text{-}CH_2{=}CH(CH_2)_nMe\}][BF_4]$		
	n = 2, 3, 4, 5	60–88	1
$FpCH(CH_2Me)_2$	$[Fp(\eta^2\text{-}MeCH{=}CHCH_2Me)][BF_4]$	55	1
$FpCH(CH_2Me)(CH_2CH_2CH_2Me)$	$[Fp\{(\eta^2\text{-}MeCH{=}CH(CH_2)_3Me\}][BF_4]$	57	1
$FpCH_2\overline{CH(CH_2)_nCH_2}$	$[Fp\{(\eta^2\text{-}CH_2{=}\overline{C(CH_2)_nCH_2}\}][BF_4]$		
	n = 3, 4, 5	50–7	1
$FpCHMe\overline{CH(CH_2)_4CH_2}$	$[Fp\{(\eta^2\text{-}CH_2{=}CH\overline{CH(CH_2)_4CH_2}\}][BF_4]$	92	1
$FpCH_2CH_2CH{=}CH_2$	$[Fp(\eta^2\text{-}CH_2{=}CHCH{=}CH_2)][BF_4]$	87	1
Fp(3-η^1-cholest-5-ene)	[Fp(3-η^2-cholesta-3,5-diene)]$[BF_4]$	57	1
Fp(endo- or exo-2-norbornyl)	[Fp(η^2-exo-norbornene)]$[BF_4]$	66	14a
$FpCH_2CH_2\overline{CHCH_2CH_2}$	$[Fp(\eta^2 - CH_2{=}CH\overline{CHCH_2CH_2})][BF_4]$	93	14b

[a] Fp = $Fe(\eta^5\text{-}Cp)(CO)_2$
[b] The corresponding η^5-idenyl complex $[Fe(\eta^5\text{-}C_9H_7)(CO)_2(\eta^2\text{-}C_2H_4)][BF_4]$ is produced[9] in 30% yield from $Fe(\eta^5\text{-}C_9H_7)(CO)_2Et$.
[c] Not reported.

other cycloalkylmethyl substrates often give mixtures of endocyclic olefin isomers[19]. Abstraction reactions of the cycloalkylmethyliron compounds are slow, probably for steric reasons, and the cyclooctylmethyl analog is inert[1]. More than one product is possible from the hydride abstraction reactions of secondary alkyliron compounds: yields are rather lower and, while most 2-alkyl compounds give only the terminal olefin products (illustrating the dominance of steric factors), the 2-butyl compound $FpCHMeCH_2Me$ gives a 2:1 ratio of the complexes of (Z)-2-butene and 1-butene[1].

Two interesting iron–olefin complexes are not mentioned in Table 1. When $FpCH_2CH_2CH_2Fp$ is treated with $[Ph_3C][PF_6]$ in CH_2Cl_2 at RT for 18 h, crystals gradually separate (62% yield)[20] that contain a cation best described[8] as having the hybrid structure $Fp(\sigma\text{-}CH_2CH{=}CH_2\text{-}\eta^2)\overset{+}{F}p \leftrightarrow \overset{+}{F}p(\eta^2\text{-}CH_2{=}CHCH_2\text{-}\sigma)Fp$. In similar fashion the salt derived from the treatment of $FpCH_2\overline{CH(CH)_5CH}$ with $[Ph_3C][SbF_6]$, namely the stable cationic complex of heptafulvene, is considered a hybrid rather than a simple η^2-alkene complex, the two structures contributing about equally[8,11]: $\overset{+}{F}p\{\eta^2\text{-}CH_2{=}\overline{C(CH)_5CH}\} \leftrightarrow Fp\{\sigma\text{-}CH_2\overset{+}{\overline{C(CH)_5CH}}\}$[8,11].

At reduced temperatures, especially in the presence of CO, migratory carbonyl insertion can compete with triphenylcarbenium-induced β-abstraction[10]:

$$FpCH_2CH_2R + [Ph_3C][BF_4] \rightarrow [Fp(\eta^2\text{-}CH_2{=}CHR)][BF_4] + FpCOCH_2CH_2R + CHPh_3 \qquad \text{(b)}$$

where R = Me; i-Pr. When the β-H is hindered and/or cannot attain an antiperiplanar conformation relative to iron, carbonyl insertion may be the exclusive reaction. Thus no π-olefinic product is formed, even after 48 h reaction at 25°C under N_2, when R = t-Bu in Eq. (b)[10]. Redox-catalyzed migratory carbonyl insertion also accompanies β-abstraction in the triphenylcarbenium-induced reactions of Fp norbornyls (see Table 1)[14a]. The $[Ph_3C]^+$ cation, besides abstracting hydride, is also capable of being a one-electron oxidant, and in the latter reaction with the Fp norbornyls it acts as both an electron-pair acceptor (hydride abstractor) and a one-electron oxidant toward the same organometallic under a single set of reaction conditions[14a].

Like the trityl cation, $[Fp(\eta^1\text{-benzocyclobutenylidene})][PF_6]$ is also a hydride abstractor, e.g.[18,21]:

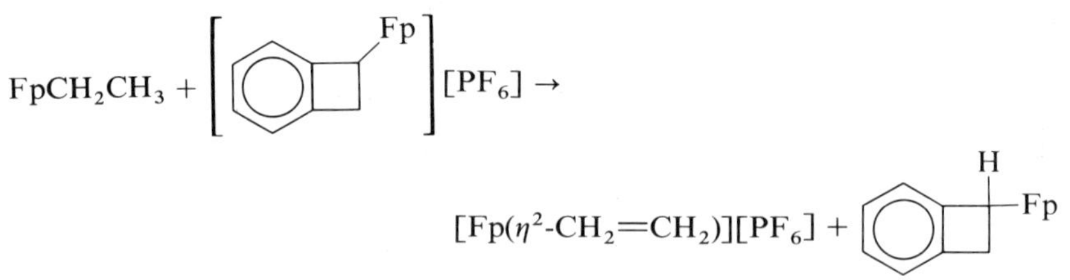

(c)

It functions as a more specific hydride abstractor, attacking the most exposed although not necessarily the most reactive hydrogen in a ligand of an organometallic complex. Thus it converts the cyclopropyl complex $Fp\overline{CHCH_2CH_2}$ or the cyclopropylmethyl complex $FpCH_2\overline{CHCH_2CH_2}$ to the respective allene $[Fp(\eta^2\text{-}CH_2{=}C{=}CH_2)][PF_6]$ or butadiene $[Fp(\eta^2\text{-}CH_2{=}CHCH{=}CH_2)][PF_6]$ complexes by respective α- or

γ-hydride abstractions concomitant with the opening of the three-membered rings, whereas reaction with $[Ph_3C]^+$ yields primarily addition products. In neither case are cyclopropene complexes formed[18,21]. Certainly other pathways than β-hydride abstraction are available for reactions of the triphenylcarbenium ion with nontypical Fp(alkyl) complexes[1,13,18,22] e.g., the σ-allyl compound $FpCH_2CH{=}CH_2$ is attacked by $[Ph_3C]^+$ at the double bond, giving[1,18] a complex of 4,4,4-triphenylbutene, $[Fp(\eta^2\text{-}CH_2{=}CHCH_2CPh_3)]^+$.

Similar reactions to those of the iron–alkyl complexes are found for alkyl complexes of other transition metals where a β-hydrogen is present. Thus, reaction in $CHCl_3$ of a stoichiometric amount of $[Ph_3C]^+$ with $Mo\eta^5\text{-}Cp(CO)_3Et$ gives the ethylene complex cation $[Mo\eta^5\text{-}Cp(CO)_3(\eta^2\text{-}C_2H_4)]^+$ as either the $[BF_4]^-$ [23] or $[PF_6]^-$ [7] salt (ca. 30% yield). The W analog $[W\eta^5\text{-}Cp(CO)_3(\eta^2\text{-}C_2H_4)][PF_6]$ is similarly accessible[7] from $W\eta^5\text{-}Cp(CO)_3Et$, while treatment of $\{W\eta^5\text{-}Cp(CO)_3CH_2CH_2\}_2NMe$ in CH_2Cl_2 with a three-fold xs of $[Ph_3C][PF_6]$ gives $[\eta^5\text{-}Cp(CO)_3W\text{-}(\sigma\text{-}CH_2CH_2NMeCH{=}CH_2\text{-}\eta^2)W(CO)_3[Cp]PF_6]$ in 72% yield[7]. Pure $Mn(CO)_5Et$ distilled under vacuum onto an excess of finely powdered $[Ph_3C][BF_4]$, followed by extraction with liquid SO_2, gives[6] a 10% yield of $[Mn(CO)_5\text{-}(\eta^2\text{-}C_2H_4)][BF_4]$ after four recrystallizations from ether–liquid SO_2. The Re(III) compound $Re(\eta^5\text{-}Cp)_2Et$, which does not undergo β-hydride elimination when heated, gives[24] a 41% yield of $[Re(\eta^5\text{-}Cp)_2(\eta^2\text{-}C_2H_4)][BF_4]$ when a β-hydride ion is abstracted by treatment with an equimolar proportion of $[Ph_3C][BF_4]$ in $CHCl_3$. Despite the available β-hydrogens in $Re\eta^5\text{-}Cp(NO)(PPh_3)R$ (R = Et, n-Pr), however, only α-hydride abstraction occurs on treatment with $[Ph_3C][PF_6]$, affording isolable cationic alkylidene complexes[25]. β-Hydride abstraction from $\eta^5\text{-}CpRu(CO)_2Et$ by $[Ph_3C][BF_4]$ gives $[\eta^5\text{-}CpRu(CO)_2(\eta^2\text{-}C_2H_4)][BF_4]$ in 25% yield after recrystallization from acetonitrile–diethyl ether[9].

Treatment of suitable monocations with trityl salts allows access to dicationic π-olefin complexes: thus the PF_6^- salt of the dicationic Ru(II) arene complex $[Ru(\eta^6\text{-}C_6H_6)(PMe_3)_2(\eta^2\text{-}C_2H_4)]^{2+}$ is prepared in quantitative yield by reaction of $[Ru(\eta^6\text{-}C_6H_6)(PMe_3)_2Et][PF_6]$ with $[Ph_3C][PF_6]$ in CH_3NO_2 at RT[26]. Similarly prepared is the dicationic Rh(III) complex $[\eta^5\text{-}CpRh(PMe_3)_2(\eta^2\text{-}C_2H_4)][PF_6]_2$ from $[Rh\eta^5\text{-}Cp(PMe_3)_2Et][PF_6]$, while the corresponding BF_4^- salt $[Rh\eta^5\text{-}Cp(PMe_3)_2(\eta^2\text{-}C_2H_4)][BF_4]_2$ is obtained in quantitative yield from $[Rh\eta^5\text{-}Cp(PMe_3)_2Et]I$ and $[Ph_3C][BF_4]$ (mole ratio 1:2.5) in nitromethane[26].

If the alkyl ligand has hydrogen atoms in the β position with respect to the metal and the metal also has an empty coordination site (which might be attained by ligand dissociation), then intramolecular β-hydrogen elimination can occur, e.g., by thermolysis, to give hydrido–olefin complexes:

$$L_nMCH_2CH_3 \rightarrow L_{n-1}MH(\eta^2\text{-}CH_2{=}CH_2) + L \qquad \text{(d)}$$

Thus the π-allyl complex $Ni(\eta^3\text{-}C_3H_5)(PPh_3)Et$ is stable below $-60°C$ but above that temperature the Ni atom abstracts a hydride ion from the alkyl group to give the hydrido–ethylene complex $Ni(\eta^3\text{-}C_3H_5)(PPh_3)H(\eta^2\text{-}C_2H_4)$. This is stable only up to $-40°C$, when ethylene is eliminated, and above $-30°C$ propene is evolved as the complex decomposes[27]. Other hydrido–olefin complexes formed by intramolecular β-hydride elimination are more thermally stable, however, and when solutions of the alkylruthenium complexes $Ru\eta^5\text{-}Cp(PPh_3)_2R$ (R = Et, n-Pr, n-Bu, i-Bu) in toluene are warmed to 50–80°C for $1\frac{1}{2}$–2 h a PPh_3 ligand dissociates and β-hydride elimination results in the formation of the Ru alkene hydride complexes Ruη^5-Cp(PPh_3)H(η^2-alkene) (alkene = ethene, propene, 1-butene, isobutene) in ca. 80% yield (32% for isobutene derivative). Higher temperatures result in the formation of Ruη^5-Cp$(PPh_3)_2$H and the orthometallation product Ru{$PPh_2C(CH)_4C$}(PPh_3)Cp-η^5 only[28]. Treatment of Os(η^6-C_6H_6)(PMe_3)EtI with $AgPF_6$ removes the iodide ligand and [Os(η^6-C_6H_6)(PMe_3)H(η^2-C_2H_4)][PF_6] is formed very quickly and in quantitative yield, presumably via the intermediate [Os(η^6-C_6H_6)(PMe_3)Et]$^+$ and subsequent β-hydride elimination[29]. Treatment of Mo(η^5-Cp)$_2$Cl$_2$ with $(EtAlCl_2)_2$ in benzene for 12 h at RT, followed by ethanolysis, gives 20% yield of Mo(η^5-Cp)$_2$ClEt and 20% yield of [Mo(η^5-Cp)$_2$H(η^2-C_2H_4)]$^+$, isolable as the PF_6 salt. This cation is also readily prepared starting from the chloroethyl derivative Mo(η^5-Cp)$_2$ClEt by treatment with $TlPF_6$ in acetone or with xs $(EtAlCl_2)_2$ in benzene. The tungsten analog [W(η^5-Cp)$_2$H(η^2-C_2H_4)][PF_6], however, is prepared in 60% yield from the dichloride W(η^5-Cp)$_2$Cl$_2$ and $(EtAlCl_2)_2$ in toluene with no evidence for the formation of the W analog of the neutral ethyl derivative. Treatment of W(η^5-Cp)$_2$Cl$_2$ in toluene with xs i-PrMgBr in ether, followed after 48 h at RT by ethanolysis and addition of [NH_4][PF_6] to the aqueous extract of the hydrolyzed reaction product gives the hydridopropene salt [W(η^5-Cp)$_2$H(η^2-CH_2=CHMe)][PF_6] in 30% yield[30]. These cations react further with aqueous base to form the neutral olefin derivatives M(η^5-Cp)$_2$(η^2-CH_2=CHR) (M = Mo, R = H; M = W, R = H, Me)[30,31].

The tungsten hydridoethylene or -propene cations, [W(η^5-Cp)$_2$H(η^2-CH_2=CHR)]$^+$ (R = H, Me), are also formed and isolated in ca. 90% yield as the BF_4 salts when the neutral dimethyl or methylethyl complexes W(η^5-Cp)$_2$MeR (R = Me, Et) are treated with equimolar quantities of [Ph_3C][BF_4] in CH_2Cl_2 at 25°C. If the reaction of W(η^5-Cp)$_2$MeEt with [Ph_3C][PF_6] is carried out at −78°C, then [W(η^5-Cp)$_2$MeEt][PF_6] is isolated instead in 94% yield. Both this and [W(η^5-Cp)$_2$Me$_2$][PF_6] then react[32] in CH_2Cl_2 or THF at RT with the trityl radical $Ph_3C\cdot$[31] to give [W(η^5-Cp)$_2$H(η^2-CH_2=CHR)][PF_6] (R = H, Me) in ca. 90% yield[33,34]. This reaction proceeds via hydrogen loss from the methyl group of W(η^5-Cp)$_2$MeR (R = Me, Et) to form the methylidene cation [W(η^5-Cp)$_2$(CH_2)R]$^+$ (R = Me, Et); this gives rise to the π-olefinic product by intramolecular insertion of the methylidene ligand into the W—R (R = Me, Et) bond and subsequent β-hydride elimination[34]. A similar reaction occurs for Ta(η^5-Cp)$_2$(CHMe)Me, where despite the presence of β-hydrogen atoms the ethylidene ligand does not rearrange to ethylene to give the expected Ta(η^5-Cp)$_2$(η^2-C_2H_4)Me on decomposition. Instead, on warming to 70°C in benzene, the complex rearranges to a 1.0:0.8 mixture of the exo and endoisomers of Ta(η^5-Cp)$_2$H(η^2-CH_2=CHMe) (32% isolated yield), presumably by migration of the alkyl to the alkylidene ligand followed by β-hydride elimination[35]. The methylidene complex Ta(η^5-Cp)$_2$(CH_2)Me, when treated with equimol MeI in benzene at RT (12 h), evolves methane and Ta(η^5-Cp)$_2$(η^2-C_2H_4)I is isolated, probably formed via the intermediate Ta(η^5-Cp)$_2$I(CH_2Me)Me and loss of a β-hydrogen (CH_3D is evolved when CD_3I is used as reagent). The iodo-ethylene complex reacts further with KH in THF to give the hydrido-ethylene complex Ta(η^5-Cp)$_2$H(η^2-C_2H_4)[36]. The nicest example from the early transition elements, however, is Nb(η^5-Cp)$_2$Et(η^2-C_2H_4), which on loss of ethylene by thermolysis regenerates crystallizable Nb(η^5-Cp)$_2$H(η^2-C_2H_4) by β-hydride elimination[37]. The same mechanism probably operates[38,39] in the formation of the Rh(III) hydrido-ethylene complex [η^5-CpRh(PMe_3)H(η^2-C_2H_4)][BF_4] from [η^5-CpRh(PMe_3)Et(η^2-C_2H_4)][BF_4].

More recently intramolecular β-hydrogen elimination has been effected by photolysis[40]. A well-studied example is the photoinduced dealkylation of the thermally inert complexes $W\eta^5$-$Cp(CO)_3R$ (R = Et, n-Pr, n-Bu, n-pentyl, CH_2CH_2Ph) to give, following photoejection of a CO ligand and β-hydrogen transfer to the metal, the trans complexes $W\eta^5$-$Cp(CO)_2H(\eta^2$-olefin) (olefin = ethene[41,43], propene[42], 1-butene[42], 1-pentene[43,44], styrene[45]). The cis isomer is generated initially but this rapidly converts into the thermodynamically more favorable form in which the hydrido ligand and the olefin occupy trans positions in the tetragonal pyramid opposite the η^5-Cp ring. Typically, photolysis of the alkyl complexes in pentane solution at $-30°C$ using a Hg high-pressure lamp ($\lambda > 300$ nm) proceeds rapidly and after 10 min the hydrido–olefin complexes represent the main products, separation and isolation being effected by column chromatography on silica at $-30°C$ and fractional crystallization at $-78°C$. Small yields (6–11%) of yellow crystals are obtained. The pentamethyl derivatives $W(\eta^5$-$C_5Me_5)(CO)_2H(\eta^2$-olefin) (olefin = ethene[43], propene[42]), obtained by photolysis of $W(\eta^5$-$C_5Me_5)(CO)_3R$ (R = Et, n-Pr), are thermally and kinetically more stable than the corresponding η^5-Cp complexes, while the η^5-Cp alkene hydrides of molybdenum are labile with respect to alkene loss, η^5-$CpMo(CO)_2H(\eta^2$-$C_2H_4)$ decomposing above ca. $-20°C$[41]. The photolysis of $Ta(\eta^5$-$Cp)_2(CO)Et$ in solution also proceeds via CO loss and subsequent β-hydrogen elimination to give $Ta(\eta^5$-$Cp)_2H(\eta^2$-$C_2H_4)$[46]. Photochemistry thus allows isolation and characterization of thermally inaccessible species; e.g., the thermolysis of $W(\eta^5$-$Cp)(CO)_3R$ requires temperatures well above 25°C to obtain a significant rate of reaction, and the product formed is mainly $W_2(\eta^5$-$Cp)_2(CO)_6$, not an alkene–hydride[43]. The temperatures necessary to promote dissociative loss of CO from $WCp(CO)_3R$ are apparently sufficiently great to lead to good rates for decomposition of the alkene–hydride or to other thermal processes of η^5-$CpW(CO)_3R$.

In general, the lanthanide and actinide σ-bonded organometallics have surprisingly high thermal stability and are not normally considered prone to β-hydride elimination—there is very little difference in stability between compounds with or without β-hydrogens[47,48]. However, ultraviolet photolysis of such organoactinides as $M(\eta^5$-$Cp)_3R$ (M = Th, U; R = i-Pr, n-Bu) does result in facile photoinduced β-hydride elimination and provides an efficient route to organoactinide hydrides and to actinide organometallics in unusual oxidation states, but no hydrido–olefin intermediates can be isolated[46,49].

(A. T. HUTTON)

1. D. E. Laycock, J. Hartgerink, M. C. Baird, *J. Org. Chem.*, *45*, 291 (1980).
2. A. Rosan, M. Rosenblum, J. Tancrede, *J. Am. Chem. Soc.*, *95*, 3062 (1973).
3. P. Lennon, M. Madhavarao, A. Rosan, M. Rosenblum, *J. Organomet. Chem.*, *108*, 93 (1976).
4. P. Lennon, A. M. Rosan, M. Rosenblum, *J. Am. Chem. Soc.*, *99*, 8426 (1977).
5. M. L. H. Green, P. L. I. Nagy, *Proc. Chem. Soc., London*, 74 (1962).
6. M. L. H. Green, P. L. I. Nagy, *J. Organomet. Chem.*, *1*, 58 (1963).
7. W. H. Knoth, *Inorg. Chem.*, *14*, 1566 (1975).
8. R. C. Kerber, D. J. Ehntholt, *J. Am. Chem. Soc.*, *95*, 2927 (1973).
9. J. W. Faller, B. V. Johnson, *J. Organomet. Chem.*, *88*, 101 (1975).
10. R. S. Bly, G. S. Silverman, M. M. Hossain, R. K. Bly, *Organometallics*, *3*, 642 (1984).
11. D. J. Ehntholt, G. F. Emerson, R. C. Kerber, *J. Am. Chem, Soc.*, *91*, 7547 (1969).
12. A. Cutler, D. Ehntholt, W. P. Giering, P. Lennon, S. Raghu, A. Rosan, M. Rosenblum, J. Tancrede, D. Wells, *J. Am. Chem. Soc.*, *98*, 3495 (1976).
13. A. Sanders, T. Bauch, C. V. Magatti, C. Lorenc, W. P. Giering, *J. Organomet. Chem.*, *107*, 359 (1976).
14a. R. S. Bly, G. S. Silverman, R. K. Bly, *Organometallics*, *4*, 374 (1985); (b) R. S. Bly, R. K. Bly, M. M. Hossain, C. S. Silverman, E. Wallace, *Tetrahedron*, *42*, 1093 (1986).

15. Kabir-ud-Din, P. H. Plesch, *J. Chem. Soc., Perkin Trans. 2*, 937 (1978). THF was used in early hydride abstraction experiments as the best solvent for dissolution of $[Ph_3C][ClO_4]$[5,6].
16. Solutions of many of the iron–olefin complexes in MeCN are observed to undergo substitution to give free olefin and the cationic acetonitrile complex $[Fp(MeCN)]^+$; see P. M. Treichel, R. L. Shubkin, K. W. Barnett, D. Reichard, *Inorg. Chem.*, *5*, 1177 (1966).
17. H. J. Dauben, L. R. Honnen, K. M. Harmon, *J. Org. Chem.*, *25*, 1442 (1960).
18. L. Cohen, W. P. Giering, D. Kenedy, C. V. Magatti, A. Sanders, *J. Organomet. Chem.*, *65*, C57 (1974).
19. M. Schlosser, in *Methoden der organischen Chemie (Houben-Weyl)*, Vol. V/1b, Georg Thieme Verlag, Stuttgart, 1972.
20. R. B. King, M. B. Bisnette, *J. Organomet. Chem.*, *7*, 311 (1967).
21. A. Sanders, L. Cohen, W. P. Giering, D. Kenedy, C. V. Magatti, *J. Am. Chem. Soc.*, *95*, 5430 (1973).
22. A. R. Cutler, *J. Am. Chem. Soc.*, *101*, 604 (1979).
23. M. Cousins, M. L. H. Green, *J. Chem. Soc.*, 889 (1963).
24. R. I. Mink, J. J. Welter, P. R. Young, G. D. Stucky, *J. Am. Chem. Soc.*, *101*, 6928 (1979).
25. W. A. Kiel, G.-Y. Lin, J. A. Gladysz, *J. Am. Chem. Soc.*, *102*, 3299 (1980); S. Georgiou, J. A. Gladysz, *Tetrahedron*, *42*, 1109 (1986).
26. H. Werner, R. Feser, R. Werner, *J. Organomet. Chem.*, *181*, C7 (1979).
27. H. Bönnemann, C. Grard, W. Kopp, G. Wilke, *Proc. 23rd Int. Congr. Pure Appl. Chem.*, Butterworths, London, 1971, Vol. 6, p. 265.
28. H. Lehmkuhl, J. Grundke, R. Benn, G. Schroth, R. Mynott, *J. Organomet. Chem.*, *217*, C5 (1981); H. Lehmkuhl, J. Grundke, R. Mynott, *Chem. Ber.*, *116*, 159 (1983).
29. R. Werner, H. Werner, *Chem. Ber.*, *116*, 2074 (1983).
30. F. W. S. Benfield, B. R. Francis, M. L. H. Green, *J. Organomet. Chem.*, *44*, C13 (1972).
31. F. W. S. Benfield, M. L. H. Green, *J. Chem. Soc., Dalton Trans.*, 1324 (1974).
32. Added as 4-(triphenylmethyl)-1-(diphenylmethylidene)-2,5-cyclohexadiene (the dimer of the trityl radical). See J. W. Copenhaver, M. F. Roy, C. S. Marvel, *J. Am. Chem. Soc.*, *57*, 1311 (1935); J. M. McBride, *Tetrahedron*, *30*, 2009 (1974) and refs. therein.
33. J. C. Hayes, G. D. N. Pearson, N. J. Cooper, *J. Am. Chem. Soc.*, *103*, 4648 (1981).
34. J. C. Hayes, N. J. Cooper, *J. Am. Chem. Soc.*, *104*, 5570 (1982).
35. P. R. Sharp, R. R. Schrock, *J. Organomet. Chem.*, *171*, 43 (1979).
36. R. R. Schrock, P. R. Sharp, *J. Am. Chem. Soc.*, *100*, 2389 (1978).
37. F. N. Tebbe, G. W. Parshall, *J. Am. Chem. Soc.*, *93*, 3793 (1971).
38. H. Werner, R. Feser, *Angew. Chem., Int. Ed. Engl.*, *18*, 157 (1979).
39. R. Feser, H. Werner, *J. Organomet. Chem.*, *233*, 193 (1982).
40. H. G. Alt, *Angew. Chem., Int. Ed. Engl.*, *23*, 766 (1984). Review of the photochemistry of alkyl transition-metal complexes.
41. H. G. Alt, M. E. Eichner, *Angew. Chem., Int. Ed. Engl.*, *21*, 78 (1982).
42. K. A. Mahmoud, A. J. Rest, H. G. Alt, M. E. Eichner, B. M. Jansen, *J. Chem. Soc., Dalton Trans.*, 175 (1984).
43. R. J. Kazlauskas, M. S. Wrighton, *J. Am. Chem. Soc.*, *104*, 6005 (1982).
44. R. J. Kazlauskas, M. S. Wrighton, *J. Am. Chem. Soc.*, *102*, 1727 (1980).
45. B. Klein, R. J. Kazlauskas, M. S. Wrighton, *Organometallics*, *1*, 1338 (1982).
46. M. E. Eichner, H. G. Alt, M. D. Rausch, *J. Organomet. Chem.*, *264*, 309 (1984).
47. See, e.g., T. J. Marks, R. D. Ernst, in *Comprehensive Organometallic Chemistry*, G. Wilkinson, F. G. A. Stone, E. W. Abel, eds., Pergamon, Oxford, 1982, Vol. 3, p. 173.
48. W. J. Evans, in *The Chemistry of the Metal-Carbon Bond*, F. R. Hartley, S. Patai, eds., Wiley (Interscience), Chichester, 1982, Vol. 1, p. 489.
49. J. W. Bruno, D. G. Kalina, E. A. Mintz, T. J. Marks, *J. Am. Chem. Soc.*, *104*, 1860 (1982).

5.8.2.2.6. Alkylidene Complexes by α-Hydride Elimination.

Isolated, well-characterized alkylidene complexes of Ta, Nb, W, Mo, Re, Ir and Fe have been obtained by the α-hydride elimination reaction. In some cases these and other alkylidene complexes have been synthesized by other reactions, such as hydrogenation of alkylidyne complexes, alkylidene transfer or protonation of alkoxy complexes. Discussion is limited to the α-hydride elimination reaction. Only IR evidence has been obtained for the product shown[1]:

$$\eta^5\text{-CpCr(CO)}_3\text{CH}_3 \xrightarrow[12\,K]{h\nu} \text{CO} + \eta^5\text{-CpCr(CO)}_2\text{(H)CH}_2 \quad \text{(a)}$$

The α-hydride elimination reactions of Ta and Nb have been investigated most extensively. The driving force for the reaction is provided by the interaction of an empty metal orbital with the α C—H bonding electrons. Steric crowding on the metal as well as the absence of β hydrogens on the ligands are important for successful synthesis. The same principles governing the alkyl-to-alkylidene transformation apply to the alkylidene-to-alkylidyne transformation and a number of alkylidyne complexes have been made by the same methods described here for alkylidene synthesis[2].

Relative ease of proton removal from the alkyl ligands is consistently $CH_2CMe_3 > CH_2SiMe_3 > CH_2Ph >> CH_3$ [3]. The alkylidene complexes of Ta and Nb are air and moisture sensitive; the Ta complexes are more stable than the Nb complexes.

The trisneopentyl neopentylidene complexes of Ta and Nb can be prepared[4]:

$$\text{M(CH}_2\text{CMe}_3)_3\text{Cl}_2 + 2\,\text{LiCH}_2\text{CMe}_3 \xrightarrow{\text{Pentane}} \text{M(CH}_2\text{CMe}_3)_3\text{(CHCMe}_3) + \text{CMe}_4. \quad \text{(b)}$$

where M = Ta, Nb. The Ta complex can be sublimed at 71°C in vacuo but the Nb complex decomposes on attempted sublimation. Mechanistic studies suggest that the elimination reaction occurs from $M(CH_2CMe_3)_4Cl$, not from $M(CH_2CMe_3)_5$.

Benzylidene complexes of Ta have been prepared similarly:

$$\text{Ta(CH}_2\text{Ph)}_3\text{Cl}_2 + 2\,\text{TlC}_5\text{H}_5 \rightarrow \text{Ta}(\eta^5\text{-Cp})\text{(CHPh)(CH}_2\text{Ph)} \quad \text{(c)}$$

$$\text{Ta(CH}_2\text{PH)}_3\text{Cl}_2 + \text{LiC}_5\text{Me}_5 \rightarrow \text{Ta}(\eta^5\text{-Cp})\text{(CHPh)(CH}_2\text{Ph)Cl} \quad \text{(d)}$$

Equation (e) describes the synthesis of a methylidene complex[7]:

$$\text{Ta}(\eta^5\text{-Cp})_2\text{Me}_3 + [\text{Ph}_3\text{C}]^+ \rightarrow [\text{Ta}(\eta^5\text{-Cp})_2\text{Me}_3]_3^+ \xrightarrow{\text{Me}_3\text{P=CH}_2} \text{Ta}(\eta^5\text{-Cp})_2(\text{Me}_3\text{P})(\text{CH}_2) \quad \text{(e)}$$

The neopentylidene and benzylidene complexes shown above are stable but the methylidene complex decomposes at RT in solution.

Disubstituted phenoxides have proved to be synthetically useful in preparing Ta alkylidenes[8]:

$$\text{Ta(OAr)}_2\text{Cl}_3 + 3\,\text{LiCH}_2\text{SiMe}_3 \rightarrow \text{Ta(OAr)}_2(\text{CH}_2\text{SiMe}_3)(\text{CHSiMe}_3) \quad \text{(f)}$$

where OAr = 2,6-di-t-butylphenoxide. Treatment of Ta tris-phenoxide dichloride with 3 equiv alkyl lithium or organomagnesium halide reagent yields tris alkylated products that produce Ta alkylidenes photochemically[8,9]:

$$\text{Ta(OAr)}_3\text{Cl}_2 + 2\,\text{LiCH}_2\text{SiMe}_3 \rightarrow \text{Ta(OAr)}_2(\text{CH}_2\text{SiMe}_3)_3 \quad \text{(g)}$$

$$\text{Ta(OAr)}_2(\text{CH}_2\text{SiMe}_3)_3 \rightarrow \text{Me}_4\text{Si} + \text{Ta(OAr)}_2(\text{CH}_2\text{SiMe}_3)(\text{CHSiMe}_3) \quad \text{(h)}$$

where OAr = 2,6-bis-i-propylphenoxide;

$$Ta(OAr)_2Me_3 \rightarrow Ta(OAr)_2(Me)(CH_2) + CH_4 \quad \text{(i)}$$

where OAr = 2,6-di-t-butylphenoxide. Iridium and Re alkylidenes have been made photochemically[10,11].

Bis-alkylidene complexes can be made by a second α elimination from a monoalkylidene complex, or directly from a tetraneopentyl complex or by what appears to be a Ta-mediated hydrogen transfer from a neopentyl ligand to a neopentylidyne ligand[12]:

$$M(CH_2Me_3)_3(CHCMe_3) + 2\,L \rightarrow M(CH_2CMe_3)(CHCMe_3)_2L_2 \quad \text{(j)}$$

where M = Ta, Nb; L = PMe_3, PMe_2PH;

$$M(CH_2CMe_3)_4Cl + 2\,L \rightarrow M(CHCMe_3)_2L_2Cl \quad \text{(k)}$$

where M = Ta, Nb; L = PMe_3, PMe_2Ph;

$$\eta^5\text{-}C_5Me_5Ta(Cl)(PMe_3)_2(CMe_3) + LiCH_2CMe_3 \rightarrow (\eta^5\text{-}C_5Me_5)Ta(CHCMe_3)_2PMe_3. \quad \text{(l)}$$

Electron pair bases have been used extensively to induce α-hydride elimination, probably by bringing about the steric crowding necessary to initiate the reaction[13]. Equation (m) exemplifies this process:

$$MR_2X_3 + 2L \rightarrow M(CHR')L_2X_3 \quad \text{(m)}$$

where X = Cl, Br; M = Ta, Nb; R = CH_2CMe_3, CH_2SiMe_3, CH_2Ph; L = PMe_3, $PPhMe_2$, PPh_2Me, dmpe, diphos, THF. The conditions necessary to induce the reaction are increasingly stringent from CH_2CMe to CH_2SiMe to CH_2Ph. In a related reaction the first cyclic alkylidene complex has been made by treating $TaCl_5$ with Na in trimethylphosphorane solvent[14]:

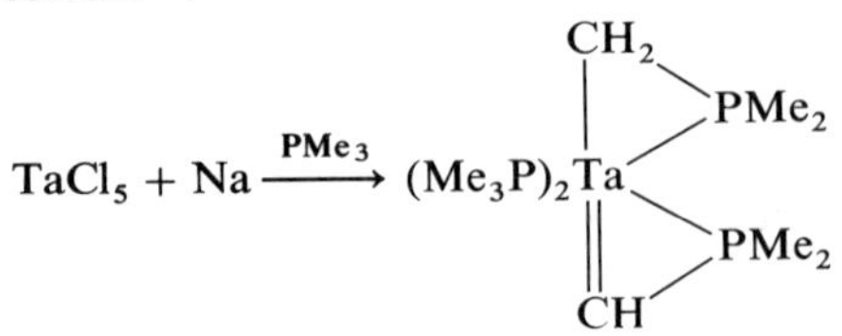

(n)

Whereas treatment of $TaCl_5$ with neopentyllithium does give a moderate yield of alkylidene Ta compound, the analogous reaction with W(VI) chloride or Mo(V) chloride yields the trisneopentylmetalalkylidyne complex. A mixed alkylidene–alkylidyne complex can then be made[15]:

$$W(CH_2CMe_3)_3(CCMe_3) + 2\,PMe_3 \rightarrow W(CH_2CMe_3)(PMe_3)_2(CHCMe_3)(CCMe_3) \quad \text{(o)}$$

which is again an intramolecular hydrogen transfer. In general, it seems that good π-donating ligands are necessary to stabilize high-oxidation, state W and Re alkylidene complexes. Thus alkoxy and phenylimido W and Re complexes may be prepared by methods similar to those already described for the Ta alkylidenes:

$$W(OAlCl_3)(OCH_2CMe_3)_2(CH_2CMe_3)_2 \xrightarrow[25°C]{CH_2Cl_2} W(OCH_2CMe_3)_2(Cl_2)(CMe_3) \quad \text{(p)}$$ [16]

$$W(NPh)(CH_2CMe_3)_3Cl + LiCH_2CMe_3 \rightarrow W(NPh)(CHCMe_3)(CH_2CMe_3)_2 \qquad (q)^{17}$$

$$Re(NCMe_3)_2CL_3 + 3\ Me_3CCH_2MgCl \rightarrow Re(NCMe_3)_2(CHCMe_3)(CH_2CMe_3) \qquad (r)^{18}$$

There has been discussion in the literature of 1,2-hydrogen shifts in alkyltungsten compounds[15]: An early example of a 1,2-hydrogen shift that produces a stable alkylidene–Ir product is[20]:

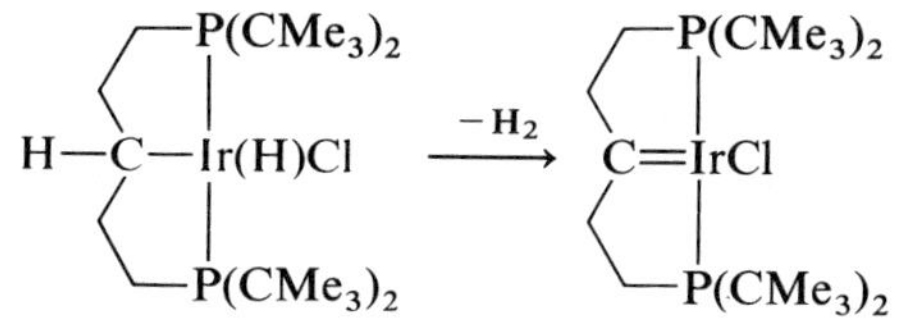

(s)

Chemical reduction of neopentyltantalum (V) species produces isolable alkylidene hydride products[21]:

$$Ta(CH_2CMe_3)Cl_4 \xrightarrow[PMe]{Na-Hg} Ta(CHCMe_{3e})(H)(PMe_3)Cl_2 \qquad (t)$$

It is also possible to make alkylidene hydride complexes of W by hydrogenation of tungsten alkylidyne complexes[22].

The trityl cation is a useful deprotonating agent to produce lower oxidation state metal alkylidenes:[23,24]

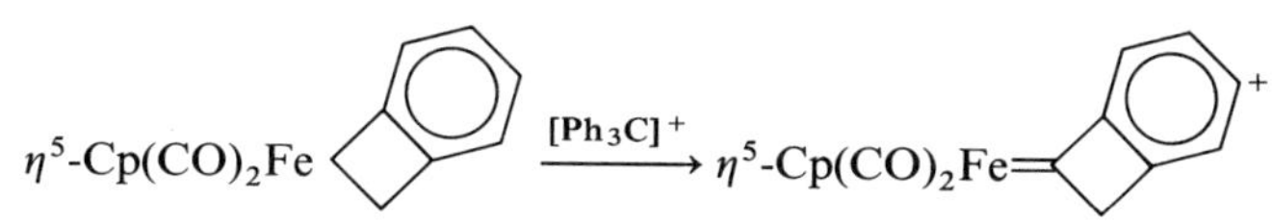

(u)

$$\eta^5\text{-}CpW(CO)_3CH_3 \xrightarrow{[Ph_3C]^+} \eta^5\text{-}CpW(CO)_3CH_2^+ \qquad (v)$$

The corresponding reaction with η^5-CpRe(NO)(PPh$_3$)R has been more thoroughly studied. When R is a straight-chain alkyl group, α-hydride elimination occurs exclusively. Even β-hydride elimination is possible when R is ethyl or larger. When R is β-phenylethyl, α and β elimination occur and when R is a branched alkyl group β elimination occurs exclusively. This remarkable selectivity toward α-hydride elimination is explained in terms of anchimeric assistance by the LUMO which is a Re 5d orbital. Deuterium labeling studies on the benzyl derivative show that only one rotamer reacts, even though there is unrestricted rotation, and the hydrogen (deuterium) antiperiplanar to the triphenyl phosphine ligand is selectively removed[25,26]:

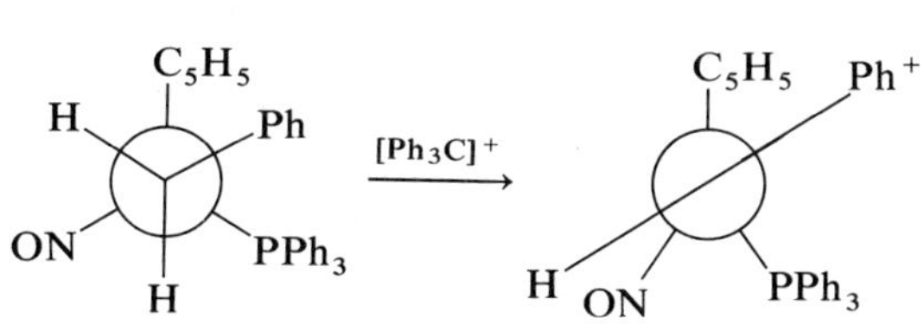

(w)

Detailed investigations of all these reactions show they proceed by an electron-transfer hydrogen-atom-abstraction process[27–29]

Evidence for a U–C double bond has been presented[30] for the compound $(\eta^5\text{-Cp})_3UCHPMe_2Ph$:

$$(\eta^5\text{-Cp})_3UCl + Li(CH_2)(CH_2)PMePh \xrightarrow[-50°C]{THF} (\eta^5\text{-Cp})_3UCHPMe_2Ph \qquad (x)$$

While this compound would have to be classified as a carbene rather than an alkylidene complex, the synthesis involves an intramolecular H atom transfer, which is a recurring phenomenon in metal–alkylidene complex synthesis.

(J. W. CONNOLLY)

1. A. Mahmoud, A. J. Rost, H. G. Alt, *J. Chem. Soc., Chem. Commun*, 1011 (1983).
2. R. R. Schrock, D. N. Clark, J. Sancho, J. H. Wengrovius, S. M. Rocklage, S. F. Pederson, *Organometallics*, *1*, 1645 (1982).
3. R. R. Schrock, *Accts. Chem. Res.*, *12*, 98 (1979).
4. R. R. Schrock, J. D. Fellman, *J. Am. Chem. Soc.*, *100*, 3359 (1978).
5. R. R. Schrock, L. W. Messerle, C. D. Wood, L. L. Guggenberger, *J. Am. Chem. Soc.*, *100*, 3793 (1978).
6. L. W. Messerle, P. Jennische, R. R. Schrock, G. Stucky, *J. Am. Chem. Soc.* *102*, 6744 (1980).
7. R. R. Schrock, P. R. Sharp, *J. Am. Chem. Soc.*, *102*, 2389 (1978).
8. L. R. Chamberlain, I. P. Rothwell, K. Folting, J. C. Huffman, *J. Chem. Soc., Dalton Trans.*, 155, (1987).
9. L. Chamberlain, I. P. Rothwell, J. C. Huffman, *J. Am. Chem. Soc.*, *104*, 7338 (1982).
10. M. D. Fryzuk, P. A. MacNeil, S. J. Rettig, *J. Am. Chem. Soc.*, *107*, 6708 (1985).
11. D. C. Edwards, L. V. Biondi, J. W. Zillner, M. R. Churchill, R. R. Schrock, *Organometallics*, *2*, 1505 (1983).
12. J. D. Fellman, G. A. Ruprecht, C. D. Wood, R. R. Schrock, *J. Am. Chem. Soc.*, *100*, 5964 (1978).
13. G. A. Ruprecht, L. W. Messerle, J. D. Fellman, R. R. Schrock, *J. Am. Chem. Soc.*, *102*, 6236 (1980).
14. V. C. Gibson, P. D. Grebenic, M. L. H. Green, *J. Chem. Soc., Chem. Commun.*, 1101 (1983).
15. D. N. Clark, R. R. Schrock, *J. Am. Chem. Soc.*, *100*, 6774 (1978).
16. J. Kress, M. Wesolek, J. A. Osborn, *J. Chem. Soc., Chem. Commun.*, 514 (1982).
17. S. F. Pederson, R. R. Schrock, *J. Am. Chem. Soc.*, *104*, 7483 (1982).
18. D. S. Edwards, R. R. Schrock, *J. Am. Chem. Soc.*, *104*, 6806 (1982).
19. M. Canestrari, M. L. H. Green, *J. Chem. Soc., Dalton Trans.*, 1789 (1982).
20. H. D. Empsall, E. M. Hyde, R. Markham, W. S. McDonald, M. C. Norton, B. L. Shaw, B. Weeks, *J. Chem. Soc., Chem. Commun.*, 589 (1977).
21. H. W. Turner, R. R. Schrock, J. D. Fellman, S. J. Holmes, *J. Am. Chem. Soc.*, *105*, 4942 (1983).
22. J. H. Wengrovius, R. R. Schrock, M. R. Churchill, H. J. Wasserman, *J. Am. Chem. Soc.*, *104*, 1739 (1982).
23. A. Sanders, T. Bauch, C. V. Magatti, W. P. Giering, *J. Organomet. Chem.*, *107*, 359 (1976).
24. J. C. Hayes, G. Pearson, N. J. Cooper, *J. Am. Chem. Soc.*, *103*, 4648 (1981).
25. W. A. Kiel, G-Y Lin, G. S. Bodner, J. A. Gladysz, *J. Am. Chem. Soc.* *105*, 4958 (1983).
26. W. A. Keil, W. E. Buhro, J. A. Gladysz, *Organometallics*, *3*, 879 (1984).
27. J. C. Hayes, N. J. Cooper, *J. Am. Chem. Soc.*, *104*, 5570 (1982).
28. C. Roger, L. Toupet, C. Lapinete, *J. Chem. Soc., Chem. Commun.*, 713 (1988).
29. G. S. Bodner, J. A. Gladysz, M. F. Nielsen, V. D. Parker, *J. Am. Chem. Soc.*, *109*, 1757 (1987).
30. R. E. Cramer, R. B. Maynard, J. C. Paw, J. W. Gilje, *Organometallics*, *2*, 1336 (1983).

5.8.2.3. from Monoolefins

5.8.2.3.1. η^2 π-Olefin Complexes by Reaction of Metal Complexes with Monoolefins by Ligand Displacement.

The preparation of olefin complexes is considered here in three parts: first the displacement of anionic ligands, then via neutral ligands and finally from coordinatively unsaturated metal complexes where no ligands are displayed by the incoming olefins.

(i) Displacement of Anionic Ligands. The displacement of anionic ligands, particularly halide ions, has proved a valuable route to metal–olefin complexes. Thus K_2PtCl_4 reacts in aqueous solution with both water-soluble olefins, such as allylamine[1], and water-insoluble and even gaseous olefins, such as ethylene, with[2] or without[3] pressure. The reaction usually takes a number of days to go to completion[3] but can be accelerated to take 2–4 h by a trace of tin(II) chloride[2]:

$$K_2PtCl_4 + C_2H_4 + HCl \xrightarrow{H_2O} K[PtCl_3(C_2H_4)]\cdot H_2O + KCl \qquad \text{(a)}$$

Full experimental details have been given[4,5]. The formation of Pt(II) complexes:

$$[PtX_4]^{2-} + L \rightarrow [PtX_3L]^- + X^- \qquad \text{(b)}$$

normally follows the two-term rate law:

$$\text{Rate} = \{k_1 + k_2[L]\}[PtX_4]^{2-} \qquad \text{(c)}$$

where k_1, the first-order rate constant, corresponds to the rate for the solvent-dependent path and k_2, the second-order rate constant, corresponds to the rate for the solvent-independent path[6]. However, $k_1 = 0$ when the entering ligand is an olefin[7–9], which implies that the aquospecies is not involved in the rate-determining step. However recent work[10–13] has shown that ethylene reacts more rapidly with $[PtCl_3(H_2O)]^-$ than with $[PtCl_4]^{2-}$ and that $[PtCl_3(C_2H_4)]^-$ catalyzes the aquation of $[PtCl_4]^{2-}$. The failure of earlier workers to observe a k_1 term is ascribed to their having a large concentration of the leaving chloride group present[12], so that the overall mechanism is:

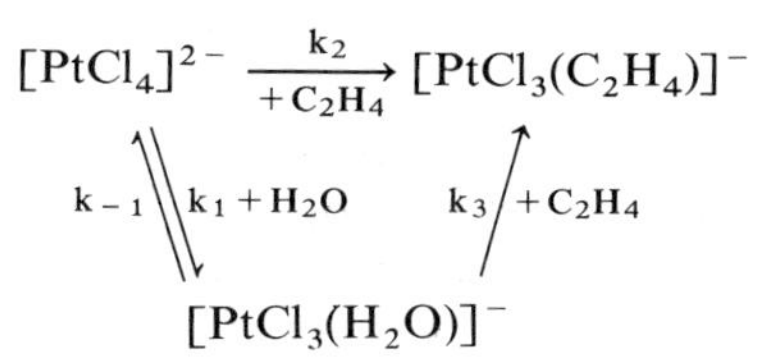

(d)

Although the charge on the olefin makes relatively little difference to the rate, positively charged olefins, such as $[CH_2{=}CHCH_2NH_3]^+$, react slightly faster than neutral olefins, such as C_2H_4 or $CH_2{=}CHCH_2OH$ which in turn react faster than anionic olefins, such as $[CH_2{=}CHCH_2SO_3]^-$ [8,12]. Tin(II) chloride is used as a catalyst; however, when the displacement is carried out in DMSO the latter is incorporated in the final product.

$$K_2PtCl_4 + ol + DMSO + HCl \xrightarrow{acetone} cis\text{-}PtCl_2(ol)(DMSO) \qquad \text{(e)}$$

where ol = C_2H_4, $CH_3CH{=}CH_2$, $C_2H_5CH{=}CH_2$, cis- or trans-$CH_3CH{=}CHCH_3$.

Not all olefins react with Pt(II) complexes to form the expected product. Thus while allyl alcohol displaces chloride from $K_2[PtCl_4]$ to form the expected olefin complex initially, this reacts further if held at 50°C for a couple of days to form a diolefin complex through a dehydration reaction[15]:

$$K_2[PtCl_4] + CH_2{=}CHCH_2OH \xrightarrow[50^\circ C]{10\ min} K[PtCl_3(CH_2{=}CHCH_2OH)] + KCl$$

$$\downarrow + CH_2{=}CHCH_2OH \quad 2\ d,\ 50^\circ C$$

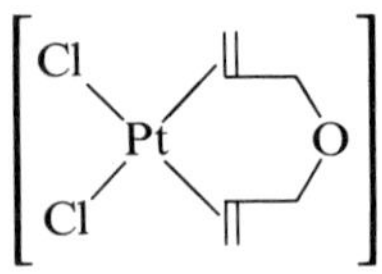

$+ H_2O + KCl$ (f)

Platinum(II) complexes containing both η^1-alkyl and η^2-olefins form by removal of the chloride ligand from the metal by Ag(I) salt[16,17]:

$$\text{trans-}Pt(CH_3)Cl(PMe_2Ph)_2 + AgPF_6 + ol \rightarrow \text{trans-}[Pt(CH_3)(ol)(PMe_2Ph)_2][PF_6] \quad (g)$$

where ol = C_2H_4, $CH_3CH{=}CH_2$, $CH_2{=}CHCH_2OH$ or vinyl ether.

Many Rh(I)–olefin complexes are formed by displacing ethylene from $[RhCl(C_2H_4)_2]_2$. The parent ethylene complex is prepared by simultaneous reduction and chloride displacement from $RhCl_3$ [18,19]:

$$2\ RhCl_3{\cdot}H_2O + 6\ CH_2{=}CH_2 \xrightarrow{MeOH-H_2O}$$

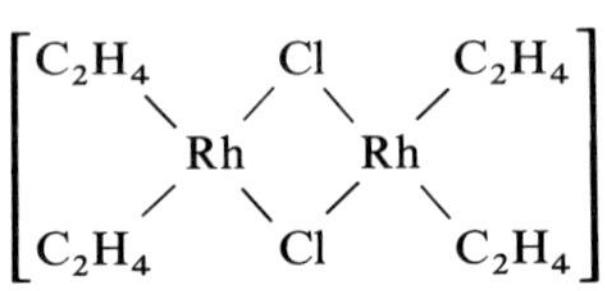

$+ 2\ CH_3CHO + 4\ HCl$ (h)

The reducing agent is ethylene, which is itself oxidized to acetaldehyde. An analogous reaction occurs with $IrCl_3$ and cyclooctene, although in this case isopropanol is used as the reducing agent as well as being the major component of the solvent[20,21]:

$$2\ IrCl_3 + 4\ C_8H_{14} + 2\ (CH_3)_2CHOH \xrightarrow{H_2O} [IrCl(C_8H_{14})_2]_2 + 4\ HCl + 2\ (CH_3)_2CO \quad (i)$$

An alternative, less convenient preparation of $[IrCl(C_8H_{14})_2]_2$ starts with $[NH_4]_3[IrCl_6]$ [22]:

$$2\ [NH_4]_3[IrCl_6] + 4\ C_8H_{14} + 2\ (CH_3)_2CHOH \xrightarrow{H_2O} [IrCl(C_8H_{14})_2]_2 + 6\ NH_4Cl + 2\ (CH_3)_2CO + 4\ HCl \quad (j)$$

When the Ir(IV) complex, H_2IrCl_6, is treated with cyclooctene in aqueous ethanol or n-propanol the Ir(I) carbonyl olefin complex $IrCl(CO)(C_8H_{14})_3$ is formed[23,24]:

$$H_2IrCl_6 + 4\ C_4H_{18} + C_2H_5OH \rightarrow [IrCl(C_8H_{14})_2]_2 + CH_3CHO + 4\ HCl \qquad (k)$$

The osmium complex $OsCl(CO)(NO)(PPh_3)_2$ can be converted into the corresponding cationic ethylene complex by a halide displacement promoted by Ag(I)[25]:

$$OsCl(CO)(NO)(PPh_3)_2 + C_2H_4 + AgPF_6 \rightarrow [Os(CO)(NO)(C_2H_4)(PPh_3)_2][PF_6] \qquad (l)$$

(ii) Displacement of Neutral Ligands: Olefins. Chromium–olefin complexes form by displacement of cyclooctene from $Cr(\eta^5\text{-Cp})(CO)(NO)(C_8H_{14})$[26]:

$$[Cr(\eta^5\text{-Cp})(CO)(NO)(C_8H_{14}] + ol \xrightarrow{C_6H_6} Cr(\eta^5\text{-Cp})(CO)(NO)(ol) + C_8H_{14} \qquad (m)$$

where ol = C_2H_4 or maleic anhydride. Displacement of norbornadiene from $[M(NBD)(CO)_4]$ (where M = Cr and Mo) by o-styryldiphenylphosphine yields the Cr(0) and Mo(0) olefin complexes of the unsaturated phosphine[27]:

$M(NBD)(CO)_4$ + 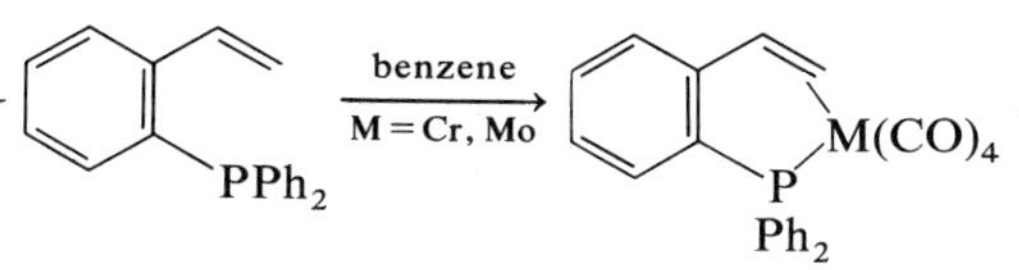 + nbd (n)

where M = Cr, Mo.

Cationic iron–olefin complexes have been prepared[28,29] by displacing isobutene from $[Fe(\eta^5\text{-Cp})(CO)_2(isobutene)][BF_4]$:

$$[Fe(\eta^5\text{-Cp})(CO)_2(isobutene)][BF_4] + ol \xrightarrow[ol = C_2H_4,\ cycloheptene,\ cyclooctene]{CH_2Cl_2} [Fe(\eta^5\text{-Cp})(CO)_2(ol)][BF_4] + isobutene \qquad (o)$$

The reaction is rapid, being complete in about 10 min, and may be monitored by the evolution of isobutene. The products involve the formation of only one iron–olefin bond so the $[Fe(\eta^5\text{-Cp})(CO)_2]^+$ cation may be used as a protecting group for one double bond of a polyolefin[30]:

$[Fe(\eta^5\text{-Cp})(CO)_2(isobutene)]^+$ + 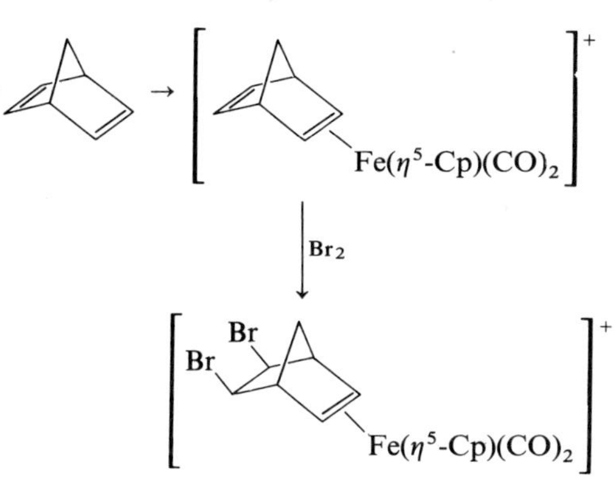 (p)

For compounds containing both double and triple bonds, such as $CH_2{=}CHCH_2C{\equiv}CCH_2CH{=}CH_2$, $[Fe(\eta^5\text{-}Cp)(CO)_2]^+$ selectively coordinates to one of the double bonds.

Many Rh(I)-olefin complexes have been prepared by displacing ethylene from $[RhCl(C_2H_4)_2]_2$ (see above) and $[Rh(acac)(C_2H_4)_2]$[31]:

$$[RhCl(C_2H_4)_2]_2 + 4\ ol \rightarrow [RhCl(ol)_2]_2 + 4\ C_2H_4 \qquad (q)$$

Iridium(I) complexes may be prepared similarly[31,32]:

$$M(acac)(C_2H_4)_2 + 4\ ol \rightarrow M(acac)(ol)_2 + 2\ C_2H_4 \qquad (r)$$

where M = Rh, Ir;

$$[IrCl(cyclooctene)_2]_2 + 8\ C_2H_4 \rightarrow 2\ IrCl(C_2H_4)_4 + 4\ cyclooctene \xrightleftharpoons{-4\ C_2H_4} [IrCl(C_2H_4)_2]_2 \qquad (s)$$

Nickel(0)-olefin complexes are often prepared by displacement reactions on $[Ni(1,5\text{-}COD)_2]$[33-36]:

$$Ni(1,5\text{-}COD)_2 + 2\ PPh_3 + PhCH{=}CHPh \xrightarrow{pentane} Ni(PPh_3)_2(PhCH{=}CHPh) + 2\ 1,5\text{-}COD \qquad (t)$$

These displacements are so facile that $Ni(1,5\text{-}COD)_2$ has been described as naked nickel[37]. In some cases the reactions are carried out in two stages with intermediate formation of, for example, an isonitrile complex[35]:

$$Ni(1,5\text{-}COD)_2 + 2\ t\text{-}BuNC \xrightarrow{Et_2O} Ni(CNBu\text{-}t)_2 \xrightarrow[\text{CHCN}]{+\ \text{CHCN}\ |\ \text{in}\ Et_2O} Ni(CNBu\text{-}t)(NCCH{=}CHCN) \qquad (u)$$

In many cases molecules of solvent remain coordinated to the nickel in the products[34]:

$$Ni(1,5\text{-}COD)_2 + ol + 1,5\text{-}COD \xrightarrow{CH_3CN} Ni(CH_3CN)(ol)_2 + 3\ 1,5\text{-}COD \qquad (v)$$

where ol = dimethylmalate or dimethyl + fumarate. The tridentate ligand trans,trans, trans-1,5,9-cyclododecatriene can be displaced from Ni(0) by ethylene to form $Ni(C_2H_4)_3$[38]:

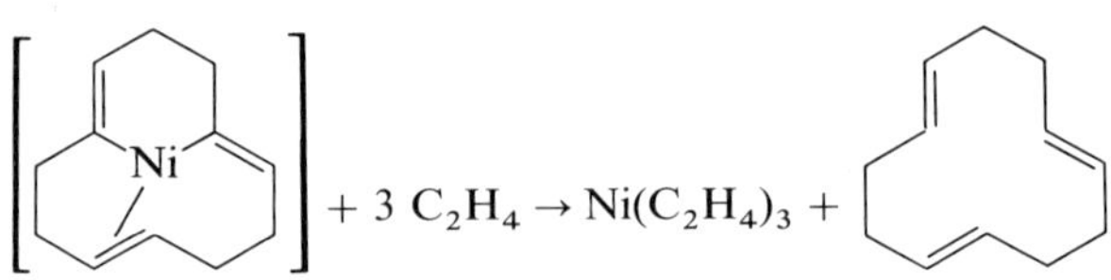

(w)

Treatment with phosphines results in ethylene displacement to form $Ni(PR_3)(C_2H_4)_2$, R = C_6H_{11} or C_6H_5; xs phosphine displaces all the ethylene ligands to form $Ni(PR_3)_4$.

Displacement of ethylene from both its Pt(II) and Pt(0) complexes is a useful route to platinum complexes of less volatile olefins[39–52]:

$$K[PtCl_3(C_2H_4)] + PhCH{=}CH_2 \rightarrow K[PtCl_3(PhCH{=}CH_2)] + C_2H_4 \qquad (x)$$

$$[PtCl_2(C_2H_4)]_2 + 2\ CH_3CH{=}CHCH_3 \rightarrow [PtCl_2(CH_3CH{=}CHCH_3)]_2 + 2\ C_2H_4 \qquad (y)$$

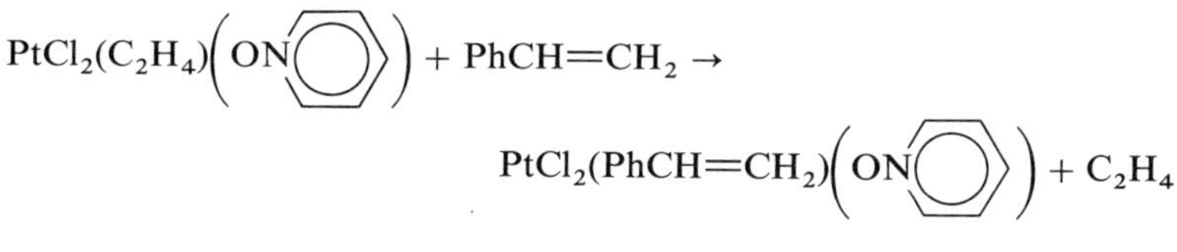

(z)

$$PtCl(acac)(C_2H_4) + CH_2{=}CHOSiMe_3 \xrightarrow{C_6H_6} PtCl(acac)(CH_2{=}CHOSiMe_3) + C_2H_4 \qquad (aa)$$

$$Pt(PPh_3)_2(C_2H_4) + ol \rightarrow Pt(PPh_3)_2(ol) + C_2H_4 \qquad (ab)$$

The rate of displacement of ethylene from trans-$PtCl_2(py)(C_2H_4)$ by free olefins is markedly reduced by the presence of ortho substituents on the pyridine, suggesting an associative mechanism[53]. The great advantage of displacing ethylene is its high volatility which enables mild reaction and workup conditions to be used. However, Pt(0) complexes of phenylacetylene[54], fumaronitrile[55] and trans-stilbene[47,56] are also used as starting materials.

Platinum(0) and Pd(0) complexes containing one cyclooctatetraene and one 1,5-cyclooctadiene ligand:

$$M(1{,}5\text{-}COD)Cl_2 + Li_2C_8H_8 \xrightarrow[\text{low T}]{Et_2O} M(C_8H_8)(1{,}5\text{-}COD) + 2\ LiCl \qquad (ac)$$

(M = Pd, Pt) have enabled a new homoleptic olefin complexes of Pt(0) and Pd(0) to be prepared by olefin displacement reactions (Scheme 1)[57,58]. The dibenzylideneacetone (DBA) complexes (DBA = PhCH=CHCOCH=CHPh) of Pt(0) and Pd(0), which are formed by treating Na_2MCl_4 with DBA in methanol in the presence of sodium acetate[59,60], readily lose dibenzylideneacetone and so form valuable starting materials for Pt(0) and Pd(0) complexes of weakly bonded ligands, such as olefins[61–65]:

$$Pt(DBA)_2 + C_2X_4\ (xs) + 2\ PPh_3 \xrightarrow{\text{benzene}} Pt(PPh_3)_2(C_2X_4) \qquad (ad)$$

where X = Cl, F;

$$[Pd_2(DBA)_3 + ol + LL \xrightarrow{\text{acetone}} Pd(LL)(ol) \qquad (ae)$$

where ol = maleic anhydride, dimethylmalate, dimethylfumarate, acrylonitrile; LL = bipyl or o-phen;

$$Pd_3(DBA)_2(bipy)_3 + ol \xrightarrow{\text{acetone}} Pd(bipy)(ol) \qquad (af)$$

where ol = maleic anhydride, dimethylmaleate, dimethylfumarate. Recently, the Pt(0) complex, $Pt_2(dipDBA)_3$, dip = pi-PrC_6H_4CH=CH-COCH=CHC_6H_4Pr-i-p has been prepared by the same route used to prepare the dibenzylideneacetone complexes. It has

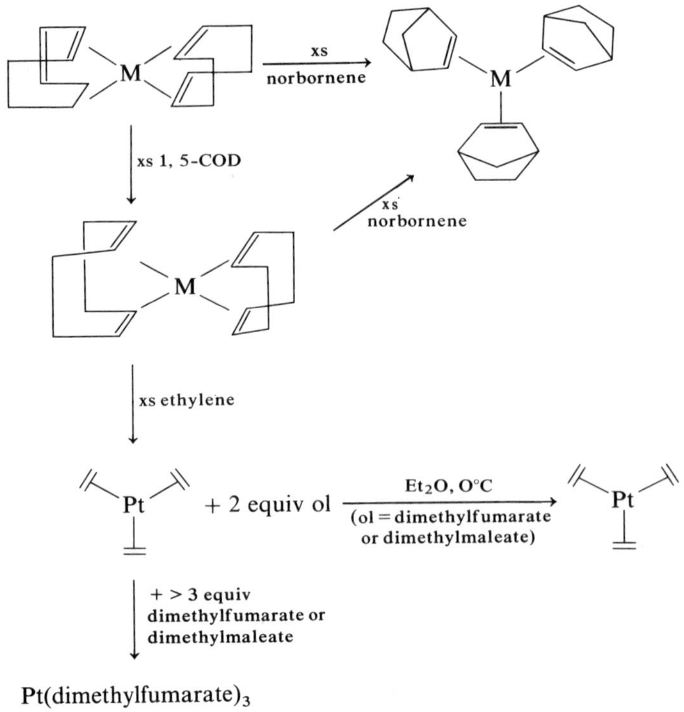

Scheme 1

greater solubility in organic solvents and should be of synthetic value as well as being helpful in enabling NMR to be used to characterize the species formed in solution[66].

(iii) Displacement of Neutral Ligands: Carbon Monoxide. Displacement of carbon monoxide from carbonyl complexes is an important route to η^2-olefin complexes. The reaction involves initial loss of carbon monoxide followed by reaction of olefin with the coordinatively unsaturated intermediate:

$$M(CO)_n \xrightarrow{-CO} M(CO)_{n-1} \xrightarrow{+ol} M(CO)_{n-1}(ol) \tag{ag}$$

The initial activation may be accomplished using either thermal or photochemical[67–71] energy to cleave the metal–carbonyl bond.

Few olefin complexes of the group-VA metals are known. However, o-CH_2=$CHCH_2C_6H_4PPh_2$ displaces two carbonyl ligands from $V(CO)_6$ to form V(o-CH_2=$CHCH_2C_6H_4PPh_2$)$(CO)_4$ in which the bidentate ligand bonds through both olefin and phosphine[72].

Thermal displacement of carbonyl ligands has rarely been used to prepare olefin complexes of group-VIA metals, although $Mo(CO)_6$ and acrylonitrile form the diamagnetic, insoluble and probably polymeric $Mo(CH_2$=$CHCH)_2(CO)_2$ [79]. Where the group-VIA hexacarbonyl is to be refluxed with an olefin problems arise because the unreacted $M(CO)_6$ sublimes out of the reaction flask. This may be partially alleviated using special glassware[67,74]. The principle method for the displacement of CO from group-VIA hexacarbonyls by olefins has been photochemical[75]:

$$M(CO)_6 + ol \xrightarrow{h\nu} M(CO)_5(ol) + M(CO)_4(ol)_2 \quad (ah)$$

where ol = C_2H_4, $CH_3CH{=}CH_2$, cis-$CH_3CH{=}CHCH_3$; M = Mo, W. Complexes with π-ligands are more stable than those with solely carbonyl supporting ligands[76–79]:

$$Mo(mesitylene)(CO)_3 + C_2H_4 \xrightarrow{h\nu} Mo(mesitylene)(CO)_2(CH_2{=}CH_2) + CO \quad (ai)$$

$$Cr(C_6Ph_6)(CO)_3 + ol \xrightarrow{h\nu} Cr(C_6Ph_6)(CO)_2(ol) + CO \quad (aj)$$

where ol = cyclopentene, maleic anhydride, maleic acid, fumaric acid;

$$W(\eta^5\text{-}Cp)(CO)_3CH_3 + C_2H_4 \xrightarrow{h\nu} W(\eta^5\text{-}Cp)(CO)_2(CH_2{=}CH_2)CH_3 + CO \quad (ak)$$

$$W(CO)_5(EMe_3) + ol \xrightarrow[E=P,\,As]{h\nu} cis\text{-}W(CO)_4(EMe_3)(ol) \quad (al)$$

where ol = C_2H_4, dimethylmaleate, dimethylfumarate.

Activated and unactivated olefin complexes of Mn are prepared from $Mn(\eta^5\text{-}Cp)(CO)_3$, in a two-step sequence; THF is used to displace one carbonyl ligand, probably forming $Mn(\eta^5\text{-}Cp)(CO)_2(THF)$ as an intermediate, which reacts rapidly with an olefin to yield an air-sensitive product[80]:

$$Mn(\eta^5\text{-}Cp)(CO)_3 \xrightarrow[(2)\ RR'C{=}CH_2,\ heat]{(1)\ THF,\ heat} Mn(\eta^5\text{-}Cp)(CO)_2(RR'C{=}CH_2) + CO \quad (am)$$

where R = R' = H, CH_3O; R = H, R' = CH_3O. The two-step mechanism of Eq. (ak) illustrates a very common difference between the thermal and photochemical displacement of carbon monoxide. The thermal displacement occurs via an S_{N2} mechanism in which loss of carbon monoxide is accompanied by gain of a new donor, whereas the photochemical reaction involves loss of carbon monoxide from the excited molecule in an S_{N1} reaction[67]. These mechanisms are supported by the fact that neither $Mn(\eta^5\text{-}Cp)(CO)_3$ nor $Mn_2(CO)_{10}$ undergoes thermal ^{14}CO exchange at room temperature[81]; this would be a necessary prerequisite for an S_{N1} mechanism. Olefin complexes of manganese, $Mn(\eta^5\text{-}Cp)(CO)_2(ol)$, where ol = C_2H_4, $CH_3CH{=}CH_2$, 1-pentene, cyclopentene, cycloheptene and cis-cyclooctene, are readily prepared by the photolysis of $Mn(\eta^5\text{-}Cp)(CO)_3$ in hexane[82,83].

The iron-olefine complexes $Fe(ol)(CO)_4$ can be prepared by the thermal reaction of the olefin with $Fe_2(CO)_9$ in benzene at modest temperatures. The reaction does not formally involve carbonyl displacement; it is a bridge splitting reaction in which $Fe_2(CO)_9$ is cleared into two moieties $Fe(CO)_5$ and $Fe(CO)_4$:

$$Fe_2(CO)_9 + ol \rightarrow Fe(CO)_5 + Fe(CO)_4(ol) \quad (an)$$

The $Fe(CO)_5$ may be removed by sublimation in vacuum[82]. Olefins that may be used include ethylene, maleic anhydride, maleic acid, fumaric acid and $PhCH{=}CHCOMe_4$ [84–87]. Olefins with strained rings may also be used to form $Fe(CO)_4$ derivatives. The extent to which ring opening occurs depends upon the reaction conditions. Thus $Fe_2(CO)_9$ reacts with substituted methylenecyclopropanes to form η^2 complexes, with no equilibration between cis and trans isomers[88]:

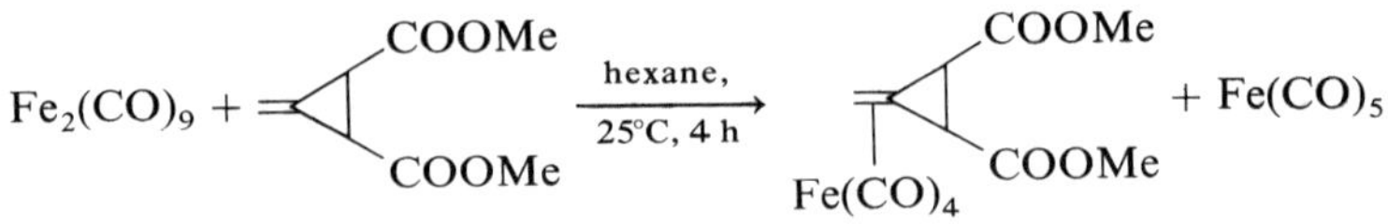

(ao)

However, products of ring opening are obtained[88]:

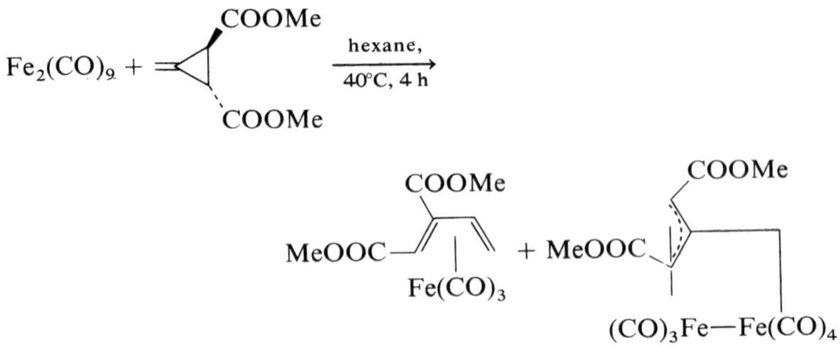

(78%) (3%)

(ap)

The substituted methylenecyclopropane $CH_2{=}\overline{CCH(CH_2OH)C}H(CH_2OH)$ reacts with Fe_2CO_9 to give ring-opened products[89]:

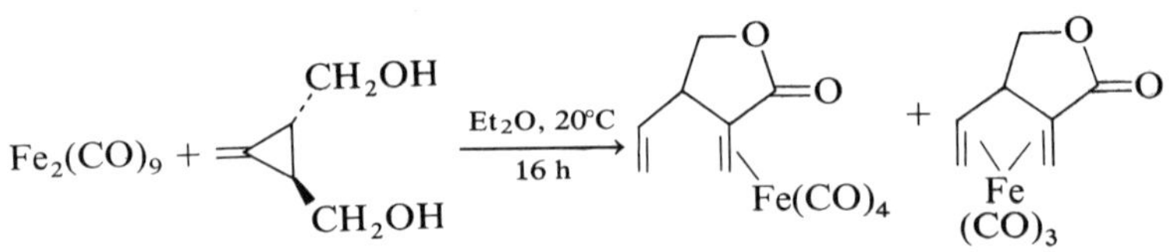

(aq)

Hexafluoro-dewar-benzene forms both mono- and bis-$Fe(CO)_4$ η^2 complexes[90]:

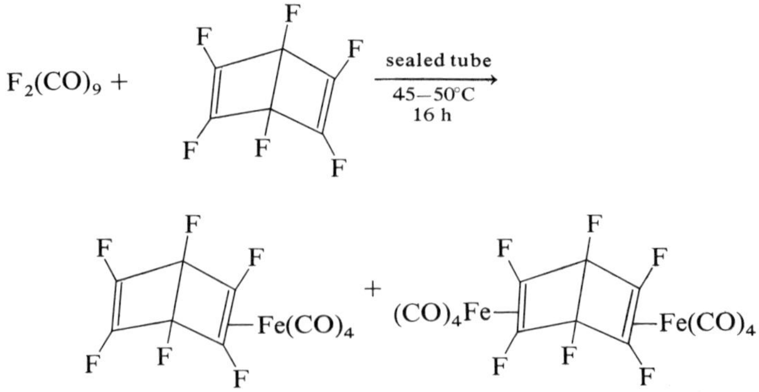

(ar)

although it may be converted to hexafluorobenzene at higher temperatures.

Iron-olefin complexes have been prepared photochemically[91,92] from $Fe(CO)_5$, $Fe_2(CO)_9$ and $Fe_3(CO)_{12}$:

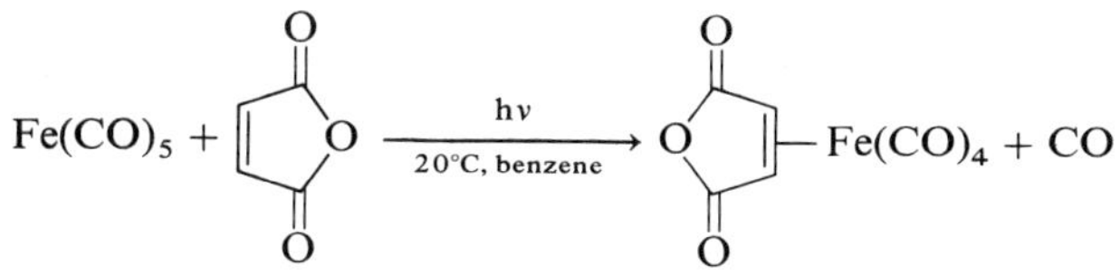

(as)

$$Fe_2(CO)_9 + ol \xrightarrow[\text{pet. ether}]{h\nu,\ 20^\circ C} Fe(CO)_4(ol) + Fe(CO)_5 \tag{at}$$

where ol = $CH_2{=}CHF$, $CH_2{=}CF_2$, $CF_2{=}CHX$ (X = F, Cl, Br), $CF_2CF{=}CFCF_3$, $CF_3CH{=}CHCF_3$, $(CF_3)_2C{=}C(CF_3)_2$, $CCl_2{=}CCl_2$. If the photolyses are carried out at $-50°C$, iron complexes of strained ring olefins may be isolated[93–95]:

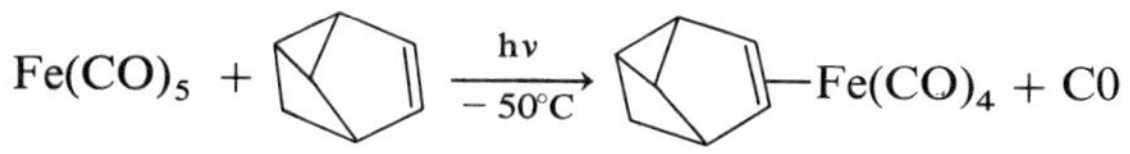

(au)

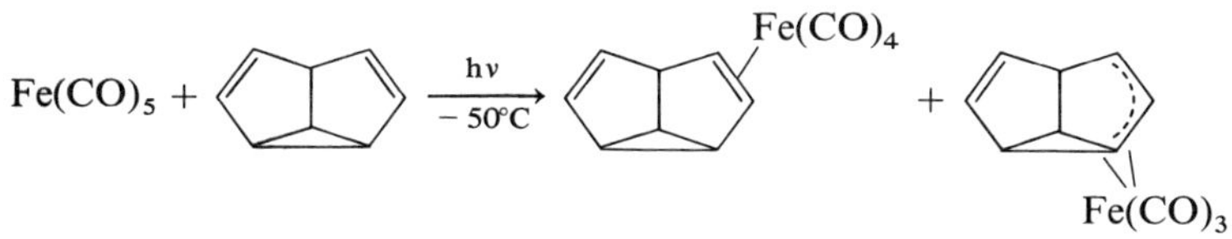

(av)

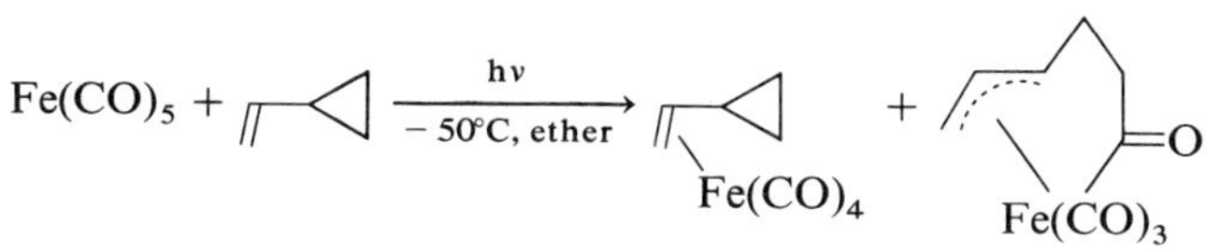

(aw)

The olefin complex is not always the sole product; ring opening and subsequent η^3-allyl complex formation is a competing reaction. It is possible to cleave selectively a carbonyl ligand photochemically without dissociation of any other π-coordinated ligands. Thus irradiation of $[Fe(cyclobutadiene)(CO)_3]^-$ in the presence of dimethylmaleate or dimethylfumarate yields the corresponding $[Fe(cyclobutadiene)(olefin)(CO)_2]^-$ complex[96].

Ruthenium-olefin complexes have been prepared by photochemical carbonyl displacement reactions[97]:

$$Ru(CO)_3L_2 + ol \xrightarrow[\text{hexane}]{h\nu} Ru(CO)_2L_2(ol) + CO \tag{ax}$$

where $L = P(OMe)_3$, $P(OEt)_3$, PMe_2Ph; ol = C_2F_4, $CClF{=}CF_2$, $CHF{=}CF_2CF_3CF{=}CF_2$, hexafluorocyclobutene. To obtain isolable complexes, phosphine ligands should be present on the Ru and the olefins should carry fluorine substituents.

Cobalt forms few complexes of monoolefins, although diolefins react with $Co_2(CO)_8$ without cleavage of the $Co(\mu\text{-}CO)_2Co$ bridge[98]. Similarly, monoolefins do not displace CO from rhodium carbonyl, although diolefins react with $Rh_6(CO)_{16}$ to form a complex series of products[99].

(iv) Displacement of Neutral Ligands: Tertiary Phosphine. Tertiary phosphines are strong ligands in transition-metal complexes, which olefins are not[100], so that the preparation of olefin complexes by phosphine displacement is rare. However, $Fe(PF_3)_5$ does undergo olefin substitution upon irradiation in the presence of olefins such as styrene or methacrylate[101]:

$$Fe(PF_3)_5 + ol \rightleftarrows Fe(PF_3)_4(ol) + PF_3 \tag{ay}$$

where ol = $CH_3CH{=}CHCN$, $PhCH{=}CH_2$, $CH_2{=}CHCOOCH_3$. The resulting trifluorophosphine complexes are more stable than the equivalent carbonyl complexes, $Fe(CO)_4(ol)$. Since PF_3 is a gas, reaction (ay) is reversible. This is an important factor in making the reaction go to completion.

Ruthenium–hydridophosphine complexes are well-known homogeneous hydrogenation catalysts[102,103]. In some cases stable η^2-olefin complexes can be isolated. Thus ethylene and styrene react with $RuH_2(PPh_3)_4$ to yield the Ru(0) complexes, $Ru(PPh_3)_3$(olefin). These reactions are accompanied by olefin hydrogenation[104]:

$$RuH_2(PPh_3)_4 + RCH{=}CH_2 \xrightarrow{R=H,\ Ph} Ru(PPh_3)_3(RCH{=}CH_2) + PPh_3 + RCH_2CH_2 \tag{az}$$

Simple phosphine displacement occurs when $Ru(CO)_2(PPh_3)_3$ is treated with ethylene in acetonitrile[105]:

$$Ru(CO)_2(PPh_3)_3 + CH_2{=}CH_2 \xrightarrow{MeCN} Ru(CO)_2(PPh_3)_2(CH_2{=}CH_2) + PPh_3 \tag{ba}$$

Although the ethylene complex $RhCl(PPh_3)_2(CH_2{=}CH_2)$ is unstable with respect to ethylene dissociation followed by Rh dimerization, ethylene will displace triphenylphosphine from[106]:

$$RhCl(PPh_3)_3 + C_2H_4 \rightarrow RhCl(PPh_3)_2(CH_2{=}CH_2) + PPh_3 \tag{bb}$$

The reaction occurs by initial dissociation of triphenylphosphine. For isolation of the product the reaction must be carried out in an ethylene-saturated solution under an atmosphere of ethylene. Other olefins generally form too weak a complex with Rh(I) to be able to displace triphenylphosphine. Olefins such as dimethylmaleate and tetracyanoethylene displace triphenylphosphine from Ir(I)[197]:

$$IrH(CO)(PPh_3)_3 + ol \xrightarrow{benzene} IrH(CO)(PPh_3)_2(ol) + PPh_3 \tag{be}$$

where ol = fumaric acid, dimethylfumarate, fumaronitrile, dimethylmaleate, tetracyanoethylene, maleic acid, cinnamic acid, cinnamonitrile.

Refluxing $Pt(PPh_3)_4$ with activated olefins gives platinum(0)–olefin complexes with chloroolefins[108,109], fluoroolefins[110], tetracyanoethylene[111] and fumaronitrile[112]:

$$Pt(PPh_3)_4 + ol \xrightleftharpoons{benzene} Pt(PPh_3)_2(ol) + 2\ PPh_3 \tag{bd}$$

If the benzene is replaced by ethanol, however, the chloroolefin–platinum(0) complex isomerizes to a Pt(II)–η^1-vinyl product[113]:

$$Pt(PPh_3)_2(CCl_2{=}CCl_2) \xrightarrow{EtOH} Pt(CCl{=}CCl_2)Cl(PPh_3)_2 \tag{be}$$

Substantial dissociation to $Pt(PPh_3)_3$ occurs when $Pt(PPh_3)_4$ is dissolved in benzene[114]. This reacts reversibly with ethylene:

$$Pt(PPh_3)_3^- + C_2H_4 \overset{K_{bf}}{\rightleftharpoons} Pt(PR_3)_2(C_2H_4) + PR_3 \qquad \text{(bf)}$$

Values of K_{bf} for $R = C_6H_5$ and $p\text{-}CH_3C_6H_4$ in benzene at 25°C are 0.12 and 0.21, respectively, emphasizing the importance of an electron donating phosphine in promoting the formation of the olefin complex[114]. When $Pt(PEt_3)_3$ is treated with hexa(trifluoromethyl)benzene in toluene at −50°C a smooth displacement of triethylphosphine occurs to form an η^2-olefin complex in which the benzene ring is nonplanar[115]:

$Pt(PEt_3)_3 + C_6(CF_3)_6 \rightarrow$ 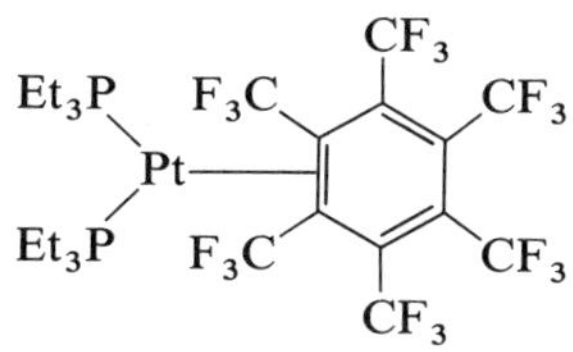 $+ PEt_3$ (bg)

A number of olefins that are too weak to displace triphenylphosphine from $Pt(PPh_3)_4$ [Eq. (bd) is reversible] can form[116] complexes with $Pt(PPh_3)_2$ which can be prepared in situ, by sodium borohydride or hydrazine reduction of $Pt(PPh_3)_2O_2$ [117], from $Pt(PPh_3)_2CO_3$ or by irradiation of $Pt(PPh_3)_2(C_2O_4)$ [118]. Platinum(0) forms only biscomplexes with bulky phosphines such as $PPhBu_2$-t and these react with olefins such as dimethylfumarate or maleic anhydride to form platimum(0)-olefin complexes[119]:

$$(t\text{-}Bu_2PhP)_3Pt + ol \rightarrow (t\text{-}Bu_2PhP)_2P(ol) \qquad \text{(bh)}$$

where ol = dimethylfumarate or maleic anhydride.

(v) Displacement of Neutral Ligands: Nitriles. Although the acetonitrile complexes $M(CO)_3(CH_3CN)_3$ (where M = Cr, Mo and W) have proved useful starting materials for the preparation of acetylene, diene and triene complexes, they have scarcely been exploited for monoolefins. Three carbonyl ligands of $M(CO)_6$, M = Cr, Mo or W, may be replaced on refluxing in acetonitrile to form[120,121] $M(CO)_3(CH_3CN)_3$. When $M(CO)_3(CH_3CN)_3$ is treated with acrylonitrile the η^2-olefin complexes $M(CO)_3(CH_2{=}CHCN)_3$ are obtained[121]. Treating $W(CO)_3(CH_3CN)_3$ with methylvinylketone gives not the expected $W(CO)_3(olefin)_3$ complex, but $W(CH_3COCH{=}CH_2)_3$, in which all six of the original ligands have been displaced[122].

The palladium complex $Pd(PhCN)_2Cl_2$, which is readily prepared by reaction of $PdCl_2$ in benzonitrile, has proved a valuable source of olefin complexes through displacement of benzontrile[123]:

$$2\ Pd(PhCN)_2Cl_2 + 2\ ol \rightarrow Pd(ol)Cl_{2\,2} + 4\ PhCN \qquad \text{(bi)}$$

where ol = C_2H_4, $PhCH{=}CH_2$, $Me_2C{=}CH_2$, cyclohexene.

(vi) Displacement of Solvents. Solvents often form relatively weak complexes with transition metals[124] and as such are attractive ligands for displacement by olefins, which are themselves relatively weak donor ligands[100]. In some cases solvent coordination is so weak that the complex is effectively coordinatively unsaturated, with the equilibrium in Eq. (bj) lying far to the right:

$$[ML_m(solvent)] \rightleftharpoons [ML_m] + solvent \qquad \text{(bj)}$$

Ethers form weak complexes with many metals and can be displaced by olefins to give thermodynamically more stable complexes[125]:

$$[Fe(\eta^5\text{-Cp})(CO)_2(THF)][BF_4] + ol \xrightarrow{CH_2CL_2} [Fe(\eta^5\text{-Cp})(CO)_2(ol)][BF_4] + THF \qquad (bk)$$

where ol = C_2H_4, cyclohexene, cycloheptene. Ethers form weak complexes with many metals and can be displaced by olefins:

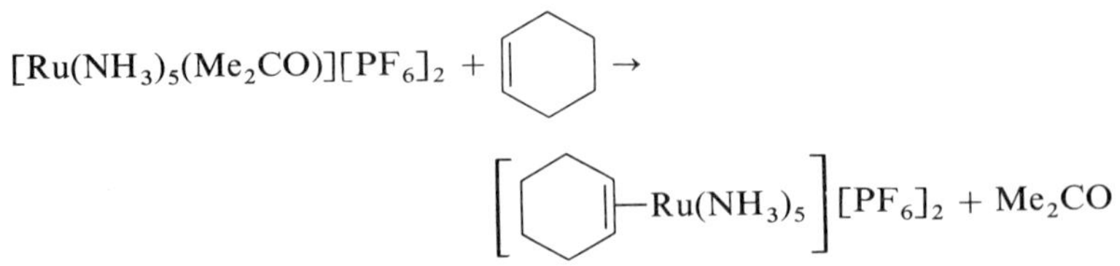

(bl)

(vii) Displacement of Hydrogen. Elimination of molecular hydrogen has been used to prepare several complexes of group-VA metals. Thus Nb- and Ta-olefin complexes can be obtained by reaction of a limited amount of the olefin with the metal trihydride[127]:

$$M(\eta^5\text{-Cp})_2H_3 + 2\ CH_2{=}CH_2 \rightarrow M(\eta^5\text{-Cp})_2(CH_2{=}CH_2)C_2H_5 + H_2 \qquad (bm)$$

where M = Nb, Ta. Two hydride ligands are displaced as gaseous hydrogen, while the third reacts with ethylene to form an ethyl group.

The osmium dihydride cluster complex $Os_3(CO)_{10}H_2$ loses hydrogen in the presence of olefins to form $Os_3(\text{olefin})_2(CO)_{10}$[128]:

$$Os_3(CO)_{10}H_2 + 2\ \text{cyclooctene} \rightarrow Os_3(\text{cyclooctene})_2(CO)_{10} \qquad (bn)$$

The fate of the original hydride ligands is not reported. Cyclooctene is replaced by other ligands so that $Os_3(\text{cyclooctene})_2(CO)_{10}$ is a very convenient source of the $Os_3(CO)_{10}$ cluster group.

Hydrogen displacement syntheses often lead to some hydrogenation of the entering olefin as in in Eqs. (az) and (bm). Thus ethylene reacts with $[Ir(PPr\text{-}i_3)_2H_5]^-$ to form 1 mol of a bis(ethylene) complex and 1 mol of ethane for every mole of iridium[129]:

$$Ir(PPr\text{-}i_3)_2H_5 + 3\,C_2H_4 \xrightarrow{\text{benzene}} \overbrace{CH_2{-}P(Pr\text{-}i)_2{-}Ir(PPr\text{-}i)_3(CH_2{=}CH_2)_2{-}CHMe{-}}^{\text{ring}} + C_2H_4 \qquad (bo)$$

(viii) Displacement of Methane. Cationic cobalt(I)–olefin complexes have been prepared by treating $H_3CCo[P(OMe)_3]_4$ with acid in the presence of an olefin[130]:

$$H_3CCo[P(OMe)_3]_4 + HBF_4 + ol \xrightarrow{-78^\circ C} [Co\{P(OMe)_3\}_4(ol)][BF_4] + CH_4 \qquad (bp)$$

The reaction involves initial oxidative addition of the proton to Co(I), followed by reductive elimination of methane to yield the coordinatively unsaturated $[Co\{P(OMe)_3\}_4]^+$ intermediate:

$$H_3CCo[P(OMe)_3]_4 + H^+ \rightarrow [H_3CCo(H)[P(OMe)_3]_4]^+ \rightarrow [Co\{P(OMe)_3\}_4]^+ + CH_4 \quad \text{(bq)}$$

(ix) Addition to Coordinatively Unsaturated Complexes (Displacement of No Ligands). Some of the most facile syntheses of olefin complexes start with coordinatively unsaturated complexes. With such complexes activation is usually unnecessary.

Activated olefins such as dimethylmaleate and diethylfumarate add to $V(\eta^5\text{-Cp})_2$ to yield paramagnetic $V(\eta^5\text{-Cp})_2$(olefin); olefins lacking strongly electron-withdrawing groups are inactive[131]. Olefin complexes of iron are readily prepared[132] from the commercially available $[Fe(\eta^5\text{-Cp})(CO)_2]_2$:

$$[Fe(\eta^5\text{-Cp})(CO)_2]_2 + \text{ol} + [Ph_3C][BF_4] \rightarrow [Fe(\eta^5\text{-Cp})(CO)_2(\text{ol})][BF_4] + C_2Ph_6 \quad \text{(br)}$$

where ol = 1-heptene, 1-octene, 4-octene, cycloheptene, cyclooctene. An oxidizing agent, $[Ph_3C][BF_4]$, must be present to split the starting dimer to form the coordinatively unsaturated $[Fe(\eta^5\text{-Cp})(CO)_2]^+$ cation. Equation (br) provides a useful route to $[Fe(\eta^5\text{-Cp})(CO)_2(\text{isobutene})][BF_4]$ used in Eqs. (bo) and (bp) above.

Trans-$[Ir(PPh_3)_2(CO)Cl]$ undegoes reversible olefin addition to form[133–135] the 5-coordinated $Ir(PPh_3)_2(CO)Cl$(olefin), whose stability increases when electron-withdrawing substituents are present on the olefin. Other 4-coordinated Ir(I) complexes that add an olefin to form 5-coordinated complexes are[136,137]:

$$[Ir(NN)(\text{diol})]X + \text{ol} \rightarrow [Ir(NN)(\text{diol})(\text{ol})]X \quad \text{(bs)}$$

where NN = bipy, phen; X = PF_6^-, ClO_4^-; diol = 1,5-COD, NBD; ol = C_2H_4, 1-hexene, acrylonitrile, fumaronitrile, tetracyanoethylene, maleic anhydride.

$$[Ir(NN)(CO)_2]X + \text{ol} \rightarrow [Ir(NN)(CO)_2(\text{ol})]X \quad \text{(bt)}$$

where NN = bipy, phen; X = PF_6^-, ClO_4^-; ol = cyclohexene, cis- and trans-2-butene.

Dialkyl(bipyridyl)nickel(II) forms unstable olefin addition complexes NiR_2-(bipy)(ol) with acrylonitrile and acrolein[138–140]. The stability of the product depends on the alkyl group present in the order R = Et > n-Pr > i-Bu > Me[140].

Hydrotris(pyrazolylborate) $[HB(pz)_3]$ stabilizes labile 5-coordinated complexes of Pt(II)[141–143]; $Pt(CH_3)\{HB(pz)_3\}$ is probably polymeric with bridging pyrazolyl groups. In dichloromethane it reacts with olefins to form 5-coordinated $Pt(CH_3)\{HB(pz)_3\}(\text{ol})$[141–143]:

$[Pt(CH_3)\{HB(pz)_3\}]_n$ + n ol → n

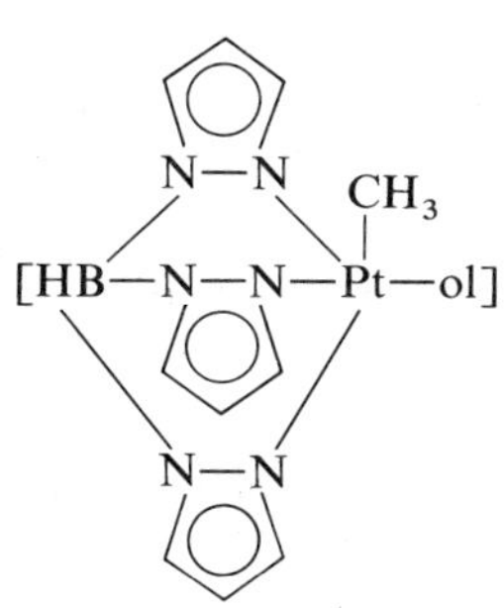

(bu)

where ol = C_2F_4, maleic anhydride, dimethylmaleate, dimethylfumarate, $CH_2{=}CHCN$, $PhCH{=}CHCHO$, $CH_2{=}CHCOOMe$, $CH_2{=}C(CH_3)CN$.

Copper(I)-, Ag(I)- and Au(I)-olefin complexes may be prepared either by dissolving the metal salt in the liquid olefin or by passing a stream of gaseous olefin through a solution of the metal salt[144]. In this way olefin complexes of Ag(I) salts with nitrate[145,146], perchlorate[145], tetrafluoroborate[147], hexafluoroacetylacetonate[148], β-diketonate[149] and trifluoromethylsulfonate[150] anions have been prepared. Similarly Cu(I) complexes with trifluoroacetate[151] and trifluoromethylsulfonate[152] anions have been prepared. These η^2-olefin complexes have varying stabilities and compositions. The Cu(I) complexes disproportionate to Cu(0) and Cu(II). The most stable complexes of both Cu(I) and Ag(I) are those with the trifluoromethylsulfonate anions[150,152]. The $[Cu(olefin)][CF_3SO_3]$ complexes are prepared by a two-step method:

$$Cu_2O + C_6H_6 + 2\ CF_3SO_3H \rightarrow [Cu_2(C_6H_6)][CF_3SO_3]_2 \quad \text{(bv)}$$

$$[Cu_2(C_6H_6)][CF_3SO_3]_2 + 2\ ol \rightarrow 2\,[Cu(ol)][CF_3SO_3] + C_6H_6 \quad \text{(bw)}$$

and show no tendency to disproportionate[153]. Gold(I) complexes with chloride and trifluoromethylsulfonate anions have also been prepared[154]:

$$AuCl + \text{trans-cyclooctene} \rightarrow [Au(\text{trans-cycloocetene})]Cl \xrightarrow{AgCF_3SO_3} [Au(\text{trans-cyclooctene})]CF_3SO_3 + AgCl \quad \text{(bx)}$$

The η^2-olefin complexes of Cu(I), Ag(I) and Au(I) dissociate in solution. The loss of the olefinic ligand can be prevented by incorporating it into a multidentate phosphine or arsine ligand[154]. Thus o-$CH_2{=}CHCH_2C_6H_4PPh_2$ reacts with Cu(I) halides to form a 1:1 complex:

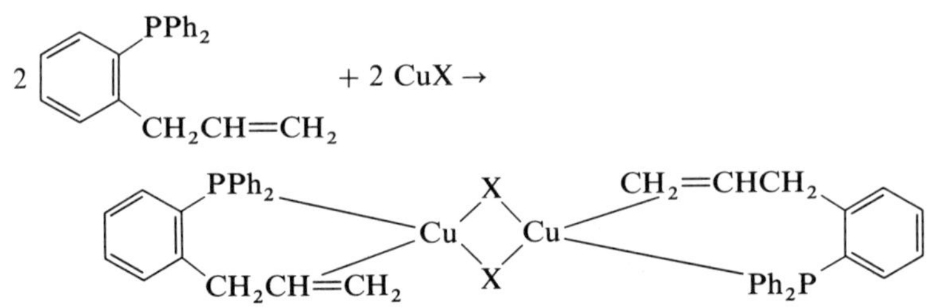

(by)

With Ag(I) and Au(I), however, this ligand forms bis complexes in which only the phosphorus donor is coordinated.

(F. R. HARTLEY)

1. R. G. Denning, L. M. Venanzi, *J. Chem. Soc.*, 3241 (1963).
2. R. D. Cramer, E. L. Jenner, R. V. Lindsey, U. G. Stolberg, *J. Am. Chem. Soc.*, *85*, 1691 (1963).
3. J. Chatt, M. L. Searle, *Inorg. Synth.*, *5*, 210 (1957).
4. F. R. Hartley, *Organomet. Chem. Rev.*, *A6*, 119 (1970).
5. P. B. Chock, J. Halpern, F. E. Paulik, *Inorg. Synth.*, *14*, 90 (1973).
6. F. Basolo, R. G. Pearson, *Mechanisms of Inorganic Reactions*, John Wiley, New York, 2nd edn., p. 377 (1967).
7. D. Banerjea, F. Basolo, R. G. Pearson, *J. Am. Chem. Soc.*, *79*, 4055 (1979).
8. R. M. Milburn, L. M. Venanzi, *Inorg. Chim. Acta*, *2*, 97 (1968).
9. M. Green, C. J. Wilson, *Disc. Faraday Soc.*, *47*, 110 (1969).

10. M. Green, C. J. Wilson, *J. Chem. Soc., Dalton Trans.*, 2302 (1977).
11. M. Green, C. J. Wilson, *J. Chem. Res. (S)*, 175 (1978).
12. L. I. Elding, A. B. Gröning, *Inorg. Chim. Acta.*, *38*, 59 (1980).
13. M. Green, M. G. Swanwick, *J. Chem. Soc., Dalton Trans.*, 158 (1978).
14. H. Boucher, B. Bosnich, *Inorg. Chem.*, *16*, 717 (1977).
15. J. Hubert, T. Theophanides, *J. Organomet. Chem.*, *93*, 265 (1975).
16. M. H. Chisholm, H. C. Clark, *Inorg. Chem.*, *10*, 2557 (1971).
17. M. H. Chisholm, H. C. Clark, *Inorg. Chem.*, *12*, 991 (1973).
18. R. Cramer, *Inorg. Chem.*, *1*, 722 (1962).
19. R. Cramer, *Inorg. Synth.*, *15*, 14 (1974).
20. J. L. Herdé, C. V. Senoff, *Inorg. Nucl. Chem. Lett.*, *7*, 1029 (1971).
21. J. L. Herdé, J. C. Lambert, C. V. Senoff, *Inorg. Synth.*, *15*, 18 (1974).
22. A. van der Ent, A. L. Onderlinden, *Inorg. Synth.*, *14*, 92 (1973).
23. G. Winkhaus, H. Singer, *Chem. Ber.*, *99*, 3610 (1966).
24. B. L. Shaw, E. Singleton, *J. Chem. Soc. A*, 1683 (1967).
25. J. A. Segal, B. F. G. Johnson, *J. Chem. Soc., Dalton Trans.*, 677 (1975).
26. M. Herberhold, H. Alt, *J. Organomet. Chem.*, *42*, 407 (1972).
27. M. A. Bennett, R. S. Nyholm, J. D. Saxby, *J. Organomet. Chem.*, *10*, 301 (1967).
28. W. P. Giering, M. Rosenblum, *J. Chem. Soc., Chem. Commun.*, 441 (1971).
29. M. Rosenblum, *Acc. Chem. Res.*, *7*, 122 (1974).
30. K. M. Nicholas, *J. Am. Chem. Soc.*, *97*, 3254 (1975).
31. M. Green, J. A. K. Howard, R. P. Hughes, S. C. Kellett, P. Woodward, *J. Chem. Soc., Dalton Trans.*, 2007 (1975).
32. A. L. Onderlinden, A. van der Ent, *Inorg. Chim. Acta*, *6*, 420 (1972).
33. R. A. Schunn, *Inorg. Synth.*, *15*, 5 (1974).
34. F. Guerrieri, G. Salerno, *J. Organomet. Chem.*, *114*, 339 (1976).
35. S. D. Ittel, *Inorg. Synth.*, *17*, 117 (1977).
36. C. S. Cundy, M. Green, F. G. A. Stone, *J. Chem. Soc. A*, 1647 (1970).
37. G. Wilke, *Angew. Chem., Int. Ed. Engl.*, *2*, 105 (1963).
38. K. Fischer, K. Jonas, G. Wilke, *Angew. Chem., Int. Ed. Engl.*, *12*, 565 (1973).
39. J. S. Anderson, *J. Chem. Soc.*, 1042 (1936).
40. H. B. Jonassen, W. B. Kirsch, *J. Am. Chem. Soc.*, *79*, 1279 (1957).
41. J. R. Joy, M. Orchin, *J. Am. Chem. Soc.*, *81*, 305 (1959).
42. L. Garcia, S. I. Shupack, M. Orchin, *Inorg. Chem.*, *1*, 893 (1962).
43. P. Schmidt, M. Orchin, *Inorg. Chem.*, *6*, 1260 (1967).
44. E. M. Haschke, J. W. Fitch, *J. Organomet. Chem.*, *57*, C93 (1973).
45. P. P. Ponti, J. C. Baldwin, W. C. Kaska, *Inorg. Chem.*, *18*, 873 (1979).
46. M. Tsutsui, M. Ori, J. Francis, *J. Am. Chem. Soc.*, *94*, 1414 (1972).
47. J. P. Birk, J. Halpern, A. L. Pickard, *Inorg. Chem.*, *7*, 2672 (1968).
48. J. P. Visser, C. G. Leliveld, D. N. Reihoudt, *J. Chem. Soc., Chem. Commun.*, 178 (1972).
49. J. P. Visser, A. J. Schipperijn, J. Lukas, *J. Organomet. Chem.*, *47*, 433 (1973).
50. M. Green, J. A. K. Howard, R. P. Hughes, S. C. Kellett, P. Woodward, *J. Chem. Soc., Dalton Trans.*, 2007 (1975).
51. J. W. Fitch, J. A. Froelich, *J. Organomet. Chem.*, *128*, C1 (1977).
52. J. W. Fitch, K. C. Chan, J. A. Froelich, *J. Organomet. Chem.*, *160*, 477 (1978).
53. D. Mansuy, J. F. Bartoli, J. C. Chottard, *J. Organomet. Chem.*, *73*, C39 (1974).
54. W. H. Baddley, L. M. Venanzi, *Inorg. Chem.*, *5*, 33 (1966).
55. G. L. McClure, W. H. Bradley, *J. Organomet. Chem.*, *27*, 155 (1971).
56. J. Bowning, M. Green, J. L. Spencer, F. G. A. Stone, *J. Chem. Soc., Dalton Trans.*, 97 (1974).
57. M. Green, J. A. K. Howard, J. L. Spencer, F. G. A. Stone, *J. Chem. Soc., Dalton Trans.*, 271 (1977).
58. M. T. Chicote, M. Green, J. L. Spencer, F. G. A. Stone, J. Vicente, *J. Chem. Soc., Dalton Trans.*, 536 (1979).
59. Y. Takahashi, T. Ito, S. Sakai, Y. Ishii, *J. Chem. Soc., Chem. Commun.*, 1065 (1970).
60. P. M. Maitlis, K. Moseley, *J. Chem. Soc., Chem. Commun.*, 982 (1971).
61. W. J. Cherwinski, B. F. G. Johnson, J. Lewis, *J. Organomet. Chem.*, *52*, C61 (1973).
62. H. Tanaka, K. Yamada, H. Kawazura, *J. Chem. Soc., Perkin Trans. II*, 231 (1978).
63. W. J. Cherwinski, B. F. G. Johnson, J. Lewis, *J. Chem. Soc., Dalton Trans.*, 1405 (1974).
64. T. Ito, S. Hasegawa, K. Takahashi, Y. Ishii, *J. Organomet. Chem.*, *73*, 401 (1974).

65. Y. Ishii, S. Hasegawa, S. Kimura, K. Itoh, *J. Organomet. Chem.*, *73*, 411 (1974).
66. A. Keasley, B. E. Mann, A. Yates, P. M. Maitlis, *J. Organomet. Chem.*, *152*, 117 (1978).
67. W. Stroheimer, *Angew. Chem.*, *Int. Ed. Engl.*, *3*, 730 (1964).
68. E. K. von Gustorf, F. W. Grevels, *Fortschr.*, *Chem. Forsch.*, *13*, 365 (1969).
69. M. Wrighton, *Chem. Rev.*, *74*, 401 (1974).
70. A. Vogler, in *Concepts of Inorganic Photochemistry*, A. W. Adamson and P. D. Fleischauer, Eds., Wiley Interscience, New York, 1975, pp. 269–298.
71. G. L. Geoffroy, M. S. Wrighton, *Organometallic Photochemistry*, Academic Press, New York, 1979.
72. L. V. Interrante, G. V. Nelson, *J. Organomet. Chem.*, *25*, 153 (1970).
73. A. G. Massey, L. E. Orgel, *Chem. Ind.*, 436 (1961).
74. W. Strohmeier, *Chem. Ber.*, *94*, 2490 (1961).
75. I. Stolz, G. R. Dobson, R. K. Sheline, *Inorg. Chem.*, *2*, 1264 (1963).
76. E. O. Fischer, P. Kuzel, *Z. Naturforsch.*, *166*, 475 (1961).
77. R. J. Angelici, L. Busetto, *Inorg. Chem.*, *7*, 1935 (1968).
78. H. G. Alt, J. A. Schwarzle, C. G. Kreiter, *J. Organomet. Chem.*, *153*, C7 (1978).
79. U. Koem, C. G. Kreiter, H. Strack, *J. Organomet. Chem.*, *148*, 179 (1978).
80. M. Herberhold, G. O. Wiedersatz, C. G. Kreiter, *J. Organomet. Chem.*, *104*, 209 (1976).
81. A. Wojcicki, F. Basolo, *J. Inorg. Nucl. Chem.*, *17*, 77 (1961).
82. E. O. Fischer, M. Herberhold, in *Essays in Coordination Chemistry*, Birkhauser Verlag, Basel, 1965.
83. I. S. Butler, N. J. Coville, A. E. Fenster, *Inorg. Synth.*, *16*, 53 (1976).
84. H. D. Murdoch, E. Weiss, *Helv. Chim. Acta*, *46*, 1588 (1963).
85. A. J. P. Domingos, J. A. S. Howell, B. F. G. Johnson, J. Lewis, *Inorg. Synth.*, *16*, 103 (1976).
86. E. Weiss, K. Stark, J. E. Lancaster, H. D. Murdoch, *Helv. Chim. Acta*, *46*, 288 (1963).
87. M. Bigorgne, *J. Organomet. Chem.*, *127*, 55 (1977).
88. T. H. Whitesides, R. W. Slaven, *J. Organomet. Chem.*, *67*, 99 (1974).
89. B. M. Chisnall, M. Green, R. P. Hughes, A. J. Welch, *J. Chem. Soc.*, *Dalton Trans.*, 1899 (1976).
90. D. J. Cook, M. Green, N. Mayne and F. G. A. Stone, *J. Chem. Soc. A*, 1771 (1968).
91. G. O. Schenck, E. K. von Gustorf, M-J. Jun, *Tetrahedron Lett.*, *23*, 1059 (1962).
92. R. Fields, G. L. Godwin, R. M. Haszeldine, *J. Chem. Soc.*, *Dalton Trans.*, 1867 (1975).
93. R. Aumann, H. Wormann, C. Kruger, *Angew. Chem.*, *Int. Ed. Engl.*, *15*, 609 (1976).
94. R. Aumann, *J. Organomet. Chem.*, *66*, C6 (1974).
95. R. Aumann, *J. Am. Chem. Soc.*, *96*, 2631 (1974).
96. P. C. Reeves, J. Henery, R. Pettit, *J. Am. Chem. Soc.*, *91*, 5888 (1969).
97. R. Burt, M. Cooke, M. Green, *J. Chem. Soc. A*, 2975 (1970).
98. P. McArdle, A. R. Manning, *J. Chem. Soc. A*, 2123 (1970).
99. J. A. J. Jarvis, R. Whyman, *J. Chem. Soc.*, *Chem. Commun.*, 562 (1975).
100. F. R. Hartley, *Chem. Rev.*, *73*, 163 (1973).
101. K. Fischer, K. Jonas, G. Wilke, *Angew. Chem.*, *Int. Ed. Engl.*, *12*, 565 (1973).
102. B. R. James, *Homogeneous Hydrogenation*, John Wiley, New York, 1973.
103. C. Masters, *Homogeneous Transition-Metal Catalysis*, Chapman and Hall, London, 1981.
104. S. Komiya, A. Yamamoto, S. Ikeda, *J. Organomet. Chem.*, *42*, C65 (1972).
105. B. E. Cavit, K. R. Grundy, W. R. Roper, *J. Chem. Soc.*, *Chem. Commun.*, 60 (1972).
106. J. A. Osborn, F. H. Jardine, J. F. Young, G. Wilkinson, *J. Chem. Soc. A*, 1711 (1966).
107. W. H. Baddley, M. Shirley-Frazer, *J. Am. Chem. Soc.*, *91*, 3661 (1969).
108. N. J. Bland, R. D. W. Kemmitt, *J. Chem. Soc. A*, 1278 (1968).
109. K. Suzuki, H. Okuda, *Bull. Chem. Soc. Jpn.*, *45*, 1938 (1972).
110. D. M. Roundhill, G. Wilkinson, *J. Chem. Soc. A*, 506 (1968).
111. W. H. Baddley, L. M. Venanzi, *Inorg. Chem.*, *5*, 33 (1966).
112. G. L. McClure, W. H. Baddley, *J. Organomet. Chem.*, *27*, 155 (1971).
113. W. J. Bland, J. Burgess, R. D. W. Kemmitt, *J. Organomet. Chem.*, *14*, 201 (1968).
114. C. A. Tolman, W. C. Seidel, D. H. Gerlach, *J. Am. Chem. Soc.*, *94*, 2669 (1972).
115. J. Browning, M. Green, B. R. Penfold, J. L. Spencer, F. G. A. Stone, *J. Chem. Soc.*, *Chem. Commun.*, 31 (1973).
116. S. Cenini, R. Ugo, G. La Monica, *J. Chem. Soc. A*, 409 (1971).
117. C. D. Cook, G. S. Jauhal, *Inorg. Nucl. Chem. Lett.*, *3*, 31 (1967).
118. D. M. Blake, C. J. Nyman, *J. Am. Chem. Soc.*, *92*, 5359 (1970).
119. T. Yoshida, S. Otsuka, *J. Am. Chem. Soc.*, *99*, 2134 (1977).

120. D. P. Tate, W. R. Knipple, J. M. Augl, *Inorg. Chem.*, *1*, 433 (1962).
121. B. L. Ross, J. G. Grasselli, W. M. Ritchey, H. D. Kaesz, *Inorg. Chem.*, *2*, 1023 (1963).
122. R. B. King, A. Fronzaglia, *Inorg. Chem.*, *5*, 1837 (1966).
123. M. S. Kharasch, R. C. Seyler, F. R. Mayo, *J. Am. Chem. Soc.*, *60*, 882 (1938).
124. J. A. Davies, F. R. Hartley, *Chem. Rev.*, *81*, 79 (1981).
125. D. L. Reger, C. Coleman, *J. Organomet. Chem.*, *131*, 153 (1977).
126. B. P. Sullivan, J. A. Baumann, T. L. Meyer, D. J. Salmon, H. Lehmann, A. Ludi, *J. Am. Chem. Soc.*, *99*, 7368 (1977).
127. F. N. Tebbe, G. W. Parshall, *J. Am. Chem. Soc.*, *93*, 3793 (1971).
128. M. Tachikawa, J. R. Shapley, *J. Organomet. Chem.*, *124*, C19 (1977).
129. M. G. Clerici, S. Di Gioacchino, F. Maspero, E. Perrotti, A. Zanobi, *J. Organomet. Chem.*, *84*, 379 (1975).
130. E. L. Muetterties, P. L. Watson, *J. Am. Chem. Soc.*, *98*, 4665 (1976).
131. G. Fachinetti, S. DelNiro, C. Floriani, *J. Chem. Soc.*, *Dalton Trans.*, 1046 (1976).
132. P. F. Boyle, K. M. Nicholas, *J. Organomet. Chem.*, *114*, 307 (1976).
133. W. H. Baddley, *J. Am. Chem. Soc.*, *88*, 4545 (1966).
134. J. P. Collman, J. W. Kang, *J. Am. Chem. Soc.*, *89*, 844 (1967).
135. W. H. Baddley, *J. Am. Chem. Soc.*, *90*, 3705 (1968).
136. G. Mestroni, A. Camus, *Inorg. Nucl. Chem. Lett.*, *9*, 261 (1973).
137. G. Mestroni, A. Camus, G. Zassinovich, *J. Organomet. Chem.*, *73*, 119 (1974).
138. T. Yamamoto, A. Yamamoto, S. Ikeda, *J. Am. Chem. Soc.*, *93*, 3350 (1971).
139. A. Yamamoto, S. Ikeda, *J. Am. Chem. Soc.*, *89*, 5989 (1967).
140. T. Yamamoto, A. Yamamoto, S. Ikeda, *J. Am. Chem. Soc.*, *93*, 3360 (1971).
141. H. C. Clark, L. E. Manzer, *J. Chem. Soc.*, *Chem. Commun.*, 870 (1973).
142. H. C. Clark, L. E. Manzer, *J. Am. Chem. Soc.*, *95*, 3812 (1973).
143. H. C. Clark, L. E. Manzer, *Inorg. Chem.*, *13*, 1996 (1974).
144. C. D. M. Beverwijk, G. J. M. van Der Kerk, A. J. Leusink, J. C. Noltes, *Organomet. Chem. Rev.*, *A5*, 215 (1970).
145. A. E. Comyns, H. J. Lucas, *J. Am. Chem. Soc.*, *79*, 4339 (1957).
146. A. C. Cope, D. C. McLean, N. A. Nelson, *J. Am. Chem. Soc.*, *77*, 1628 (1955).
147. J. Solodar, J. P. Petrovich, *Inorg. Chem.*, *10*, 395 (1971).
148. W. Partenheimer, E. H. Johnson, *Inorg. Chem.*, *11*, 2840 (1972).
149. W. Partenheimer, E. H. Johnson, *Inorg. Synth.*, *16*, 117 (1976).
150. G. S. Lewandos, D. K. Gregston, F. R. Nelson, *J. Organomet. Chem.*, *118*, 363 (1976).
151. M. B. Dines, *Inorg. Chem.*, *11*, 2949 (1972).
152. R. G. Salomon, J. K. Kochi, *J. Am. Chem. Soc.*, *95*, 1889 (1973).
153. S. Komiya, J. K. Kochi, *J. Organomet. Chem.*, *135*, 65 (1977).
154. D. I. Hall, J. H. Ling, R. S. Nyholm, *Struct. Bonding* (*Berlin*), *15*, 3 (1973).

5.8.2.3.2. η^2 π-Olefin Complexes by Reaction of Metal Vapors with Monoolefins.

Codeposition of ethylene with Ni atoms yields tris(ethylene)Ni(0), which is stable to ~0°C[1,2,3]

$$CH_2{=}CH_2 + Ni \rightarrow \underset{\text{isolable}}{Ni(CH_2{=}CH_2)_3} \tag{a}$$

A similar reaction with styrene yields tris(styrene)Ni(0), which is stable to ~ −20°C[4]:

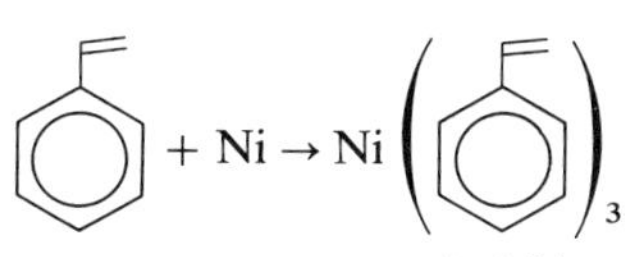

(b)

isolable

A more stable tris(alkene) complex of Pd uses norbornene tris(bicyclo[2.2.1]heptene)[1]:

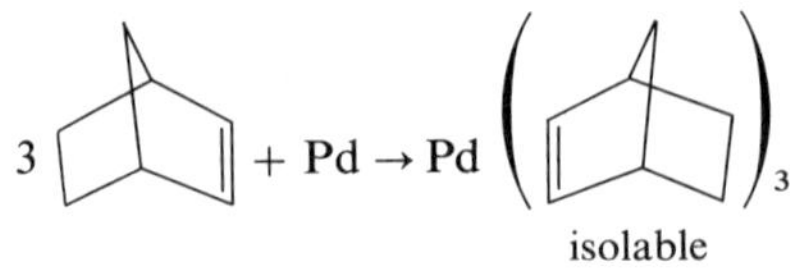

(c)

The norbornene ligand is unique in yielding more stable homoleptic alkene complexes, apparently because the strain energy in the C=C bond allows stronger π–M interactions.

Propene and 1-butene codeposited with Ni yield unstable complexes. Upon warming the 1-butene is efficiently (catalytically) isomerized to 2-butene, with π-allyl-nickel hydrides as short-lived intermediates[5,6].

Propene–PF_3–Co depositions yield a π-allyl complex where one hydrogen of the propene is transferred to Co to yield $HCo(PF_3)_4$ as a byproduct[7]:

$$\mathrm{Co + CH_3CH{=}CH_2 + PF_3 \rightarrow H{-}C(CH_2)_2{-}Co(PF_3)_3 + HCo(PF_3)_4} \quad \text{(d)}$$

Perfluoroalkenes codeposited with Ni, Pd or Pt atoms yield tris(olefin)–metal complexes that are stable to $\sim -30°C$. Above this temperature decomposition to release the olefin and metal is efficient. However, when the tris(olefin)–metal complex is treated at low temperature with $P(CH_2CH_3)_3$ or py, two olefin molecules are released and ligand-stabilized metallocyclopropane complexes are isolable[8,9]:

$CF_3CF{=}CFCF_3 + Pd \rightarrow (olefin)_3Pd$

not isolated

↓ $P(CH_2CH_3)_3$

$2CF_3CF{=}CFCF_3 +$ 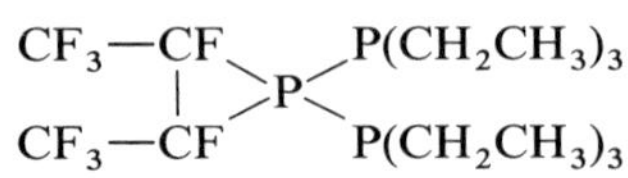

isolated

(e)

Microscale (matrix isolated only) metal atom–olefin codepositions yield unstable π-olefin–metal complexes that can be investigated by matrix isolation spectroscopy. By varying the concentrations of alkene and metal in the inert gas (at 10–50K), mono-, bis- and tris-olefin–metal complexes can be observed. Table 1 lists ethylene and tetrafluoroethylene complexes prepared for Co, Ni and Pd[1,3,10,11,12] and Co_2 and Ni_2[13,14]. If the assumption is correct that $\nu_{C\ C}$ reflects binding strength, then: (1) the metal dimers interact more strongly with C_2H_4 than the atoms, which is not expected, (2) Co and Ni atoms interact more strongly than Pd and (3) for Ni the C_2H_4 interaction decreases as more C_2H_4 molecules are added.

The only isolable complex is the tris$(C_2H_4)_3$Ni formulation. The D_{3h} planar configuration sterically impedes addition of a fourth C_2H_4 ligand, which would yield an 18-electron complex.

TABLE 1. ETHYLENE–METAL ATOM AND –METAL DIMER COMPLEXES PREPARED FOR MATRIX ISOLATION SPECTROSCOPY

Complex	$\nu_{C=C}$	Comments	Thermal stability	Ref.
$(C_2H_4)_3Co$	1499	π–Co interaction > π–Ni > π–Cu (D_{3h} planar)		3
$(C_2H_4)Co$	1504	(C_{2v})		
$(C_2H_4)Co_2$	1484	(C_{2v})		13, 14
$(C_2H_4)_2Co_2$	1508			13, 14
$(C_2H_4)_3Ni$	1512	(D_{3h} planar)	0°C	3
$(C_2H_4)_2Ni$	1465	Most π–Ni interactions	20–60K	3
$(C_2H_4)Ni$	1496	(C_{2v})	25K	3
$(C_2H_4)Ni_2$	1488			13, 14
$(C_2H_4)_2Ni_2$	1504			13, 14
$(C_2H_4)_3Pd$	1513, 1522		−70°C	1, 10, 11
$(C_2F_4)_3Ni$		Two π-bonded C_2F_4, one σ-metallocyco-propane bond		12

The spectroscopic properties of ethylene -Cu, -Ag, and -Au matrices correlate with corresponding Ni, Pd and Pt systems[15].

(K. J. KLABUNDE)

1. R. M. Atkins, R. MacKenzie, P. L. Timms, T. W. Turney, *J. Chem. Soc., Chem. Commun.*, 764 (1975).
2. P. L. Timms, private communications (1974).
3. H. Huber, G. A. Ozin, W. J. Power, *J. Am. Chem. Soc., 98*, 6508 (1976).
4. J. R. Blackboro, R. Grubbs, A. Miyashita, A. Scrivanti, *J. Organomet. Chem., 120*, C49 (1976).
5. P. S. Skell, M. J. McGlinchey, *Angew. Chem., Int. Ed. Engl., 14*, 195 (1975).
6. P. S. Skell, J. J. Havel, D. L. Williams-Smith, M. J. McGlinchey, *J. Chem. Soc., Chem. Commun.*, 1098 (1972).
7. P. L. Timms, *Angew. Chem., Int. Ed. Engl., 14*, 273 (1975).
8. K. J. Klabunde, *Acc. Chem. Res., 8*, 393 (1975).
9. K. J. Klabunde, J. Y. F. Low, H. F. Efner, *J. Am. Chem. Soc., 96*, 1984 (1974).
10. H. Huber, G. A. Ozin, W. J. Power, *Inorg. Chem., 16*, 979 (1977).
11. G. A. Ozin, W. J. Power, *Inorg. Chem., 16*, 212 (1977).
12. G. A. Ozin, W. J. Power, *Inorg. Chem., 16*, 2864 (1977).
13. A. J. Hanlan, G. A. Ozin, W. J. Power, *Inorg. Chem., 17*, 3648 (1978).
14. G. A. Ozin, W. J. Power, T. H. Upton, W. A. Goddard, *J. Am. Chem. Soc., 100*, 4750 (1978).
15. D. F. McIntosh, G. A. Ozin, R. P. Messmer, *Inorg. Chem., 19*, 3321 (1980).

5.8.2.3.3. η^1 σ-Alkyl Complexes by Addition of Metal Hydrides to Monoolefins.

The formation of transition-metal σ-alkyls by the addition of a transition-metal hydride to an alkene is a common process:

$$\text{Metal—H} + \text{C=C} \rightarrow \text{—}\overset{\overset{\text{H}}{|}}{\underset{|}{\text{C}}}\text{—}\overset{\overset{\text{Metal}}{|}}{\underset{|}{\text{C}}}\text{—} \qquad \text{(a)}$$

This reaction is an elementary step in catalytic processes such as hydrogenation, hydrosilylation, hydrocyanation, hydroformylation and olefin dimerization and isomerization. This addition is also useful as a stoichiometric route to transition-metal σ-alkyls from monoolefins, although it does have some limitations. The reaction of transition-metal hydrides with alkenes in stoichiometric synthesis of transition-metal σ-alkyls also exhibits certain features. Transition-metal hydride addition to an alkene often proceeds through initial formation of an alkene–transition-metal hydride π-complex[1]. Exceptions include reactions that proceed through a free-radical mechanism and reactions in which an intermolecular hydride transfer occurs[2,3]. When a transition-metal–hydridoalkene complex is an intermediate, the transition-metal–σ-alkyl product results from either insertion of the coordinated alkene into the metal–hydrogen bond or migration of hydride to the coordinated alkene[4]. The reverse of this addition or insertion is β elimination and is a kinetically facile thermal decomposition pathway for many σ-alkyl transition-metal compounds that contain β hydrogens (see § 5.8.2.2.5). Thus, the reversal of Eq. (a) is a complication in the synthesis of transition-metal σ-alkyls from alkenes and transition metal-hydrides. Sometimes Eq. (a) is an equilibrium that favors the hydriodoalkene complex. Then the presence of a small equilibrium concentration of transition-metal–σ-alkyl complex can be deduced by isotopic labeling[5]. In other cases, both the alkyl and the alkene complexes can be seen. In still other cases the equilibrium favors the σ-alkyl complex.

Whether products or reactants are favored in Eq. (a) is the subject of theoretical studies. Increased electron density of the metal stabilizes the hydrido–alkene complex.[6] The energy differences between metal–hydrogen and metal–carbon bonds (M—H bonds are usually strong bonds) and the possible electronic changes associated with insertion [going from an n-coordinated metal to an (n − 1)-coordinated metal][7] are also factors that determine the position of equilibrium (a). Insertion of ethylene into Rh—H[4], Co–H[8], Cr—H [9] and Pt—H [1,10] bonds are studied theoretically. The regiochemistry and the regioselectivity of the addition and the stereochemistry of the addition process are important and vary with different metal hydrides. Although many metal hydride additions to alkenes occur with the hydrogen attaching to the carbon with the fewest hydrogens to form the less substituted σ-alkyl transition-metal compound, there are also cases with mixed or altered regiochemistry in the addition process. The stereochemistry of the addition of metal hydride to an alkene is commonly assumed to be cis. Addition typically occurs from the least hindered face of the alkene.

An important example of transition-metal σ-alkyl synthesis by hydrometalation is the synthesis of bis(cyclopentadienyl)alkylzirconium chlorides $(\eta^5\text{-Cp})_2ZrRCl$ from alkenes and biscyclopentadienylchlorozirconium hydride $(\eta^5\text{-Cp})_2ZrHCl$. The facility and general applicability of this hydrometalation, coupled with the synthetic versatility of the resultant alkylzirconium reagents makes this hydrometalation an important synthetic method in organic chemistry[11].

Hydrozirconation is carried out in a hydrocarbon such as benzene or in an ether such as tetrahydrofuran at 25–80°C. The starting $(\eta^5\text{-Cp})_2ZrHCl$ used is prepared separately from $(\eta^5\text{-Cp})_2ZrCl_2$ and an aluminum hydride[12] or obtained commercially. The hydride $(\eta^5\text{-Cp})_2ZrHCl$ can also be prepared in situ with or without the alkene which is to be hydrozirconated[13]. Most hydrozirconations using preformed $(\eta^5\text{-Cp})_2ZrHCl$ involve small scales. Large-scale reactions can be carried out using zirconium hydride prepared in situ. Hydrozirconation in the most favorable instances can be accomplished by simply adding a benzene solution of an alkene to a benzene suspension of $(\eta^5\text{-Cp})_2ZrHCl$ at RT. The insoluble zirconium hydride dissolves to give a solution of the product alkylzirconium compound within several hours. Modest heating accelerates hydrozirconation of less reactive alkenes. Often the alkylzirconium compound is not isolated but directly transformed into the desired functional group.

Most alkenes can be employed in hydrozirconation reactions as shown in the equations below and in Table 1. When internal alkenes such as 4-octene are used as substrates, only the primary alkylzirconium is obtained $(\eta^5\text{-Cp})_2Zr(Cl)C_8H_{17-n}$:

$$CH_3(CH_2)_2CH{=}CH(CH_2)_2CH_3 \xrightarrow[C_6H_6]{(\eta^2\text{-Cp})_2ZrHCl} (\eta^5\text{-Cp})_2Zr[CH_2(CH_2)_6CH_3]Cl \quad \text{(b)}$$

This isomerization is attributed to rapid hydrozirconation followed by β eliminations and readditions of zirconium hydride until the more stable primary alkylzirconium is formed. No intermediate sec-alkylzirconium or isomerized alkene is detected by NMR, implying that exchange of alkene coordinated to zirconium with alkene in solution is less facile than hydrozirconation. No isomerization occurs past branches in the alkyl chain. Hydrozirconation of 2,2-disubstituted alkenes leads exclusively to the primary alkylzirconium. sec-Alkylzirconiums that cannot isomerize to a primary alkylzirconium can be

TABLE 1. BIS(CYCLOPENTADIENYL)ZIRCONIUM(ALKYL) CHLORIDES PREPARED FROM ALKENES AND $(\eta^5\text{-Cp})_2ZrHCl$

Alkene	Product	Ref.
$CH_3(CH_2)_5CH{=}CH_2$	$CH_3(CH_2)_7ZrCl(\eta^5\text{-Cp})_2$[a]	14
(Z)-$CH_3(CH_2)_2CH{=}CH(CH_2)_2CH_3$	$CH_3(CH_2)_7ZrCl(\eta^5\text{-Cp})_2$[a]	14
(E)-$CH_3(CH_2)_2CH{=}CH(CH_2)_2CH_3$	$CH_3(CH_2)_7ZrCl(\eta^5\text{-Cp})_2$[a]	14
$\overline{CH_2CH_2CH{=}CHCH_2CH_2}$	$\overline{CH_2(CH_2)_4CH}ZrCl(\eta^5\text{-Cp})_2$[b]	14
$\overline{CH_2(CH_2)_4C}{=}CH_2$	$\overline{CH_2(CH_2)_4CH}CH_2ZrCl(\eta^5\text{-Cp})_2$	14
$(CH_3)_2C{=}CHCH_3$	$(CH_3)_2CHCH_2CH_2ZrCl(\eta^5\text{-Cp})_2$[b]	14
(E)-$(CH_3)_2CHCH{=}CHCH_3$	$(CH_3)_2CH(CH_2)_2CH_2ZrCl(\eta^5\text{-Cp})_2$	11
$\overline{CH_2CH_2CH{=}CHCH_2}$	$\overline{CH_2(CH_2)_3CH}ZrCl(\eta^5\text{-Cp})_2$	14
$CH_3(CH_2)_{10}CH{=}CH(CH_2)_{10}CH_3$	$CH_3(CH_2)_{22}CH_2ZrCl(\eta^5\text{-Cp})_2$[c]	13
$CH_3(CH_2)_{11}CH{=}CH(CH_2)_{11}CH_3$	$CH_3(CH_2)_{24}CH_2ZrCl(\eta^5\text{-Cp})_2$[c]	13
$CH_3(CH_2)_{12}CH{=}CH(CH_2)_{12}CH_3$	$CH_3(CH_2)_{26}CH_2ZrCl(\eta^5\text{-Cp})_2$[c]	13
$CH_3(CH_2)_{13}CH{=}CH(CH_2)_{13}CH_3$	$CH_3(CH_2)_{28}CH_2ZrCl(\eta^5\text{-Cp})_2$[c]	13
$CH_3(CH_2)_{15}CH{=}CH(CH_2)_{15}CH_3$	$CH_3(CH_2)_{32}CH_2ZrCl(\eta^5\text{-Cp})_2$[c]	13
$CH_3(CH_2)_{19}CH{=}CH(CH_2)_{19}CH_3$	$CH_3(CH_2)_{40}CH_2ZrCl(\eta^5\text{-Cp})_2$[c]	13
$CH_2CHCH{=}CHCHCH_2CH_2$ (bicyclic)	$CH_2CH_2CHCH_2CHCH_2CHZrCl(\eta^5\text{-Cp})_2$[d] (bicyclic)	17
$CH_2CHCH_2CH_2CHC(CH_3)_2C{=}CH_2$ (bicyclic)	$CH_2CHCH_2CH_2CHC(CH_3)_2CHCH_2ZrCl(\eta^5\text{-Cp})_3$[d] (bicyclic)	17
$C(CH_3)_2CHCH_2CHCH_2CH_2C{=}CH_2$ (bicyclic)	$C(CH_3)_2CHCH_2CHCH_2CH_2CHCH_2ZrCl(\eta^5\text{-Cp})_2$[d] (bicyclic)	17
$CH_2CHC(CH(CH_3)_2)CH_2CH_2C{=}CH_2$ (bicyclic)	$CH_2CHC(CH(CH_3)_2)CH_2CH_2CHCH_2ZrCl(\eta^5\text{-Cp})_2$[d] (bicyclic)	17

[a] Characterized as alkane after quenching with H^+.
[b] Characterized as the alkyl bromide after treatment of the alkylzirconium compound with bromine.
[c] Characterized as the iodide after iodination of the alkylzirconium compound.
[d] Characterized as the alcohol after $(CH_3)_3COOH$ oxidation of the alylzirconium compound.

prepared; e.g., cyclohexene and bicyclo(2.2.1)hept-2-ene are both hydrozirconated to yield sec-alkylzirconiums. Some trisubstituted alkenes such as 3-methyl-2-butene also react with $(\eta^5\text{-Cp})_2ZrHCl$ to yield $(\eta^5\text{-Cp})_2Zr[(CH_3)_2CHCH_2CH_2]Cl$, while others such as methylcyclohexene are unreactive. Tetrasubstituted alkenes such as tetramethylethylene do not react with $(\eta^5\text{-Cp})_2ZrHCl$.

The relative reactivity is measured qualitatively. At RT the rate of hydrozirconation of various alkenes is 1-octene > cis-4-octene > trans-4-octene > methylenecyclohexane > cyclopentene > cyclohexene > 3-methyl-2-butene. In competitive studies, a terminal alkene reacts 70 times faster that a disubstituted alkene. These relative rates are attributed to steric effects[14]. Double bonds γ to zirconium are less reactive than dienes since $(\eta^5\text{-Cp})_2Zr(CH_2CH_2CH{=}CH_2)Cl$ can be isolated from the reaction of $(\eta^5\text{-Cp})_2ZrHCl$ and butadiene[15].

Hydrozirconation of alkenes as a route to σ-alkylzirconiums is limited by other reactive functional groups; e.g., alkynes react more readily than alkenes with $(\eta^5\text{-Cp})_2ZrHCl$, and other reactive functional groups with $(\eta^5\text{-Cp})_2ZrHCl$ include alcohols, aldehydes, ketones, carboxylic acids, carboxylic acid esters, nitriles and thiocarbonyls[11,16]. Alcohols may be protected either as alkyl ethers or as trialkylsilyl ethers.

The stereochemistry of hydrozirconation involves addition of hydrogen and Zr to the least hindered face of an alkene[17]:

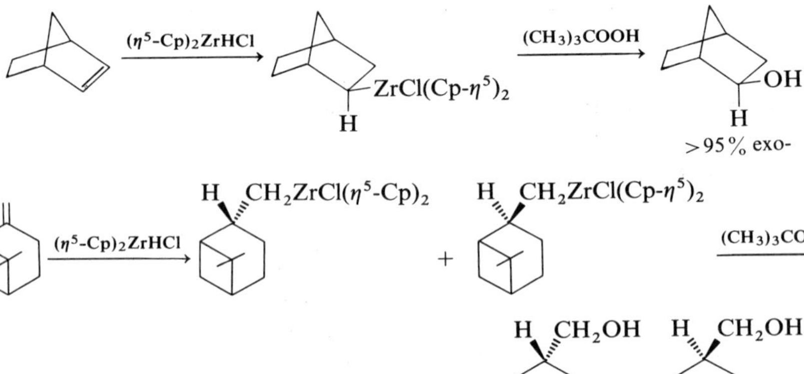

(c)

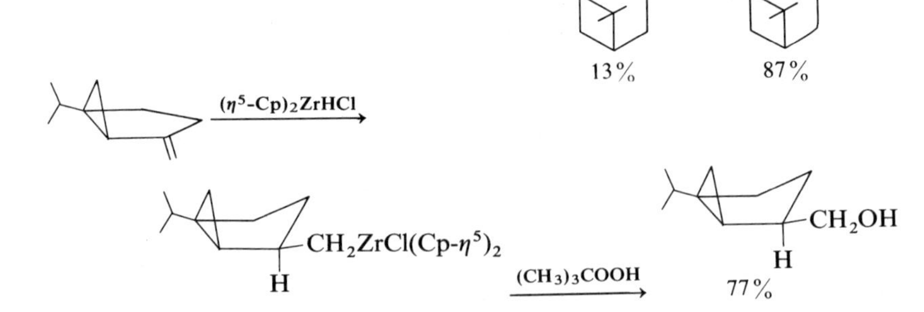

(d)

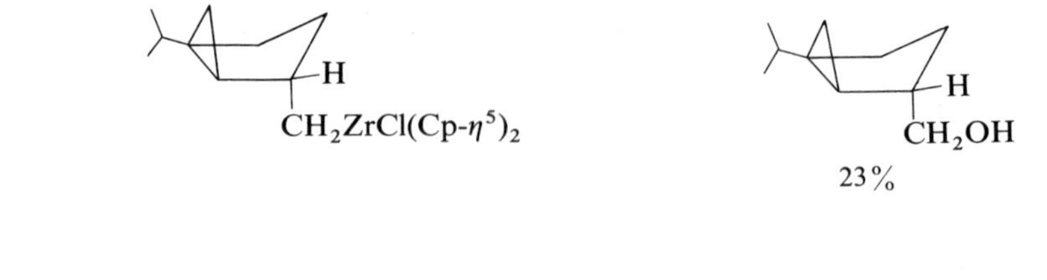

(e)

Isomerization of diastereomeric alkylzirconiums formed from reaction of $(\eta^5\text{-Cp})_2ZrHCl$ and α-pinene is implied in one instance[17], but such epimerizations are not confirmed in other systems[18]. A vicinal cyclopropyl group influences the stereoselectivity of hydrozirconation by intramolecular electronic effects (Eq. e).

Hydrozirconation occurs by cis addition of Zr and hydrogen to terminal alkenes[18]. Both the (E) and (Z) diastereomers of 1-deuterio-3,3-dimethylbutene react with $(\eta^5\text{-Cp})_2ZrHCl$ to give primary alkylzirconiums characterized by deuterium-decoupled ^1H NMR:

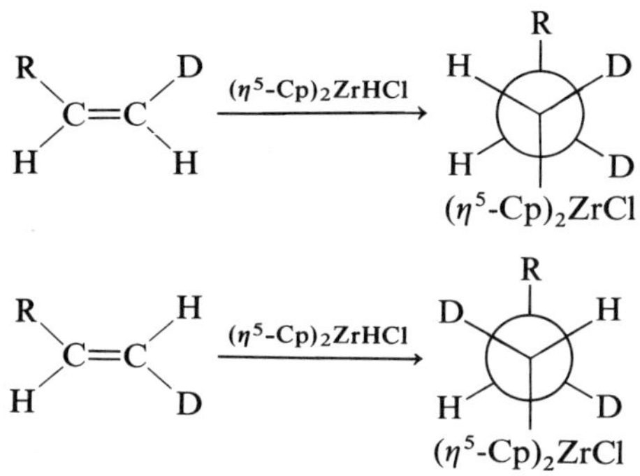

(f)

Stereochemical assignments based on the characteristic vicinal coupling constants of 12.9 and 4.2 Hz are seen for the erythro- and threo-diastereomeric alkylzirconiums, respectively. These diastereomeric primary alkylzirconiums are configurationally stable once formed.

Heterosubstituted alkenes undergo hydrozirconation. Hydrozirconation of alkenylzirconiums is a suggested mechanism for isomerization of alkenylzirconiums, and 1,2-dizirconiums are known[11,19]. Hydrozirconation of alkenylboranes and alkenylaluminums leads to both 1,1- and 1,2-dimetalated derivatives[20]:

$$(\eta^5\text{-Cp})_2ZrHCl + RCH{=}CHAl(CH_3)_2 \rightarrow$$

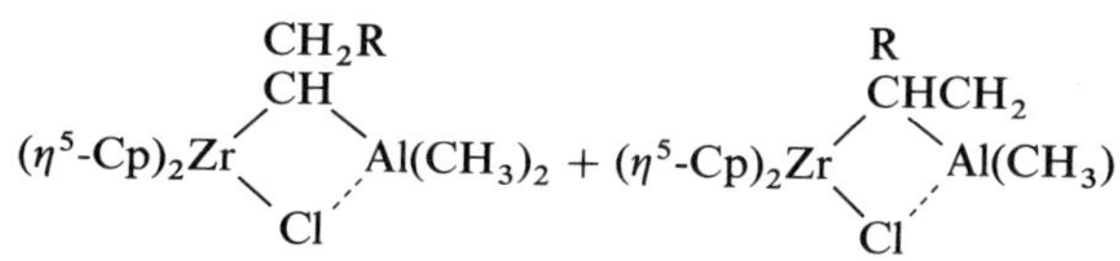

(g)

where if $R = -C(CH_3)_3$, respective product yields are >20 and $<1\%$; if $R = -(CH_2)_3CH_3$, respective product yields are 3 and 2%; and if $R = -CH_3$, respective product yields are 1 and 2%. These dimetalated species are synthetically useful as alternatives to conventional reagents for forming carbon–carbon bonds in reactions with ketones and aldehydes:

$$(\eta^5\text{-Cp})_2ZrHCl + (CH_3)_3CCH{=}CHBR_2 \rightarrow (\eta^5\text{-Cp})_2Zr\begin{matrix} \diagup CH[CH_2C(CH_3)_3]BR_2] \\ \diagdown Cl \end{matrix} \qquad \text{(h)}$$

where $-BR_2 = -\overline{B}CH(CH_2)_3\overline{C}HCH_2CH_2\underline{C}H_2$.

Hydrohafnation, the reaction of alkenes with $(\eta^5\text{-Cp})_2HfHCl$ to form σ-alkylhafniums, is similar to hydrozirconation[21] and uses the same procedures. High yields of $(\eta^5\text{-Cp})_2HfRCl$ can be obtained using alkenes similar to those used in hydrozirconation. Electrophilic substitution reactions of these alkylhafniums are also similar in scope. As in hydrozirconation, rapid isomerization leads to primary alkylhafniums from both terminal and internal alkenes if there are no branches in the carbon chain and if isomerization can occur:

$$CH_3(CH_2)_4CH{=}CH_2 \xrightarrow[20\text{ min}, -20°C]{(\eta^5\text{-Cp})_2HfHCl} CH_3(CH_2)_5CH_2HfCl(Cp\text{-}\eta^5)_2$$

$$CH_3CH{=}CHCH_2CH_3 \xrightarrow[1\text{ h}, 40°C]{(\eta^2\text{-Cp})_2HfHCl} CH_3(CH_2)_3CH_2HfCl(Cp\text{-}\eta^5)_2 \qquad \text{(i)}$$

The stereochemistry, however, is different than that of hydrozirconation; e.g., the stereochemistry of hydrohafnation of several bicyclic terpenes, examined by analysis of the product alcohols formed after oxidation of the intermediate alkylhafnium compounds with butyl hydroperoxide, corresponds to hydrohafnation from the more hindered face of the bicyclic alkene:

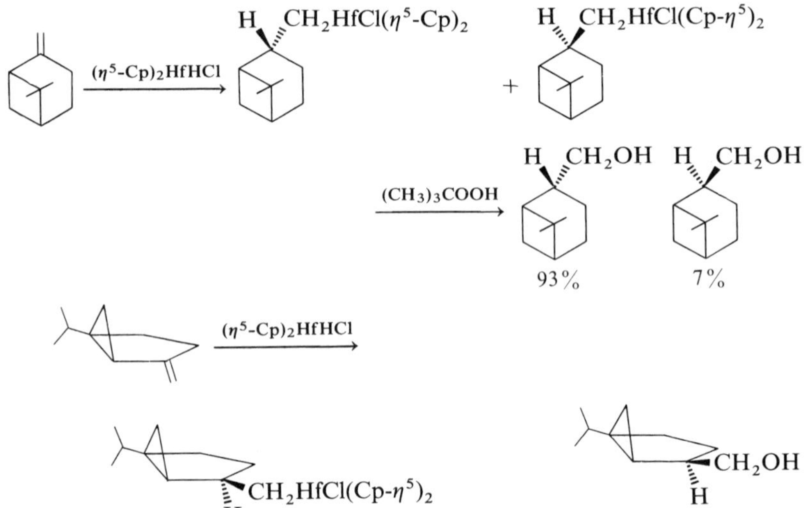

(j)

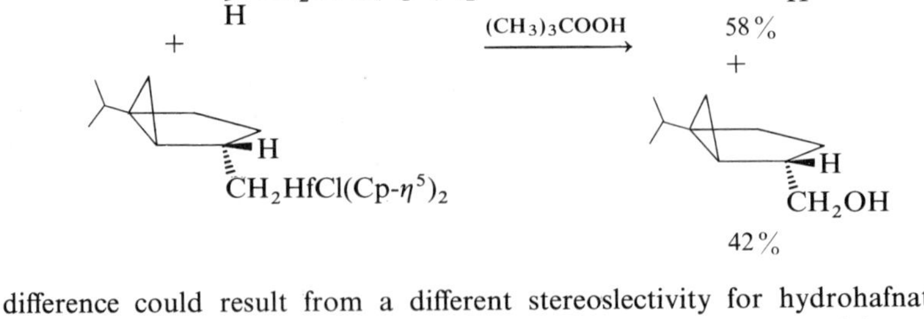

(k)

This difference could result from a different stereoslectivity for hydrohafnation vs. hydrozirconation or from epimerization of the diastereomeric alkylhafniums. Direct observation of epimerization in diastereomeric alkylhafnium compounds is not yet accomplished, but hydrohafnation is more sensitive to steric effects than hydrozirconation, since cyclohexene, camphene and β-pinene do not react with $(\eta^5\text{-Cp})_2HfHCl$ but do react with $(\eta^5\text{-Cp})_2ZrHCl$.

Hydrometalation leads to an ethylniobium from $(\eta^5\text{-Cp})_2NbH_3$ and ethylene[22]:

$$(\eta^5\text{-Cp})_2NbH_3 \xrightarrow{C_2H_4} Cp_2Nb(C_2H_4)H \xrightarrow{C_2H_4} Cp_2Nb(C_2H_4)(C_2H_5) \qquad (1)$$

This reaction proceeds through a hydriodoethylene complex, which is also isolable. The structure of this insertion product, $(\eta^5\text{-Cp})_2Nb(C_2H_4)(C_2H_5)$, is known[23]. This hydrometalation may be a step in the $(\eta^5\text{-Cp})_2NbH_3$-catalyzed hydrogenation of ethylene to ethane. Addition of $(\eta^5\text{-Cp})_2NbH_3$-catalyzed hydrogenation of ethylene to ethane. Addition of $(\eta^5\text{-Cp})_2NbH_2$ to propylene to give a paramagnetic product, $(\eta^5\text{-Cp})_2Nb(C_3H_7)_2$, also occurs, although the dialkylniobium formed is characterized by ESR spectroscopy but is not isolated.

The kinetics for alkene insertion into Nb—H bonds using permethylniobocene (III)-olefine hydride complexes to form σ-alkylniobium compounds have been measured by ^{1}H NMR spectroscopy using magnetization transfer and coalescence techniques[24] The starting $(\eta^5\text{-Cp})_2Nb(CH_2{=}CHR)H$ complexes can be converted into stable σ-alkylniobium complexes, but only by addition of CO or methyl isocyanide at 25°C. In these cases, the final products are complexes of formula $(\eta^5\text{-Cp})_2Nb(CH_2CH_2R)L$, where L is CO or $NCCH_3$. Whereas the additional ligand permits isolation of the product σ-alkylniobium, it complicates the kinetics. Thus, rates measured by NMR are for formation of $(\eta^5\text{-Cp})_2NbCH_2CH_2R$ from $(\eta^5\text{-Cp})_2Nb(CH_2{=}CHR)H$. In the case of the simplest olefin complex (R = H), this rate is 5.03 s^{-1} at 59°C. The reverse of this reaction (β-hydride elimination) was much faster with an estimated rate $\geq 100\ s^{-1}$. Rates and activation parameters have been measured for a series of alkenes in these reactions, where R included H, CH_3, Ph, p-$Me_2NC_6H_5$, p-$MeOC_6H_4$, p-MeC_6H_4 and p-$CF_3C_6H_4$. The E_a values range from 14.1 to 16.7 kcal mol^{-1} in these cases. A correlation was shown to exist with σ_{para} with a ρ value of -0.64. Thus, in this system electron-donating substituents at the para position of the phenyl group on the β-carbon accelerate the formation of a σ-alkylniobium from the niobium olefin–hydride complex.

Bercaw[25] has extended these studies of σ-alkyl formation from niobium olefin–hydride complexes to include additional examples of meta-substituted styrene complexes of permethylniobocene. As noted in earlier work[25], electron-donating substituents on the β-carbon accelerate this insertion reaction, whereas electron-withdrawing substituents retard the rate based on a σ_{meta} correlation with a ρ of -1.2 [25].

Kinetic studies have also been carried out with the sterically less constrained endo-$(\eta^5\text{-Cp})_2Nb(CH_2{=}CHR)H$ system. Again, a correlation of insertion rate with σ_{para} is seen $\rho = -1.1$). However, unlike the $(\eta^5\text{-Cp})_2Nb(CH_2{=}CHR)H$ systems studied earlier, the ground-state energies of the niobium olefin–hydride complexes also vary with the electronic nature of the substituent.

Formation of σ-alkylniobium complexes from the exo-$(\eta^5\text{-Cp})_2Nb(CH_2{=}CHR)H$ leads to a different σ-alkylniobium species, $(\eta^5\text{-Cp})_2Nb(CH_3CHR)$ and has different substituent effects. Here para-substituted aryl substituents ($R = C_6H_4Z$) are attached to the α carbon and electron-withdrawing substituents are found to accelerate the insertion reactions ($\rho = +0.7$). This is interpreted to mean that electron-withdrawing groups at C_α either stabilize a modest negative charge in the insertion reaction or increase the ionic character and bond energy of the incipient Nb—C σ bond[92].

TABLE 2. PRODUCTS OF LIGAND-INDUCED INSERTION INTO $(\eta^5\text{-Cp})_2Ta(H)L$

Alkene	Ligand	Product	Refs.
$CH_3CH{=}CH_2$	CO	$[CH_2(CH_2)_2]Ta(CO)(\eta^5\text{-Cp})_2$	26
$CH_3CH_2CH{=}CH_2$	CO	$[CH_3(CH_2)_3]Ta(CO)(\eta^5\text{-Cp})_2$	26
$CH_3(CH_2)_2CH{=}CH_2$	CO	$[CH_3(CH_2)_4]Ta(CO)(\eta^5\text{-Cp})_2$	26
$CH_3CH{=}CH_2$[a]	2,6-$(CH_3)_2C_6H_3NC$	$[CH_3(CH_2)_2]Ta[2,6\text{-}(CH_3)_2C_6H_3NC](\eta^5\text{-Cp})_2$	27
$CH_3CH_2CH{=}CH_2$[a]	2,6-$(CH_3)_2C_6H_3NC$	$[CH_3(CH_2)_3]Ta[2,6\text{-}(CH_3)_2C_6H_3NC](\eta^5\text{-Cp})_2$	27
$CH_3CH_2{=}CH_2$[a]	c-$C_6H_{11}NC$	$[CH_3(CH_2)_2]Ta(c\text{-}C_6H_{11}NC)(\eta^5\text{-Cp})_2$	26
$CH_3CH_2CH{=}CH_2$[a]	c-$C_6H_{11}NC$	$[CH_3(CH_2)_3]Ta(c\text{-}C_6H_{11}NC)(\eta^5\text{-Cp})_2$	26
$CH_3CH{=}CH_2$[a]	CH_3NC	$[CH_3(CH_2)_2]Ta(CH_3NC)(\eta^5\text{-Cp})_2$	26
$CH_3CH_2CH{=}CH_2$[a]	CH_3NC	$[CH_3(CH_2)_3]Ta(CH_3NC)(\eta^5\text{-Cp})_2$	26
$CH_3CH{=}CH_2$[a]	$(CH_3)_3CNC$	$[CH_3(CH_2)_2]Ta[(CH_3)_3CNC](\eta^5\text{-Cp})_2$	26
$CH_3CH_2CH{=}CH_2$[a]	$(CH_3)_3CNC$	$[CH_3(CH_2)_3]Ta[(CH_3)_3CNC](\eta^5\text{-Cp})_2$	26

[a] Only the endo complex undergoes the insertion reaction.

Hydrometalation reactions of Ta hydrides and alkenes are important in hydrogenations using $(\eta^5\text{-Cp})_2TaH_3$, which are analogous to those reported for the corresponding Nb compounds. Although alkyltantalums are not isolated from hydrogenation mixtures, hydridoalkene complexes of Ta undergo a ligand-induced hydrometalation. Thus, both CO and isonitriles can be used to form σ-alkyltantalum products from $(\eta^5\text{-Cp})_2TaH$(alkene) complexes (see Table 2)[26,27]:

$$(\eta^5\text{-Cp})_2Ta(C_2H_4)H + L \rightarrow (\eta^5\text{-Cp})_2Ta(C_2H_5)L \tag{m}$$

where L = CO or R′NC.
This hydrometalation reaction is regiospecific, forming only the primary alkyltantalum. These insertions proceed under mild conditions (e.g., 75°C and 100 kPa CO for 7–8 h) on the endo-hydridoalkene complexes. However, while the exo-hydridoalkene complexes do react with CO to form Ta alkyls, alkyltantalums are not formed from the exo-hydridoalkene complex and an isonitrile.

Rates of formation of σ-alkylniobium and σ-alkyltantalum compounds have been compared using $(\eta^5\text{-Cp})_2Nb(CH_2{=}CH_2)H$, $(\eta^5\text{-Cp})_2Nb(CH_2{=}CH_2)$,- $(\eta^5\text{-Cp})_2Ta(CH_2{=}CH_2)H$ and $(\eta^5\text{-Cp})_2Ta(CH_2{=}CH_2)H$. Insertion is fastest with the niobium complexes and the permethylniobocene and permethyltantalocenes are both faster than their H-substituted analogs[25].

Insertion of ethylene into a Cr—H bond can be studied using CNDO/2 and INDO calculations[9]. This insertion of ethylene into a cationic cyclopentadienyl(siloxane)chromium hydride is a model for ethylene polymerization by chromocene–silica catalysts.

Hydrometalation of alkenes with Mo and W hydrides occurs on addition of PPh_3 to cationic hydridoalkene complexes such as $[(\eta^5\text{-Cp})_2Mo(C_2H_4)H]^+$ and $[(\eta^5\text{-Cp})_2W(C_2H_4)H]^+$ and with the corresponding propylene complexes[7]:

$$[(\eta^5\text{-Cp})_2M(C_2H_4)H][PF_6] \xrightarrow{PPh_3} (\eta^5\text{-Cp})_2M(C_2H_5)(PPh_3) \tag{n}$$

where M = Mo or W. This insertion does not occur without added phosphine because the metal's two d electrons would occupy nearly degenerate orbitals, and the metal alkyl might consequently have a triplet ground state if phosphine were not present. The regiochemistry of metal hydride addition to propylene is high, and no 2-propylmolyben-

TABLE 3. PRODUCTS OF ADDITION OF $(\eta^5\text{-Cp})_2MoH_2$ TO ACTIVATED ALKENES (XCH=CHY).

Alkene substituent		Product	Refs.
X	Y		
H	CO_2CH_3	$(\eta^5\text{-Cp})_2MoH(CH(CH_3)CO_2CH_3H$	30
H	CN	$(\eta^5\text{-Cp})_2MoH(CH(CH_3)CN$	30
CO_2CH_3	CO_2CH_3	$(\eta^5\text{-Cp})_2MoH(CH(CO_2CH_3)CH_2CO_2CH_3$	30
CN	CN	$(\eta^5\text{-Cp})_2MoH(CH(CN)CH_2CN$	30

dums or -tungstens are formed. The only products are $(\eta^5\text{-Cp})_2Mo(CH_2CH_2CH_3)PPh_3$ and $(\eta^5\text{-Cp})_2W(CH_2CH_2CH_2)PPh_3$ for Mo and W, respectively. Hydrometalation of fluorinated alkenes is also known for both Mo and W tricarbonyl hydrides[28]:

$$\eta^5\text{-CpMo(CO)}_3H + C_2F_4 \rightarrow \eta^5\text{-CpMo(CO)}_3C_2F_4H$$
$$\eta^5\text{-CpW(CO)}_3H + C_2F_4 \rightarrow \eta^5\text{-CpW(CO)}_3C_2F_5H \qquad \text{(o)}$$

Electron-withdrawing substituents on the substrate alkene such as carbonyl or nitrile groups also facilitate hydrometalation of alkenes by Mo hydrides. The synthesis of several hydrido σ-alkyl complexes of Mo from alkenes containing various X and Y groups as electron acceptors is described (Table 3)[29,30]:

$$(\eta^5\text{-Cp})_2MoH_2 + YCH{=}CHX \rightarrow (\eta^5\text{-Cp})_2Mo(H)CHXCH_2Y \qquad \text{(p)}$$

where X, Y = electron-accepting groups. The stereochemistry of this addition can be examined using dimethylfumarate and dimethylmaleate:

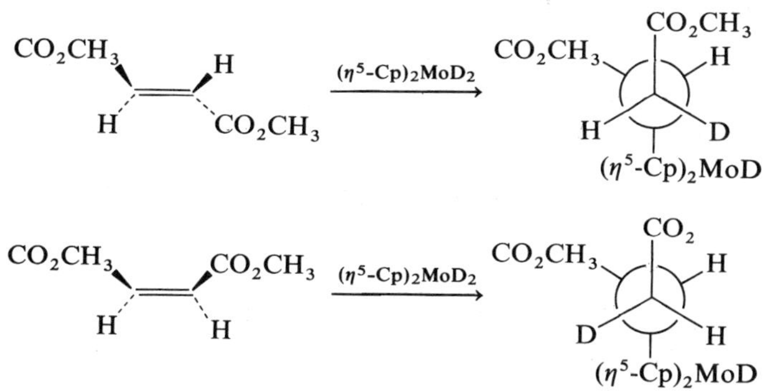

(q)

The addition of Mo and H to these diastereomeric alkenes is stereospecifically cis, based on the observed NMR coupling constants of 10.5 and 5.5 Hz for the threo- and erythro-Mo hydrido(alkyl) complexes formed, respectively. No deuterium scrambling is observed in these reactions, showing that the hydrometalation is not reversible. Protonation of alkenylmolybdenum hydrides by HCl also leads to σ-alkylmolybdenums[30]:

$$(\eta^5\text{-Cp})_2Mo(H)C(CO_2CH_3){=}CHCO_2CH_3 \xrightarrow{HCl} (\eta^5\text{-Cp})_2Mo(Cl)CH(CO_2CH_3)CH_2CO_2CH_3 \qquad \text{(r)}$$

Metaloheterocycles are formed by addition of $[(\eta^5\text{-Cp})_2MoH_3][PF_6]$ and $(\eta^5\text{-Cp})_2MoHI$ to 2-vinylpyridine, 1-allylpyrazole and 2-allylpyridine[31]:

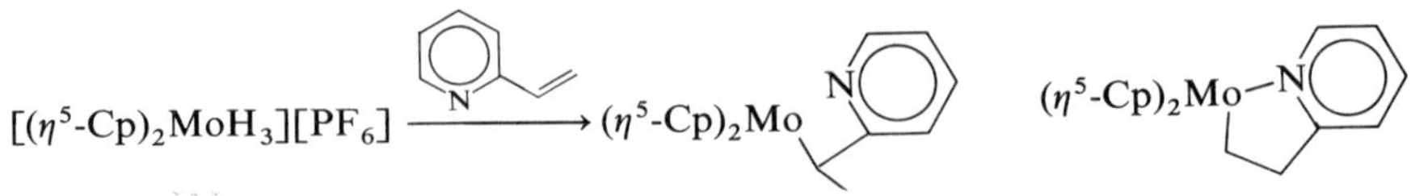

(s)

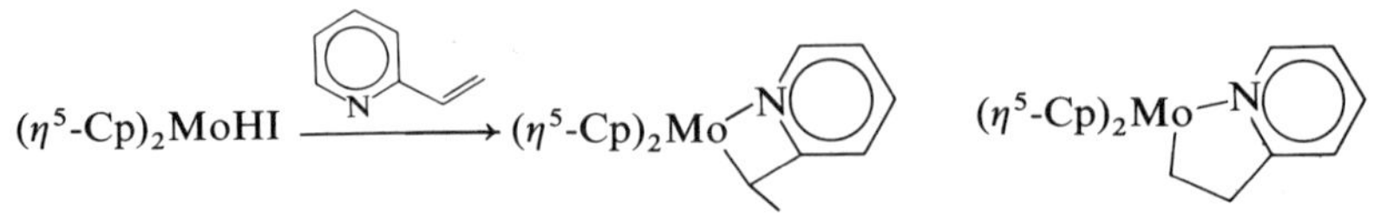

(t)

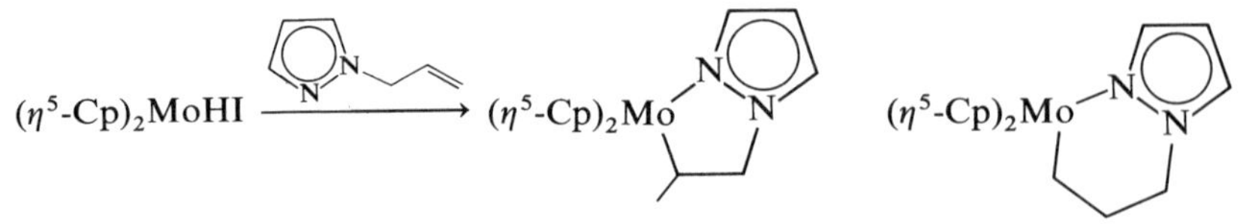

(u)

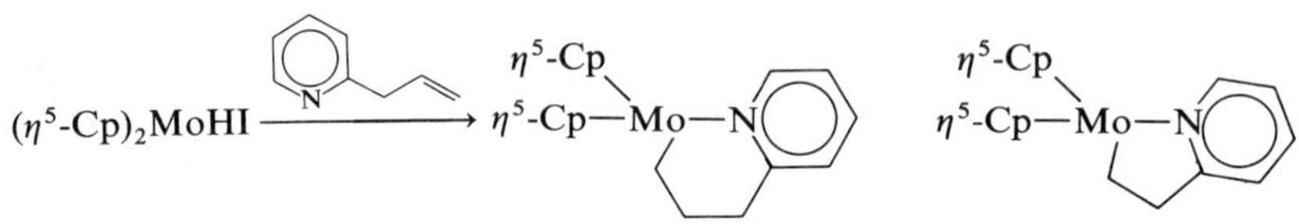

(v)

This intramolecular hydrometalation is not regiospecific.

Alkylmolybdenums and tungstens are sometimes formed in low concentrations in an equilibrium hydrometalation; e.g., both $[HMo(C_2H_4)(Ph_2PCH_2CH_2PPh_2)]$-$[CF_3CO_2]^-$ and $W_4(\mu\text{-H})_2[OCH(CH_3)_2]_{14}$ exchange H with ethylene as determined by ^1H NMR and labeling studies[32,33].

σ-Alkylmanganeses can be formed from $HMn(CO)_5$ and alkenes. Halogenated alkenes such as chlorotrifluoroethylene react to form mixtures of regiochemical isomers with either irradiation or heat[34]:

$$HMn(CO)_5 + FClC{=}CF_2 \xrightarrow{\text{heat or } h\nu} HCF_2CFClMn(CO)_5 + HCFClCF_2Mn(CO)_5 \qquad \text{(w)}$$

Styrene also reacts with $HMn(CO)_5$ to form 1-phenethylmanganese pentacarbonyl as a detectable intermediate in styrene hydrogenations using $HMn(CO)$[35]. Hydrometalation of a chelated alkene by a hydridomanganese species is also involved in ring contractions of Mn phosphino heterocycles[36].

Alkene insertions into Re—H bonds to form σ-alkylrheniums occurs for halogenated alkenes[37]. Reaction of $HRe(CO)_5$ with C_2F_4, C_2F_3Cl and CCl_2CF_2 occurs to give $(C_2F_4H)Re(CO)_5$, $(CHClFCF_2)Re(CO)_5$ and $(CHCl_2CF_2)Re(CO)_5$, respectively, with the metal becoming attached to the carbon bearing the greatest number of fluorines. Another approach to a σ-ethylrheniumpentacarbonyl that is a different way of preparing σ-alkyl transition-metal compound from a hydridoalkene complex is [38]:

$$[(PPh_3)_2Ir(H)(Cl)(CO)(C_2H_4)]^+ + [Re(CO)_5]^- \rightarrow Re(CO)_5(C_2H_5) + ClIr(CO)(PPh_3)_2 \qquad \text{(x)}$$

where pentacarbonylrhenate in THF attacks a coordinated alkene nucleophilically in a cationic iridium hydride complex to give a 1,2-dimetalated ethane. Subsequent reductive elimination then forms the products shown.

σ-Alkylirons form by insertion of both activated and unactivated alkenes into Fe—H bonds. Hydrometalation and β elimination both occur during decomposition and rearrangement of σ-alkylirons[39,40]. Both ethylene and cyclohexene insert into phosphine complexes of iron dihydride and tetrahydride[40]:

$$RCH{=}CHR + FeH_2N_2(EtPPh_2)_3 \rightarrow FeH(R')(EtPPh_2)_2(solvent) \quad (y)$$

where R = —$CH_2CH_2CH_2CH_2$— or —H; R′ = C_6H_{11} or C_2H_5. Reduction of activated alkenes containing electron-withdrawing groups such as carbonyls or nitriles can be effected using various Fe hydridocarbonyl complexes and is a synthetically useful reduction[40,43]. Where the intermediate σ-alkyliron is isolated, the Fe atom is σ-bonded to the carbon atom α to the carbonyl or nitrile, and the reaction is regiospecific (Table 4): The mechanistic details using $[HFe(CO_4]^{2-}$ and $[HFe_2(CO)_8]^-$ are known, as are the effects of ion pairing, isotopic substitution, stereochemistry, and steric effects[43,44]:

$$\eta^5\text{-Cp Fe(CO)}_2H + CH_2{=}CHCN \rightarrow \eta^5\text{-Cp Fe(CO)}_2[CH(CH_3)CN] \quad (z)$$

Coordination of alkenes by $[\eta^5\text{-CpFe(CO)}_2]^+$ activates alkenes toward nucleophilic substitution. For cationic alkene complexes η^5-CpFe(CO)(PPh_3)H can serve as a source of nucleophilic hydride[3]. Thus, the transformations:

$$\eta^5\text{-CpFe(CO)}_2\text{(alkene)}]^+ + \eta^5\text{-CpFe(CO)(PPh}_3\text{)H} \rightarrow \eta^5\text{-CpFe(CO)}_2(\sigma\text{-alkyl}) \quad (aa)$$

Insertion of alkenes into Ru—H bonds is step in many Ru-catalyzed reactions of alkenes such as homogeneous hydrogenation. Stoichiometric formation of isolable σ-alkyl complexes is also observed. The reaction of ethylene with Ru hydrides such as $RuHCl(PPh_3)_3$ or $RuH(OCOCF_3)(PPh_3)_3$ is an equilibrium process[45,46]:

$$RuHX(PPh_3)_3 + C_2H_4 \rightleftharpoons Ru(X)(C_2H_5)(PPh_3)_3 \quad (ab)$$

where X = Cl^- or $[OCOF_3]^-$. Although ethylrutheniums are detectable with ethylene as the alkene, other σ-alkyl intermediates are not detected with $RuHCl(PPh_3)_3$ and other alkenes. For X = $[OCOCF_3]^-$, the ethylruthenium species is favored at 600 kPa of CO and the ethylene complex at 100 kPa of CO and 35°C.

Insertion of cyclohexene into a Ru—H bond occurs at a rate that can be studied by ^{1}H NMR for the reaction of a hydrido(η^3-cyclohexadienyl)cyclohexene complex of ruthenium(II)[47]. Insertion of both activated and unactivated alkenes into the Ru—H bond of $K[(PPh_3)_2Ph_2PC_6H_4RuH_2]\cdot C_{10}H_8\cdot(C_2H_5)_2O$ also occurs, although isolable σ-alkylruthenium species are not yet characterized[48].

Formation of σ-alkyl complexes from ruthenium hydrides and alkenes containing a potential chelating group occurs regioselectively such that the product σ-alkyl contains a five-membered chelate; e.g., reaction of $RuCl(CO)H(PPh_3)_3$ with $CH_2{=}CHCO_2Me$ produces $Ru(CH_2CH_2CO_2Me)Cl(CO)(PPh_3)_2$. Similarly, such alkenes as $CH_2{=}CHCONMe_2$, 2-vinylpyridine and 2-vinyl-5-ethylpyridine react with this same Ru hydride to yield σ-alkyls in which either the O of the carbonyl group or the N of the Py is coordinated to the metal in a five-membered chelate ring.[49] This same regioselectivity however, is not seen in reactions of $RuCl(CO)H(PPh_3)_3$ with $CH_2{=}CHC{\equiv}N$ or

TABLE 4. REACTIONS OF IRON HYDRIDES AND ACTIVATED ALKENES

Activated alkene	Iron hydride	Product	Ref.
$CH_2{=}CHCN$	$(\eta^5\text{-Cp})Fe(CO)_2H$	$(\eta^5\text{-Cp})Fe(CO)_2[CH(CH_3)CN]$	42
trans-$CH_3CH{=}CHCO_2Et$	$Na[HFe_2(CO)_8]$	$CH_3CH_2CH[Fe(CO)_4]CO_2Et$	46
trans-$CH_3CH{=}CHCO_2Et$	$Na[HFe(CO)_4]$	$CH_3CH_2CH[Fe(CO)_4]CO_2Et$	43, 44
$CH_2{=}CHCO_2Et$	$Na[DFe(CO)_4]$	$CH_2DCH[Fe(CO)_4]CO_2Et$	43, 44
$[\eta^5\text{-Cp})Fe(CO)_2(C_2H_4)]^+$	$(\eta^5\text{-Cp})Fe(CO)(PPh_3)H$	$(\eta^5\text{-Cp})Fe(CO)_2(C_2H_5)$	3
$[\eta^5\text{-Cp})Fe(CO)_2(CH_2{=}CHCH_3)]^+$		$(\eta^5\text{-Cp})Fe(CO)_2(CH_2CH_2CH_3)$	3
$[\eta^5\text{-Cp})Fe(CO)_2(CH_2{=}C(CH_3)_2)]^+$		$(\eta^5\text{-Cp})Fe(CO)_2(CH_2CH(CH_3)_2)$	3
$[\eta^5\text{-Cp})Fe(CO)_2(CH_2{=}CHOEt)]^+$		$(\eta^5\text{-Cp})Fe(CO)_2(CH_2CH_2OCH_2CH_3)$	3
$[\eta^5\text{-Cp})Fe(CO)(PPh_3)(C_2H_4)]^+$	$(\eta^5\text{-Cp})Fe(Ph_2PCH_2CH_2PPh_3)H$	$(\eta^5\text{-Cp})Fe(CO)(PPh_3)(C_2H_5)$	3

N≡CCH=CHC≡N; with these nitriles, the σ-alkyl products are the chloro-bridged dimers

$$\{Ru(MeCHC{\equiv}N)Cl(CO)(PPh_3)_2\}_2$$

and

$$\{Ru(N{\equiv}CCH_2CHC{\equiv}N)Cl(CO)(PPh_3)_2\}_2,$$

respectively[50].

Cationic pentaaminoruthenium hydride complexes also undergo facile formation of alkylrutheniums in reactions with both simple and halogenated alkenes (see Table 5)[51]:

$$[RuH(NH_3)_5][SO_4] + \text{alkene} \rightarrow [Ru(NH_3)_5(\sigma\text{-alkyl})][SO_4] \qquad \text{(ac)}$$

These insertions are regiospecific, forming the least substituted ruthenium carbon σ bond. The relative rates for additions of these ruthenium hydrides to various alkenes are:

$$C_2H_4 > C_3H_6 > C_4H_8, C_2F_4 > C_3F_6 > C_4F_8.$$

Ethylosmiums form by insertion of ethylene into the Os—H bond of a cationic Os hydridoethylene complex[52]. The Os hydride intermediate is not isolated, and the reaction may proceed by an alternative pathway of protonation of coordinated ethylene to form the product ethylosmium:

$$Os(C_2H_4)(CO)_2(PPh_3)_2 \xrightarrow{HX} Os(C_2H_5)(X)(CO)_2(PPh_3)_2 \qquad \text{(ad)}$$

where X = Cl, I, O_2CCF_3 and OClO. Insertion of activated alkenes such as ethyl maleate, ethylfumarate, maleic anhydride and ethylacrylate into an Os—H bond in $H_2Os_3(CO)_{10}$ also occurs[53]:

$$H_2Os_3(CO)_{10} + XCH{=}CHX \rightarrow HOs_3(CO)_{10}[CH(CH_2X)X] \qquad \text{(ae)}$$

where X = $-CO_2CH_3$ or —C(O)OC(O)—.

The synthesis of σ-alkylcobalts from the reaction of Co hydrides and alkenes is relevant to the use of Co hydrides in alkene hydroformylation and is of interest because of the biological significance of carbon-cobalt bonds in vitamin B_{12} chemistry. The reaction of $HCo(CO)_4$ with simple alkenes involves hydrometalation as a key step in hydrogenation, isomerization or hydroformylation of alkenes[54]. σ-Alkylcobalt compounds are detectable intermediates[55] in reactions of $HCo(CO)_4$ and alkenes by Ir, but

TABLE 5. σ-ALKYLRUTHENIUM COMPLEXES[48] FROM INSERTION OF ALKENES INTO THE Ru—H BOND OF $[RuH(NH_3)_5][SO_4]$.

Alkene	σ-Alkylruthenium product
$CH_2{=}CH_2$	$[Ru(C_2H_5)(NH_3)_5][SO_4]$
$CH_2{=}CHCH_3$	$[Ru(CH_2CH_2CH_3)(NH_3)_4][SO_4]$
$CH_2{=}CHCH_2CH_3$	$[Ru(CH_2(CH_2)_2CH_3)(NH_3)_5][SO_4]$
$CF_2{=}CF_2$	$[Ru(CF_2CF_2H)(NH_3)_5][SO_4]$
$CF_2{=}CFCF_3$	$[Ru(CF_2CFHCF_3)(NH_3)_5][SO_4]$
$CF_2{=}CFCF_2CF_3$	$[Ru(CF_2CFHCF_2CF_3)(NH_3)_5][SO_4]$

such compounds are too reactive to isolate and either isomerize, revert to alkenes, hydrogenate or react to form an acylcobalt. The high reactivity of these alkylcobalt may be due to a weak C—Co bond. Alternatively, the paucity of isolated σ-alkylcobalts from catalytic processes may be a reflection of the possible radical mechanism for these catalyses[54]. Insertion of alkenes into H—Co bonds and the regiochemistry of this process are treated theoretically using CNDO/2 and ab initio MO-SCF methods[8], which show that electronic factors are important in determining the regiochemistry of this insertion, and that a more basic apical ligand such as R_3P favors linear products in the insertion and subsequent hydroformylation.

σ-Alkylcobalts form stoichiometrically in reactions of $HCo(CO)_4$ and activated halogenated alkenes[56]:

$$C_2F_4 + HCo(CO)_4 \xrightarrow[\text{pentane}]{-70^\circ C} HCF_2CF_2Co(CO)_4 \xrightarrow{PPh_3} HCF_2CF_2Co(CO)_3PPh_3 \quad \text{(af)}$$

Alkylcobalts form in stoichiometric reactions of pentacyanocobalt(III) hydride and alkenes. This reaction occurs both for halogenated alkenes such as tetrafluoroethylene[57] and for alkenes that contain other electron-withdrawing groups such as carbonyls, nitriles and arenes as substituents (see Table 6)[2,58,59]. The addition is regiospecific, forming the more substituted alkylcobalt. Prior coordination of alkene to cobalt to form an alkene(hydrido)cobalt complex, an intermediate in hydrometalation reactions, is not important[2]. This reaction is a radical process[2,58,59]; however, by ^1H NMR, additions of $[HCo(CN)_5]^{3-}$ to diastereomeric alkenes such as fumaric and maleic acid salts lead to a σ-alkylcobalt by stereospecific cis addition of Co and H to the double bond[58]. The overall reduction is not stereospecific. σ-Alkylcobalt bond formation proceeds by either a concerted addition or a rapid collapse of a radical cage.

σ-Alkylcobalt compounds can be obtained[60] from the reaction of simple alkenes with netural cobalt hydrides such as $HCo(N_2)(PPh_3)_2$:

$$RCH{=}CH_2 + HCo(N_2)(PPh_3)_2 \rightarrow \xrightarrow{R'COCl} RCH_2CH_2C(O)R' \quad \text{(ag)}$$

where R = H, $n\text{-}C_4H_9$ or $n\text{-}C_5H_{11}$. The σ-alkylcobalt formed in this reaction are characterized by their transformation into ketones on reaction with an acid chloride. The hydrogen adds to the carbon atom with least hydrogen.

TABLE 6. PRODUCTS OF ADDITION OF $[HCo(CN)_5]^{3-}$ TO ACTIVATED ALKENES.

Substrate alkene	Product	Ref.
$[CH_2{=}CHCO_2]^-$	$[Co(CN)_5(CH(CH_3)CO_2^-)]^{3-}$	60
$CH_2{=}CHCN$	$[Co(CN)_5(CH(CH_3)CN)]^{3-}$	2
$CH_2{=}CH\text{-}2\text{-}C_5H_4N$	$[Co(CN)_5(CH(CH_3)\text{-}2\text{-}C_5H_4N)]^{3-}$	2
$CH_2{=}CHC_6H_5$	$[Co(CN)_5(CH(CH_3)C_6H_5)]^{3-}$	2
$CH_2{=}C(CH_3)CN$	$[Co(CN)_5(CH(CH_3)_2CN)]^{3-}$	2
cis-$^-[O_2CCH{=}CHCO_2]^-$	$[Co(CN)_5(CH(CO_2^-)CH_2CO_2^-)]^{3-}$	60
trans-$^-[O_2CCH{=}CHCO_2]^-$	$[Co(CN)_5(CH(CO_2^-)CH_2CO_2^-)]^{3-}$	60
trans-$[CH_3CH{=}CHCO_2]^-$	$[Co(CN)_5(CH(CH_3)CH_2CO_2^-)]^{3-}$	65

Cobalt hydrides formed in situ from cobalt(dimethylglyoximato) (DMG) complexes containing amines such as pyridine or water as axial bases undergo hydrometalation with alkenes as do the more stable $HCo(DMG)_2[P(C_4H_9)_3]$ complexes:

$$CH_2{=}C(CH_3)CN + HCo(DMGH)_2(C_5H_5N) \rightarrow Co(DMGH)_2(C_5H_5N)(CH_2CH(CH_3)CN) \quad \text{(ah)}$$

$$CH_2{=}CHCN + HCo(DMGH)_2(PPh_3) \rightarrow Co(DMGH)_2(PPh_3)(CH(CH_3)CN) \quad \text{(ai)}$$

Similar chemistry is seen with hydridocobalamin[62]:

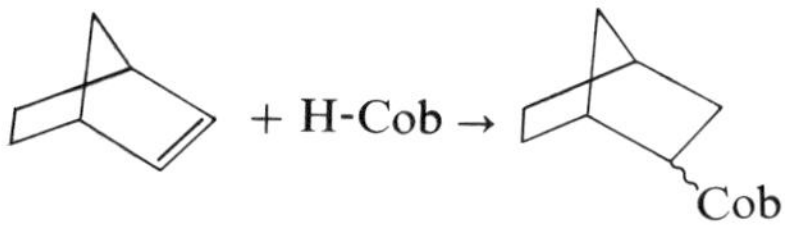

(aj)

where H-Cob = cobalamine. Substituted alkenes react with these Co hydrides (see Table 7), leading to formation of the more substituted alkylcobalt compounds.

Alkylrhodiums form from the reaction of Rh hydrides and alkenes. This reaction is important in hydrogenation, asymmetric hydrogenation, alkene isomerization and hydroformylation and other catalytic processes. The regiochemistry seen in this reaction is the subject of theoretical study that rationalizes the formation of the less substituted σ-alkyrhodium intermediate on electronic grounds[4]. Several Rh complexes form stable σ-alkylrhodiums on reaction of Rh hydrides with fluorinated alkenes (see Table 8)[62–66].

TABLE 7. ORGANOCOBALT COMPOUNDS FORMED FROM ALKENES AND HYDRIDOCOBALAMIN AND HYDRIDOCOBALOXIME.

Alkene	σ-Alkylcobalt	Refs.
$CH_2{=}CHOCOCH_3$	$CH_3CH(OCOCH_3)Co(DMGH)_2(C_5H_5N)$	61
$CH_2{=}CHCO_2H$	$CH_3CH(CO_2H)Co(DMGH)_2(H_2O)$	61
$CH_2{=}CHCO_2H$	$CH_3CH(CO_2H)Co(DMGH)_2(C_5H_5N)$	61
$CH_2{=}CHCO_2CH_3$	$CH_3CH(CO_2CH_3)Co(DMGH)_2(H_2O)$	61
$CH_2{=}CHCO_2CH_3$	$CH_3CH(CO_2CH_3)Co(DMGH)_2(C_5H_5N)$	61
$CH_2{=}C(CH_3)CO_2CH_3$	$CH_3O_2CCH(CH_3)CH_2Co(DMGH)_2(C_5H_5N)$	61
cis-$CO_2EtCH{=}CHCO_2Et$	$EtO_2CCH_2CH(CO_2Et)Co(DMGH)_2(C_5H_5N)$	61
$CH_2{=}CHCN$	$CH_3CH(CN)Co(DMGH)_2(C_5H_5N)$	61
$CH_2{=}CHCN$	$CH_3CH(CN)Co(DMGH)_2(C_6H_5NH_2)$	61
$CH_2{=}C(CH_3)CN$	$CH_3CH(CN)CH_2Co(DMGH)_2(C_6H_5NH_2)$	61
$CH_3CH{=}CHCN$	$CH_3CH_2CH(CN)Co(DMGH)_2(C_5H_5N)$	61
$CNCH{=}CHCN$	$CNCH_2CH(CN)Co(DMGH)_2(C_5H_5N)$	61
$CH_2{=}CHCO_2Et$	$CH_3CH(CO_2Et)Co(DMGH)_2(PBu_3)$	63
$CH_2{=}CH_2$	$CH_3CH_2{-}Cob^a$	62
$CH_3CH{=}CH_2$	$(CH_3)_2CH{-}Cob^a$	62
$CH_2{=}CHCO_2Et$	$CH_3CH(CO_2Et){-}Cob^a$	62
$\overline{CH_2CH_2CH{=}CHCH_2CH_2}$	C_6H_{11}-Cob^a	62

[a] Cob = cobalamine.

TABLE 8. ALKYLRHODIUM COMPOUNDS FROM INSERTION OF ALKENES INTO Rh—H BONDS

Alkene	Rhodium hydride	Product	Refs.
$CF_2{=}CF_2$	$[RhH(CN)_4(H_2O)]^{2-}$	$[Rh(Cn)_5(C_2F_4H)]^{3-}$	73
$CF_2{=}CF_2$	$RhHCl_2(PPh_3)_2$	$Rh(C_2F_4H)(Cl)_2(PPh_3)_2$	66
$CF_2{=}CF_2$	$RhH(PF_3)(PPh_3)_3$	$Rh(C_2F_4H)(PF_3)(PPh_3)_3$	65
$CF_2{=}CF_2$	$RhH(CO)(PPh_3)_2$	$Rh(C_2F_4H)(CO)(PPh_3)_2$	64
$CH_2{=}CHF$	$(\eta^5\text{-Cp})Rh(C_2H_4)(CH_2{=}CHF)$[a]	$(\eta^5\text{-Cp})Rh(C_2H_4)(CH_2CH_2F)(Cl)$[b]	71
$CH_2{=}CHOCOCH_3$	$(\eta^5\text{-Cp})Rh(C_2H_4)(CH_2{=}CHOCOCH_3)$[a]	$(\eta^5\text{-Cp})Rh(C_2H_4)(CH_2CH_2OCOCH_3)(Cl)$[c]	71
$CH_2{=}CHCN$	$(\eta^5\text{-Cp})Rh(C_2H_4)(CH_2{=}CHCN)$[a]	$(\eta^5\text{-Cp})Rh(C_2H_4)(CH_2CH_2CN)(Cl)$[d]	71
$CH_2{=}CHCH_3$	$(\eta^5\text{-Cp})Rh(CH_2{=}CHCH_3)_2$[a]	$(\eta^5\text{-Cp})Rh(CH_2{=}CHCH_3)(CH_2CH_2CH_3)(Cl)$	71

[a] The hydride is formed by protonation in situ.
[b] A minor product is $(\eta^5\text{-Cp})Rh(CH_2{=}CHF)(C_2H_5)(Cl)$.
[c] A minor product is $(\eta^5\text{-Cp})Rh(CH_2{=}CHOCOCH_3)(C_2H_5)(Cl)$.
[d] A minor product is $(\eta^5\text{-Cp})Rh(CH_2{=}CHCN)(C_2H_5)(Cl)$.

The hydrides also react with alkenes such as acryonitrile to give isolable σ-alkyl complexes[67,68]. These additions are regiospecific and give the more substituted alkylrhodium:

$$HRh(DMGH)_2PPh_3 + CH_2{=}CHCN \rightarrow Rh(DMGH)_2(PPh_3)[CH(CH_3)CN]$$

$$(MePh_2A_s)_3RhHCl_2 + CH_2{=}CHCN \rightarrow RhCl_2[CH(CH_3)CN] \xrightarrow{C_5H_5N} (C_5H_5N)_3RhCl_2[CH(CH_3)CN] \quad \text{(ak)}$$

Chlororhodium phosphine complexes also react with simple alkenes such as ethylene to form isolable ethyl complexes similar to those proposed as intermediates in alkene hydrogenations catalyzed by Rh(I)[66,69]. Ethylrhodium(III) is also formed on protonation of η^5-$CpRh(C_2H_4)_2$ [70]:

$$\eta^5\text{-}CpRh(C_2H_4)_2 + HCl \rightarrow \eta^5\text{-}CpRh(C_2H_4)(C_2H_5)(Cl) \quad \text{(al)}$$

The ethylrhodium(III) formed in this reaction results from insertion of ethylene into an intermediate cationic Rh(III) hydride. This is an initial step in ethylene dimerization catalyzed by Rh(I). Protonation of other η^5-CpRh(alkene)$_2$ complexes can be studied by ^{1}H NMR (see Table 8)[71].

Equilibration between a rhodium(hydrido)(ethylene) and ethylrhodium complex is directly observed by ^{1}H NMR[72]:

$$[(\eta^5\text{-}Cp)RhH[P(CH_3)_3](C_2H_4)][BF_4] \rightarrow [(\eta^5\text{-}Cp)Rh(C_2H_5)(P(CH_3)_3)][BF_4] \quad \text{(am)}$$

Although coalescence of the hydride and ethylene signals is not observed, the doublet owing to H—Rh coupling broadens, and the ethylene signal also broadens as the hydridoethylene complex is warmed from -20 to 60°C.

Anionic Rh hydrides add to halogenated alkenes (see Table 8)[57,73]:

$$[RhH(CN)_4(H_2O)]^{2-} + CF_2{=}CF_2 \rightarrow [Rh(CN)_5(C_2F_5H)]^{3-} \quad \text{(an)}$$

However, nonhalogenated alkenes such as ethylene or propylene are unreactive toward $[RhH(CN)_4(H_3O)]^{2-}$.

Formation of a σ-alkylrhodium by insertion of an alkene into a Rh—H bond is a key step in homogeneous asymmetric hydrogenation of alkenes by Rh(I) catalysts. Such intermediates are characterized using multinuclear NMR both in hydrogenation of (Z)-α-acetamidocinnamate by $[Rh(Ph_2PCH_2CH_2PPh_2)]^+$ and in the asymmetric hydrogenation of α-benzamidocinnamic acid or its methyl ester by (R,R)-1,2-bis(o-methoxyphenylphenylphosphino)ethane complexes of Rh(I)[74,75].

The suitability of in situ trapping of intermediate transition-metal hydrides formed by C—H bond activation by a coordinated alkene to form a σ-alkyl has been demonstrated to be feasible with η^3-allyl hydridoiridium complexes. However, insertion was also shown to occur in an intramolecular sense to yield methyl-substituted iridacyclopentanes[76]:

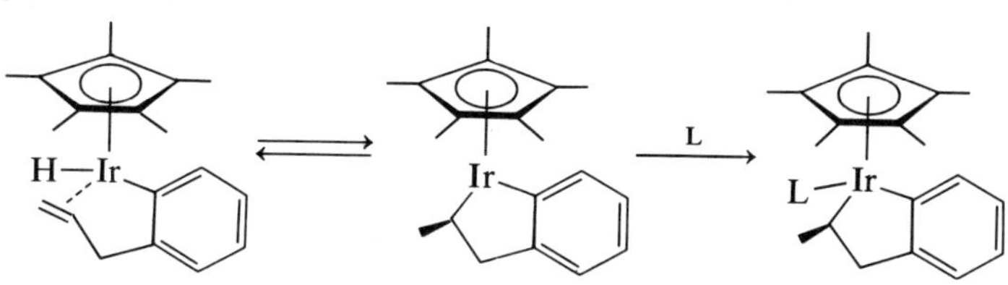

(ao)

Activated alkenes insert into Ir—H bonds. These insertions occur so that the less substituted alkyliridium compounds are formed[77–79]:

$$IrH(CO)(PPh_3)_2(CH_2{=}CHCN) + CH_2{=}CHCN \rightarrow Ir(CH_2CH_2CN)(CO)(PPh_3)_2(CH_2{=}CHCN) \quad \text{(ap)}$$

Thus, the hydride complex $IrHCl_2(DMSO)_3$ [DMSO = $(CH_3)_2SO$] reacts with trans-$C_6H_5CH{=}CHCOC_6H_5$ to give $IrCl_2(DMSO)_3[CH(C_6H_5)CH_2COC_6H_5]$ and acryonitrile reacts with an Ir hydride as shown in Eq. (ap). Although the origin of this regiochemistry is unclear for acryonitrile, an X-ray structure suggests that the regiochemistry seen in reaction of Ir—H with α,β-unsaturated carbonyls may be the result of chelation of the carbonyl oxygen and formation of a five-membered ring[79]. σ-Alkyliridium complexes form from insertion of alkenes into Ir—H bonds in $HIr(CO)_3I_2$ and $HIr(CO)_2I_2(H_2O)$. Ethylene also inserts into $IrH(CO)(PPh_3)_2$ to give $Ir(C_2H_5)(CO)(PPh_3)$[63].

Formation of σ-alkylnickels by addiction of Ni hydride to an alkene is important in many nickel-catalyzed reactions such as alkene dimerization and isomerization. However, stoichiometric formation of alkylnickels from alkenes and Ni hydrides is not synthetically useful because of the reactivity of alkylnickels. The regiochemistry of the addition of $HNiX(PR_3)_2$ to propene is affected by the nature of the trialkylphosphine group used in the Ni hydride complex. The observed ratios of n-propyl- to isopropyl-nickel found after the insertion step vary from 82:18 to 82:20 to 19:81 as the phosphine is changed from $P(OC_6H_5)_3$ to $P(c\text{-}C_6H_{11})$ to $P(t\text{-}C_4H_9)_2(i\text{-}C_3H_7)$.[81]

Insertion of ethylene into Pt—H bonds forms an alkyl platinum (see Table 9). Examples that illustrate important features are discussed below. The theoretical aspects of this insertion are also discussed[1,10].

Ethylene inserts[82] into the Pt—H bond of trans-$PtCl(C_2H_4)(PEt_3)_2$:

$$\text{trans-}HPtCl(PEt_2) + C_2H_4 \rightleftharpoons C_2H_5PtCl(PEt_3)_2 \quad \text{(aq)}$$

This insertion reverses on heating. The addition is slow in the absence of additives; e.g., 1% $SnCl_2$, which forms a good π-acceptor ligand $[SnCl_3]^-$, accelerates the reaction[83]. Cationic Pt hydride complexes form σ-alkylplatinum compounds from ethylene more readily than neutral Pt hydrides[84]. The mechanism by which trans-$[PtH(PR_3)_2(acetone)]^+X^-$ forms an ethylplatinum product[85] does not require the hydride and alkene to adapt a mutually cis orientation for the insertion[86]. Kinetic studies reveal that hydrometalation of methyl acrylate by $HPt(PEt_3)_2(NO_3)$ proceeds by initial dissociation of NO_3 prior to formation of a σ-alkylplatinum[87]. The regiochemistry of addition of Pt and H to an alkene can be determined by electronic and steric effects[88–90]. When there is a good chelating group present, addition can occur to give the more subsituted product:

$$\text{trans-}[(PtH(PPh_3)_2(acetone)]^+X^- + CH_2{—}CHCH_2OCOCH_3 \rightarrow (Ph_3P)_2Pt[CH(CH_3)CH_2OCOCH_3] \quad \text{(ar)}$$

where X = $[BF_4]^-$ or $[PF_6]^-$. The alternative regiochemistry is seen in several cases (Table 9). Rearrangements of the initially formed σ-alkylplatinum can be seen when cyclopropylmethylplatinums are formed after the initial hydrometalation[91,92]. Tetracyanoethylene does not undergo insertion with trans-$[PtH(X)(PPh_3)_2]$ (X =

TABLE 9. σ-ALKYLPLATINUM COMPOUNDS FROM INSERTION OF ALKENES INTO Pt—H BONDS

Alkene	Platinum hydride	Product	Ref.
$CH_2{=}\overline{CCH_2CH_2}$	trans-$\{PtH(NO_3)[P(C_6H_5)_3]_2\}$	$CH_2{=}C(CH_3)CH_2Pt(NO_3)[P(C_6H_5)_3]_2$	92
$CH_2{=}C\overline{CH(CO_2CH_3)CH}(CO_2CH_3)$	trans-$\{PtH(NO_3)[P(C_2H_5)_3]_2\}$	$CH_2{=}CHCH(CO_2CH_3)CH(CO_2CH_3)Pt(NO_3)[P(C_2H_5)_3]_2$	91
$CH_2{=}CHCH_3$	trans-$\{PtHCl[P(C_2H_5)_3]_2\}$	trans-$CH_3CH_2CH_2PtCl(P(C_2H_5)_3)_2$	90
$CH_2{=}CH_2$	trans-$\{PtHCN[P(C_2H_5)_3]_2\}$	trans-$PtCN(C_2H_5)[P(C_2H_5)_3]_2$	90
$CH_2{=}CH_2$	trans-$\{PtH(C_2H_4)[P(C_2H_5]_3\}_2][B(C_6H_5)_4]$	$\{Pt(C_2H_5)[P(C_2H_5)_3]_2\}[B(C_6H_5)_4]$	89
$CH_2{=}CHCOCH_3$	trans-$\{(PtH(NO_3)[P(C_2H_5)_3]_2\}$	$[CH_3COCH_2CH_2Pt(C_2H_5)_3]_2(NO_3)$	89
$CH_2{=}CHCH_2NHCOCH_3$	trans-$\{PtH(PCH_3[C_6H_5]_2\}_2((CH_3)_2CO)][BF_4]$	$\{CH_3CH_2CH(NHCOCH_3)Pt[PCH_3(C_6H_5)_3]_2\}[BF_4]$	88
$CH_2{=}CHCO_2C_2H_5$	trans-$\{PtH(NO_3)[P(C_2H_5)_3]_2\}$	$CH_3O_2CCH_2CH_2Pt(NO_3)[P(C_2H_5)_3]_2$	87
$CH_2{=}CH_2$	trans-$\{PtHCl[P(C_2H_5)_3]_2\}$	trans-$PtCl(C_2H_5)(P(C_2H_5)_3)_2$	80
$CH_2{=}CHCH_2CH_3$	trans-$\{PtH[PCH_3(C_6H_5)_2]_2\}(CH_3)_2CO)][PF_6]$	trans-$\{Pt(CH_2[CH_2]_2CH_3][PCH_3(C_6H_5)_2]]\}^+$	84
$CH_2{=}CH_2$	trans-$\{PtH[PCH_3(C_6H_5)_2]_2\}(CH_3)_2CO)][PF_6]$	trans-$\{Pt(C_2H_5)[PCH_3(C_6H_5)_2](C_2H_4)\}^+$	84
$CH_2{=}CH_2$	trans-$\{PtH(NO_3)[PCH_3(C_6H_5)_2]_2\}$	trans-$Pt(C_2H_5)(NO_3)[PCH_3(C_6H_5)_2]_2$	84

halogen)[93]. In the decomposition of $(PPh_3)_2PtR_2$ (R = n-C_4H_9 or C_2H_5) the hydrido(alkene)platinum complex presumed to be an intermediate in hydrometalation does not exchange coordinated alkene with alkenes in solution at a rate competitive with Pt—H addition to coordinated alkene[94,95].

Copper hydrides do not react readily with unactivated alkenes such as ethylene or 1-butene; however, they will add to activated alkenes such as α,β-unsaturated carbonyls[96]. The organometallic products from such addition reactions, however, are not completely characterized.

Addition of Au hydride to alkenes coordinated to Au is an essential step in the isomerization of t-butyldimethylgold(III) to the less hindered isobutyldimethylgold(III)[97]:

$$(CH_3)_3CAu(CH_3)_2(PPh_3) \rightleftharpoons (CH_3)_2CHCH_2Au(CH_3)_2(PPh_3) \qquad \text{(as)}$$

However, the direct formation of σ-alkylgolds by the addition of Au hydride to alkenes is not an important procedure for the synthesis of σ-alkylgolds.

(D. E. BERGBREITER)

1. D. L. Thorn, and R. Hoffmann, *J. Am. Chem. Soc., 100*, 2079 (1978).
2. J. Halpern, L.-Y. Wong, *J. Am. Chem. Soc., 90*, 6665 (1968).
3. T. Bodnar, S. J. LaCroce, A. R. Cutler, *J. Am. Chem. Soc., 102*, 3292 (1980).
4. A. Dedieu, *Inorg. Chem., 20*, 2803 (1981).
5. R. A. Schunn, *Inorg. Chem., 9*, 2567 (1970).
6. J. W. Lauher, R. Hoffmann, *J. Am. Chem. Soc., 98*, 1729 (1976).
7. F. W. S. Benfield, M. L. H. Green, *J. Chem. Soc., Dalton Trans.*, 1324 (1974).
8. J. P. Grima, F. Choplin, G. Kaufmann, *J. Organomet. Chem., 129*, 221 (1977).
9. D. R. Armstrong, R. Fortune, P. G. Perkins, *J. Catal., 42*, 435 (1970).
10. S. Sakaki, H. Kato, H. Kanai, K. Tarama, *Bull. Chem. Soc. Jpn., 48*, 813 (1975).
11. J. Schwartz, J. A. Labinger, *Angew. Chem., Int. Ed. Engl., 15*, 333 (1976).
12. P. C. Wailes, H. Weigold, *Inorg. Synth., 19*, 223 (1979).
13. T. Gibson, L. Tulich, *J. Org. Chem., 46*, 1821 (1981).
14. D. W. Hart, J. Schwartz, *J. Am. Chem. Soc., 96*, 8115 (1974).
15. C. A. Bertelo, J. Schwartz, *J. Am. Chem. Soc., 98*, 262 (1976).
16. D. E. Laycock, H. Alper, *J. Org. Chem., 46*, 289 (1981).
17. M. S. Miftakhov, N. N. Sidorov, G. A. Tolstikov, *Akad. Nauk SSSR, Bull., Div. Chem. Sci. (Engl. Trans.), 28*, 2556 (1979).
18. J. A. Labinger, D. W. Hart, W. E. Siebart III, J. Schwartz, *J. Am. Chem. Soc., 97*, 3851 (1975).
19. P. C. Wailes, H. Weigold, A. P. Bell, *J. Organomet. Chem., 27*, 373 (1971).
20. F. W. Hartner, J. Schwartz, *J. Am. Chem. Soc., 103,*, 4979 (1981).
21. G. A. Tolstikov, M. S. Miftahov, F. A. Valeev, *Akad. Nauk SSSR, Bull., Div. Chem. Sci. (Engl. Trans.), 28*, 2392 (1979).
22. F. N. Tebbe, G. W. Parshall, *J. Am. Chem. Soc., 93*, 3793 (1971).
23. L. S. Guggenberger, P. Meakin, F. N. Tebbe, *J. Am. Chem. Soc., 96*, 5420 (1974).
24. N. M. Doherty, J. E. Bercaw, *J. Am. Chem. Soc., 103*, 2670 (1985).
25. B. J. Burger, B. D. Santarsiero, M. S. Trimmer, J. E. Bercaw, *J. Am. Chem. Soc., 110*, 3134 (1988).
26. A. H. Klazinga, J. H. Teuben, *J. Organomet. Chem., 165*, 31 (1978).
27. A. H. Klazinga, J. H. Teuben, *J. Organomet. Chem., 192*, 75 (1980).
28. P. M. Treichel, J. H. Morris, F. G. A. Stone, *J. Chem. Soc.*, 720 (1973).
29. A. Nakamura, S. Otsuka, *J. Am. Chem. Soc., 94*, 1886 (1972).
30. A. Nakamura, S. Otsuka, *J. Am. Chem. Soc., 95*, 7262 (1973).
31. M. J. Calhorda, A. R. Dias, *J. Organomet. Chem., 198*, 41 (1980).
32. J. W. Byrne, H. U. Blazer, J. A. Osborn, *J. Am. Chem. Soc., 97*, 3872 (1975).
33. M. Akiyama, M. H. Chisholm, F. A. Cotton, M. W. Extine, D. A. Haitko, J. Leonelli, D. Little, *J. Am. Chem. Soc., 103*, 779 (1981).

34. J. B. Wilford, P. M. Treichel, F. G. A. Stone, *J. Organomet. Chem.*, *2*, 119 (1964).
35. R. L. Sweany, J. Halpern, *J. Am. Chem. Soc.*, *99*, 8335 (1977).
36. E. Lindner, G. Funk, S. Hoehne, *J. Organomet. Chem.*, *214*, C7 (1981).
37. J. B. Wilford, F. G. A. Stone, *Inorg. Chem.*, *4*, 93 (1965).
38. B. Olgemoler, W. Beck, *Angew. Chem., Int. Ed. Engl.*, *19*, 834 (1980).
39. D. L. Reger, E. C. Culbertson, *J. Am. Chem. Soc.*, *98*, 2789 (1976); *Inorg. Chem.*, *16*, 3104 (1977).
40. P. L. Bock, D. J. Boschetto, J. R. Rasmussen, J. P. Demers, G. M. Whitesides, *J. Am. Chem. Soc.*, *96*, 2814 (1974).
41. V. D. Bianco, S. Doronzo, N. Gallo, *J. Organomet. Chem.*, *124*, C43 (1977).
42. J. K. P. Ariyaratne, M. L. H. Green, *J. Chem. Soc.*, 2976 (1963).
43. J. P. Collman, R. G. Finke, P. L. Matlock, R. Wahren, R. G. Komoto, J. I. Brauman, *J. Am. Chem. Soc.*, *100*, 1119 (1978).
44. T. Mitsudo, Y. Watanabe, M. Yamashita, Y. Takegami, *Chem. Lett.*, 1385 (1974).
45. D. Rose, J. D. Gilbert, R. P. Richardson, G. Wilkinson, *J. Chem. Soc., A*, 2610 (1969).
46. P. S. Hallman, B. R. McGarvey, G. Wilkinson, *J. Chem. Soc., A*, 3143 (1968).
47. B. N. Chaudret, D. J. Cole-Hamilton, G. Wilkinson, *Acta Chem. Scand.*, *A32*, 763 (1978).
48. R. A. Gray, G. P. Pez, A. Wallo, *J. Am. Chem. Soc.*, *103*, 7536 (1981).
49. K. Hiraki, N. Ochi, Y. Sasada,H. Hayashida, Y. Fuchita and S. Yamanaka, *J. Chem. Soc., Dalton Trans.*, 873 (1985).
50. K. Hiraki, N. Ochi, T. Kitamura, Y. Sasada and S. Shinoda, *Bull. Chem. Soc. Jpn.*, *55*, 2356 (1982).
51. K. Thomas, J. A. Osborn, A. R. Powell, G. Wilkinson, *J. Chem. Soc., A*, 1801 (1968).
52. W. R. Grundy, W. R. Roper, *J. Organomet. Chem.*, *216*, 255 (1981).
53. J. B. Keister, J. R. Shapely, *J. Am. Chem. Soc.*, *98*, 1056 (1976).
54. M. Orchin, *Acc. Chem. Res.*, *9*, 259 (1981).
55. R. F. Heck, D. S. Breslow, *J. Am. Chem. Soc.*, *83*, 4023 (1961).
56. J. B. Wilford, A. Forster, F. G. A. Stone, *J. Chem. Soc.*, 6519 (1965).
57. M. J. Mays, G. Wilkinson, *J. Chem. Soc.*, 6629 (1965).
58. L. M. Jackman, J. A. Hamilton, J. M. Lawlor, *J. Am. Chem. Soc.*, *90*, 1914 (1968).
59. J. Kwiatek, *Catal. Rev.*, *1*, 37 (1967).
60. J. Schwartz, J. B. Cannon, *J. Am. Chem. Soc.*, *96*, 4721–4723 (1974).
61. G. N. Schrauzer, R. J. Windgassen, *J. Am. Chem. Soc.*, *89*, 1999 (1967).
62. G. N. Schrauzer, R. J. Holland, *J. Am. Chem. Soc.*, *93*, 4062 (1971).
63. G. N. Schrauzer, R. J. Holland, *J. Am. Chem. Soc.*, *93*, 1505 (1971).
64. G. Yagupsky, C. K. Brown, G. Wilkinson, *J. Chem. Soc., A*, 1392 (1970).
65. J. F. Nixon, J. R. Swain, *J. Organomet. Chem.*, *72*, C15 (1974).
66. M. C. Baird, J. T. Mague, J. A. Osborn, G. Wilkinson, *J. Chem. Soc., A*, 1347 (1967).
67. K. C. Dewhirst, *Inorg. Chem.*, *5*, 319 (1966).
68. J. H. Weber, G. N. Schrauzer, *J. Am. Chem. Soc.*, *92*, 726 (1970).
69. M. C. Baird, D. N. Lawson, J. T. Mague, J. A. Osborn, G. Wilkinson, *J. Chem. Soc., Chem. Commun.*, 129 (1966).
70. R. Cramer, *J. Am. Chem. Soc.*, *87*, 4717 (1965).
71. L. P. Seiwell, *Inorg. Chem.*, *15*, 2560 (1976).
72. H. Werner, and R. Feser, *Angew. Chem., Int. Ed. Engl.*, *18*, 157 (1979).
73. D. N. Lawson, M. J. Mays, G. Wilkinson, *J. Chem. Soc., A*, 52 (1966).
74. A. S. C. Chan, J. Halpern, *J. Am. Chem. Soc.*, *102*, 838 (1980).
75. J. M. Brown, P. A. Chaloner, *J. Chem. Soc., Chem. Commun.*, 344 (1980).
76. W. D. McGhee and R. G. Bergman, *J. Am. Chem. Soc.*, *110*, 4246 (1988).
77. W. H. Baddley, M. S. Fraser, *J. Am. Chem. Soc.*, *91*, 3661 (1969).
78. M. McPartlin, R. Mason, *J. Chem. Soc., A*, 2206 (1970).
79. H. B. Henbest, J. Trocha-Grimshaw, *J. Chem. Soc., Perkin Trans. I*, 601 (1974).
80. D. Forster, A. Hershman, D. E. Morris, *Catal. Rev.*, *23*, 89 (1981).
81. P. W. Jolly, G. Wilke, *The Organic Chemistry of Nickel*, Vol. I, Academic Press, New York, 1974.
82. J. Chatt, B. L. Shaw, *J. Chem. Soc.*, 5075 (1962).
83. R. Cramer, R. V. Lindsey, Jr., *J. Am. Chem. Soc.*, *88*, 3534 (1966).
84. H. C. Clark, H. Kurosawa, *Inorg. Chem.*, *11*, 1275 (1972).
85. H. C. Clark, C. R. Jablonski, C. S. Wong, *Inorg. Chem.*, *14*, 1332 (1974).
86. C. Bracher, P. S. Pregosin, L. M. Venanzi, *Angew. Chem., Int. Ed. Engl.*, *14*, 563 (1975).

87. H. C. Clark, C. S. Wong, *J. Am. Chem. Soc.*, *96*, 7213 (1974).
88. H. C. Clark, H. Kurosawa, *Inorg. Chem.*, *12*, 357 (1973).
89. A. J. Deeming, B. F. G. Johnson, J. Lewis, *J. Chem. Soc., Dalton Trans.*, 1848 (1973).
90. J. Chatt, R. S. Coffey, A. Gough, D. T. Thompson, *J. Chem. Soc., A*, 190 (1968).
91. T. G. Attig, *Inorg. Chem.*, *17*, 3097 (1978).
92. R. L. Phillips, R. J. Puddephatt, *J. Chem. Soc., Dalton Trans.*, 1736 (1978).
93. R. Uguagliati, W. H. Baddley, *J. Am. Chem. Soc.*, *90*, 5446 (1968).
94. G. M. Whitesides, J. F. Gaasch, E. R. Stedronsky, *J. Am. Chem. Soc.*, *94*, 5428 (1972).
95. T. J. McCarthy, R. G. Nuzzo, G. M. Whitesides, *J. Am. Chem. Soc.*, *103*, 3396 (1981).
96. E. C. Ashby, J. J. Lin, A. B. Goel, *J. Org. Chem.*, *43*, 183 (1978) and references therein.
97. J. K. Kochi, *Organometallic Mechanisms and Catalysis*, Academic Press, New York, 1978.

5.8.2.3.4. η^1 σ-Alkyl Complexes from Nucleophilic Attack on π-Olefin Metal Complexes.

Coordination of olefins to a wide variety of transition metals activates the olefin for nucleophilic attack, the result of which is the formation of a carbon–nucleophile bond and a carbon–transition-metal σ-bond:

$$\underset{\displaystyle\mathrm{M}}{\mathrm{H_2C{=}CH_2}} + \mathrm{Nuc^-} \rightarrow \mathrm{Nu{-}CH_2{-}CH_2{-}M} \tag{a}$$

The stability of such σ-alkyl–transition-metal complexes is dependent on the particular transition metal and its associated ligands. A number of stable η^1-alkyl–metal complexes can be made in this fashion, and a larger number of unstable σ-alkyl–metal complexes made in the fashion are reactive intermediates in synthetic processes of some importance[1].

Complexes derived from nucleophilic attack on $[\eta^5\text{-}C_5H_5Fe(CO)_2(\text{olefin})]^+$ complexes, hereafter abreviated $[Fp(\text{olefin})]^+$, are among the most common. Heteroatomic nucleophiles including methanol, dimethylamine, trimethylamine, benzyl amine, t-butylmercaptan, triphenylphosphine and triethylphosphite all react with $[Fp(\text{olefin})]^+$ complexes to produce stable η^1-alkyliron complexes[2]. Monosubstituted olefin complexes react with methanol with high regiospecificity to produce air-sensitive amber oils, which, however, are well characterized:

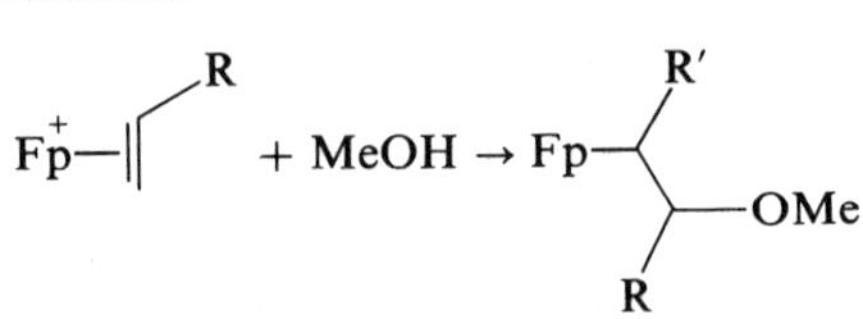

(b)

where $Fp = \eta^5\text{-}C_5H_5Fe(CO)_2$; and R = R′ = H; R = Me; R′ = H; R = Ph; R′ = H; R = H; R′ = CHO. Secondary and tertiary amines react exclusively at the substituted position of the olfefin to give the corresponding ammonium salts, while with primary amines, addition to a second mole of $[Fp(\text{olefin})]^+$ complex occurs, resulting in the formation of the corresponding 2:1 adduct:

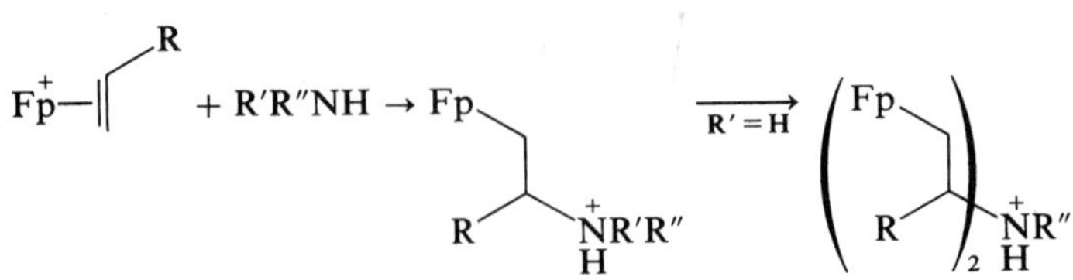

(c)

These amine adducts are air-stable solids that melt in the range of 100–150°C. Phosphines and phosphites react in a similar fashion, attacking unsymmetrical olefins at the most substituted position to produce phosphonium salts. t-Butylmercaptide is much less regioselective and reacts at both olefinic carbons to give unseparable mixtures of positional isomers.

Many carbanions react with $[Fp(olefin)]^+$ complexes to form relatively stable η^1-alkyliron complexes[3]. Stabilized enolates derived from cyanoacetates, malonates, acetoacetates and nitromethane react with the $[Fp(styrene)]^+$ complex almost exclusively at the benzylic carbon but react with $[Fp(propene)]^+$ complexes with low regioselectivity:

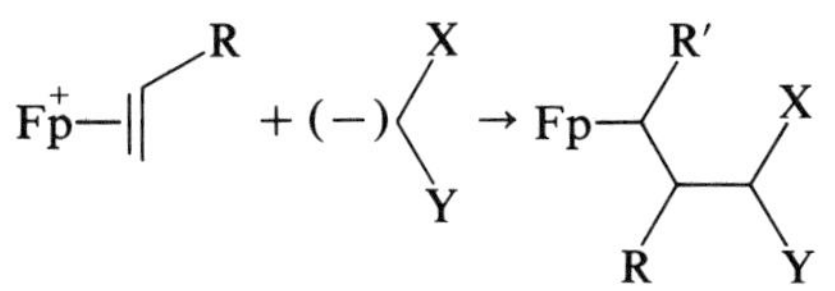

(d)

where R = Me, Ph; X = Y = COOR; X = CH_3CO, Y = COOR; X = CN, Y = COOR; X = H, Y = NO_2; R = Ph, R′ = H; R = H, R′ = Me; R = Me, R′ = H. This reaction proceeds with Fp complexes of ethene, propene, styrene, butadiene (1,2 addition), cyclopentene, cyclohexene, 1-methylallene and 3-substituted cyclopentenes in 60–90% yield to produce complexes that are air-stable low-melting solids or oils. The stereochemistry of addition is trans in all cases, characteristic of attack by nucleophile without prior coordination.

Enamines react with $[Fp(olefin)]^+$ complexes at 0°C in acetonitrile in a similar fashion to give carbonyl-substituted η^1-alkyliron complexes. The reaction of the pyrrolidine enamine with the $[Fp(ethylene)]^+$ complex is illustrative.

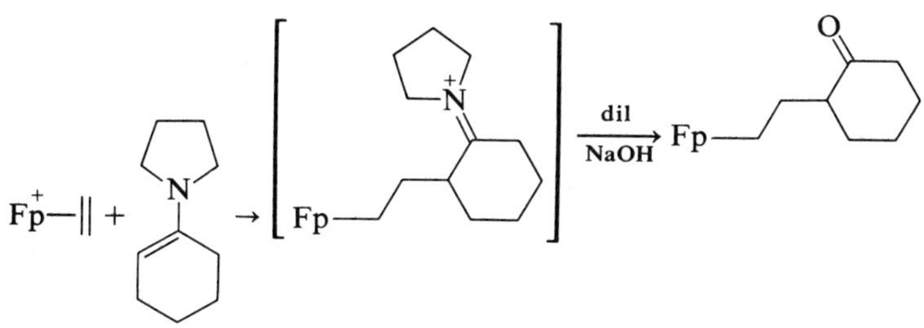

(e)

The olefin complexes studied include those of ethylene, propene, styrene, butadiene (1,2 and 1,4 addition) and cyclopentene. The pyrrolidine enamines of cyclochexanone, cyclopentanone and isobutyraldehyde all react in good yield. The regioselectivity parallels that noted with stabilized enolates.

Reaction of nonstabilized carbanions with $[Fp(olefin)]^+$ complexes generally results in either displacement of the olefin or reduction of the metal rather than formation of stable η^1-alkyliron complexes. This is especially true with simple, nonstabilized organomagnesium halide or lithio reagents. However, allylmagnesium chloride and phenylmagnesium chloride react in modest (20–40%) yield with the ethylene, propene and butadiene (1,4 addition) iron complexes. Lithium dimethylcuprate is even more efficient, reacting in up to 70% yield with Fp complexes of styrene, butadiene (1,4 addition) isoprene (1,4 addition) and allene. Complexes of cyclopentene and allene react in low yield. As might be expected with these more reactive carbanions, the regioselectivity of the process is low, reactive species being in general less selective.

The phosphonium ylid from ethyl α-bromoacetate reacts with the $[Fp(ethylene)]^+$ complex to give a stable η^1-alkyliron phosphonium salt, which reacts with base and benzaldehyde after the fashion of a normal reagent:

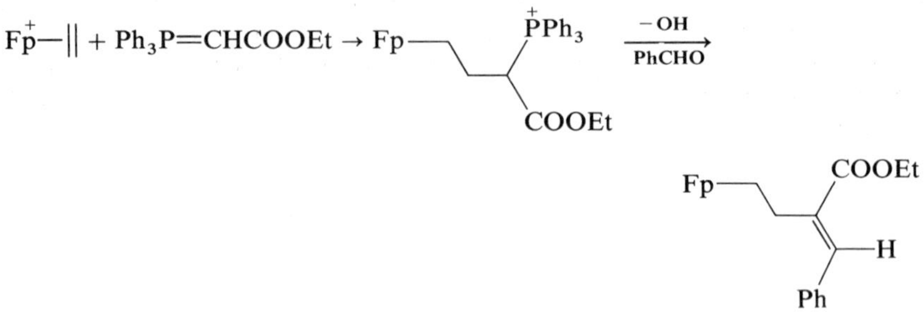

(f)

Nonstabilized ylids do not lead to isolable η^1-alkyliron complexes.

Enol ethers also form $[Fp(olefin)]^+$ complexes, which react with ketone enolates to form β-alkoxy-η^1-alkyliron complexes[4]:

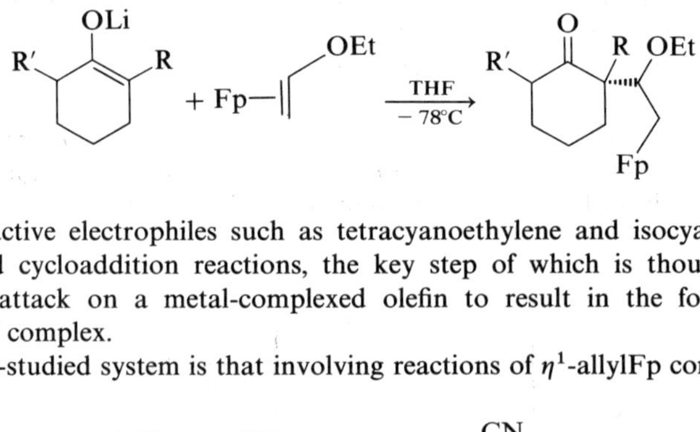

(g)

Very reactive electrophiles such as tetracyanoethylene and isocyanates undergo metal-assisted cycloaddition reactions, the key step of which is thought to involve nucleophilic attack on a metal-complexed olefin to result in the formation of an η^1-alkylmetal complex.

The best-studied system is that involving reactions of η^1-allylFp complexes[5,6]:

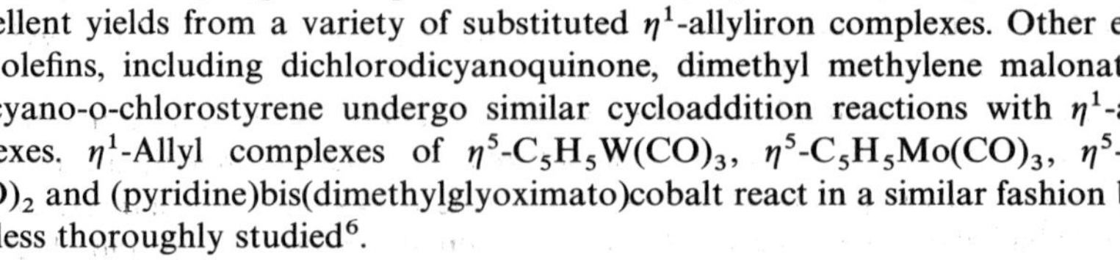

(h)

This reaction is very rapid and produces air-stable, crystalline η^1-alkyliron complexes in excellent yields from a variety of substituted η^1-allyliron complexes. Other electrophilic olefins, including dichlorodicyanoquinone, dimethyl methylene malonate, and β,β-dicyano-o-chlorostyrene undergo similar cycloaddition reactions with η^1-allylFp complexes. η^1-Allyl complexes of η^5-$C_5H_5W(CO)_3$, η^5-$C_5H_5Mo(CO)_3$, η^5-C_5H_5-$Cr(NO)_2$ and (pyridine)bis(dimethylglyoximato)cobalt react in a similar fashion but are much less thoroughly studied[6].

This cycloaddition reaction is not restricted to η^1-allylmetal complexes; rather it is a general process for many η^1-organometal complexes having incipient β-unsaturation[5]. Illustrated below are cycloaddition reactions of η^1-allenyl-, η^1-propargyl-, and

η^1-cyclopropylmethyl-Fp complexes, all of which are presumed to involve nucleophilic attack on an $[Fp(olefin)]^+$ complex as the ring-forming step:

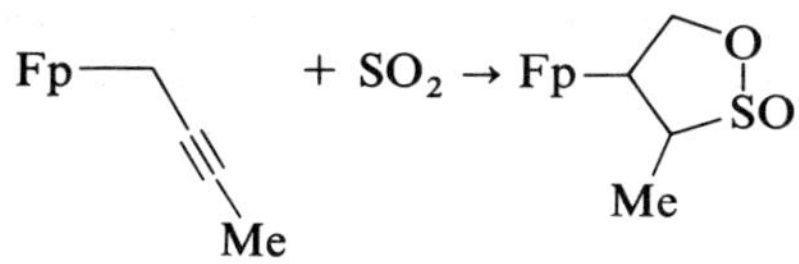

(i)

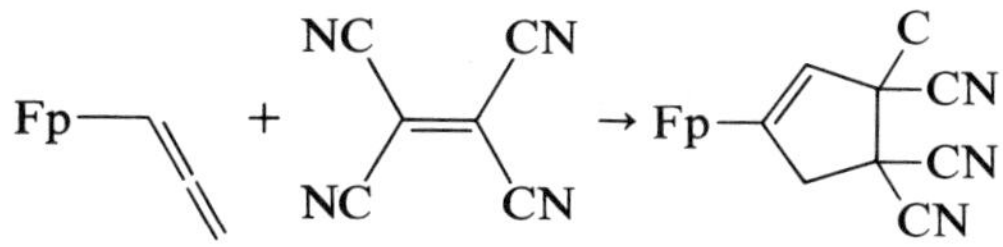

(j)

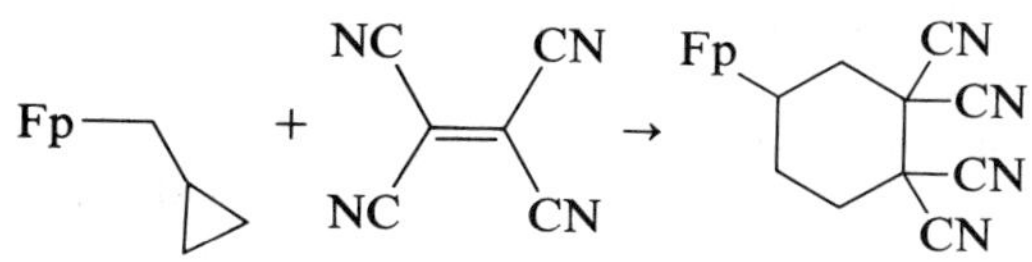

(k)

Finally, the olefin in $[Fp(olefin)]^+$ complexes is sufficiently electrophilic to react with η^1-allylFp complexes to produce a new η^1-alkylFp complex containing an $[Fp(olefin)]^+$ group as well[7]:

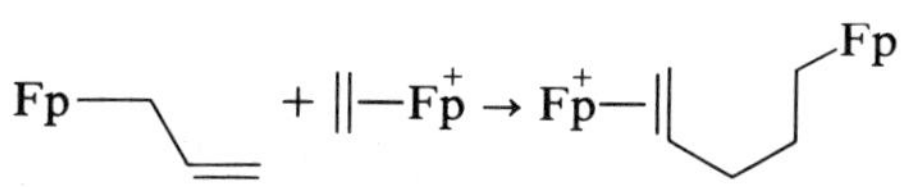

(l)

The reaction proceeds more rapidly, in higher yield, regiospecifically β if the olefin in the $[Fp(olefin)]^+$ complex contains an electron-withdrawing group such as ester or aldehyde group.

The cationic olefin complex of Pd(II), $[Pd(\eta^5\text{-}C_5H_5)(PR_3)(ethylene)]^+$ also reacts with nucleophiles (Nuc) to form stable η^1-alkylpalladium(II) complexes of the type $[Pd(\eta^5\text{-}C_5H_5)(PPh_3)(CH_2CH_2Nuc)]$[8]. Both alkoxides and sodium 2,4-pentandionate react in high yield to produce quite stable complexes. The stereochemistry of this addition has been shown to be clearly trans, as is the case with the related $Fp(olefin)^+$ complexes discussed above. (Both types of complexes are coordinatively saturated, and coordination of the nucleophile prior to addition, a requirement for cis stereochemistry, is not possible.) Amines do not add to the coordinated olefin but displace it to produce $[Pd(\eta^5\text{-}C_5H_5)(PPh_3)(amine)]^+$ complexes.

The η^1-alkylmetal complexes discussed above are undoubtedly stable because they are coordinatively saturated and, lacking a vacant coordination site, are unable to undergo a facile β-hybride elimination reaction. Since β-hydride elimination is the most common way for η^1-alkylmetal complexes to decompose, coordinatively unsaturated complexes are often considerably less stable.

Reaction of η^2-ethylenetetracarbonyliron with dimethylsodiomalonate produces a modestly stable anionic η^1-iron carbonyl complex[9]:

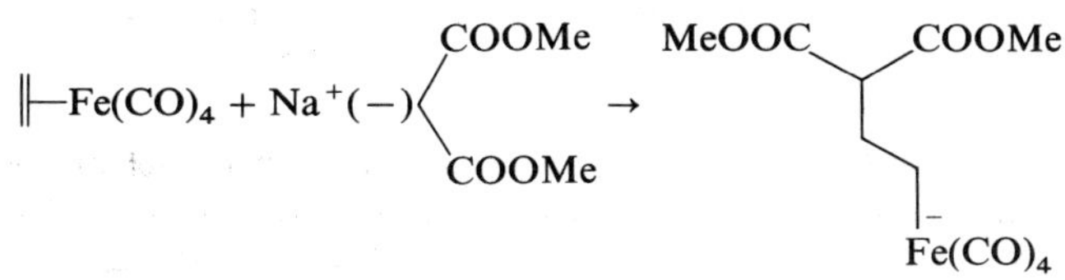

(m)

The tetracarbonyl iron complex of methyl acrylate, methyl 2-chloroacrylate[10] and methyl crotonate[11] react in a similar fashion. Although the resulting η^1-alkyliron complexes are stable (requiring treatment with trifluoroacetic acid followed by a oxidant to free the organic group from the iron), they are not well characterized.

η^2-Olefinplatinum(II) complexes react with secondary amines to form quite stable zwitterionic η^1-alkylPt(II) complexes:

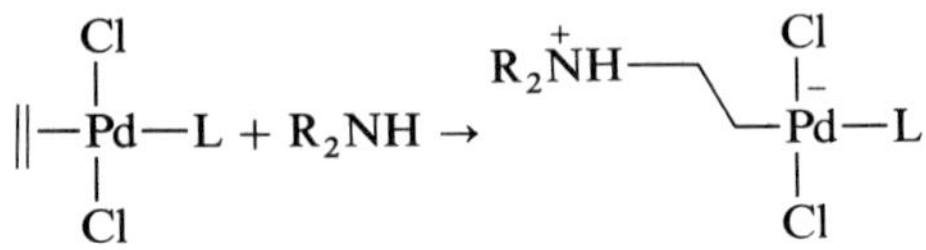

(n)

With higher olefins, such as propene or 1-butene, amination occurs predominantly at the substituted olefin position, although both regioisomers are obtained[12]. The steric course of this reaction was shown to involve a trans (addition to the olefin face opposite the metal) addition of the amine to the olefin by carrying out the reaction on a resolved diastereoisomer complex containing a chiral amine as an ancillary ligand, and a porchiral α-olefin[13]. The structure of the η^1-alkylPt complex resulting from the reaction of diethylamine with trans-[(diethylamine)dichloro(ethylene)Pt(II)] has been shown to be that in Eq. (n) by single-crystal X-ray diffraction[14]. Pyridine reacts with η^2-olefinPt(II) complexes in a similar fashion to produce zwitterionic η^1-alkylPt(II) complexes[15].

The corresponding η^2-olefinPd(II) complexes undoubtedly react with amines in an analogous manner. However, the resulting η^1-alkylPd(II) complexes undergo a facile β-hydride elimination, thus precluding their isolation and characterization. That they form has been demonstrated by trapping the unstable η^1-alkylPd(II) complex with CO to produce the corresponding β-aminoacyl complex[16]:

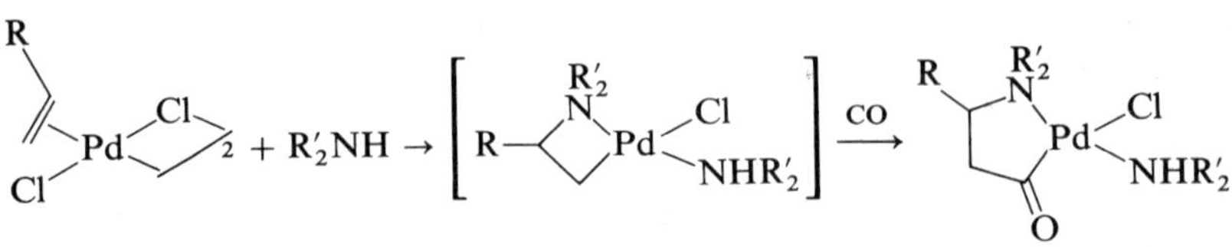

(o)

The stability of this η^1-acylPd(II) complex undoubtedly owes to the inherent stability of a five-membered palladiacycle.

Another series of η^1-alkyl–metal complexes that owe their stabilities to the formation of a five-membered chelate ring is those produced in the reaction of allylic and homoallylic amine and sulfide complexes of Pd(II) with nucleophiles. Lithium tetrachloropalladate reacts with N,N-dimethylallylamine in methanol to produce the β-methoxy-η^1-alkylPd complex[17]:

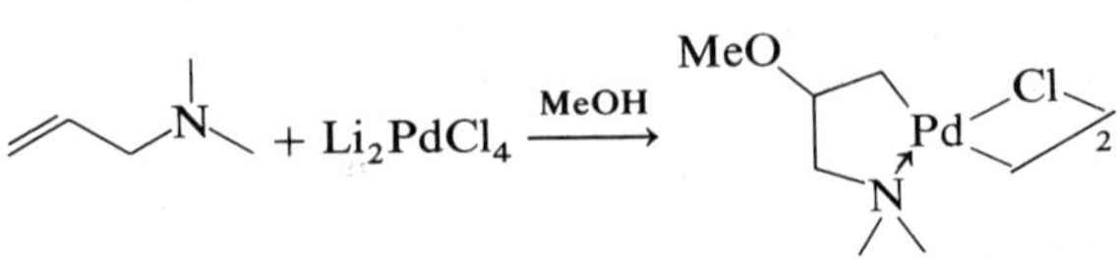

(p)

This complex is clearly formed by nucleophilic attack of methanol on the Pd-bound olefin. The η^1-alkyl complex is stabilized by its incorporation into a five-membered chelate ring[18]. Allyl methylsulfide reacts in a similar fashion, forming a sulfur-chelated metallacycle. These same olefins react with stabilized carbanions such as malonates, and acetoacetates, to give similar η^1-alkylPd(II) complexes[19]:

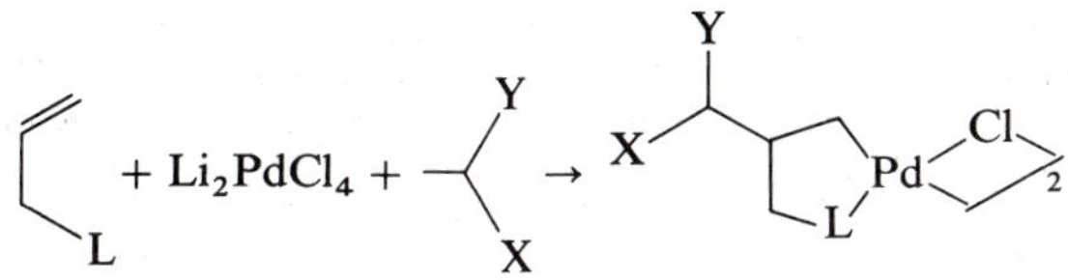

(q)

Chelation is clearly necessary, since allyl alcohol, allyl phenyl ether, ethyl acrylate and 1-octene fail to produce addition products.

The importance of five-membered chelate ring formation in the stabilization of these η^1-alkylPd(II) complexes is seen in the reactions of homoallylic amines and sulfides with lithium tetrachloropalladate and stabilized enolates[20]:

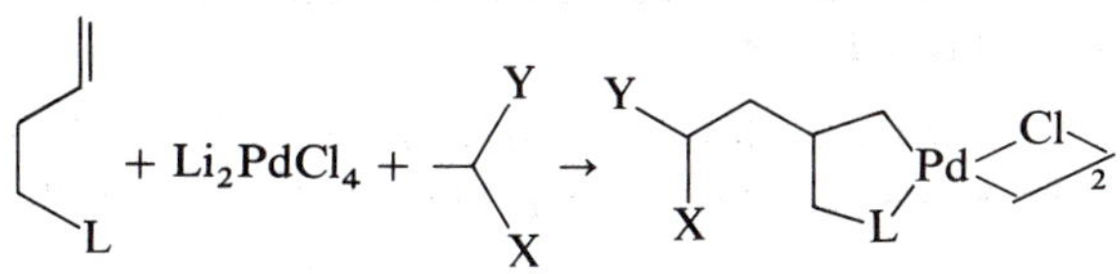

(r)

where L = Me_2N, i-PrS. In these cases, the regiochemistry of nucleophilic attack is exactly opposite that observed with allylic amines and sulfides, and reaction occurs exclusively at the terminal olefin carbon rather than the internal olefin carbon. That is both systems react to give five-membered chelate rings.

Neither the Pd(II) nor the Pt(II) complexes of allylphosphines react with nucleophiles to form η^1-alkyl–metal complexes. Instead, longer aliphatic chains are required to form chelate-stabilized ring systems. Thus, both 1-butenyl- and 1-pentenylphosphines react with $PtCl_2$ to give stable 6- and 7-metallacycles[21]:

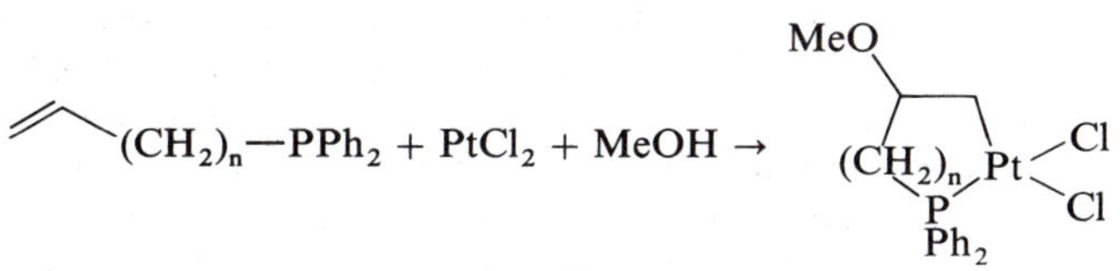

(s)

where n = 2, 3.

((L. S. HEGEDUS)

1. J. P. Collmam, L. S. Hegedus, J. R. Norton, R. G. Finke, *Principles and Applications of Organotransition Metal Chemistry*, 2nd Ed., University Science Books, Mill Valley, Calif., 1987.
2. P. Lennon, M. Madhavarao, A. Rosen, M. Rosenblum, *J. Organomet. Chem.*, *108*, 93 (1976).
3. P. Lennon, A. M. Rosan, M. Rosenblum, *J. Am. Chem. Soc.*, *99*, 8426 (1977).
4. T. C. T. Chang, M. Rosenblum, S. B. Samuels, *J. Am. Chem. Soc.*, *102*, 5930 (1980).
5. M. Rosenblum, *Acc. Chem. Res.*, *7*, 122 (1974).
6. A. Cutler, D. Ehntholt, W. P. Giering, P. Lennon, S. Ragliu, A. Rosen, M. Rosenblum, J. Tancrede, D. Wells, *J. Am. Chem. Soc.*, *98*, 3495 (1976).
7. P. J. Lennon, A. Rosan, M. Rosenblum, J. Tancrede, P. Waterman, *J. Am. Chem. Soc.*, *102*, 7033 (1980).
8. H. Kurosawa, T. Majima, N. Asada, *J. Am. Chem. Soc.*, *102*, 6996 (1980).
9. B. W. Roberts, J. Wong, *J. Chem. Soc., Chem. Commun.*, 20 (1977).
10. M. A. Baer, B. W. Roberts, *J. Chem. Soc., Chem. Commun.*, 1129 (1979).

11. B. W. Roberts, M. Ross, J. Wong, *J. Chem. Soc., Chem. Commun.*, 428 (1980).
12. A. Panunzi, A. DeRenzi, R. Palumbo, G. Paiaro, *J. Am. Chem. Soc.*, *91*, 3879 (1969).
13. A. Panunzi, A. DeRenzi, G. Paiaro, *J. Am. Chem. Soc.*, *92*, 3488 (1970).
14. E. Benedetti, A. DeRenzi, G. Paiaro, A. Panunzi, C. Pedone, *Gazz. Chim. Ital.*, *102*, 744 (1972).
15. G. Natile, L. Maresca, L. Cattalini, *J. Chem. Soc., Dalton Trans.*, 651 (1977).
16. Louis S. Hegedus, Owen P. Anderson, Krister Zetterberg, Gary Allen, Kirstii Siirala-Hansen, David J. Olsen, Allan B. Packard, *Inorg. Chem.*, *16*, 1887 (1977).
17. A. C. Cope, J. M. Kliegman, E. C. Friedrich, *J. Am. Chem. Soc.*, *89*, 287 (1967).
18. D. Medema, R. van Helden, C. F. Kohl, *Inorg. Chim. Acta*, *3*, 255 (1969).
19. R. A. Holton, R. A. Kjonaas, *J. Am. Chem. Soc.*, *99*, 4177 (1977).
20. R. A. Holton, R. A. Kjonaas, *J. Organomet. Chem.*, *142*, C15 (1977).
21. R. N. Haszeldine, R. J. Lundt, R. V. Parish, *J. Chem. Soc.*, *A*, 3705 (1971).

5.8.2.3.5. η^1 σ-Alkyl Complexes by Insertion of Monoolefins into σ-Alkyl–, σ-Aryl–, σ-Vinyl–, and σ-Acyl-Metal Bonds.

The term olefin insertion in this section means the placing of the two double-bond carbons of the olefin between the alkyl group and the metal of a metal alkyl:

$$R{-}ML_n + {>}C{=}C{<} \rightarrow R{-}\overset{|}{\underset{|}{C}}{-}\overset{|}{\underset{|}{C}}{-}ML_n \qquad \text{(a)}$$

where R = an organic group, M = a transition-metal group and L = a ligand. No particular mechanism is implied. Indeed, the mechanisms often are not yet known. Excluded from this discussion are reactions involving attack of stabilized carbon nucleophiles upon olefin π-complexes. These reactions may involve initial attack of the carbon nucleophile upon the metal followed by insertion, or the nucleophile may attack the olefinic group directly. Unless the intermediate metal alkyl can be detected, the reactions of stabilized carbon nucleophiles are such as enolate anions, with olefinic π-complexes treated in §5.8.2.3.4. Likewise, insertions that occur by electrophilic attack of carbon reagents upon olefin π-complexes are also excluded from this section.

Olefin insertion into metal alkyls is involved in many valuable catalytic and stoichiometric organometalic syntheses. The high reactivity of most transition-metal alkyls causes them to react further under the conditions required for their formation, and the intermediate alkyls often have not been identified. Nevertheless, insertions into metal alkyls are important steps in many reactions.

As a consequence of the high reactivity of metal alkyls, there are few examples of olefin insertions in which the simple insertion product (with the metal still σ bonded to the alkyl group) can be isolated. This lack of clear examples has led to alternative explanations involving formation of a carbene metal complex from the alkyl, which then forms a metallocyclobutane with the olefin[1]. A final hydrogen shift with ring opening yields the product:

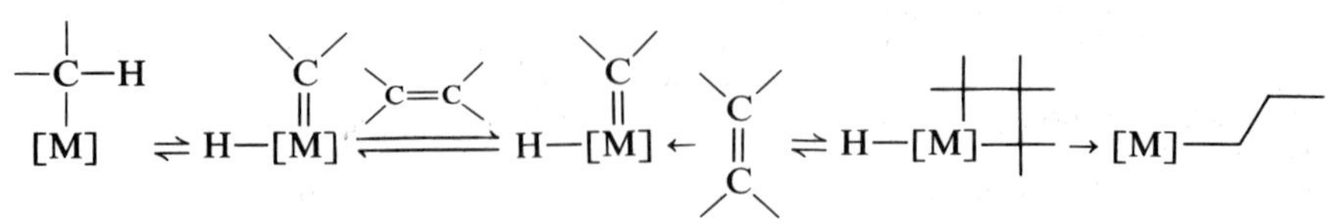

(b)

This mechanism has been disproved in one case by the use of deuterium labels[2]. The results of that study are consistent with the insertion mechanism, although the bracketed intermediates were not detected:

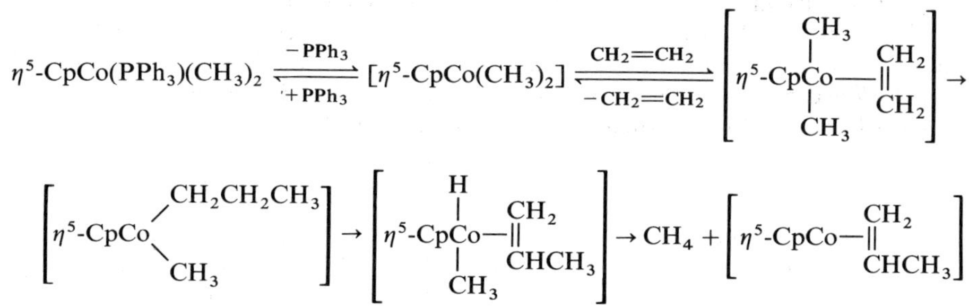

(c)

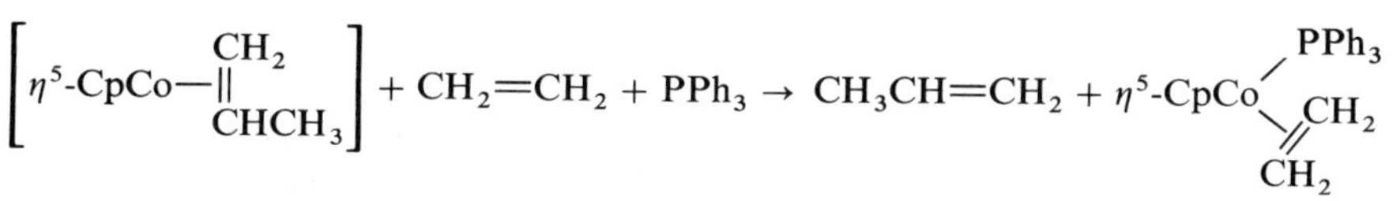

(d)

In this example the metal alkyl insertion product decomposes by a β-metal hydride elimination. This is common. The general inhibition of the insertion reaction by phosphines, as in the above case, and by other good ligands has been interpreted to mean that olefin coordination to the metal alkyl must occur before insertion can take place.

Consistent with the mechanism in Eq. (c), the more hindered the olefin is, and consequently the less stable its metal π-complexes are, the less reactive the olefin is in the insertion reaction. All olefin π-complexes of metal alkyls such as the intermediate in Eq. (c) do not undergo the insertion reaction. Several of these complexes have been prepared that do not undergo the reaction. Some of the unreactive complexes are:

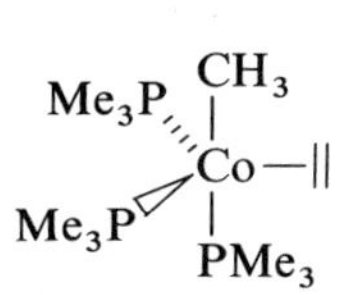

I

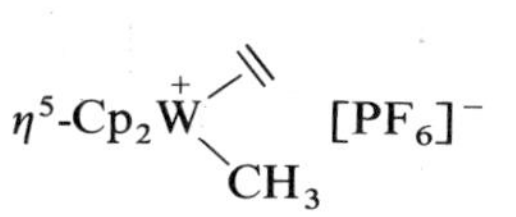

II

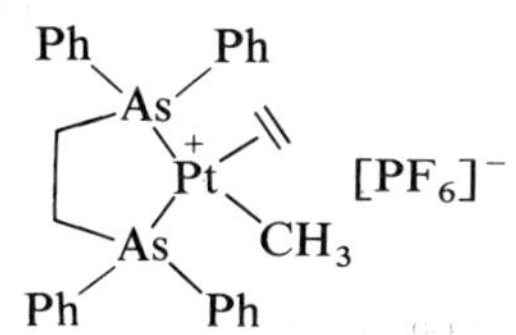

III

The unreactivity of these complexes may arise from an unfavorable preferred orientation of the double bond with respect to the alkyl group. The electronic properties of ligands on the metal may also influence the rate of insertion relative to other possible reactions.

Other mechanisms of insertion than the type shown in Eqs. (c) and (d) may exist, ESR evidence, for example, supports a radical mechanism for the reaction of alkylplatinum complexes with olefins[6].

(i) Insertion Reactions that Form Isolable Metal Alkyls. The only examples of simple olefin insertions where the products are isolable occur when the products are stabilized relative to the starting alkyl. This usually means that the β-metal hydride elimination is inhibited by the structure of the product. Where the hydride elimination is not rapid compared with insertion, and the initial insertion product is not stabilized relative to the starting alkyl, then polyinsertions of the olefin (polymerization) are observed, particularly with the early transition-metal alkyls (e.g., Ti, V, Cr).

Insertion of norbornadiene into π-allyl(hexafluoroacetylacetonato)-Pd(II) yields an adduct stabilized by both chelation of the Pd with the allylic double bond and by the lack of a cis-β-hydrogen to the metal needed for β-hydride elimination[7]. The reaction proceeds by way of the σ-allylpalladium–olefin π complex:

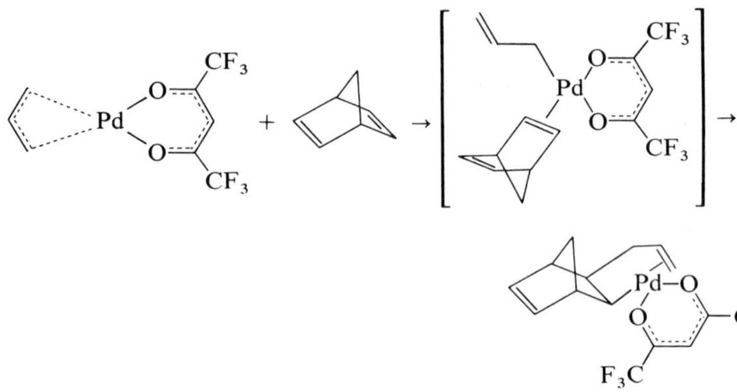

(e)

The addition occurs exo-cis since insertion takes place in a concerted fashion on the least hindered side of the double bond.

Reaction of the preformed $PdCl_4$–norbornadiene π-complex with diphenylmercury or sodium tetraphenylborate produces the endo-phenyl endo-norbornenylpalladium complex[8]. The mercurial or borate apparently exchange a phenyl for a chloro group and the phenyl migrates to one of the double-bond carbons. The second double bond remains coordinated to the metal during the reaction:

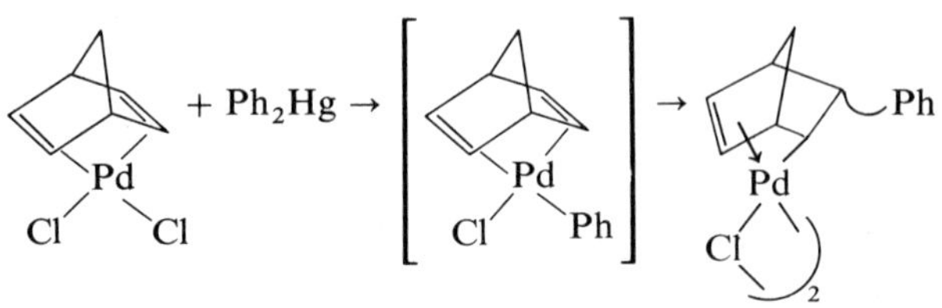

(f)

Similar chemistry occurs with 1,5-cyclooctadiene[9] and endo-dicyclopentadiene[9]:

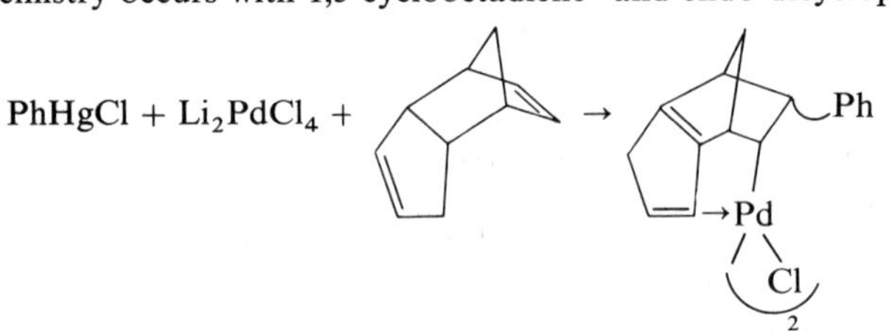

(g)

Surprisingly arylpalladium complexes generated from Na arylsulfinates and $PdCl_4$ (with deinsertion of SO_2) add to endo-dicyclopentadiene to form the exo-cis adduct[10]. The aryl group may coordinate the Pd in the adduct to stabilize it:

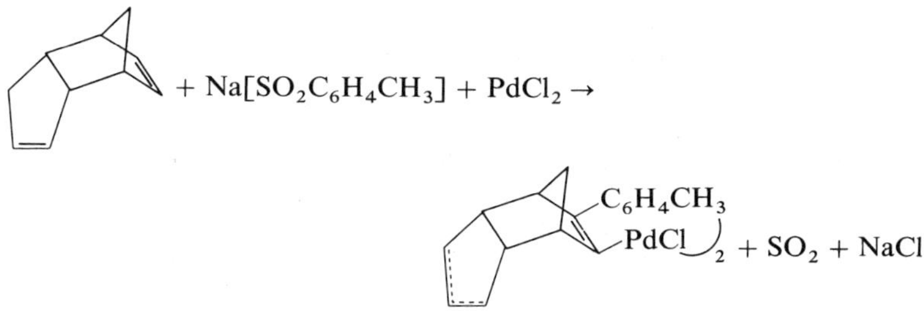

(h)

Two diasteriomers differing in the position of the double bond are produced in the reaction.

The related reaction of chloro(-π-2-methallyl)Ni(II) dimer with norbornene yields a stable, monomeric π-olefin complex that gives the expected exo-cis insertion product when treated with sodium acetate[11]:

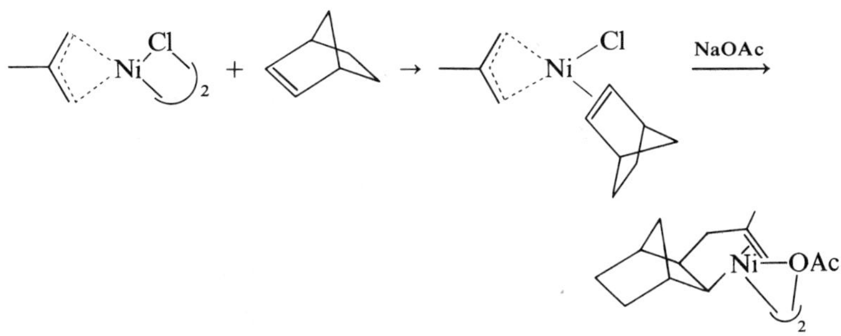

(i)

The corresponding chloro-π-allylic Pd dimer reacts similarly[12].

Allene also gives stable insertion products. The insertion occurs to place the alkyl group from the metal alkyl on the second carbon of the allene forming a stable π-allyl-metal complex as the product by way of a σ-allyl intermediate[13–17]. The reaction of allene with tetracarbonyl(acetyl)Co(I) is typical[13]:

$$CH_3COCo(CO)_4 \underset{CO}{\overset{-CO}{\rightleftharpoons}} [CH_3COCo(CO)_3] \xrightarrow{CH_2{=}C{=}CH_2}$$

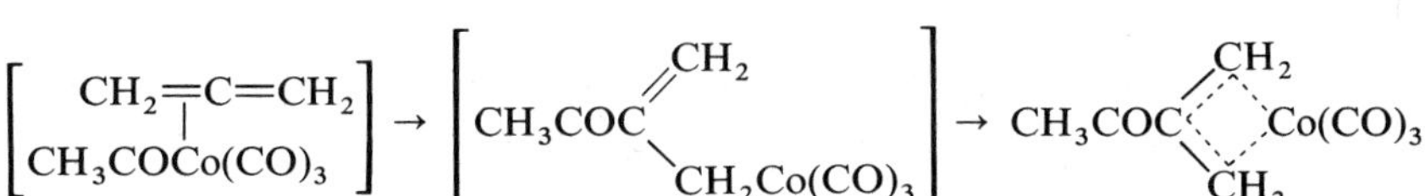

(j)

Tetracyanoethylene inserts into some metal alkyls giving isolable tetracyanoalkyl-metal complexes[18,19]:

$$\eta^5\text{-}CpFe(CO)_2CH_3 + (NC)_2C{=}C(CN)_2 \rightarrow$$

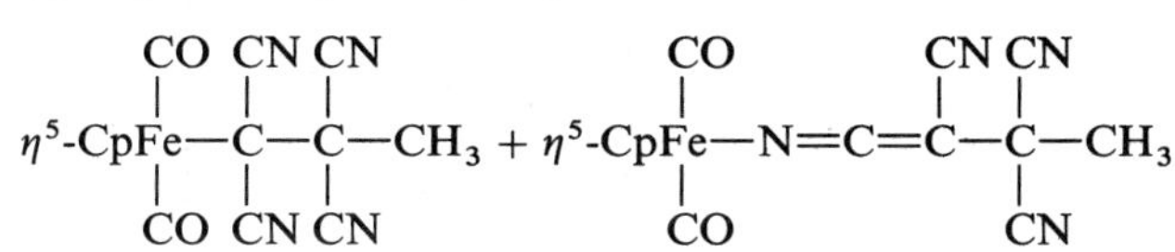

(k)

Qualitative relative reactivity data indicate that metal alkyls with alkyl groups that form the best radicals are the most reactive, suggesting a free-radical mechanism[18]. These reactions give the 1,2-addition products as well as the 1,4-keteniminate isomers.

Perfluoroolefins also give stable insertion products. Tetrafluoroethylene and pentacarbonyl(methyl)rhenium, for example, yield the tetrafluoropropylrhenium complex along with a small amount of a double insertion product[20]:

$$CH_3Re(CO)_5 + CF_2{=}CF_2 \rightarrow CH_3CF_2CF_2Re(CO)_5 + CH_3(CF_2)_4Re(CO)_5 \qquad (1)$$

Pentacarbonyl(methyl)- and phenylMn(I)[21,22] and bis(dimethylphenylphosphine) dimethylPt(II)[23] also insert tetrafluoroethylene.

π-Allylnickel[24] and -cobalt compounds[25,26] insert perfluoroolefins. These are included here because σ-alkyl complexes probably are the true reactants. Bis-(π-2-methallyl)-Ni(II) gives an insertion followed by a reductive elimination producing a Ni(0) complex with tetrafluoroethylene and the 1,7-diene product as a ligand[24]:

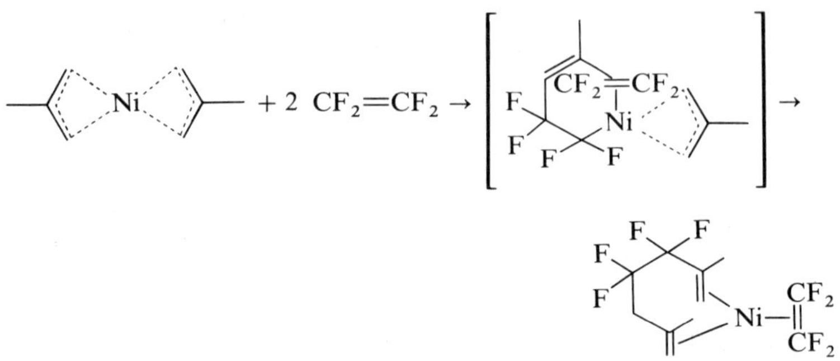

(m)

Tricarbonyl-π-allylcobalt(I) gives π-fluoroalkenylCo complexes under similar conditions[25,26]:

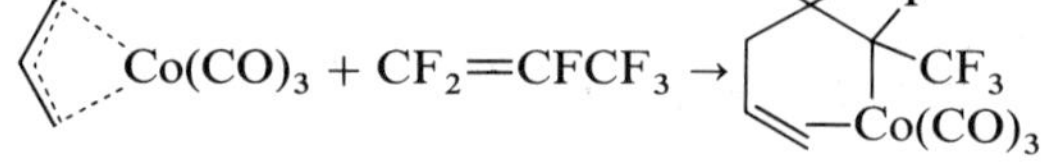

(n)[25]

Insertion of ethylene into a vinylic metal complex occurs with a cobaltacyclopentadiene complex, but the initial insertion product undergoes a rapid reductive elimination to form a π-1,3-cyclohexadiene complex and the intermediate product is not seen[27]:

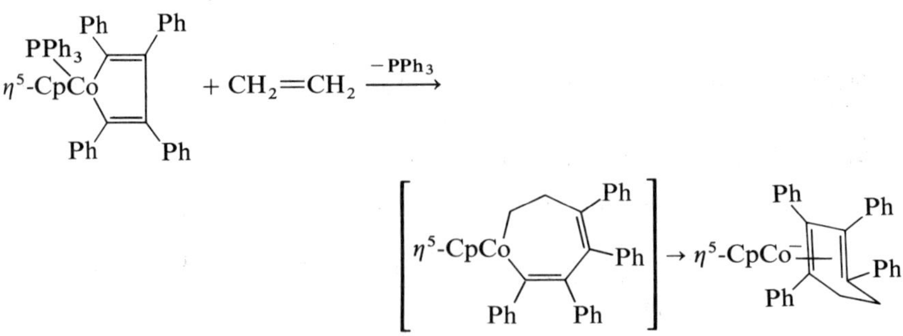

(o)

Rearrangement of the initial insertion product occurs in the reaction of various vinylic palladium halides with alkenes[28]. The vinylic palladium halides are prepared in situ from vinyl mercurials and palladium halides. The initial insertion products undergo a β-metal hydride elimination followed by a reverse readdition to form relatively stable π-allylic Pd complexes:

$$(CH_3)_3CCH{=}CHHgCl + Li_2[PdCl_4] + CH_2{=}CHCO_2CH_3 \rightarrow$$

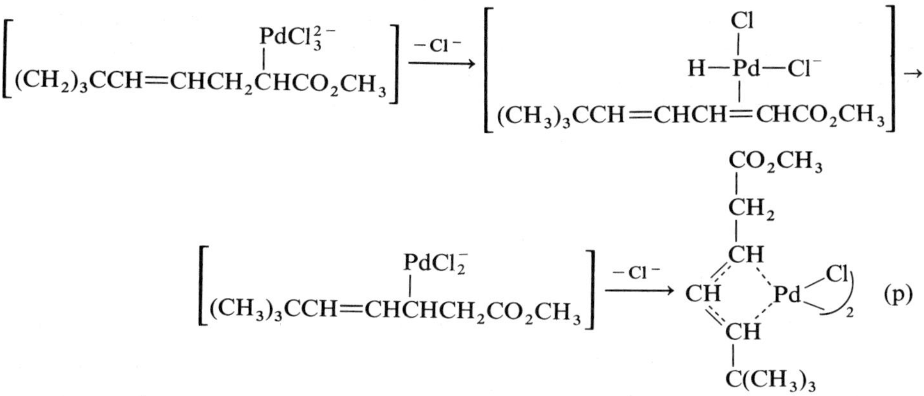

(p)

(ii) Insertion Reactions in Which Stable Organometallic Products Are not Obtained. When the insertion product is not isolated, there is uncertainty whether that reaction has occurred. The situation is even less certain where the starting alkyl is unstable or not identified and made in the reaction mixture. However, enough is known to be reasonably confident about the intermediacy of insertion steps. These reactions can be divided into groups based upon the mode of decomposition or further reaction of the insertion products.

(a) Polyinsertion. Polyinsertions of olefins (oligomerization or polymerization) occur only with the lighter transition-metal alkyls because the completing β-hydride elimination is relatively less favorable with these alkyls. Ethylene is most easily inserted for steric reasons. For example, a mixture of trichloro(methyl)Ti(IV) and dichloro(methyl)aluminum at $-50°C$ rapidly converts ethylene into a mixture of oligomeric straight- and branched-chain olefins[29]. The aluminum compound or other electron-pair acceptors are required in these reactions to complex with a halo group on the transition metal and activate the alkyl by removing electron density. Oligomers rather than high polymers are formed owing to the successful competition of β-hydride elimination with the insertion. The titanium hydride formed in the elimination may add back to the terminal olefin formed, in the reverse direction to reeliminate and produce internal olefins or to undergo further insertions to form branched-chain oligomers.

Changes in the metal ligands can change the relative rates of elimination and insertion. For example, if dichlorodicyclopentadienyltitanium instead of trichloro(methyl)-titanium is combined with chloro(diethyl)aluminum and the mixture reacted with ethylene, linear high polymers are produced[30]. Since the higher olefins insert less readily than ethylene, they are more reluctant to polymerize. High polymers can be made from propylene and higher 1-alkenes only with heterogeneous catalysts, because more electron-deficient and therefore more reactive catalysts can exist in the solid phase than can in solution[31,32]. A heterogeneous commercial catalyst for propylene polymerization is a combination of titanium trichloride and a chloro(alkyl)-aluminum compound. The polyinsertion of substituted ethylenes has the added complication of stereochemistry, since chiral centers are produced in the reaction and there is the possibility for succeeding centers to be of the same or alternate chirality, be randomly produced, or be formed in some more complex way. Organometallic catalysts vary in the stereochemistry they produce depending upon the catalyst employed and how it is prepared and used[33].

(b) Alkyl Transfer to Another Metal. Few examples of this useful reaction have been studied. An example is the Ni-catalyzed addition of phenylmagnesium bromide to propylene[34]. The reaction involves transfer of the phenyl group from Mg to Ni followed by insertion of the olefin and reexchange of the metals. The insertion occurs in both possible directions, but the primary magnesium alkyl is the major product[36]:

$$C_6H_5MgBr + NiCl_2 \rightarrow [C_6H_5NiCl] + MgBrCl \qquad \text{(q)}$$

$$C_6H_5NiCl + CH_3CH{=}CH_2 \rightarrow C_6H_5\overset{CH_3}{\overset{|}{C}}HCH_2NiCl \qquad \text{(r)}$$

$$C_6H_5\overset{CH_3}{\overset{|}{C}}HCH_2NiCl + MgBrCl \rightarrow C_6H_5\overset{CH_3}{\overset{|}{C}}HCH_2MgBr + NiCl_2 \qquad \text{(s)}$$

(c) Formation of Transition-Metal Enolates. Insertions of α,β-unsaturated carbonyl compounds yield transition-metal enolates similar to the better known reactions of organocopper compounds. Since the enolates are generally reactive and difficult to purify, they are not isolated but are hydrolyzed instead. The formation of the enolate often stops further insertions from occurring so that a single product can be obtained. Nickel and Fe are useful in these reactions. Acylnickel carbonyl anions, for example, add to α,β-unsaturated ketones to form enolates of unknown structure that form 1,4-diketones on hydrolysis and oxidation[35]. The oxidative hydrolysis is used in order to destroy nickel carbonyl present in the hydrolysis mixture:

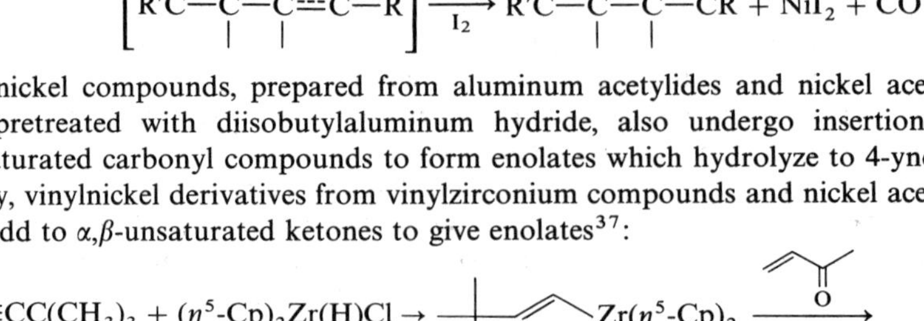

(t)

Ethynylnickel compounds, prepared from aluminum acetylides and nickel acetylacetonate pretreated with diisobutylaluminum hydride, also undergo insertions with α,β-unsaturated carbonyl compounds to form enolates which hydrolyze to 4-ynones[36]. Similarly, vinylnickel derivatives from vinylzirconium compounds and nickel acetylacetonate add to α,β-unsaturated ketones to give enolates[37]:

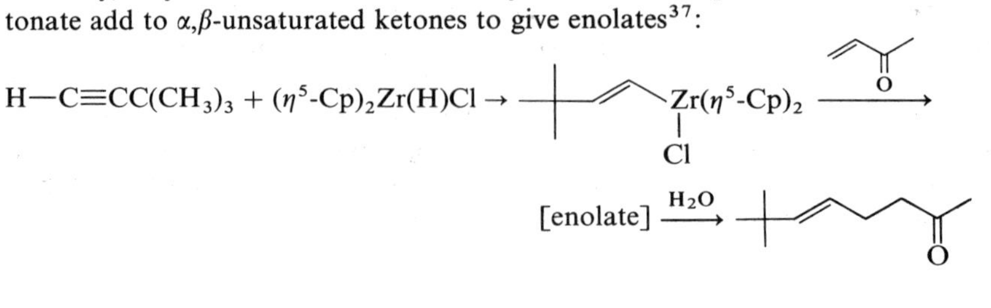

(u)

π-Allylnickel bromide dimer adds to acrylonitrile and methyl acrylate similarly to form complexes that hydrolyze to 5-hexenonitrile or methyl 5-hexenoate, respectively[38]. Tetracarbonyl(alkyl)iron anions behave like the nickel alkyls with α,β-unsaturated ketones, nitriles and esters[39], and they also add to ethylene[40]. The ethylene insertion stops with one insertion because the initial product rearranges by β-hydride elimination and readdition to form an enolate complex:

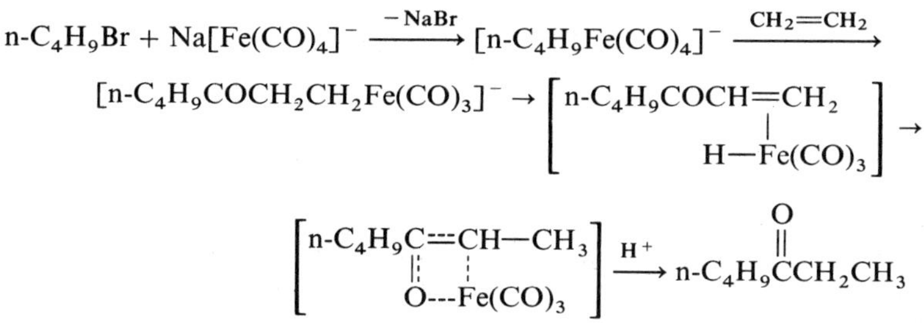

(v)

(d) Reductive Elimination of Insertion Productions. An example of this mode of reaction which produces a cyclohexadiene π-complex is given in Eq. (o). The formation of cyclopropanes from alkenes, dicarbonylcyclopentadienyl(methoxymethyl)Fe(II) and an acid, or other similar combinations which produce the methyl cation, involves similar insertions of the olefin. Insertion appears to be in the Fe–methyl carbonium ion bond here, and this is followed by reductive elimination of the cyclopropane[41,42]:

$$\eta^5\text{-CpFe(CO)}_2\text{CH}_2\text{OCH}_3 + 2\ \text{HBF}_4 \rightarrow [\eta^5\text{-CpFe(CO)}_2\text{CH}_2]\text{BF}_4 + [\text{H}_2\text{OCH}_3]\text{BF}_4 \qquad \text{(w)}$$

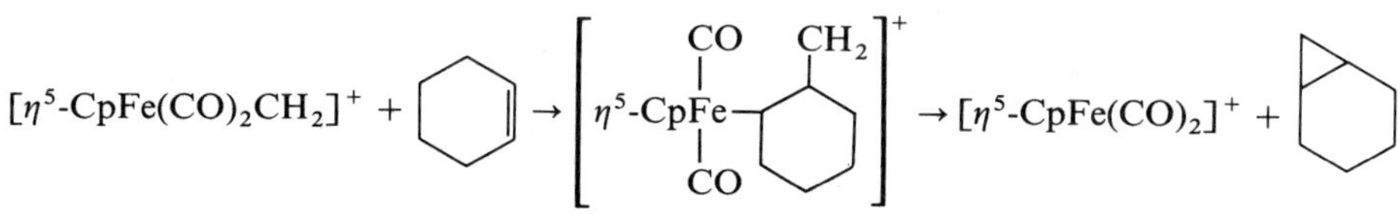

(x)

An insertion–reductive elimination sequence is involved in the rhodium-catalyzed, intramolecular hydroacylation of 4-alkenals to form cyclopentanes[43,44,45]:

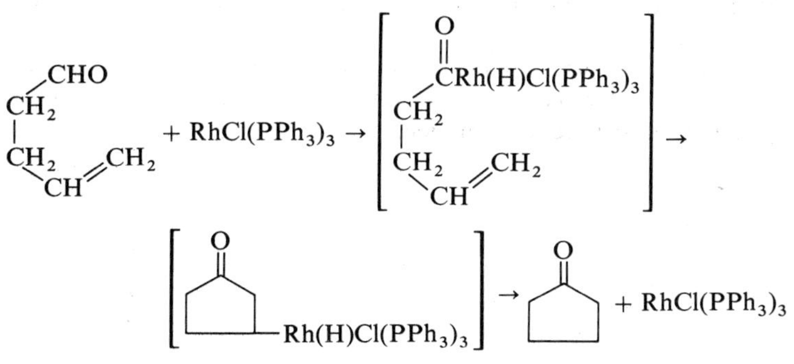

(y)[44]

This reaction also occurs intermolecularly with ethylene[44,46].

(e) Carbonylation of Insertion Products. Olefin insertion products may undergo insertion by CO followed by other reactions. For example, hydrolysis of the carbonylated intermediate to the carboxylic acid occurs in the reaction of crotyl chloride with ethylene in the presence of CO and H_2O with nickel carbonyl as catalyst[47]:

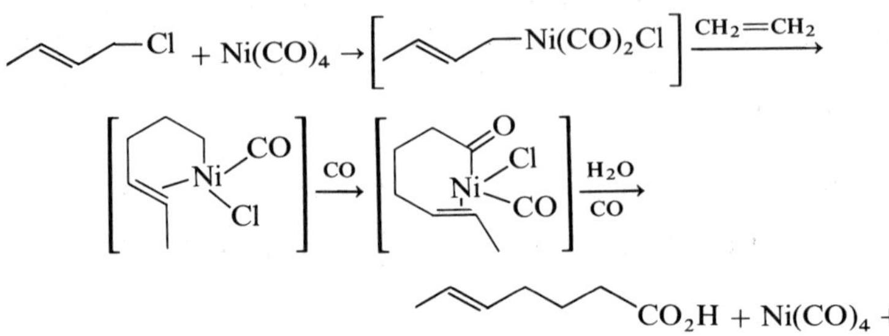

(z)

The stabilized cyclic olefin insertion product explains why the reaction stops with the insertion of only one ethylene group. The carboxylation product of crotyl chloride, 3-pentenoic acid, and 2,6-octadiene (from the coupling of two crotyl chloride molecules together) also forms.

Cyclization of the β-ketoacylmetal complexes formed by carbonylation of acylmetal complex–olefin insertion products is another commonly observed sequence that is completed by a β-hydride elimination of the cyclized product[48]:

$$PhI + Ni(CO)_4 \rightarrow [PhCONi(CO)_2I] \xrightarrow{PhCH=CH_2}$$

$$[PhCOCH_2CH(Ph)Ni(CO)_2I] \xrightarrow{CO} [PhCOCH_2CH(Ph)CONi(CO)_2I] \rightarrow$$

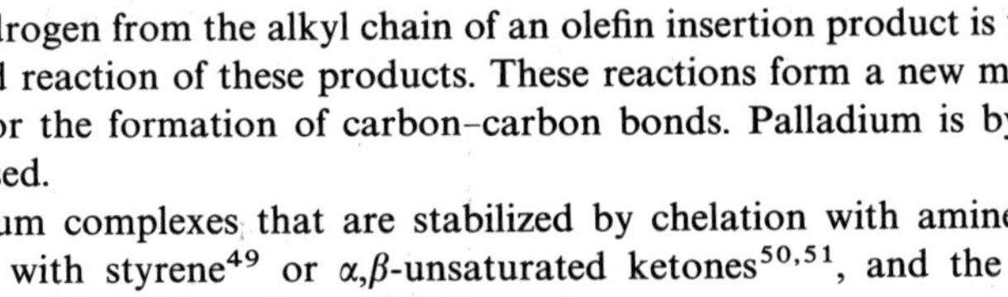

$[+HNi(CO)_2I]$ (aa)

(f) β-Metal Hydride Eliminations of Insertion Products. Elimination of the metal moiety with a β-hydrogen from the alkyl chain of an olefin insertion product is the most commonly observed reaction of these products. These reactions form a new method in organic synthesis for the formation of carbon–carbon bonds. Palladium is by far the metal most often used.

Organopalladium complexes that are stabilized by chelation with amine groups undergo insertions with styrene[49] or α,β-unsaturated ketones[50,51], and the adducts quickly eliminate metal hydride to form olefinic products:

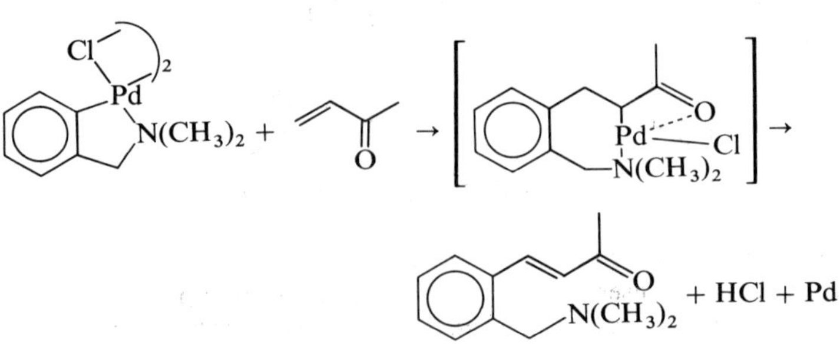

(ab)[50]

Similar reactions are involved where the organometallics only are formed in situ. Palladium acetate metallates aromatic hydrocarbons at elevated temperatures, and in the presence of olefins insertion followed by β-hydride elimination can occur, forming styrene or stilbene derivatives:

$$C_6H_6 + Pd(OAc)_2 \rightarrow [C_6H_5PdOAc] + HOAc$$

$$C_6H_5PdOAc + C_6H_5CH{=}CH_2 \rightarrow \left[C_6H_5\underset{}{\overset{\overset{\displaystyle PdOAc}{|}}{C}}HCH_2C_6H_5 \right] \rightarrow$$

$$C_6H_5CH{=}CHC_6H_5[+HPdOAc]$$

$$[HPdOAc] \rightarrow Pd + HOAc \qquad \text{(ac)}^{52}$$

Furan and thiophene heterocycles undergo similar reactions[53,54], as does ferrocene[55].

Since the direct palladation of aromatic hydrocarbons often gives isomeric mixtures of arylmetal derivatives[56], other, more specific methods of forming the organometallics for the insertion reaction are usually employed. These are also useful for the preparation of organometallics that cannot be made by direct metallation.

Exchange reactions between stable organometallics and palladium salts are useful for preparing isomerically pure organopalladium species to be used in situ for insertion reactions. Exchanges have been made with organocobalt[57,58], -iron[58], -phosphorus[59,60], -silicon[61,62], -lead[63], -tin[63], -boron[64], -thallium[65], -zinc[66], -magnesium[67] and -mercury[63,68,69]. Of these, mercurials are most often used, e.g.[63]:

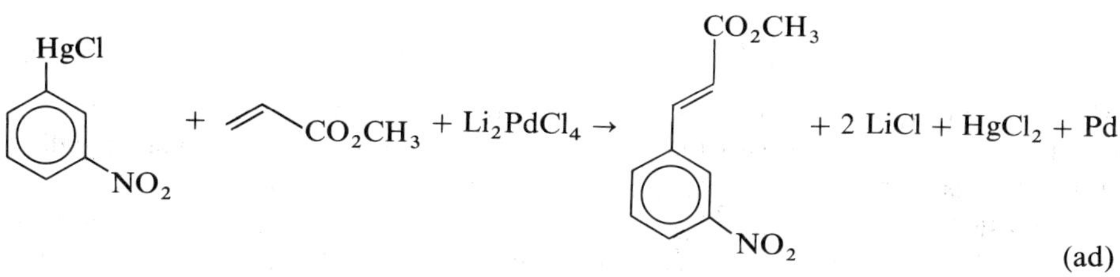

(ad)

β-Elimination of chloride instead of hydride occurs if the chloro substituent is in the proper position in the insertion product. For example, this elimination occurs with allylic and organopalladium chlorides obtained by exchange reactions from mercurials[70]:

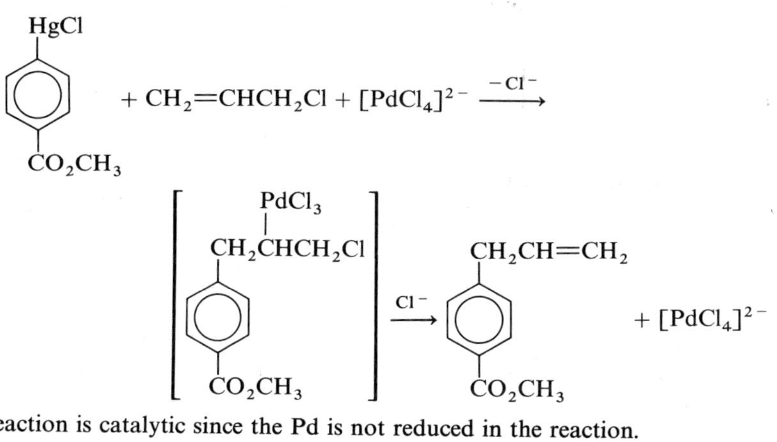

(ae)

This reaction is catalytic since the Pd is not reduced in the reaction.

Other sources of aryl groups for preparing arylpalladium complexes in situ are arylamines[71,72], arylhydrazines[72], aryldiazonium salts[73] and aryl halides[74]. Aryl halides as well as heterocyclic, benzylic and vinylic halides are used to prepare organopalladium complexes in situ for olefin insertion reactions[74]. These reactions are catalytic in Pd when a base is included in the reaction mixtures to neutralize the hydrogen halide produced. The organic halide oxidative addition–insertion—β-hydride elimination sequence is carried out simply by heating mixtures of the halide (iodides or bromides) with the olefin to be inserted in the presence of an amine as the base and a catalytic amount of palladium acetate with 2 equiv of a triarylphosphine as ligand. The palladium acetate–phosphine complex is formed in situ and reduced by the olefin to the required catalyst, a Pd(0)–phosphine complex. 2,4-Dienoic acid derivatives, for example, are obtained stereospecifically from vinylic halides and acrylate derivatives this way[75]:

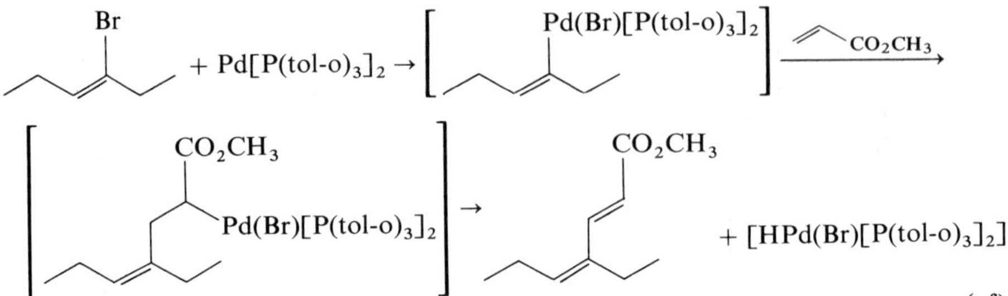

(af)

$$[HPd(Br)_2] + Et_3N \rightarrow Pd_2 + [Et_3NH]Br \qquad (ag)$$

where $Po(tol\text{-}o)_3$ = tri-o-tolylphosphine. This reaction is stereospecific because the oxidative addition occurs with retention of the stereochemistry of the vinylic halide. Intramolecular versions of the reaction produce heterocyclic products[74].

Insertions of vinylic palladium halides with olefins not conjugated with carbonyl groups is more complex. Rearrangement of the initial insertion product to a π-allylpalladium derivative usually occurs as in Eq. (p). Since the π-allylic complexes are relatively stable, catalytic reactions to form dienes with tertiary amines as bases are slow and inefficient[76]. A useful catalytic reaction occurs, however, if a nucleophilic secondary amine is used as base instead of the tertiary amine. The π-allylic palladium intermediates are attacked by the amine to form tertiary allylic amines and regenerate the catalyst[76]:

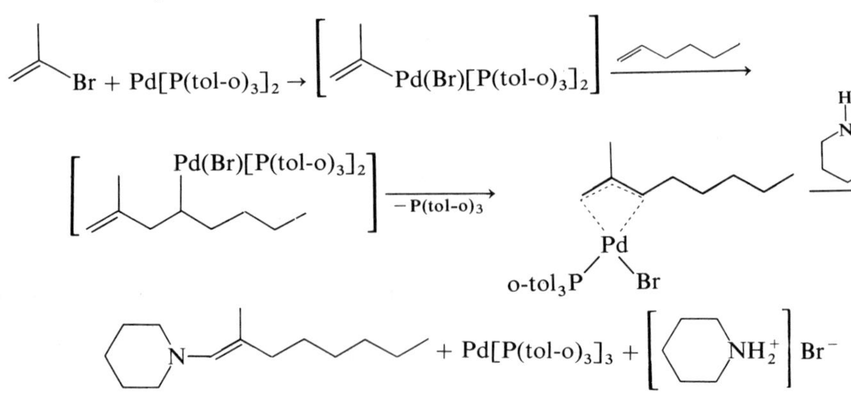

(ah)

The β-metal hydride elimination occurs to form the conjugated diene or styrene product if a β hydrogen is present in the position cis to the metal group. If only trans hydrogens are available, elimination occurs more slowly. If there is a choice of forming a conjugated olefin by a trans elimination or a non conjugated isomer by a cis elimination, the latter reaction is preferred:

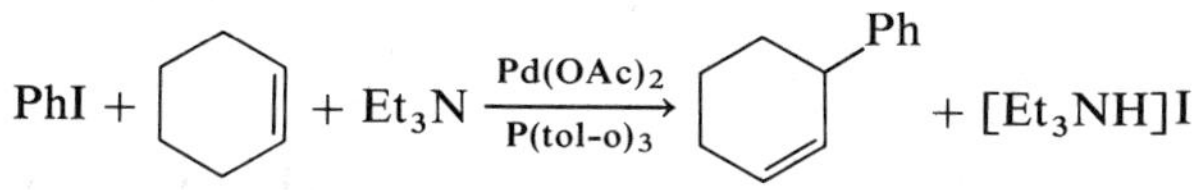

(ai)[77]

In hydroxylic solvents solvolylic displacement of the Pd has been observed to occur in preference to trans-β-hydride elimination[78].

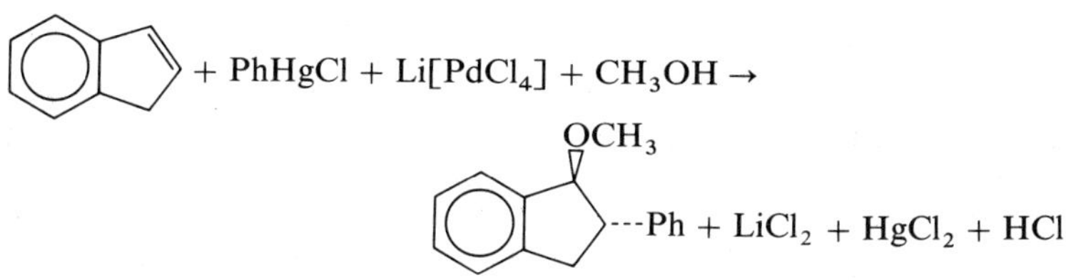

(aj)

Another exception to the elimination to form conjugated olefins occurs in the insertion reactions of allylic alcohols with aryl or heterocyclic halides where elimination of the β-hydrogen on the carbon bearing the hydroxyl group is preferred so that carbonyl products (as enols) are formed[79,80,74]:

$$PhI + CH_2{=}C(CH_3)CH_2OH + Et_3N \xrightarrow[PPh_3]{Pd(OAc)_2} PhCH_2C(CH_3)HCHO + [Et_3NH]I \qquad \text{(ak)}^{79}$$

The organometallic-olefin insertion-hydride elimination sequence occurs with other transition metals than Pd, e.g., with Ni[81,82], Ir[84], Rh[85,86,87], Ru[88], Co[2,89] and Ti[90,91]. In most cases the organometallic undergoing insertion is formed in situ by a metal hydride addition (insertion) with an olefin rather than by the exchange reaction. The olefin reacting with the hydride to form the alkyl may be the same one that undergoes the insertion with the metal alkyl, or a different one. This sequence with a single olefin produces olefin dimers after the final β-hydride elimination. A mechanism for the $RhCl_3$-catalyzed dimerization of ethylene is[87]:

$$RhCl_3 \cdot 3\,H_2O \xrightarrow[EtOH]{CH_2{=}CH_2} [Cl_2Rh(CH_2{=}CH_2)_2] \xrightarrow{HCl}$$

$$[C_2H_5RhCl_3(C_2H_4)S]^- \xrightarrow{S} [n\text{-}C_4H_9RhCl_3S_2]^- \xrightarrow[-S]{-HCl}$$

$$[Cl_2Rh(S)(CH_3CH_2CH{=}CH_2)]^- \xrightarrow[-S]{CH_2{=}CH_2}$$

$$CH_3CH_2CH{=}CH_2 + [Cl_2Rh(CH_2{=}CH_2)_2]^- \qquad \text{(al)}$$

Two different alkenes can be combined selectively with metal hydride catalysts. This is termed a codimerization and is selective when one alkene reacts only with the hydride and the other alkene selectively inserts into the metal alkyl formed from the hydride. A good example of a selective reaction is the titanium tetrachloride-dimethylaluminum chloride-catalyzed reaction of ethylene with higher terminal olefins[90] in which the titanium halide catalyst is formed from a methyltitanium trichloride–methylaluminum dichloride complex produced by an alkylation of the Ti by the aluminum alkyl. The methyltitanium complex can undergo insertion with one of the olefins and then eliminate the hydride:

$$TiCl_4 + (CH_3)_2AlCl \rightarrow [CH_3TiCl_3 \cdot Al(CH_3)Cl_2] \qquad \text{(am)}$$

$$[CH_3TiCl_3 \cdot Al(CH_3)Cl_2] + RCH{=}CH_2 \rightarrow [R\overset{CH_3}{\overset{|}{C}}HCH_2TiCl_3 \cdot Al(CH_3)Cl_2] \rightarrow [HTiCl_3 \cdot Al(CH_3)Cl_2] + R\overset{CH_3}{\overset{|}{C}}{=}CH_2 \qquad \text{(an)}$$

The Al coordinates with the halogen on the titanium and activates it for insertion by removing electron density as in the polymerization reactions described in §5.8.2.3.5.ii.a. The titanium hydride reacts more easily with ethylene than with the higher l-alkene present and forms the ethyltitanium complex. This complex then selectively inserts the l-alkene and the adduct finally eliminates the metal hydride catalyst and forms the 2-ethyl-l-alkene specifically[90]:

$$[HTiCl_3 \cdot Al(CH_3)Cl_2] + CH_2{=}CH_2 O [CH_3CH_2TiCl_3 \cdot Al(CH_3)Cl_2] \xrightarrow{RCH=CH_2} [R\overset{CH_2CH_2}{\overset{|}{C}}HCH_2TiCl_3 \cdot Al(CH_3)Cl_2] \rightarrow [HTiCl_3 \cdot Al(CH_3)Cl_2] + R{-}\overset{CH_2CH_3}{\overset{|}{C}}{=}CH_2 \qquad \text{(ao)}$$

The preferential insertion of the titanium alkyl into the l-alkene is one of few cases known where the more hindered olefin (the l-alkene) inserts more easily than the less hindered one (ethylene). Presumably, electronic factors are more important than steric ones in this case.

Insertion reactions of olefins with acylmetal complexes are also followed by β-metal hydride eliminations. Diethylketone, for example, is a by product in the hydroformulation of ethylene and is probably formed by a final hydrogenation of the vinyl group. Diethylketone is a major product if the hydrogen used is limited[92]. Intramolecular versions of the reaction proceed catalytically[93] and stoichiometrically[95]:

$$CH_2{=}CH(CH_2)_2COCl + Na[Co(CO)_4] \xrightarrow{-NaCl} CH_2{=}CH(CH_2)_2COCo(CO)_4 \xrightarrow[-CO]{\Delta}$$

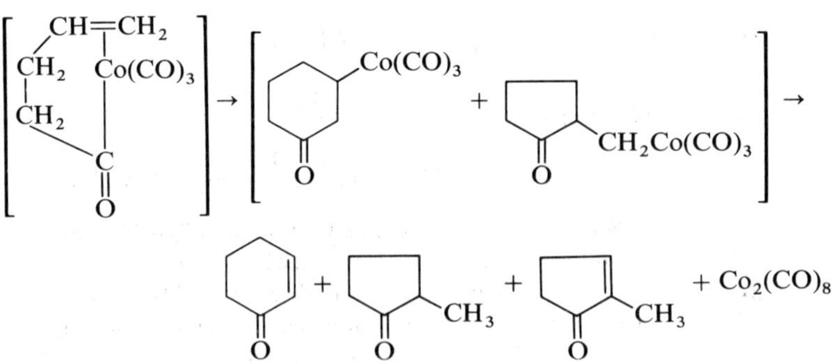

(ap)

The cobalt hydrocarbonyl formed in this elimination partially reduces the unsaturated ketones produced[95].

(R. F. HECK)

1. K. J. Ivin, J. J. Rooney, C. D. Stewart, M. L. H. Green, R. Mahtab, *J. Chem. Soc., Chem. Commun.*, 604 (1978).
2. E. R. Evitt, R. G. Bergman, *J. Am. Chem. Soc.*, *101*, 3973 (1979); *102*, 7003 (1980).
3. H-F. Klein, R. Hammer, J. Gross, U. Schubert, *Angew. Chem., Int. Ed. Engl.*, *19*, 809 (1980).
4. F. W. S. Benfield, N. J. Cooper, M. L. H. Green, *J. Organomet. Chem.*, *76*, 49 (1974).
5. H. C. Clark, C. R. Jablonski, K. von Werner, *J. Organomet. Chem.*, *82*, C51 (1974).
6. N. G. Hargreaves, R. J. Puddephatt, L. H. Sutcliffe, P. J. Thompson, *J. Chem. Soc., Chem. Commun.*, 861 (1973).
7. R. P. Hughes, J. Powell, *J. Organomet. Chem.*, *30*, C45 (1971); *60*, 387 (1973).
8. A. Segnitz, E. Kelly, S. H. Taylor, P. M. Maitlis, *J. Organomet. Chem.*, *124*, 113 (1977).
9. A. Kasahara, T. Izumi, K. Endo, T. Takeda, M. Ookita, *Bull. Chem. Soc., Jpn.*, *47*, 1967 (1974).
10. Y. Tamaru, Z. Yoshida, *Tetrahedron Lett.*, 4527 (1978).
11. M. C. Gallozzi, T. C. Hanlon, G. Vilulli, L. Porri, *J. Organomet. Chem.*, *33*, C45 (1971).
12. M. Zocchi, G. Tiegln, A. Albinate, *J. Organomet. Chem.*, *33*, C47 (1971).
13. S. Otsuka, A. Nakamura, *Inorg. Chem.*, *11*, 644 (1972).
14. D. H. Gibson, R. L. Vonnahme, J. E. McKiernan, *J. Chem. Soc., Chem. Commun.*, 720 (1971).
15. M. H. Chisholm, W. S. Johns, *Inorg. Chem.*, *14*, 1189 (1975).
16. R. P. Hughes, J. Powell, *J. Organomet. Chem.*, *34*, C51 (1972); *60*, 409 (1973).
17. H. C. Clark, C. R. C. Milne, C. S. Wong, *J. Organomet. Chem.*, *136*, 265 (1977).
18. S. R. Su, A. Wojcicki, *Inorg. Chem.*, *14*, 89 (1975).
19. J. A. Hanna, A. Wojcicki, *Inorg. Chim. Acta*, *9*, 55 (1974).
20. J. B. Wilford, F. G. A. Stone, *Inorg. Chem.*, *4*, 93 (1965).
21. J. B. Wilford, P. M. Treichel, F. G. A. Stone, *Proc. Chem. Soc.*, 218 (1963).
22. J. B. Wilford, A. Forster, F. G. A. Stone, *J. Chem. Soc.*, 6519 (1965).
23. H. C. Clark, R. J. Puddephatt, *Inorg. Chem.*, *9*, 2670 (1970).
24. J. Browning, D. J. Cook, C. S. Cundy, M. Green, F. G. A. Stone, *J. Chem. Soc., Chem. Commun.*, 929 (1968).
25. A. Greco, M. Green, F. G. A. Stone, *J. Chem. Soc., A*, 3476 (1971).
26. M. Bottrill, R. Goddard, M. Greene, P. Woodward, *J. Chem. Soc., Dalton Trans.*, 1671 (1979).
27. H. Yamazaki, N. Hagihara, *J. Organomet. Chem.*, *21*, 431 (1970).
28. R. C. Larock, M. A. Mitchell, *J. Am. Chem. Soc.*, *100*, 180 (1978).
29. H. Bestian, K. Claus, H. Jensen, E. Prinz, *Angew. Chem., Int. Ed. Engl.*, *2*, 32 (1963).
30. W. P. Long, D. S. Breslow, *J. Am. Chem. Soc.*, *82*, 1953 (1960).
31. P. Cossee, *J. Catal.*, *3*, 80, 89, 99 (1964).
32. T. Suzuki, Y. Takegami, *Bull. Chem. Soc., Jpn.*, *43*, 1484 (1970).
33. N. C. Gaylord, H. F. Mark, *Linear and Stereoregular Addition Polymers: Polymerization with Controlled Propagation*, Interscience, New York, 1959.
34. L. Faráday, L. Markó, *J. Organomet. Chem.*, *28*, 159 (1971).
35. E. J. Corey, L. S. Hegedus, *J. Am. Chem. Soc.*, *91*, 4927 (1969).
36. R. T. Hansen, D. B. Carr, J. Schwartz, *J. Am. Chem. Soc.*, *100*, 2244 (1978).
37. M. J. Loots, J. Schwartz, *J. Am. Chem. Soc.*, *99*, 8045 (1977).
38. M. Dubini, F. Montino, G. P. Chiusoli, *Chimica L'Industria (Milan)*, *47*, 839 (1965).
39. M. P. Cooke, *J. Am. Chem. Soc.*, *99*, 5222 (1977).
40. M. P. Cooke, R. M. Parlman, *J. Am. Chem. Soc.*, *97*, 6863 (1975).
41. P. W. Jolly, R. Pettit, *J. Am. Chem. Soc.*, *88*, 5044 (1966).
42. S. Brandt, P. Helquist, *J. Am. Chem. Soc.*, *101*, 6473 (1979).
43. K. Sakai, O. Oda, *Tetrahedron Lett.*, 4375 (1972).
44. C. F. Lochow, R. G. Miller, *J. Am. Chem. Soc.*, *98*, 1281 (1976).
45. R. C. Larock, K. Oerthe, G. F. Potter, *J. Am. Chem. Soc.*, *102*, 190 (1980).
46. K. P. Vora, C. F. Lochow, R. W. Miller, *J. Organomet. Chem.*, *192*, 257 (1980).
47. G. P. Chiusoli, G. Cometti, *J. Chem. Soc., Chem. Commun.*, 1015 (1972).
48. E. Yoshisato, M. Ryang, S. Tsutsumi, *J. Am. Chem. Soc.*, *34*, 1500 (1969).
49. S. W. Hart, R. Bau, C. H. Chao, R. F. Heck, *J. Organomet. Chem*, *179*, 301 (1979).

50. R. A. Holton, *Tetrahedron Lett.*, 355 (1977).
51. R. A. Holton, R. A. Kjonaas, *J. Organomet. Chem.*, *133*, C5 (1977).
52. Y. Fujiwara, I. Moritani, S. Danno, R. Asano, S. Teranishi, *J. Am. Chem. Soc.*, *91*, 7166 (1969).
53. R. Asano, I. Moritani, Y. Fujiwara, S. Teranishi, *Bull. Chem. Soc., Jpn.*, *46*, 663 (1973).
54. A. Kasahara, T. Izumi, M. Yodono, R. Saito, T. Takeda, T. Sugawara, *Bull. Chem. Soc., Jpn.*, *46*, 1220 (1973).
55. R. Asano, *J. Chem. Soc., C*, *1971*, 3691.
56. Y. Fujiwara, R. Asano, I. Moritani, S. Teranishi, *J. Org. Chem.*, *41*, 1681 (1976).
57. C-Y. Chen, T. Kwan, *Tetrahedron Lett.*, 2651 (1973).
58. M. E. Vol'pin, R. Taube, H. Drevs, L. G. Volkova, I. Y. Levitin, T. M. Ushakova, *J. Organomet. Chem.*, *39*, C79 (1972).
59. T. Yamane, K. Kikukawa, M. Takagi, T. Matsuda, *Tetrahedron*, *29*, 955 (1973).
60. R. Asano, I. Moritani, Y. Fujiwara, S. Teranishi, *Bull. Chem. Soc., Jpn.*, *46*, 2910 (1973).
61. W. P. Weber, R. A. Felix, A. K. Willard, K. E. Koenig, *Tetrahedron Lett.*, 4701 (1971).
62. I. S. Akhrem, N. M. Christovalova, E. I. Mysov, M. E. Vol'pin, *J. Organomet. Chem.*, *72*, 163 (1974).
63. R. F. Heck, *J. Am. Chem. Soc.*, *90*, 5518 (1968).
64. H. A. Dieck, R. F. Heck, *J. Org. Chem.*, *40*, 1083 (1975).
65. S. Uemura, K. Zushi, M. Okano, K. Ichikawa, *J. Chem. Soc., Chem. Commun.*, 234 (1972).
66. J. F. Fauvarque, A. Jutand, *J. Organomet. Chem.*, *132*, C17 (1977).
67. N. Luong-Thi, H. Riveriere, *J. Chem. Soc., Chem. Commun.*, 918 (1978).
68. R. F. Heck, *J. Organomet. Chem.*, *37*, 389 (1972).
69. A. Kasahara, T. Izumi, G. Saito, M. Yodono, G. Saito, M. Yodono, R. Suito, Y. Goto, *Bull. Chem. Soc., Jpn.*, *45*, 895 (1972).
70. R. F. Heck, *J. Am. Chem. Soc.*, *90*, 5531 (1968).
71. K. Kikukawa, K. Maemura, K. Nagira, F. Wada, T. Matsuda, *Chem. Lett.*, 551 (1980).
72. F. Akiyama, H. Miyazaki, K. Kaneda, S. Teranishi, *J. Org. Chem.*, *45*, 2359 (1980).
73. K. Kikukawa, K. Nagira, N. Terao, F. Wada, T. Matsuda, *Bull. Chem. Soc., Jpn.*, *52*, 2609 (1979).
74. R. F. Heck, *Org. React.*, *27*, 345 (1981).
75. J-I. Kim, B. A. Patel, R. F. Heck, *J. Org. Chem.*, *46*; 1067 (1981).
76. B. A. Patel, R. F. Heck, *J. Org. Chem.*, *43*, 3898 (1978).
77. N. A. Cortese, C. B. Ziegler, Jr, B. J. Hrnjez, R. F. Heck, *J. Org. Chem.*, *43*, 2953 (1978).
78. H. Horino, N. Inoue, *Bull. Chem. Soc., Jpn.*, *44*, 3210 (1971).
79. J. B. Melpolder, R. F. Heck, *J. Org. Chem.*, *41*, 265 (1976).
80. A. J. Chalk, S. A. Magennis, *J. Org. Chem.*, *41*, 273, 1206 (1976).
81. L. S. Hegedus, B. R. Evans, *J. Am. Chem. Soc.*, *100*, 3461 (1978).
82. G. N. Schrauzer, *Chem. Ber.*, *94*, 1891 (1961).
83. J. Ewers, *Angew. Chem.*, *78*, 593 (1966).
84. J. Blum, M. Weitzberg, R. J. Mureinik, *J. Organomet. Chem.*, *122*, 261 (1976).
85. K. Mori, T. Mizoroki, A. Ozaki, *Bull. Chem. Soc., Jpn.*, *49*, 758 (1976).
86. T. Alderson, E. L. Jenner, R. V. Lindsey Jr., *J. Am. Chem. Soc.*, *87*, 5638 (1965).
87. R. Cramer, *J. Am. Chem. Soc.*, *87*, 4717 (1965).
88. H. Umezaki, Y. Fujiwara, K. Sawara, S. Teranishi, *Bull. Chem. Soc., Jpn.*, *46*, 2230 (1973).
89. G. Hata, *Chem. Ind. (London)*, 223 (1965).
90. H. Bestian, K. Clauss, *Angew. Chem., Int. Ed. Engl.*, *2*, 704 (1963).
91. J. J. Barber, C. Willis, G. M. Whitesides, *J. Org. Chem.*, *44*, 3603 (1979).
92. J. Staib, W. Guyer, O. Slotterbeck, U.S. Patent 2,864,864 (1958); *Chem. Abstr.* *53*, 9063 (1959).
93. J. A. Bertrand, C. L. Aldridge, S. Husebye, H. B. Jonassen, *J. Org. Chem.*, *29*, 790 (1964).
94. R. F. Heck, D. S. Breslow, *J. Am. Chem. Soc.*, *83*, 1097 (1961).

5.8.2.3.6. Metallacycles from Insertion of Monoolefins into Metal–Carbene Complexes.

The interaction of metallocarbene (or alkylidene) complexes with unsaturated compounds via the formation of a metallacyclobutane intermediate plays an important role in several stoichiometric and catalytic reactions, e.g., metathesis, cycloaddition and polymerization.

The first elementary steps are:

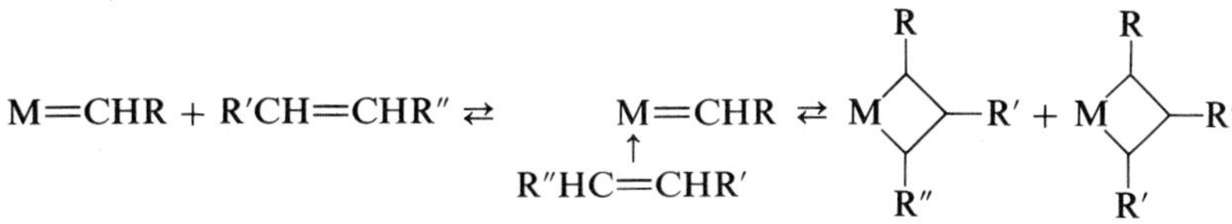

(a)

Then the metallacycles may rearrange:

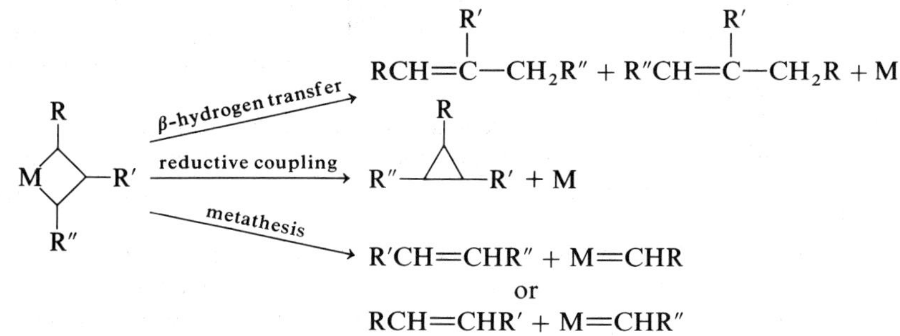

(b)

(c)

(d)

Direct evidence for this mechanism has been proved in several cases. Furthermore, the characterization of the organic and organometallic products of decomposition offers insight into the structure of the intermediate metallacycle.

(i) Titanium. α-Olefins react with the methylene-bridged compound, **I**, which may be considered as a Ti alkylidene complex stabilized by an electron-pair acceptor acid.

In a toluene or benzene solution (**I**) reacts at RT with ethylene or propylene to give the homologated olefin via a β-hydrogen transfer[1,2]:

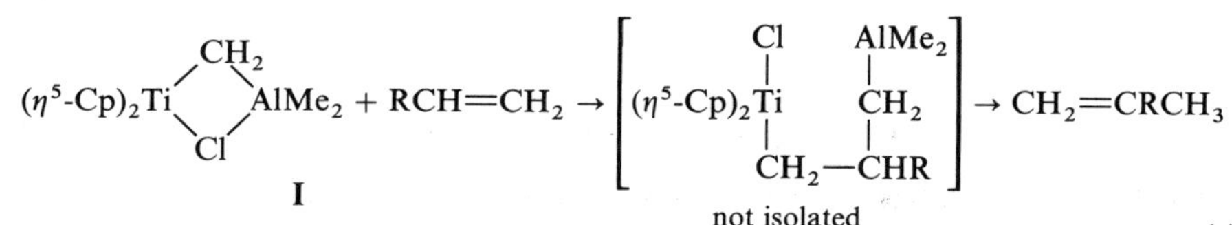

(e)

where if R = H, yield = 32% and if R = CH_3, yield = 59%. Small amounts of cyclopropane derivatives are also formed when THF or trimethylamine, Me_3N, is added to the reaction medium.

However, owing to the lack of β-hydrogen in the intermediate metallacycle, 1,1-disubstituted ethylenes (2-methylpropene, methylenecyclohexane) do not undergo β-hydrogen transfer; instead, their methylene group is catalytically exchanged by a degenerated olefin metathesis reaction[2,3]:

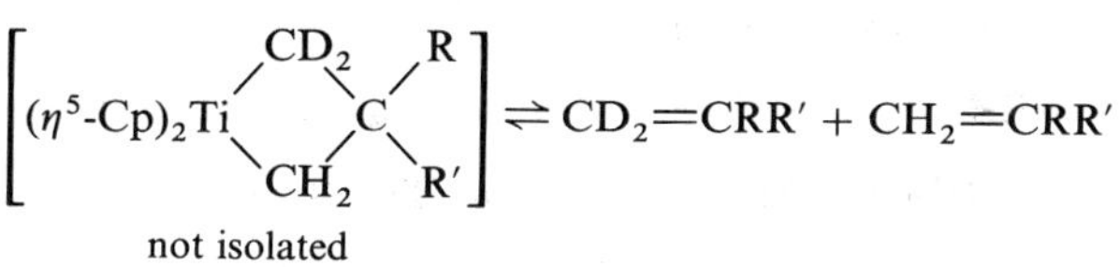

(f)

where R = R′ = CH_3.

In the presence of 1 equiv pyridine or 4-(N,N-dimethylamino)pyridine **I** reacts with terminal olefins to give fully characterized metallacyclobutane derivatives, **II**[4–6]:

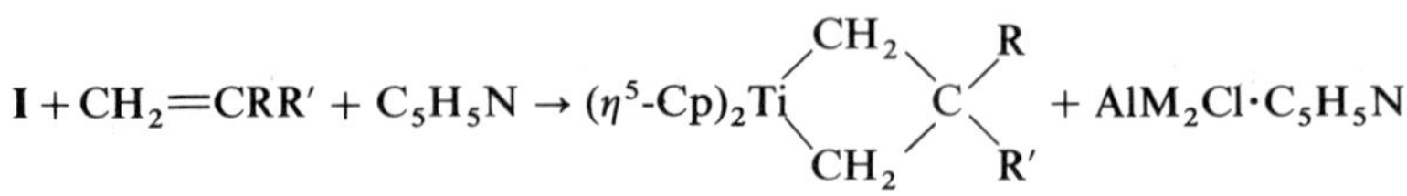
(g)

II

where R = t-Bu, R′ = H; R = i-P$_r$, R′ = H; R = Ph, R′ = H; R = H, R′ = H; R = CH_3, R′ = CH_3. In the same conditions **I** reacts with norbornene, giving the corresponding metallatricyclic compound; this latter catalyzes the living polymerization of norbornene and its copolymerization with bicyclic olefins[53,54].

Structural studies indicate that the metallacycle **II** is nearly planar and symmetrical. Thus the source of stereoselection in olefin metathesis, a question which is often disputed, must result from factors other than those arising from the conformational effects owing to the puckering of the intermediate.

The reaction of **II** with $AlMe_2Cl$ regenerates **I**, suggesting that the role of the electron-pair acceptor acid cocatalyst in metathesis reactions may be to facilitate carbene–metallacycle equilibrium. Labeled olefin, CDH=CRR′, reacts with **II** affording labeled **II**, then labeled **I** by subsequent reaction with $AlMe_2Cl$. Thus, the required metathesis reaction cycle is completed.

The mechanism of reduction coupling, Eq. (b), is discussed in ref. 55.

Compounds **I** and **II** react with acetylenes in THF to produce stable titanacyclobutene, **III**[2,4,7,8].

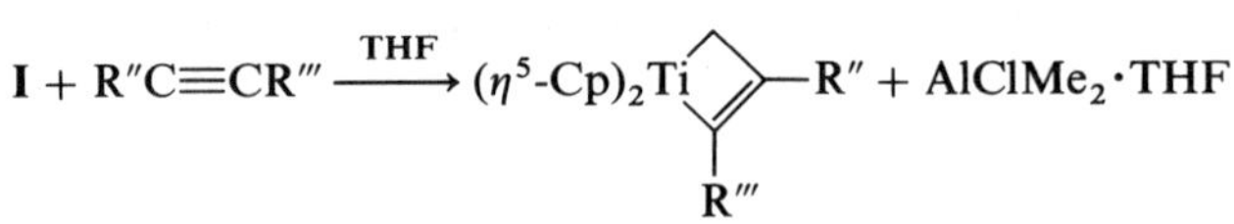
(h)

III

$$\mathbf{II} + R''C{\equiv}CR''' \rightarrow \mathbf{III} + CH_2{=}CRR' \tag{i}$$

where R″ = R‴ = Ph, **IIIa**; R″ = R‴ = $SiMe_3$, **IIIb**; R″ = Ph, R‴ = $SiMe_3$, **IIIc**.

The structure and reactivity of **III** depend on the substituents. Thus **IIIa** is quite inert to substitution reagents, and structural data are consistent with the metallocyclobutene formalism, **A**:

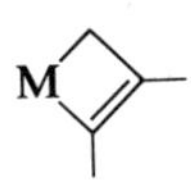

A

On the other hand the $C_2(SiMe_3)_2$ unit in **IIIb** is labile and may be substituted by C_2Ph_2 and $AlMe_2Cl$ (in the absence of THF):

$$\mathbf{IIIb} + PhC{\equiv}CPh \rightarrow \mathbf{IIIa} + Me_3SiC{\equiv}CSiMe_3 \tag{j}$$

$$\mathbf{IIIb} + AlMe_2Cl \rightarrow \mathbf{I} + Me_3SiC{\equiv}CSiMe_3 \tag{k}$$

Further, **IIIb** exchanges its methylene moiety with labeled 2-methylpropene ($^{13}CH_2{=}CMe_2$). These facts and structural data suggest for **IIIb** (and **IIIc**) the formalism **B**, in which the methylene and acetylene groups may be considered as nearly separated:

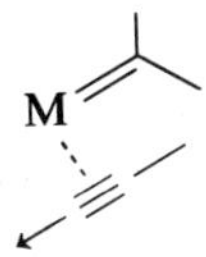

B

Thus the synthesis, detailed structural analysis and reactivity of complexes **II** and **III** make an important contribution to the mechanisms suggested previously for the metathesis of olefins and for the polymerization of acetylenes by metal–carbene complexes[9].

(ii) Niobium and Tantalum. Several Ta complexes containing both neopentylidene and ethylene ligands are known[10,11].

The alkylidene ligand can be formed from a nonisolated trisneopentyl–ethylene derivative by transfer of an α-hydrogen atom from one neopentyl group to a leaving neopentyl group:

$$Ta(C_2H_4)Cl_3(PMe_3)_2 + 1.5\ Mg(CH_2CMe_3)_2 \rightarrow (Me_3CCH_2)Ta(CHCMe_3)(C_2H_4)(PMe_3)_2 + CMe_4 \quad (l)$$

IV

Ethylene can also be formed from a nonisolated triethyl–neopentylidene derivative by transfer for a β-hydrogen atom from one ethyl group to a leaving ethyl group:

$$Ta(CHCMe_3)Cl_3(PMe_3)_2 + 1.5\ MgEt_2 \rightarrow EtTa(CHCMe_3)(C_2H_4)(PMe_3)_2 + C_2H_6 \quad (m)$$

V

Complex **V** also can be prepared by treating $Ta(CH_2CMe_3)_2Et_3$ with PMe_3:

$$Ta(CH_2CMe_3)_2Et_3 + 2\ PMe_3 \rightarrow \mathbf{V} + Ta(C_2H_4)_2(Et)(PMe_3)_2 \quad (n)$$

Both ethylene and neopentylidene ligands can be formed in the same step by transfer of a β-hydrogen atom from an ethyl group to a neopentylidyne ligand:

$$(\eta^5\text{-}Cp)Ta(CCMe_3)Cl(PMe_3)_2 + 0.5\ MgEt_2 \rightarrow (\eta^5\text{-}Cp)Ta(CHCMe_3)(C_2H_4)PMe_3 \quad (o)$$

VI

Complexes **IV** and **V** have a trigonal-bipyramidal geometry in which the neopentylidene and ethylene ligands occupy the equatorial sites:

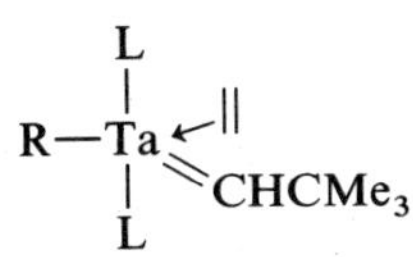

($R = Et, CH_2CMe_3$) and the relatively large angle between the two nearly perpendicular ligands prevents rearrangement to a tantalacyclobutane derivative.

Complex **VI** has a tetrahedral geometry, and its stability toward the formation of a tantalacyclobutane complex is ascribed to the large distance between the two carbon atoms involved.

Theoretical analysis defines the conditions of relative orientation that must be taken by carbene and olefin ligands to allow metallacyclobutane formation[12].

However, the neopentylidene ligand(s) of pentacoordinated Ta(V) complexes, **V**, and Ta $(CHCMe_3)_2(R)(PMe_3)_2$ (R = Cl, Me, Et, CH_2CMe_3) react slowly with free ligand, so a tantalacyclobutane may form owing to the small angle between the two ligands involved:

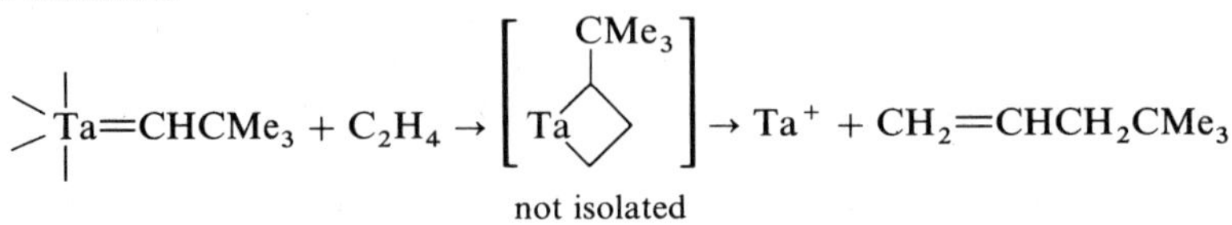
(p)

The Ta^+ complex, formulated as $Ta(C_2H_4)_2(Bu)(PMe_3)_2$ (except R = Cl, Me), catalytically dimerizes ethylene to 1-butene, confirming the tendency of these complexes to undergo β-hydrogen transfer.

The 6-coordinated complexes $Ta(CHR)Cl_3L_2$ (R = Ph, CMe_3; L = PMe_3, PPhMe, PPh_2Me_2), $Ta(CHCMe_3)(OCMe_3)Cl_2(PMe_3)_2$ and $Nb(CHCMe_3)Cl_3(PMe_3)_2$ react with ethylene, propylene or styrene to give the four isomeric products of β-hydrogen transfer for the two possible (α,β- and α,α′-disubstituted) metallacyclobutane intermediates without any metathesis products[13]:

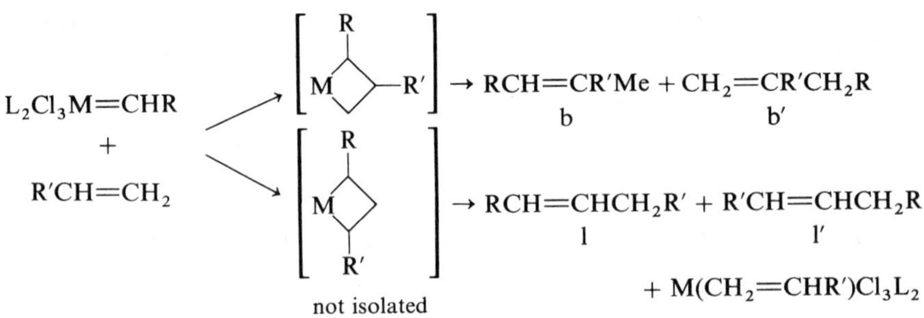
(q)

(r)

where l is linear and b is branched.

The distribution of isomeric olefins is given in Table 1. The product ratio is not

TABLE 1. OLEFINIC PRODUCTS FROM THE REACTION OF TANTALUM-ALKYLIDENE COMPLEXES WITH OLEFINS

Complexes	Olefins	Products[a] %			
		b	b′	l	l′
$Ta(CHPh)Cl_3PMe_3$	Propene	49	21	4	26
	Styrene	5	49	46	
$Ta(CHCMe_3)Cl_3(PMe_3)_2$	Propene	8	42	7	43
	Styrene	6	35	30	29

[a] See text; l = linear; b = branched.

affected by the nature of the complex, linear products being slightly favored for the bulky neopentylidene group.

Likewise, the organic products of the reaction between $Ta(CHCMe_3)Cl_3L_2$ (L = py, THF) and ethylene issue from β-hydrogen transfer (4,4-dimethylpentene) but also contain the metathesis product 3,3-dimethyl-1-butene. Unlike phosphine derivatives, these latter complexes react with cis-2-pentene to give exclusively the metathesis olefins and act as catalysts for the metathesis of cis-2-pentene. Thus, the presence of a hard ligand and the formation of an intermediate $\alpha,\alpha'\beta$-trisubstituted metallacyclobutane complex do not favor β-hydrogen transfer:

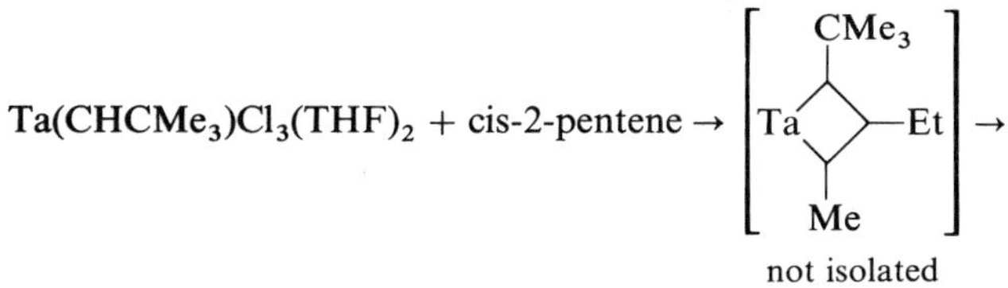

$$Me_3CCH{=}CHEt(+3\text{-hexenes} + 2\text{-butenes}) \qquad (s)$$

The corresponding benzylidene complex generated in situ from a benzyl complex reacts in the same way:

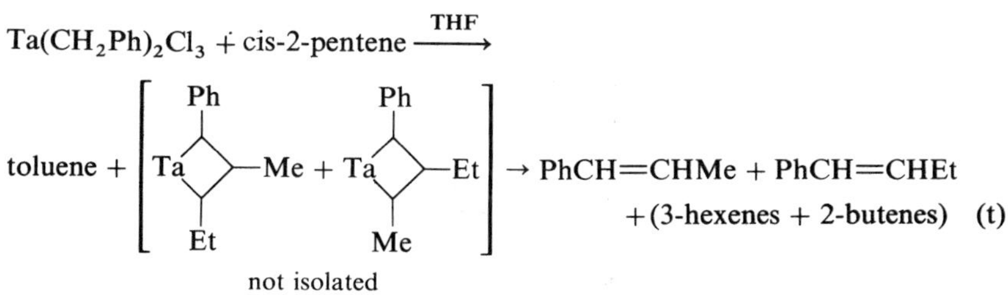

The alkoxo-neopentylidene complex, $Ta(CHCMe_3)(OCMe_3)_2Cl(PMe_3)$, reacts with ethylene to give t-butylethylene and propylene:

$$Ta(CHCMe_3)(OCMe_3)_2Cl(PMe_3) + 3\ C_2H_4 \rightarrow$$
$$Me_3CCH{=}CH_2 + CH_3CH{=}CH_2 + Ta(C_2H_4)(OCMe_3)_2(PMe_3)_2Cl \qquad (u)$$

The formation of olefinic products can be explained as follows:

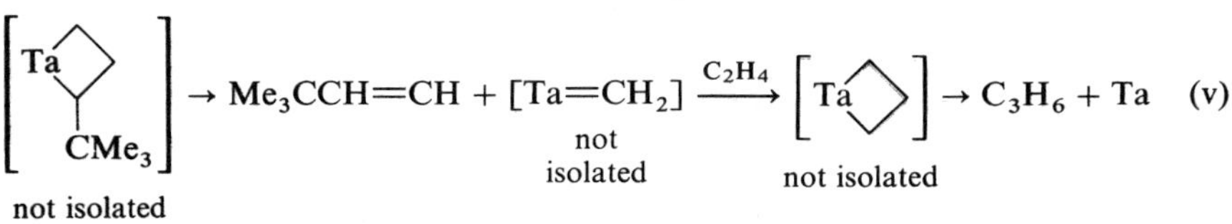

Thus the first tantalacyclobutane intermediate decomposes by a metathesis process, and the second one by a β-hydrogen transfer process.

The reaction of neopentylidene-alkoxo complexes, $M(CHCMe_3)(OCMe_3)_2$-$(PMe_3)Cl$ (M = Nb, Ta), with styrene, propenylbenzene and stilbene gives the organic products arising from the metathesis of the nonisolated metallacyclobutane complex (see Table 2), and the corresponding benzylidene complex $M(CHPh)(OCMe_3)_2(PMe_3)Cl$.

TABLE 2. ORGANIC PRODUCTS FROM THE REACTION OF $M(CHCMe_3)(OCMe_3)_2Cl(PMe_3)$ WITH STYRENE, PROPENYLBENZENE AND STILBENE

Complex	Olefin	Time (h)	Stilbene	$PhCH{=}CHCMe_3$	$Me_3CCH{=}CH_2$	$Me_3CCH{=}CHMe$
Ta	$PhCH{=}CH_2$	2	2	6	69	
Nb	$PhCH{=}CH_2$	2	1	3	69	
Ta	$PhCH{=}CHMe$	40	13	2		87
Nb	$PhCH{=}CHMe$	40	26	8		86
Nb	$PhCH{=}CHPh$	2	48	99		

These results suggest that the intermediate **C** (or its decomposition) is favored more than the intermediate **D**:

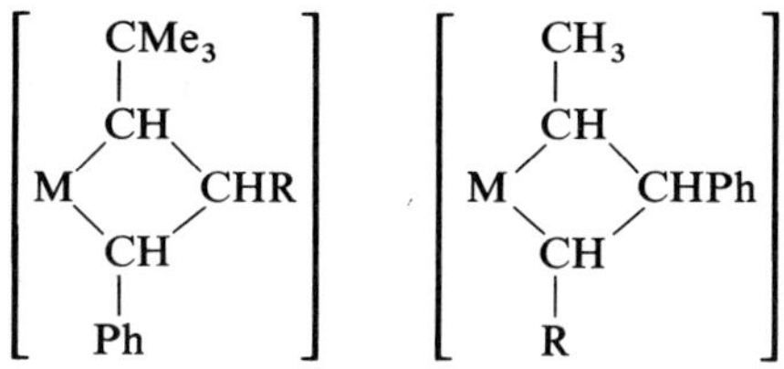

C (not isolated) **D** (not isolated)

where R = H, CH_3.

The same neopentylidene–alkoxo complexes react with cis-2-pentene to give the two initial metathesis products (4,4-dimethyl-2-pentene and 5,5-dimethyl-3-hexene) and catalyze the metathesis of cis-2-pentene to 2-butenes and 3-hexenes. Furthermore, propylene and ethylene appear in the reaction medium as the catalyst deactivates. These latter olefins are formed by rearrangement of the ethylidene and propylidene intermediates, providing the mechanism for the metathesis chain termination step:

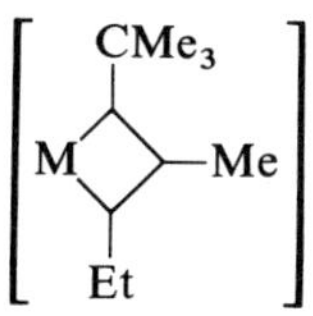 $\rightarrow Me_3CCH{=}CHMe + [M{=}CHEt] \rightarrow M + C_3H_6$ (w)

not isolated not isolated

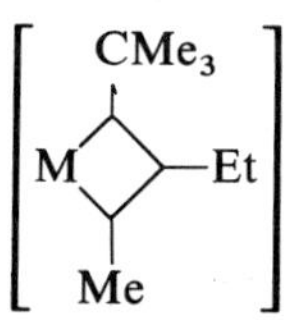 $\rightarrow Me_3CCH{=}CHEt + [M{=}CHMe] \rightarrow M + C_2H_4$ (x)

not isolated not isolated

In the same way $Ta(CHCMe_3)(OCMe_3)_2$ reacts with cis-2-pentene to give the initial and productive metathesis olefins but deactivates without formation of propylene and ethylene. In this case the chain termination steps must be a bimolecular decomposition of the alkylidene intermediates. Adding monodentate tertiary phosphine increases the lifetime of the complex, probably by a stabilizing effect of the alkylidene intermediates.

Several conclusions can be drawn[13]: (a) α,α'-substituted metallacycles form and metathesize faster than α,β-disubstituted metallacycles; (b) $\alpha,\alpha'\beta$-trisubstituted metalacycles undergo the β-hydrogen transfer process more slowly than di- or monosubstituted metallacycles and (c) replacing chloride ligands with alkoxide slows the rate of β-hydrogen transfer compared to metathesis.

The alkylidene–triphenoxo complexes $Ta(CHR)(OR')_3(THF)$ (R = CMe_3, $SiMe_3$, Ph; R' = 2,6-$C_6H_3Pr_2$-i) react reversibly with α-olefins ($CH_2{=}CHR''$), giving the fully characterized tantalacyclobutane (X-ray structures) $Ta[CHRCHR''CH_2](OR')_3$ ($R'' = R$ or H). In the same way, norbornene gives:

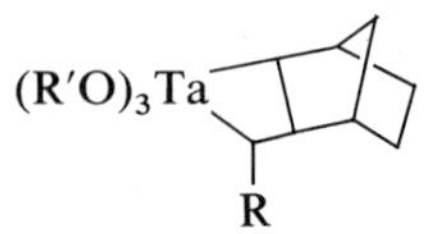

(X-ray structure)

which is in equilibrium with the carbene complex, $(R'O)_3(THF)Ta{=}CHC_5H_8CH{=}CHR$, that catalyzes the living polymerization of norbornene[56].

The cyclopentadienyl and neopentylidene derivatives $(\eta^5\text{-}Cp)M(CHCMe_3)Cl_2$ and $(\eta^5\text{-}C_5H_4Me)M(CHMe_3)Cl_2$ (M = Ta, Nb) also react with ethylene, prolylene and styrene to give the organic products of β-hydrogen transfer of the metallacyclobutane intermediate[14–16]. Propylene exclusively affords the α,β-disubstituted metallacycle, and styrene the α,α'-disubstituted one:

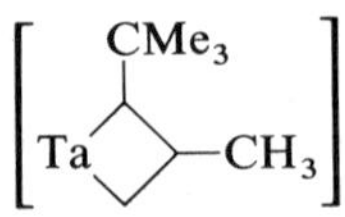

not isolated

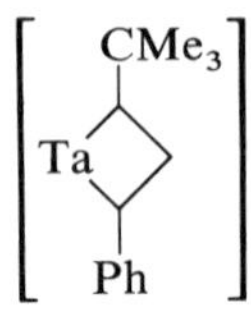

not isolated

while $(\eta^5\text{-}C_5Me_4Et)Ta(CHPh)Cl_2$ gives the same cleavage products, but the products derive from the α,β-disubstituted metallacycle:

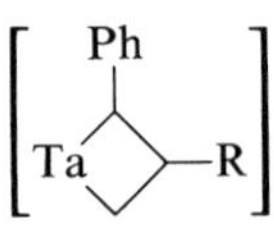

not isolated

where R = CH_3, Ph.

Acetylenes also react with alkylidene complexes to afford a new carbene derivative formed by inserting the triple bond into the alkylidene–metal bond[17]. A metallacyclobutene is the intermediate species:

$(\eta^5\text{-}Cp)Ta(CHCMe_3)Cl_2 + PhC{\equiv}CPh \rightarrow$

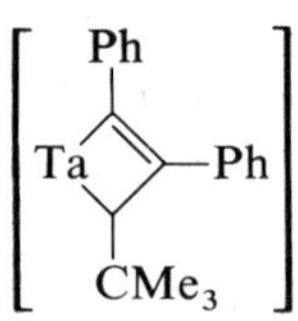

$\rightarrow (\eta^5\text{-}Cp)Ta(CPhPhC{=}CHCMe_3)Cl_2$ (y)

not isolated

A second insertion does not take place (a mechanism for the polymerization of acetylenes[9]) for steric and electronic reasons.

(iii) Chromium, Molybdenum, Tungsten. The group-VIA metal–carbene complexes are obtained from the carbonyl derivatives $M(CO)_6$, and contain five CO ligands. These compounds start the reaction of metal–carbene complexes with olefinic compounds, where it is found that, depending on the olefin structure and reaction conditions, either cyclopropanation or metathesis products are formed.

Thus heating pentacarbonylmethoxyphenylcarbene complexes under drastic conditions with α,β-unsaturated esters affords two isomeric cyclopropanes[18,19]:

$$(CO)_5M[C(OMe)Ph] + MeCH{=}CHCO_2Me$$

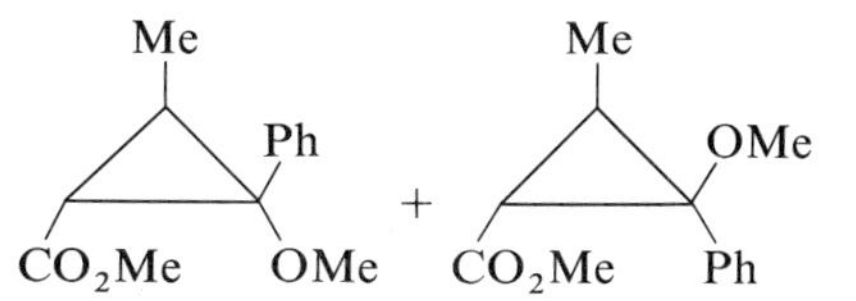

(z)

(M = Cr, Mo, W) the cis–trans ratio of which depends on the metal. Similar results are obtained starting from methylcinnamate, diethylfumarate and diethylmaleate, but simple alkenes such as cyclohexene or tetramethylethylene fail to give cyclopropane. In the same way the ferrocenylcarbene complex $(CO)_5Cr[CFc(OMe)]$ (Fc = ferrocenyl) reacts at 136°C with dimethylfumarate to give the expected cyclopropane derivative[23]. The optically active carbene complex $(CO)_4Cr[C(OMe)Ph](PMePhPr)$ obtained from (—)(R)-methylphenylpropylphosphine reacts with diethylfumarate at 60°C to afford, in low yield, the optically active cyclopropane[24]:

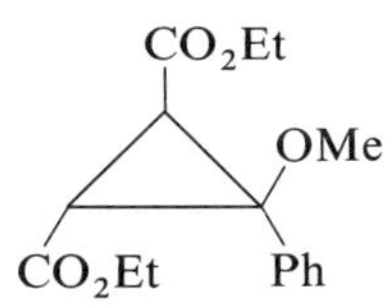

On the other hand, the Cr-carbene complex reacts with ethylvinyl ether or isobutylvinyl ether, under milder conditions, to give the metathesis product, α-methoxystyrene[20]:

$$(CO)_5Cr[C(OMe)Ph] + CH_2{=}CHOR \rightarrow CH_2{=}C(OMe)Ph \qquad \text{(aa)}$$

(R = Me, CH_2CHMe_2) but under CO pressure and higher temperature the products are cyclopropanes:

$$(CO)_5M[C(OMe)Ph] + CH_2{=}COR \xrightarrow{CO}$$

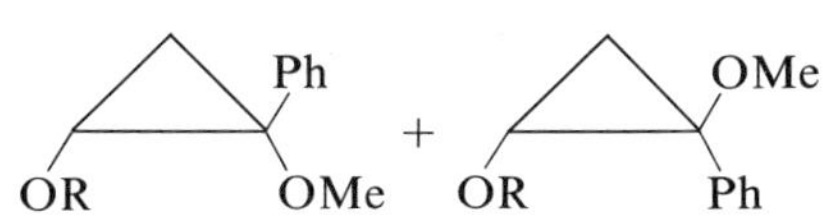

(ab)

where M = Cr, Mo, W; R = Me, CH_2CHMe_2. In this case also the cis–trans ratio of the cyclopropane depends on the metal.

The same Cr–carbene complex reacts with 1-vinyl-2-pyrrolidone to give the metathesis product α-methoxystyrene[21] and with 1-phenyl-1-piperidinoethylene to give both the cyclopropane derivative and a new carbene (the inorganic metathesis product)[22]:

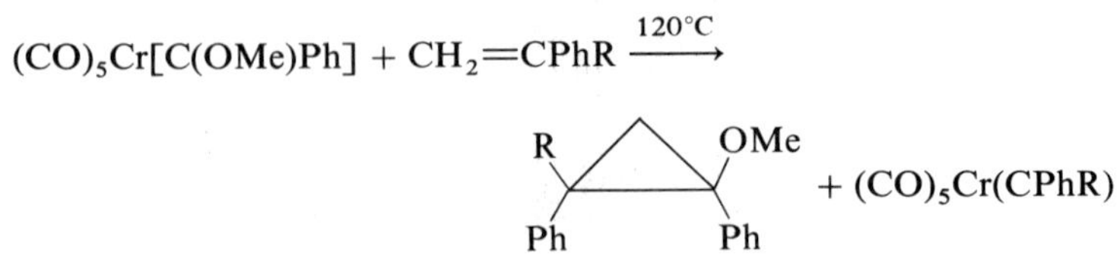

(ac)

where R = 

.

These facts indicate that no free organic carbenes are involved in the mechanism of the reaction, but owing to the difference in the experimental conditions the cyclopropanes and metathesis products do not involve the same intermediate.

A similar duality is noted with alkylidene complexes derived from the same type of carbenes, but not containing any electron-donating heteroatoms capable of stabilizing the electron-deficient carbene carbon atom. These latter complexes have been synthesized in the hope that they are more reactive toward nonactivated olefins.

Thus pentacarbonyldiphenylcarbenetungsten reacts, e.g., with 2-methylpropene or 2-butene, to give both the cyclopropane and the metathesis olefins, via a nonisolated metallacycle[25,26]:

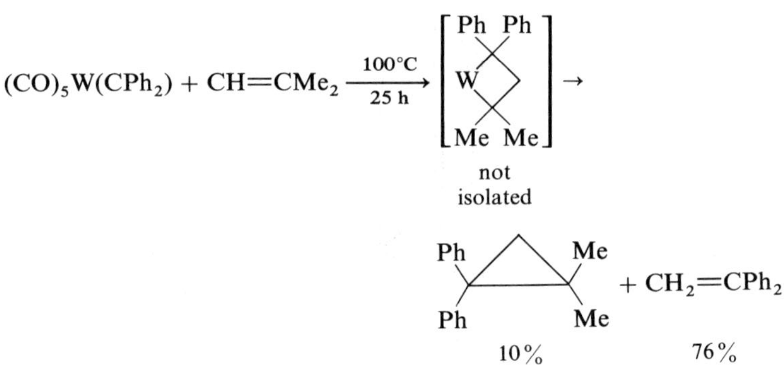

(ad)

Reacting an enol ether serves to isolate a new carbene derivative, thus confirming the tungstacyclobutane structure of the intermediate:

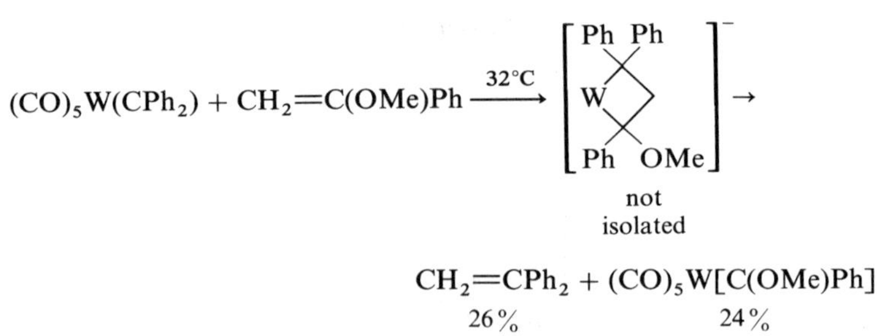

(ae)

TABLE 3. REACTION OF $(CO)_5WC(tol)_2$[a] WITH OLEFINS: ORGANIC PRODUCTS AND RELATIVE REACTIVITY OF OLEFINS

Olefins	Products (%)		Rel. reactivity
	Olefins	Cyclopropane	
1-Pentene	$CH_2{=}C(tol)_2$ 36	29	40
2-Pethylpropene	$CH_2{=}C(tol)_2$ 73	5	10
Styrene	$CH_2{=}C(tol)_2$ 61	8	
cis-2-Butene	$MeCH{=}C(Tol)_2$ 27	2	1
2-Methyl-2-hexene	—	—	0

[a] tol = p-C_6H_4Me.

The reaction of α-olefins with $(CO)_5W[C(p\text{-}C_6H_4Me)_2]$ supports the preferential formation of the α,α'-metallacycle[27] (see Table 3). Only trace amounts of the olefins coming from the α,β-substituted metallacycle are formed. Competition studies demonstrate that the relative reactivity of olefins toward metathesis is: 1-pentene > 2-methylpropene > cis-2-butene >> 2-methyl-2-hexene. This stability pattern parallels the stability of π-olefin–metal complexes.

The reaction of pentacarbonyldiphenylcarbenetungsten with xs of 2,3-dimethyl-1,3-butadiene at 70°C results in the formation of isoprene and 1,1-diphenylethylene[28]:

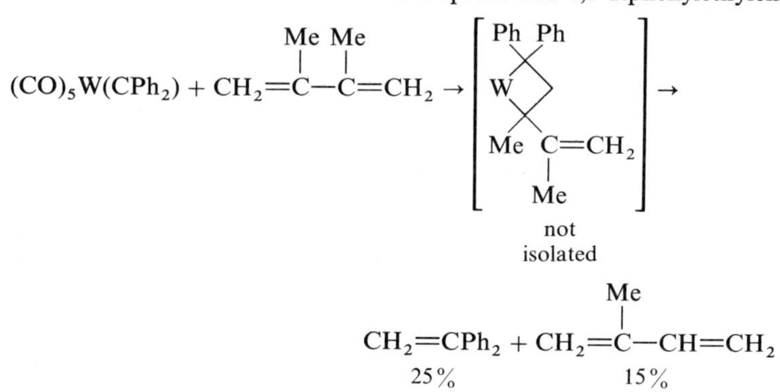

(af)

The dissymmetric pentacarbonylmethylphenylcarbenetungsten complex does not react with olefins[29] but undergoes decomposition via β-hydride elimination from the methyl group.

The reactivity of the corresponding phenylcarbene complex contrasts with the behavior of the above diphenylcarbene. Thus, the in situ reaction of olefins with the non-isolated pentacarbonylbenzylidenetungsten, obtained by protonation of this ammonium salt derivative, exclusively gives cyclopropanes without any organic metathesis products or inorganic alkylidene derivative[30,31] (see Table 4):

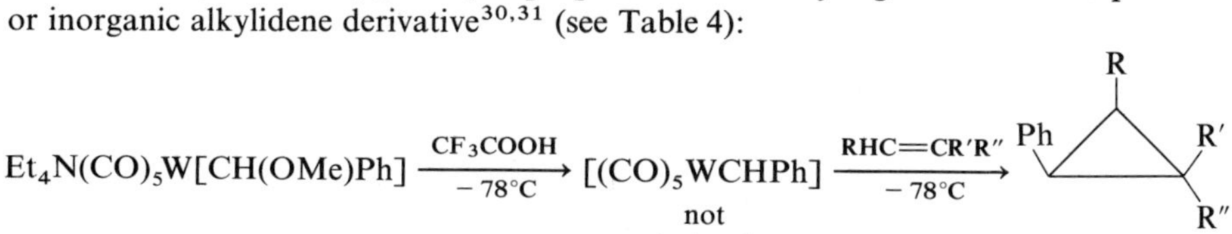

(ag)

TABLE 4. REACTION OF VARIOUS OLEFINS WITH THE NONISOLATED $(CO)_5WC(HPh)$, ORGANIC PRODUCTS, STEREOCHEMISTRY AND RELATIVE REACTIVITY OF OLEFINS

Olefin	Cyclopropane (%)	Cis–trans ratio	Rel. reactivity
2-Methylpropene	98		3500
Propene	80	2	11
Styrene	39	9	410
3,3-Dimethyl-1-butene	69	<0.01	1
cis-2-Butene	60	40	7.6
trans-2-Butene	82		3.5
Ethylene	0		0

where R,R′ = H, R″ = Me, Et, i-Pr, t-Bu, Ph, p-C_6H_4Me; R = H, R′R″ = Me, p-C_6H_4Me; R′ = H, R,R″ = Me.

Both cis- and trans-2-butene give cyclopropane with retention of the alkene geometry. Propene, styrene, 2-methyl-2-butene lead to the predominant formation of the thermodynamically less stable cis-cyclopropane (see Table 4). However, this cannot be explained, as suggested earlier, by a preferred stereochemistry of a hypothetical metalla-cyclobutane intermediate.

Competition reactions demonstrate that the reactivity of pentacarbonylbenzylidene-tungsten toward olefins is: 2-methylpropene >> styrene >> propene > n-butenes >> ethylene; i.e., the number of substituents attached to the more substituted end of the alkene controls the reaction rate.

This reactivity series and the stereochemistry of the cyclopropane adducts, contrasting with that observed in the case of pentacarbonyldiphenyl carbene, indicate that the reaction first involves an electrophilic attack of the alkylidene complex on the alkene without prior formation of an alkene–metal complex. A transition state (see Fig. 1) involves the formation of a bond from the alkylidene carbon atom to the less substituted end of the alkene and interaction of the positively polarized, more substituted end of the alkene with the carbon atom of the phenyl group.

The difference between the reaction condition for the two carbenes (40°C vs. −70°C) suggests that the reaction of diphenylcarbene proceeds via the dissociation of a molecule, and that the π-complexation of reactant and product is an essential step in the mechanism.

The neopentylidenetungsten complex, $W(O)(CHCMe_3)(PEt_3)_2Cl_2$, not containing carbon monoxide ligands, reacts in the presence of $AlCl_3$ with 1-butene to give

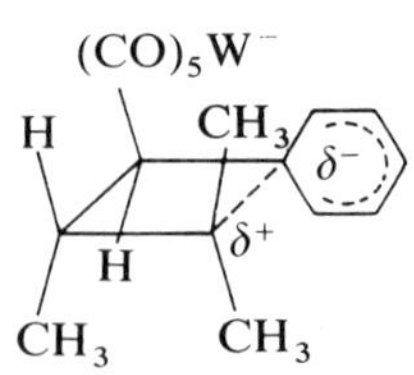

Figure 1

exclusively 3,3-dimethyl-1-butene (the metathesis product of the α,α'-disubstituted tungstacyclobutane) and a mixture of the corresponding propylidene and methylidene derivatives[32]:

$W(O)(CHCMe_3)(PEt_3)_2Cl_2 + CH_2{=}CHEt \rightarrow$

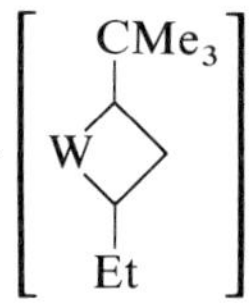

not isolated

$\rightarrow$

$$CH_2{=}CHCMe_3 + W(O)(CHEt)(PEt_3)_2Cl_2 \xrightarrow{CH_2=CHEt} W(O)(CH_2)(PEt_3)_2Cl_2 \qquad \text{(ah)}$$

Treating $W(CHR)(NR')(OR'')_2$ with ethylene, vinyltrimethyleilane or 3-hexene [so that R = t-Bu, $SiMe_2$, Et; R' = 2,6-C_6H_3Pr-i; R'' = $CMe(CF_3)_2$] gives the fully characterized metallacycle[57] $W(CHR'''CHR'''CH_2)(NR')(OR'')_2$ (R''' = R or H).

An NMR-characterized matallacycle is obtained by mixing $W(CHBu\text{-}t)(OCH_2Bu\text{-}t)_2Br_2$ with $GaBr_2$ and norbornene[58]:

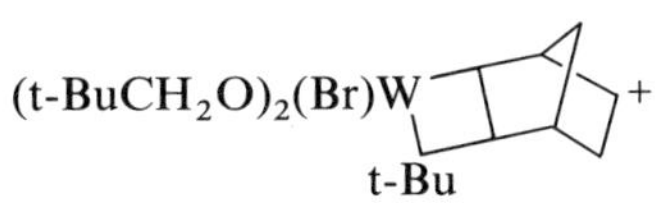

which polymerizes an excess of norbornene with an equilibrium between the metallacarbene and the metallacycle.

Just as for Ta and Nb, attempts are made to synthesize complexes containing both carbene and alkene ligands, especially where the carbene and alkene moieties are joined to form a bidentate chelating ligand. Thus the base-catalyzed exchange reaction of α,ω-unsaturated alcohols with pentacarbonylmethoxy(p-tolyl)carbenetungsten affords the expected complexes in which, however, the double bond does not interact with the metal[33]:

$$(CO)_5W[C(OMe)(p\text{-}C_6H_4Me)] + CH_2{=}CH(CH_2)_nOH \xrightarrow{[OH]^-}$$

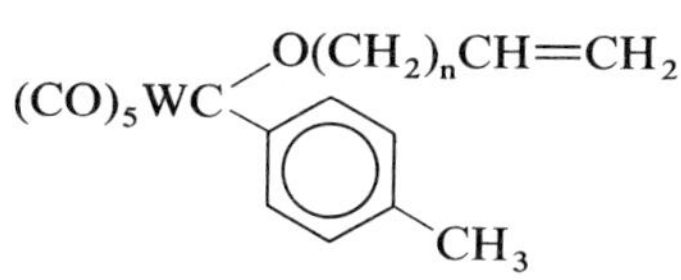

VII (ai)

where in **VIIa** n = 1; **VIIb** n = 2; **VIIc** n = 3.

Thermolysis of complexes **VIIa** and **VIIc**, respectively, gives p-tolylalkylketone and the expected cyclopropane derivative, but no π-alkene or metallacycle intermediate is characterized. However, heating **VIIb** affords a π-alkene–carbene complex which is well characterized by infrared and ^{1}H NMR spectroscopies:

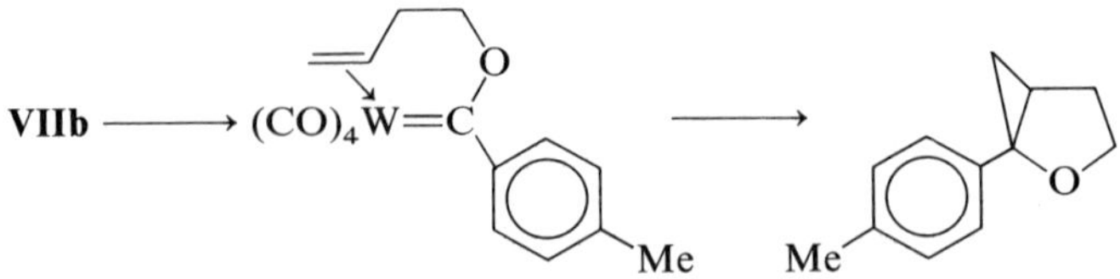

(aj)

The not very stable **VIIb** decomposes to the corresponding cyclopropane. In the same way pentacarbonyl (methoxy p-tolyl)carbenetungsten reacts with allylamine to give complex **VIII** (no interaction of the double bond with the metal), which decomposes to the stable π-alkene–carbene complex **IX**[33]:

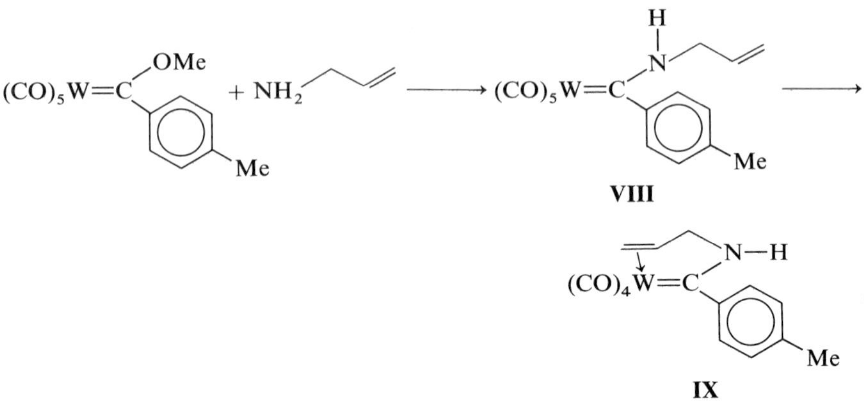

(ak)

In another approach, cyclopentanone enol ether reacts with a pentacarbonyldiphenylcarbene complex to give an unsaturated carbene, **X** (the metathesis product) and a cyclopentanone derivative formed by rearangement of the cyclopropane intermediate, both coming from the nonisolated tungstacyclobutane intermediate[34,35]:

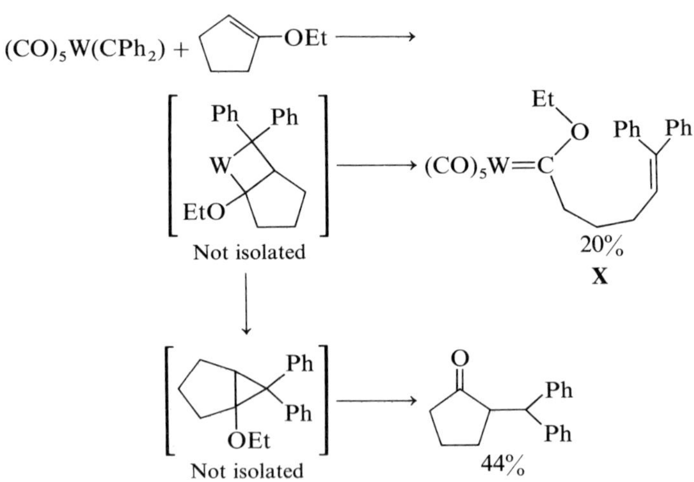

(al)

Under the same conditions the corresponding six- and eight-membered cyclic enol ethers do not react. Norcamphor enol ether also gives the inorganic metathesis product, **XI**[35]:

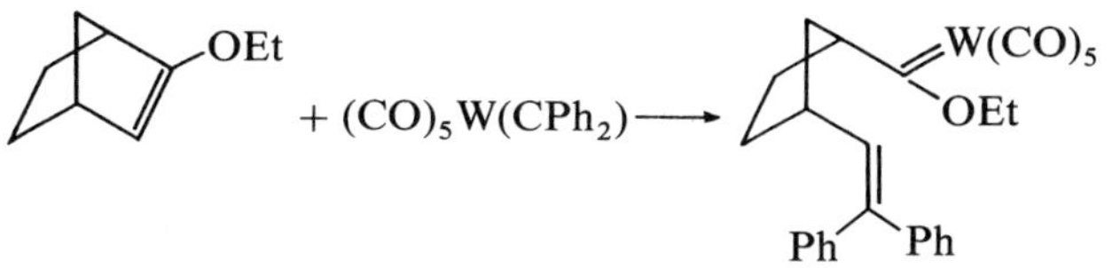

XI

(am)

There is no interaction between the double bond and the metal in either complex, ascribed to the bulkiness of the diphenylmethylene group. Similarly no interaction is observed in **XII** obtained by the reaction of cyclic eneamine 1-pyrrolidino-1-cyclopentene with a chromium–carbene complex[22]:

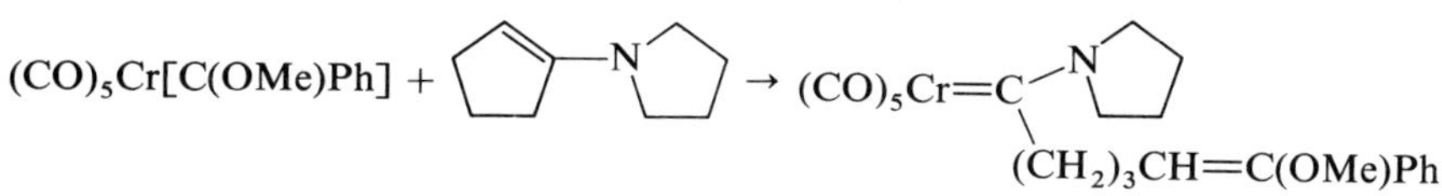

XII

(an)

In a third method, unsaturated carbenes with a free double bond, **XIII**, may be directly synthesized from $W(CO)_6$ and the corresponding unsaturated Li derivative. The π-complexation of the double bond is achieved in boiling benzene, affording thermostable **XIV**[36]:

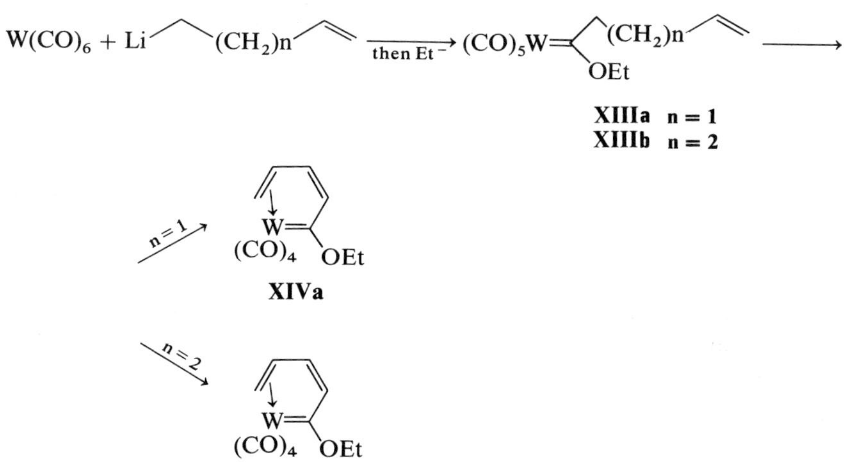

XIIIa n = 1
XIIIb n = 2

XIVa

XIVb

(ao)

Complexes **X**, **XI** and **XIVa** mimic the steps of the ring-opening polymerization of five-membered cyclic alkenes and account for the reversibility of the polymerization of cyclopentene. In the structure of compound **XIVb** the double bond shifts so that, once more, a five-membered ring is formed.

In a last route, an unsaturated μ-alkylidene complex is formed by treating pentacarbonylmethoxy methylcarbenetungsten with methyllithium at low temperature, then adding trifluoroacetic acid[37]:

$$(CO)_5W[C(OMe)Me] + MeLi \xrightarrow[H^+]{-40^\circ C}$$

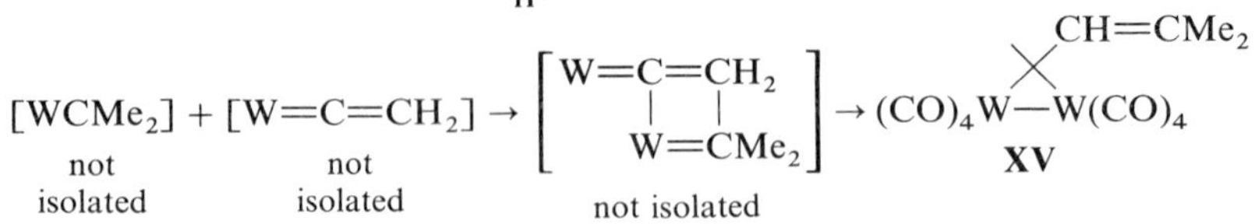

(ap)

This mechanism is consistent with the results of labeling experiments. In complex **XV** the double bond is uncomplexed. However, on standing in solution at RT, **15** gives **16** in which the double bond interacts with one W atom; **XVI** may be considered either an **XVI** alkylidene complex or a dimetallacyclopentene[38]:

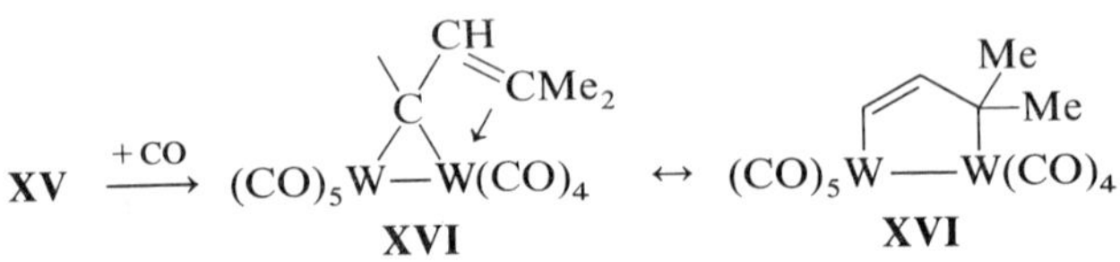

(aq)

Complex **XVI** may also be looked upon as the insertion product of acetylene into a metal–carbene bond. It is reminiscent of complexes **XXI** and **XXII**.

(iv) Reaction with Acetylenic Compounds. Acetylenic compounds bearing donating groups readily react with carbene complexes to give new carbenes resulting from the insertion into the metal–carbon double bond. A metallacyclobutene is the intermediate.

Thus, pentacarbonylmethoxymethyl-(or phenyl-)carbenechromium, -molybdenum or -tungsten react at RT with diethylaminoethyne or 1-diethylamino-1-propyne to give, in good yields, amino-stabilized carbenes[39,40]:

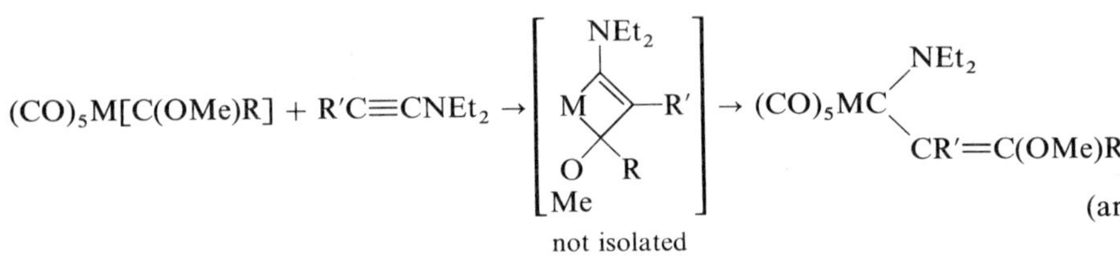

(ar)

where R = Me, R′ = H, M = Cr; R = Ph, R′ = CH_3, M = Cr, Mo, W.

Pentacarbonylmethoxyphenylcarbenechromium and -tungsten add bis(diethylamino) acetylene[41]:

$$(CO)_5M[C(OMe)Ph] + Et_2NC{\equiv}CNEt_2 \rightarrow$$

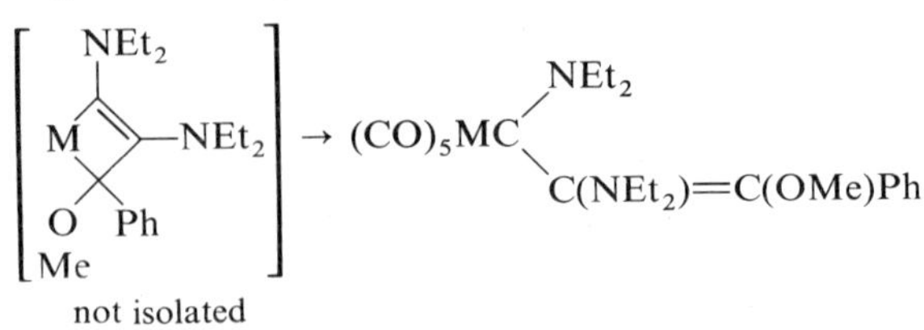

(as)

(M = Cr, W). Pentacarbonyldiphenylcarbenetungsten reacts similarly[40].

Ethoxyacetylene also undergoes an insertion reaction to the carbene-metal bond of W complexes[31]:

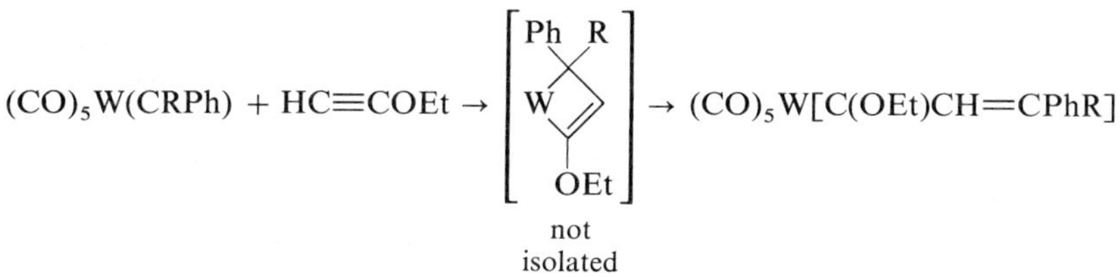
(at)

(R = H, Ph, OMe), and Mn carbene complexes add aminoacytelynes in the same way[42]:

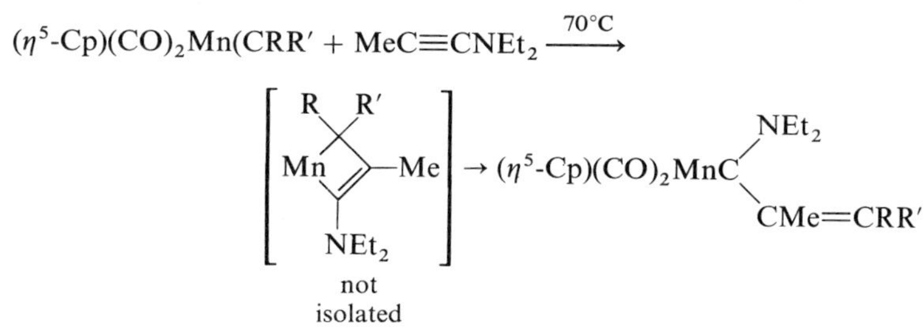
(au)

where R = Me, Ph; R′ = OMe, Ph.

In these reactions, the acetylenic carbon atom bearing the donor group bonds the metal, thus forming the more stable carbene derivative.

Metallacycles are isolated from the reaction of μ-alkylidene complexes with alkynes. Thus **XVI** reacts with 2-butyne at RT to give a mixture of **XVII** and **XVIII**[43,44]:

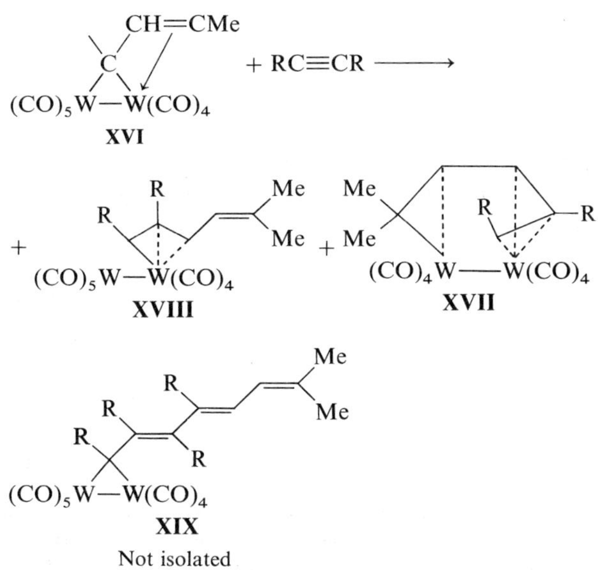
(av)

Both complexes result from the insertion of butyne into one of the tungsten-carbon bonds of the bridging alkylidene ligand. However, in **XVII** the two double bonds are coordinated to the metal atoms, whereas in **XVIII** only the double bond of the acetylene moiety is coordinated as a semibridging ligand between the two W atoms. Complex **XVIII** decomposes to give **XVII**. Furthermore, **XVIII** undergoes further insertions of 2-butyne, and **XVI** catalyzes a fast polymerization of terminal alkynes (e.g., 1-pentyne and 1-heptyne) to give a high mol wt polyalkyne that is soluble in hydrocarbon solvents. Clearly, the mechanism of polymerization involves such species as **XIX** (Fig. 7)[44].

(v) Group VIII Metals. Group VIII metal carbene complexes present diverse reactivity toward olefins.

Complexes that possess a methylene bridge between two transition-metal atoms react with olefins similarly to the Ti—CH_2—Al system. Thus treatment of octacarbonyl-μ-methylenediiron at 55°C with ethylene or propene produces homologated olefins. This reaction proceeds via the formation of a dimetallacyclopentane, then β-hydrogen transfer[45]:

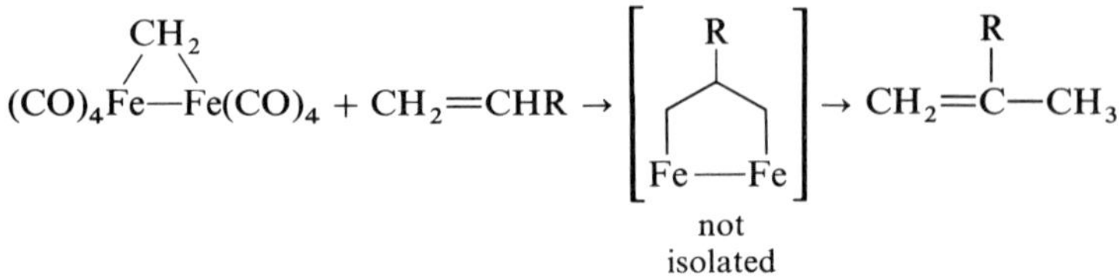

(aw)

where R = H, CH_3.

Similarly, μ-methylene bis(cyclopentadienylcarbonyl cobalt) reacts at 60°C with ethylene to give propene[46]:

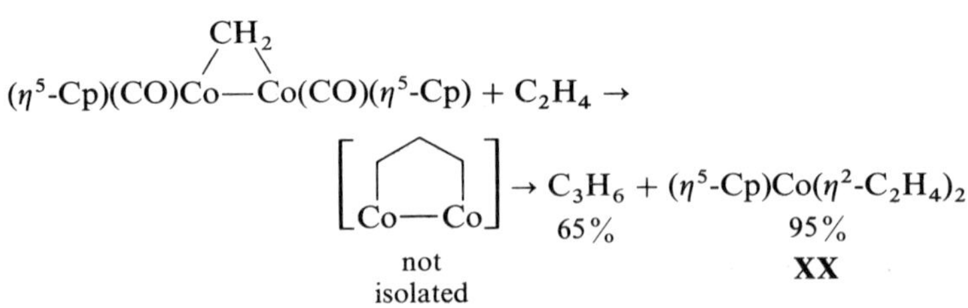

(ax)

The initial step involves CO dissociation, and **XX** has a catalytic effect on the reaction.

The methylation of the adduct of $[HFeCO_4]^-$ and dimethylacetylenedicarboxylate gives **XXI** directly without characterization of the transient carbene complex, the structure of which is reminiscent of **XVI**[47]:

(ay)

XXI

Bridging alkylidene complexes similarly react with alkynes to afford isolatable derivatives arising from the formal insertion into one metal–carbon bond. So, UV irradiation of $M_2(CO)_2(\mu\text{-}CO)(\mu\text{-}CHCH_3)(\eta^5\text{-}Cp)$, in the presence of acetylene results in the fomation of a new carbene and a three-carbon bridging system, **XXII**[48]:

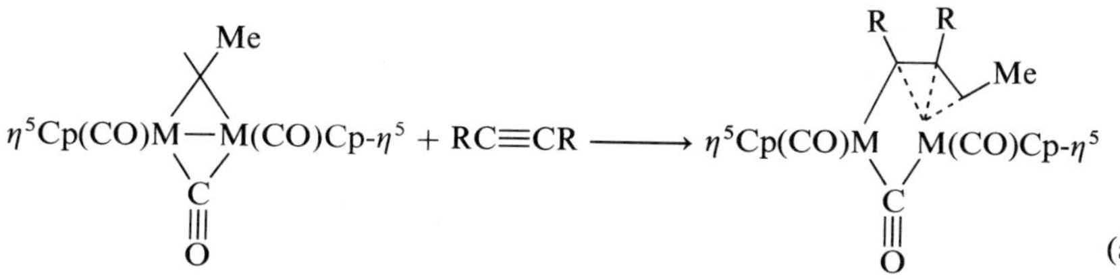

(az)

XXII

where if R = H, M = Fe, Ru; if R = CO_2Me, M = Fe. This complex is also similar to **XVI** obtained by an entirely different route and undergoes the same type of reactions.

In an attempt to prepare ferracyclobutane by the protonation of a carbene complex in the presence of ethylene, a stable compound containing both carbene and alkene ligands is obtained, which does not rearrange to a metallacycle[49]:

$$\eta^5\text{-Cp(CO)Fe[C(OMe)(OC}_6\text{H}_{11})] + \text{C}_2\text{H}_4 \rightarrow \eta^5\text{-Cp(CO)(C}_2\text{H}_4)\text{Fe[C(OMe)(OC}_6\text{H}_{11})] \quad \text{(ba)}$$

The benzylideneiron carbene complex **XXIII** is used to transfer the benzylidene ligand to unactivated alkenes to afford cyclopropane (Table 5)[50]:

$[\eta^5\text{-Cp(CO(CO)}_2\text{FeCHPh][PF}_6] + \text{RR'C=CR''R'''} \xrightarrow{-78^\circ\text{C}}$ 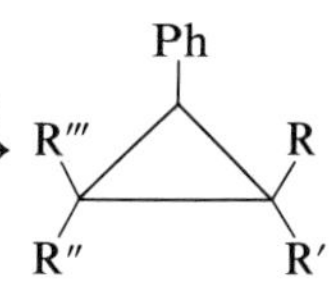 (bb)

XXIII

Phenylcyclopropanes are obtained in good yield for either ethylene or tetrasubstituted olefins. The reaction is highly stereoselective, giving thermodynamically less stable cis–isomers. The reactivity pattern is reminiscent of that of pentacarbonylbenzylidenetungsten[30,31], and the same type of intermediate is proposed (Fig. 1). Treatment of **XXIII** with 2-butyne results in the formation of a cyclopropenium ion:

XXIII + $\text{MeC}\equiv\text{CMe} \rightarrow$ 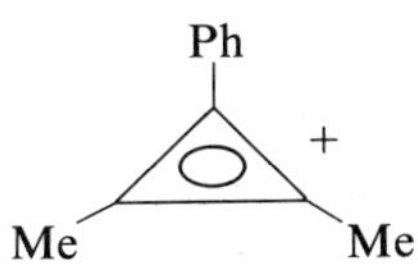 (bc)

TABLE 5. PRODUCTS OF THE REACTION OF ALKENES WITH η^5-$Cp(CO)_2Fe{=}CHPh$

Alkene	Cyclopropane (yield, %)	Cis/trans ratio
Ethylene	47	
Propene	90	7.8
1-Butene	75	6.5
Styrene	88	>100
2-Methylpropene	45	

Lastly, Rh–carbene complexes **XXIV** are intermediates in the metathesis of the electron-rich olefin **XXIV**. A transient metallacycle is proposed[51]:

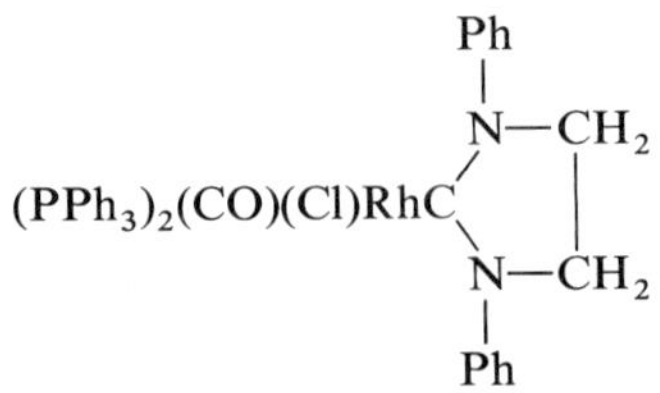

XXIV

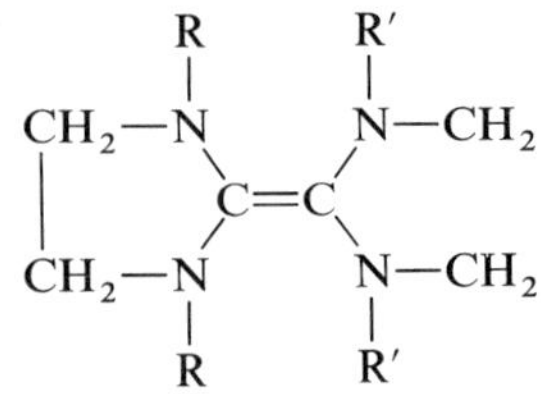

XXV

where R,R′ = Ph, p-C_6H_4Me.

(vi) Reaction of Alkylidyne (Carbyne) Complexes with Acetylenes. Unsaturated metallacycles are intermediates in the reaction between carbyne complexes and acetylenes. Thus various neopentylidynetungsten complexes such as $W(CCMe_3)Cl_3(OPEt_3)$, $(PMe_3)_3Cl_3W(CCMe_3)$ and $W(CCMe_3)(OCMe_3)_3$ react with acetylenes in toluene at 70°C to give the organic metathesis products and a new carbyne derivative:

$W(CCMe_3)(OCMe_3)_3 + PhC{\equiv}CPh \rightarrow$ 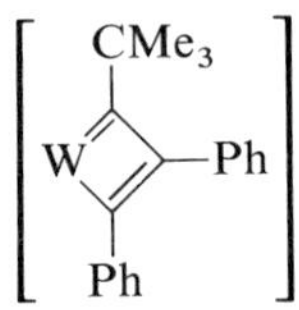 $\rightarrow$

not isolated

$PhC{\equiv}CCMe_3 + W(CPh)(OCMe_3)_3$ (bd)
90%

Furthermore, the same alkylidyne complexes readily catalyze the metathesis of acetylenic compounds (Table 6). This opens up a new area in the synthesis of metallacycles.

(vii) Outlook. The role of metallocarbenes in catalytic transformations of unsaturated compounds stimulates work to prove the existence and determine the conformation of the hypothetical intermediate metallacyclobutane complex. However, now that some metallacycles are isolated and their reactivity studied, the question becomes more

TABLE 6. METATHESIS OF ACETYLENES CATALYZED BY $W(CCMe_3)(OCMe_3)_3$ [52]

Acetylene (mmol)	Catalyst (mmol)	Equilibrium time
PhC≡CEt (1.9)	0.1	4 h
ChxC≡CEt (2.1)	0.076	18 min
PhC≡Ctol (1.9)	0.063	1 h
EtC≡CPr	0.034	10 min

complicated, since the form of the intermediates and their subsequent evolution largely depends on the structure of the complexes.

(Y. CHAUVIN)

1. F. N. Tebbe, G. W. Parshall, G. S. Reddy, *J. Am. Chem. Soc.*, *100*, 3611 (1977).
2. U. Klabunde, F. N. Tebbe, G. W. Parshall, R. L. Harlow, *J. Mol. Catal.*, *8*, 37 (1980).
3. F. N. Tebbe, G. W. Parshall, D. W. Ovenall, *J. Am. Chem. Soc.*, *101*, 5074 (1979).
4. T. R. Howard, J. B. Lee, R. H. Grubbs, *J. Am. Chem. Soc.*, *102*, 6876 (1980).
5. J. B. Lee, G. J. Gajda, W. P. Schaefer, T. R. Howard, T. Ikariya, D. A. Straus, R. H. Grubbs, *J. Am. Chem. Soc.*, *103*, 7358 (1981).
6. K. C. Ott, R. H. Grubbs, *J. Am. Chem. Soc.*, *103*, 5922 (1981).
7. F. N. Tebbe, R. L. Harlow, *J. Am. Chem. Soc.*, *102*, 6149 (1980).
8. R. J. McKinney, T. H.Tulip, D. L. Thorn, T. S. Coolbaugh, F. N. Tebbe, (to be published).
9. T. J. Katz, S. J. Lee, M. A. Shippey, *J. Mol. Catal.*, *8*, 219 (1980).
10. J. D. Fellmann, R. R. Schrock, G. A. Rupprecht, *J. Am. Chem. Soc.* *103*, 5752 (1981).
11. A. J. Schultz, R. K. Brown, J. M. Williams, R. R.Schrock, *J. Am. Chem. Soc.*, *103*, 169 (1981).
12. O. Eisenstein, R. Hoffmann, *J. Am. Chem. Soc.*, *103*, 5582 (1981).
13. S. M. Rocklage, J. D. Fellmann, G. A. Rupprecht, L. W. Messerle, R. R. Schrock, *J. Am. Chem. Soc.*, *103*, 1440 (1981).
14. J. S. McLain, C. D. Wood, R. R. Schrock, *J. Am. Chem. Soc.*, *99*, 3519 (1977).
15. J. S. McLain, C. D. Wood, R. R. Schrock, *J. Am. Chem. Soc.*, *101*, 4558 (1979).
16. L. W. Messerle, P. Jennische, R. R. Schrock, G. Stucky, *J. Am. Chem. Soc.*, *102*, 6744 (1980).
17. C. D. Wood, S. J. McLain, R. R. Schrock, *J. Am. Chem. Soc.*, *101*, 3210 (1979).
18. E. O. Fischer, K. H. Dötz, *Chem. Ber.*, *103*, 1273 (1970).
19. K. H. Dötz, E. O. Fischer, *Chem. Ber.*, *105*, 1356 (1972).
20. E. O. Fischer, K. H.Dötz, *Chem. Ber.*, *105*, 3966 (1972).
21. E. O. Fischer, B. Dorrer, *Chem. Ber.*, *107*, 1156 (1974).
22. K. H. Dötz, I. Pruskil, *Chem. Ber.*, *114*, 1980 (1981).
23. J. A. Connor, J. P. Lloyd, *J. Chem. Soc.*, *Perkin Trans. I*, 17 (1973).
24. M. D. Cooke, E. O. Fischer, *J. Organomet. Chem.*, *56*, 279 (1973).
25. C. P. Casey, T. J. Burkhardt, *J. Am. Chem. Soc.*, *96*, 7808 (1974).
26. C. P. Casey, *Chem. Tech.*, 378 (1979).
27. C. P. Casey, H. E. Tuinstra, M. C. Saeman, *J. Am. Chem. Soc.*, *98*, 608 (1976).
28. S. P. Kolesnikov, A. I. Ioffe, N. I. Okhrimenko, O. M. Nefedov, *Izv. Akad. Nauk SSSR, Ser. Khim.* 1206 (1977).
29. C. P. Casey, L. D. Albin, T. J. Burkhardt, *J. Am. Chem. Soc.*, *99*, 2533 (1977).
30. C. P. Casey, S. W. Polichnowski, *J. Am. Chem. Soc.*, *99*, 6097 (1977).
31. C. P. Casey, S. W. Polichnowski, A. J. Shusterman, C. R. Jones, *J. Am. Chem. Soc.*, *101*, 7282 (1979).
32. J. H. Wengrovius, R. R. Schrock, M. R. Churchill, J. R. Missert, W. J. Youngs, *J. Am. Chem. Soc.*, *102*, 4515 (1980).
33. C. P. Casey, A. J. Shusterman, *J. Mol. Catal.*, *8*, 1 (1980).
34. J. Levisalles, H. Rudler, D. Villemin, *J. Organomet. Chem.*, *146*, 259 (1978).
35. H. Rudler, *J. Mol. Catal.*, *8*, 53 (1980).
36. C. A. Toledano, J. Levisalles, M. Rudler, H. Rudler, *J. Organomet. Chem.*, to be published.
37. J. Levisalles, H. Rudler, F. Dahan, Y. Jeannin, *J. Organomet. Chem.*, *187*, 233 (1980).
38. J. Levisalles, H. Rudler, F. Dahan, Y. Jeannin, *J. Organomet. Chem.*, *188*, 193 (1980).
39. K. H. Dötz, C. G. Kreiter, *J. Organomet. Chem.*, *99*, 309 (1975).
40. K. H. Dötz, *Chem. Ber.*, *110*, 78 (1977).
41. K. H. Dötz, C. G. Kreiter, *Chem Ber.*, *109*, 2026 (1976).
42. K. H. Dötz, I. Pruskil, *J. Organomet. Chem.*, *132*, 115 (1977).
43. J. Levisalles, F. Rose-Munch, H. Rudler, J. C. Daran, Y. Dromzée, Y. Jeannin, *J. Chem. Soc., Chem. Commun.*, *152*, (1981).
44. J. Levisalles, F. Rose-Munch, H. Rudler, J. C. Daran, Y. Dromzée, Y. Jeannin, D. Ades, M. Fontanille, *J. Chem. Soc., Chem. Commun.*, *1055*, (1981).
45. C. E. Sumner, P. E. Riley, R. E. Davis, R. Petit, *J. Am. Chem. Soc.*, *102*, 1752 (1980).
46. K. H. Theopold, R. G. Bergman, *J. Am. Chem. Soc.*, *103*, 2489 (1981).

47. K. Nakatsu, T. Mitsudo, H. Nakanishi, T. Watanabe, Y. Takegami, *Chem. Lett.*, 1447 (1977).
48. A. F. Dyke, S. A. R. Knox P. J. Naish, G. E. Taylor, *J. Chem. Soc., Chem. Commun.*, 803 (1980).
49. W. Priester, M. Rosenblum, *J. Chem. Soc., Chem. Commun.*, 26 (1978).
50. M. Brookhart, M. B. Humphrey, H. J. Kratzer, G. O. Nelson, *J. Am. Chem. Soc., 102*, 7802 (1980).
51. D. J. Cardin, M. J. Doyle, M. F. Lappert, *J. Chem. Soc., Chem. Commun.*, 927 (1972).
52. J. H. Wengrovius, J. Sancho, R. R. Schrock, *J. Am. Chem. Soc., 103*, 3932 (1981).
53. L. R. Gilliom, R. H. Grubbs, *J. Am. Chem. Soc., 108*, 733 (1986).
54. L. F. Cinzano, R. H. Grubbs, *Macromolecules, 21*, 1961 (1988).
55. S. C. H. Ho, D. S. Stauss, R. H. Grubbs, *J. Am. Chem. Soc., 106*, 1533 (1984).
56. K. C. Wallace, A. H. Liu, J. C. Dewan, R. R. Schrock, *J. Am. Chem. Soc., 110*, 4964 (1988).
57. R. R. Schrock, R. T. DePue, J. Feldman, C. J. Schaverien, J. C. Dewan, A. H. Liu, *J. Am. Chem. Soc., 110*, 1423 (1988).
58. J. Kress, J. A. Osborn, R. M. E. Greene, K. J. Ivin, J. J. Rooney, *J. Am. Chem. Soc., 109*, 899 (1987).

5.8.2.3.7. Metallacycles by Cyclodimerization of Monoolefins.

Metallacyclopentane syntheses proceed from alkenes, by oxidative additions of fluoroolefins, from olefins activated by polar groups such as acrylic esters, from strained alkenes, and from unactivated alkenes:

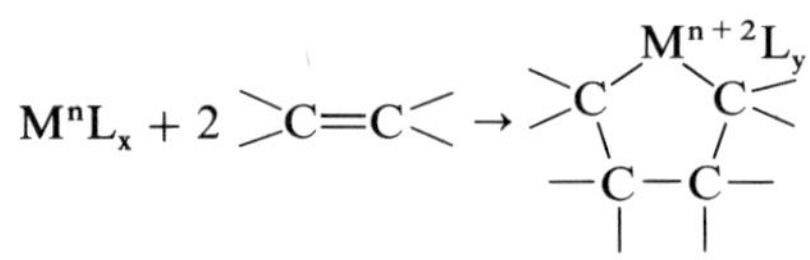

(a)

Tetrafluoroethylene undergoes oxidative cyclization as in Eq. (a) with η^5-$CpCo(CO)_2$, $Fe(CO)_5$, $Ni(CO)_2$(diphos), and NiL_4 (L = PEt_3, PPh_2Me, $P(OMe)_3$, C ≡ NBu-t) to form η^5-$CpCo(CO)(C_4F_8)$, $Fe(CO)_4(C_4F_8)$, Ni(diphos)(C_4F_8) and $NiL_2(C_4F_8)$ at > 100°C and ca. 10 atm of C_2F_4. Furthermore, (trans,trans,trans-1,5,9-cyclododecatriene)Ni and C_2F_4 at −78°C form Ni(CDT)(C_2F_4), which yields $Ni(PPh_3)_2(C_2F_4)$ upon treatment with PPh_3. Reaction with C_2F_4 then leads to $Ni(PPh_3)_2$ (C_4F_8), supporting a stepwise mechanism for formation of these metallacyclopentanes. Treatment of $Ni(AsPhMe_2)_4$ with $CHFCF_2$ yields only $L_2\overline{NiCF_2CHFCHFCF_2}$[1].

For starting complexes from which thermal dissociation of ligands to generate coordinative unsaturation is too high in energy, the reaction can be significantly facilitated by irradiation, typically by a Hg arc lamp through borate glass, CB. Under photolytic conditions $Fe[P(OEt)_3]_2(CO)_3$ undergoes reaction with $CHFCF_2$ forming $L_2\overline{FeCHFCF_2CF_2CHF}$ in 40% yield[2], notably with the opposite regiochemistry to that observed for Ni above. Similarly, irradiation of $Fe(CO)_4$ (trans-cinnamaldehyde) in xs C_2F_4 for 2 d leads to a 31% yield of $(CO)_4\overline{Fe[cis\text{-}CF_2CF_2CH(CHO)CHPh]}$ in a 4:5 molar ratio[3]. Reaction of $Fe(CO)_4(CH_2{=}CHCO_2Me)$ with xs $CH_2{=}CHCO_2Me$ under irradiation at −30°C in hexane yields $Fe(CO)_3(CH_2{=}CHCO_2Me)_2$ which is stable for long periods at −78°C, but photolysis under otherwise identical conditions except at 20°C gives the metallacycle $(CO)_4\overline{Fe[CH(CO_2Me)CH_2CH(CO_2Me)]}$ as a mixture of diastereomers[4]. The structure of this material is established by reaction with CO and with H_2, as well as by X-ray crystallography.

Strained alkenes can also undergo cyclization with low-valent metal complexes:

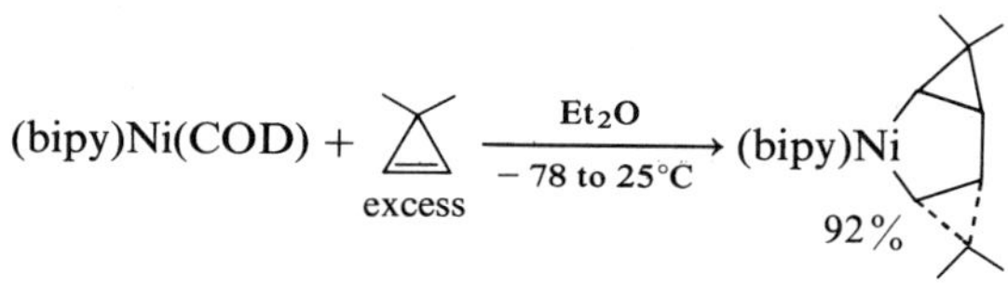

(b)

Product structure was shown by X-ray crystallography[5]. Treatment of either $[(1,5\text{-COD})IrCl]_2$ or $[(\text{cyclooctene})_2IrCl]_2$ with excess norbornadiene (NBD) in acetone at 25°C yields an.oligomeric material that, in turn, forms monomeric products of formulas $Ir(NBD)_3(acac)$ and $Ir(NBD)_3(PMe_3)Cl$ when treated with acetylacetone–Na_2CO_3 or PMe_3, respectively. The structure of the former, established by X-ray crystallography, in addition to η^2-acac and η^4-NBD, contains an exo-trans-exo metallacycle, **I**[6].

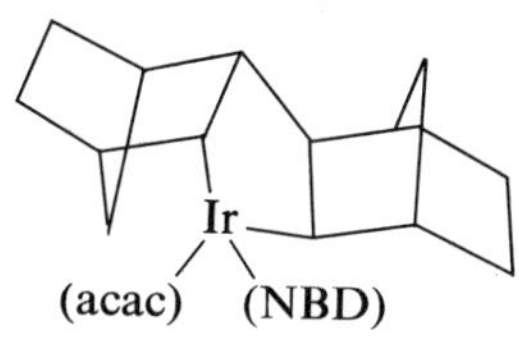

I

Chelation also aids in metallacycle formation, e.g.[7]:

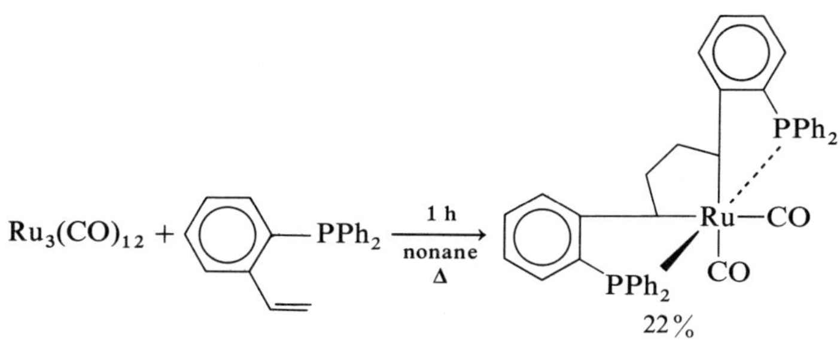

(c)

The structure of this product was also established by X-ray crystallography.

Low-valent Ti in $[(\eta^5Cp)_2Ti]_2N_2$ is capable of reductively cyclizing alkenes, albeit in low yield[8]; e.g., in reaction with ethylene at −30°C in toluene it forms $(\eta^5\text{-Cp})_2\overline{TiCH_2CH_2CH_2CH_2}$; for, although it cannot be directly characterized, reaction with CO, Br_2, or heat leads to the characteristic products, cyclopentanone, dibromobutane and butene, respectively, but never in more than ca. 10–15% yield based on Ti. Norbornene and benzonorbornadiene similarly give titanacyclopentanes in reaction with $[(\eta^5\text{-Cp})_2Ti]_2N_2$. The reaction:

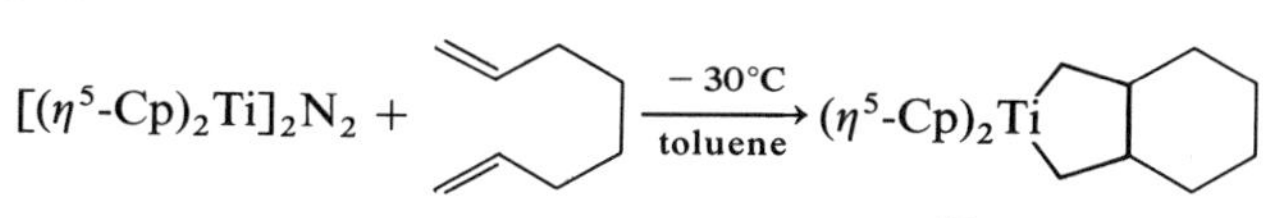

(d)

II ~2%

proceeds in low yield.

The best worked-out system for reductive dimerization is that of some η^5-$CpTaCl_2$ derivatives wherein the reaction is reversible and also forms the basis for a catalytic alkene dimerization[9]. The metallacycle can be prepared in several ways. Treatment of Ta carbene complexes bearing a C_5H_5 ligand such as **III** with either C_2H_4 or C_2H_3Me affords nearly quantitative yields of metallacycle **IV**:

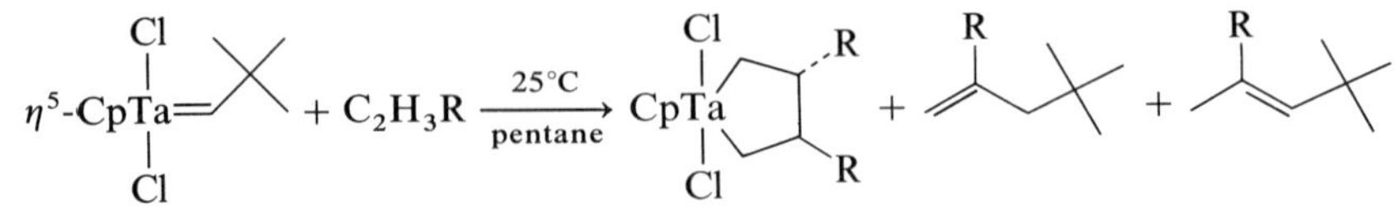

(e)

III **IV**

where R = H, Me.

Of the two possible isomers where R = Me, only the transmetallacycle is detectable by NMR. Other alkenes fail to give metallacycles by this method. The reaction also fails with the Nb analog of **III**. The metallacycles, **IV**, are unstable at 25°C and are sensitive to air and water. If 1,7-octadiene is used in place of ethylene in Eq. (e), then metallacycle **V**, stable below 100°C, is obtained[10]. The fused carbocycle thus aids stability.

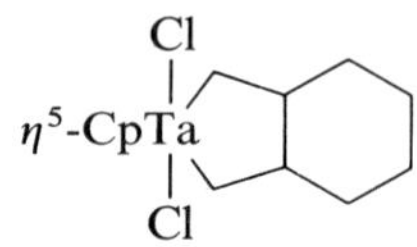

V cis + trans

More extensive and useful chemistry arises with use of η-C_5Me_5 in place of η^5-Cp; e.g., **VI** undergoes reversible cyclization with excess ethylene:

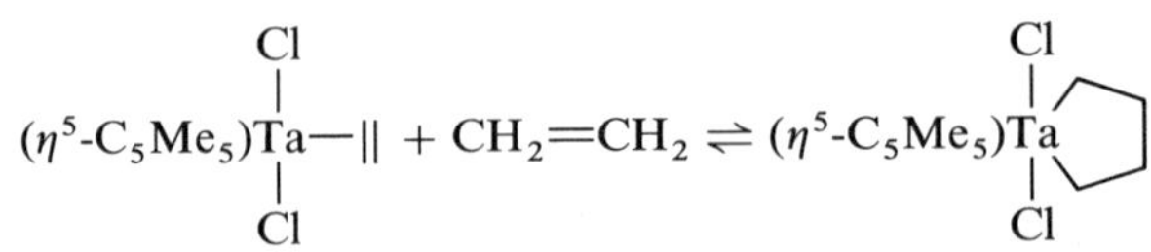

(f)

VI **VII**

Although the equilibrium lies to the right, **VI** can be regenerated under vacuum. The propylene analog of **VI** similarly forms **VIII** with excess propylene. In the case of neopentylethylene, NMR reveals the presence of both the olefin complex and the metallacycle in detectable amounts in the equilibrium. An additional indication that metallacycle formation is reversible arises by submitting **VI** to reaction with propylene:

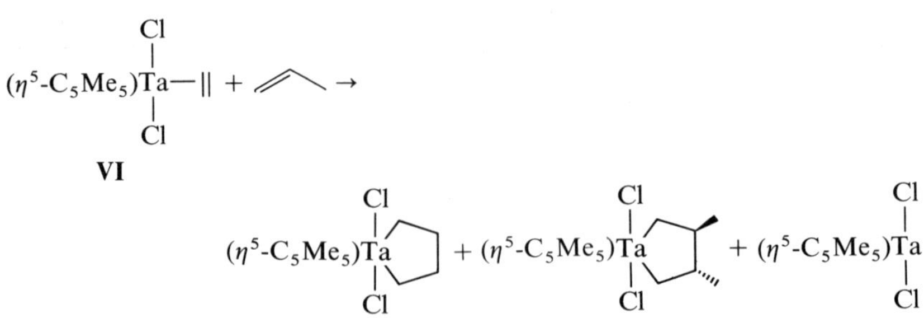

(g)

VI

VII **VIII** **IX**

This reaction forms the basis of a catalytic dimerization of these alkenes which has an unusual mechanism[11]. At $-35°C$ **IX** is formed quantitatively. Upon warming the solution of **IX** to 10°C for 2 h, a 1:2:1 mixture of **VII**:**IX**:**VII** is formed.

Tantalacyclopentanes decompose to the corresponding simple olefin complexes in solution at 25°C in 1 d in an open system. The other product is a pair of isomeric dimers of the alkene:

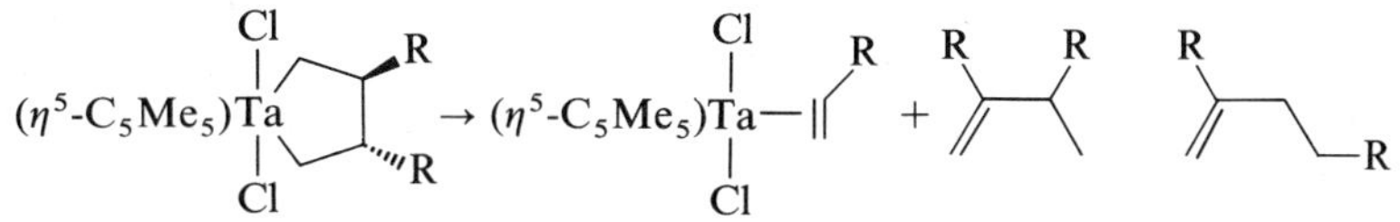
(h)

Reversible alkene cyclization also occurs with Ni and Ti[12]. Treatment of $(PPh_3)_2NiCH_2CH_2CH_2CH_2$, **X** or $(\eta^5\text{-Cp})_2TiCH_2CH_2CH_2CH_2$ (each generated from 1,4-dilithiobutane and the metal dihalide) with 1,7-octadiene leads to the formation of **XI** and **II**, respectively, with concomitant ethylene liberation.

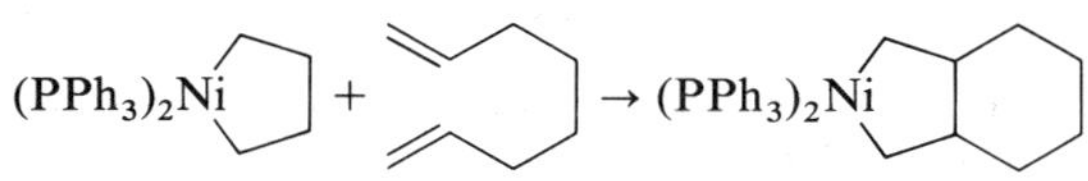
(i)

X **XI**

Neither **II** nor **XI** is isolated, but each is characterized by protonolysis and by reaction with CO. If cleavage products are an indication of the preferred stereochemistry of the metallacycles, then **II** is ca. 70% trans and **XI** is entirely trans. Also, **X** itself can be prepared by a direct reaction of ethylene[13]:

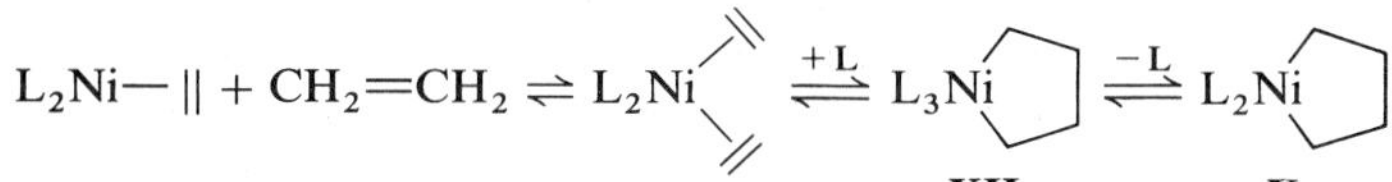
(j)

XII **X**

This last reaction necessarily proceedes through a pentacoordinated intermediate, **XII**[14]. The reaction sequence of Eq. (j), coupled with the thermal decomposition of **X** to cyclobutane, provides a path for the unique catalytic dimerization of ethylene to cyclobutene[13].

Analogous to the $[(\eta^5\text{-Cp})_2Ti]_2N_2$ reaction mentioned above, $[(\eta^5\text{-C}_5Me_5)_2ZrN_2]_2N_2$ absorbs C_2H_4 at 25°C in toluene to form $(\eta^5\text{-C}_5Me_5)_2ZrCH_2CH_2CH_2CH_2$, **XIII**, except that this reaction proceeds nearly quantitatively[15], and the product is easier to work with than its Ti analog. Reductive elimination of isobutane from $(\eta^5\text{-C}_5Me_5)_2Zr(H)CH_2Me_2$ in the presence of C_2H_4 also leads to formation of **XIII**[16]. A more reactive form of Zr(II), $(\eta^5\text{-Cp})_2Zr(PMePh_2)_2$, can also be used as a cyclization substrate, as demonstrated by its reaction with 1,7-octadiene to produce **XIV** in ca. 90% yield. Protonolysis affords 1,2-dimethylcyclohexane which is ca. 1:1, cis:trans, but it is unknown whether this reflects the stereochemistry of **XIV**.

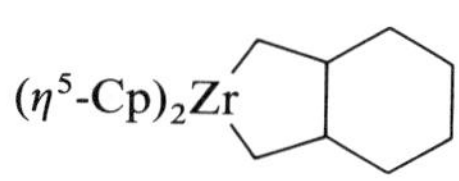

XIV

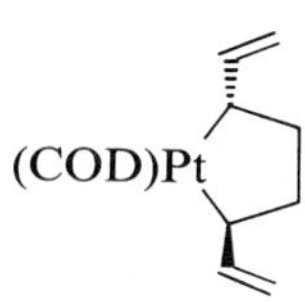

XV

1,3-Dienes and alkenes can function formally as monoenes in metallacycle formation, e.g., $(COD_2Pt$ in reaction with butadiene in hexane at $-30°C$ affords **XV** for which the transconfiguration is established by X-ray diffraction. Treatment of **XV** with t-BuNC forms $(t\text{-}BuNC)_2Pt[CH(CH{=}CH_2)CH_2CH_2CH(CH{=}CH_2)]$ in 90% yield, and PMe_3 displaces COD from **XV**, producing a similarly high yield of the analogous bisphosphene complex, $(Me_3P)_2Pt[CH(CH{=}CH_2)CH_2CH_2CH(CH{=}CH_2)]$, **XVI**[18]. Compound **XVI** can also be prepared via intermediate **XVII**, whose structure is established by X-ray diffraction, indicating that the second olefin of the diene is not innocent in these cyclizations[18]:

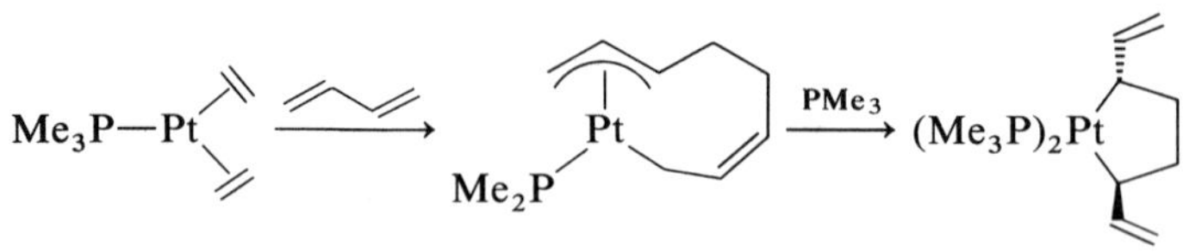

(k)

XVII **XVI**

Allenic unsaturation can participate in cyclization with metals, yielding materials containing partial structure **XVIII** or its isomers:

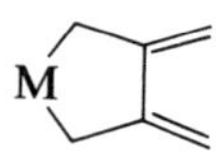

XVIII

For example, (β-diketonide)$Rh(C_2H_4)_2$ reacts with allene at $-78°C$ to form a material of empirical formula (β-diketonide)$Rh(allene)_3$ which, when treated with pyridine, is converted to (β-diketonide)$Rh(py)_2(C_6H_8)$. This last material contains the metalcyclic structure **XVIII**, by X-ray diffraction[19]. An analogous sequence leads to $(acac)Ir(py)(allene)(C_6H_8)$, which also contains partial structure **XVIII** according to its crystal structure[20].

The bis(1,5-COD) complexes of Ni and Pt are suitable starting materials for metallacycle preparation from allene. For Ni, concurrent treatment of $Ni(COD)_2$ with allene and bis(dicyclohexylphosphino)ethane yielded (chelate)$Ni(C_6H_8)$ (X-ray structure determination)[21], while for Pt, $(COD)Pt(C_6H_8)$ itself could be isolated[18]. Both of these species have structure **XVIII** for the organic ligand.

Reduced Zr complexes also cyclize with allenes, except that here the regiochemistry is different. Allene itself reacts with $[(\eta^5\text{-}C_5Me_5)_2Zr(N_2)]_2N_2$ to afford **XIX** nearly quantitatively[22].

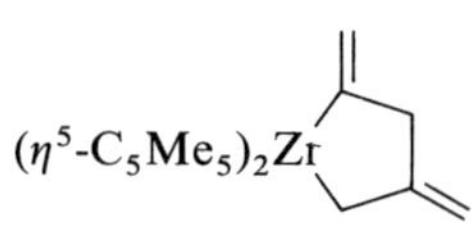

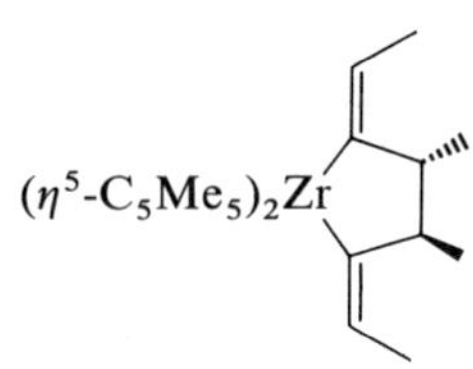

XIX **XX**

Dimethylallene forms a symmetrical product assigned structure **XX** based on its NMR spectrum[22], and methylallene yields two isomeric forms of the bisethylidene analog of

XIX[23], although exactly which two is not certain. At 80°C in C_6H_6 $(\eta^5\text{-}C_5H_5)_2ZrPh_2$ reacts with cis- or trans-2,3-diphenylzirconaindans.

(T. C. FLOOD)

1. F. G. A. Stone, *Pure Appl. Chem., 30*, 551 (1972), and refs. therein.
2. R. Burt, M. Cooke, M. Green, *J. Chem. Soc., A.*, 2975 (1970).
3. A. Bond, B. Lewis, M. Green, *J. Chem. Soc., Dalton Trans.*, 1109 (1975).
4. F. Grevels, D. Schulz, E. Koerner von Gustorf, *Angew. Chem., Int. Ed. Engl., 13*, 534 (1974).
5. P. Binger, M. Doyle, J. McMeeking, C. Kruger, Y. Tsay, *J. Organomet. Chem., 135*, 405 (1977).
6. S. Bezman, P. Bird, A. Fraser, J. Osborn, *Inorg. Chem., 19*, 3755 (1980).
7. M. Bennett, R. Johnson, I. Tomkins, *J. Am. Chem. Soc., 96*, 61 (1974).
8. J. McDermott, M. Wilson, G. Whitesides, *J. Am. Chem. Soc., 98*, 6529 (1976).
9. S. McLain, C. Wood, R. Schrock, *J. Am. Chem. Soc., 101*, 4558 (1979).
10. S. McLain, R. R. Schrock, *J. Am. Chem. Soc., 100*, 1315 (1978).
11. S. McLain, J. Sancho, R. Schrock, *J. Am. Chem. Soc., 101*, 5451 (1979).
12. R. Grubbs, A. Miyashita, *J. Chem. Soc., Chem. Commun.*, 864 (1977).
13. R. Grubbs, A. Miyashita, *J. Am. Chem. Soc., 100*, 7416 (1978).
14. R. Grubbs, A. Miyashita, M. Lin, P. Burk, *J. Am. Chem. Soc., 100*, 2418 (1978).
15. J. Manriquez, D. McAlister, R. Sanner, J. Bercaw, *J. Am. Chem. Soc., 100*, 2716 (1978).
16. D. McAlister, D. Erwin, J. Bercaw, *J. Am. Chem. Soc., 100*, 5966 (1978).
17. K. Gell, J. Schwartz, *J. Am. Chem. Soc., 103*, 2687 (1981).
18. G. Barker, M. Green, J. Howard, J. Spencer, G. Stone, *J. Chem. Soc., Dalton Trans.*, 1839 (1978).
19. G. Ingrosso, L. Porri, G. Pantini, P. Racanelli, *J. Organomet. Chem., 84*, 75 (1975).
20. P. Diversi, G. Ingrosso, A. Immirzi, M. Zocchi, *J. Organomet. Chem., 104*, C1(1976).
21. P. Jolly, G. Krüger, R. Salz, J. Sekutowski, *J. Organomet. Chem., 165*, C39 (1979).
22. J. R. Schmidt, D. M. Duggan, *Inorg. Chem., 20*, 318 (1981).
23. M. Duggan, *Inorg. Chem., 20*, 1164 (1981), and refs. therein.
24. K. Kropp, G. Erker, *Organometallics, 1*, 1246 (1982).

5.8.2.3.8. π-Allyl Complexes from Allylic C—H Bond Cleavage in Olefins by Metal Complexes

π-Allyl complexes of transition metals occur as intermediates in catalytic reactions[1–3], e.g., isomerization of olefins[4–6], oxidation of olefins[7–9] and conjugated dienes[10–13] and oligomerization and telomerization of conjugated dienes[14,15]. In these reactions the intermediate π-allyl complex can be isolated and characterized. Allylic carbon–hydrogen bonds are activated by metals to give π-allyl–metal complexes of interest in catalytic transformations of olefins. This process may formally be an oxidative addition where the metal cleaves the carbon–hydrogen bond with the formation of a π-allyl–metal hydride complex. Two principal pathways may be depicted (Scheme 1). The process may either be a concerted reaction (path A) or it may take place via a base-induced proton abstraction (path B). The former (path A) operates in the cleavage of nonactivated allylic carbon–hydrogen bonds.

A completely different way of obtaining π-allyl complexes from olefins is by hydride abstraction, e.g.:

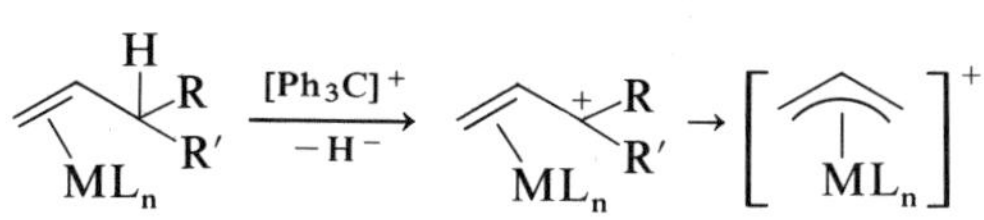

(a)

This procedure results in a formal oxidation of the metal.

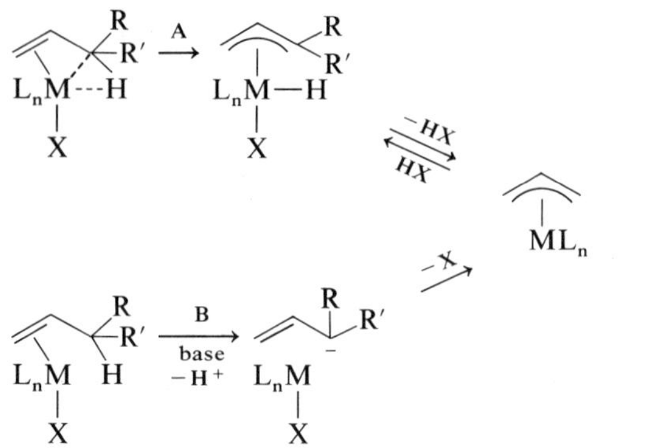

Scheme 1

(i) Nickel, Palladium and Platinum. Isobutene and cyclohexene on treatment with palladium chloride give a mixture of an olefin complex **I** and a π-allyl complex **II**[16–18]:

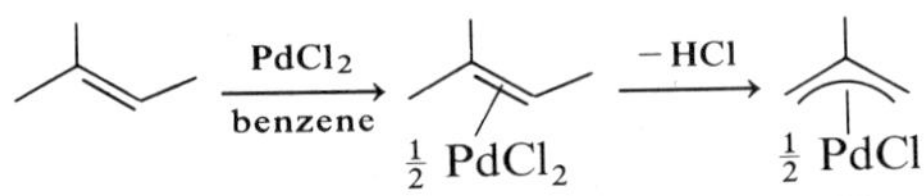

Ia **IIa** (b)

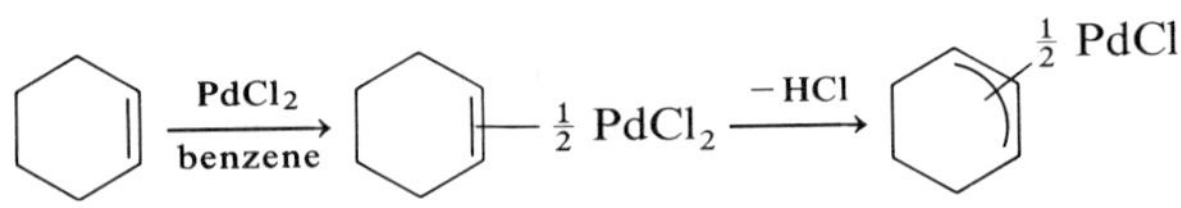

Ib **IIb** (c)

This reaction is a preparative method for the synthesis of π-allylpalladium complexes from olefins. Thus the π-allyl complex **III** can be prepared in 90% yield from 2,3-dimethyl-1-butene[19]:

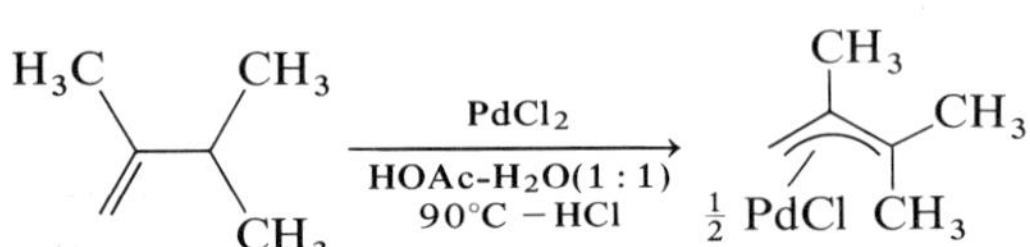

III (d)

In these reactions the metal cleaves the allylic carbon–hydrogen bond to eliminate HCl. Since HCl is formed a base may be added to improve the yield of the π-allylpalladium complex[20–26] and allow less substituted olefins to be transformed into π-allylpalladium complexes. Reaction of olefins with $PdCl_2$ in acetic acid in the presence of sodium acetate[20–23], or with $PdCl_2(PhCN)_2$ in the presence of sodium carbonate[24] gives the corresponding π-allyl complexes in good yield:

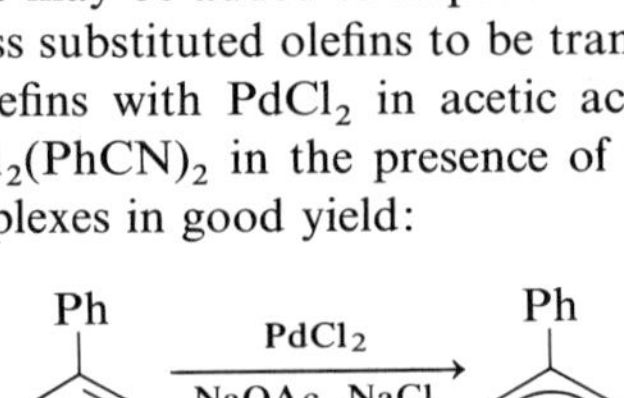

(e)[20]

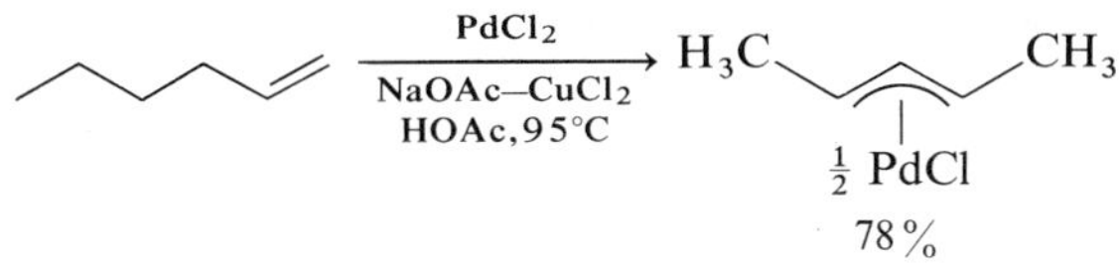

(f)[21]

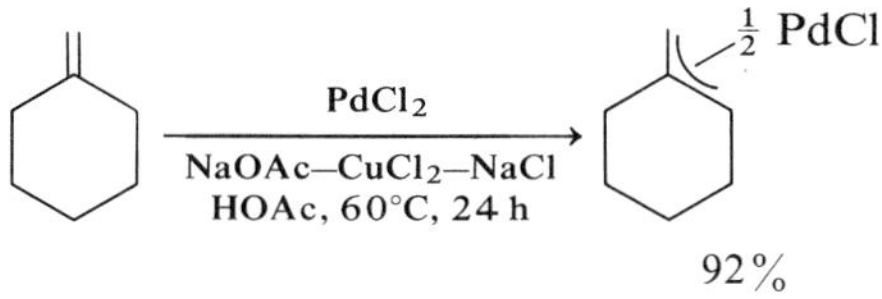

(g)[22,23]

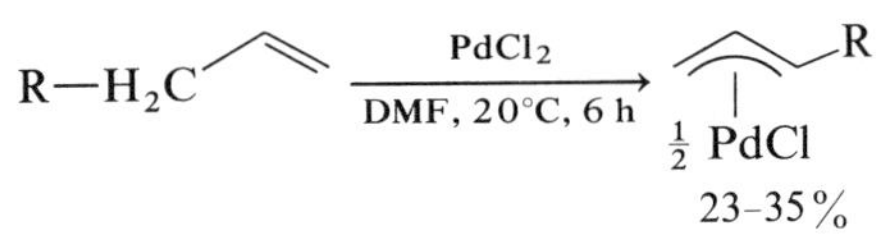

(h)[24]

By using a basic solvent, e.g., DMF, π-allyl complexes can be prepared under mild conditions, but yields are moderate. The promoting effect of the base may favor a mechanism in which the base abstracts a hydrogen in the allylic position (cf. Scheme 1). The facile formation of π-allylpalladium complexes from olefins with slightly acidic allylic hydrogens[27,28]:

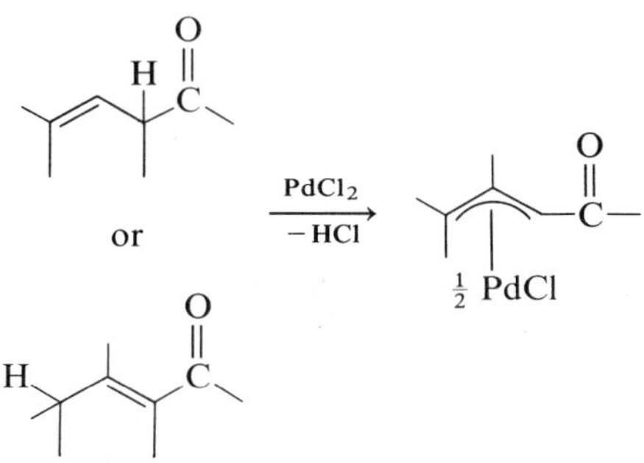

(i)

is consistent with such a mechanism. However, formation of π-allyl complexes from π-olefinpalladium acetate complexes is catalyzed by perchloric acid, suggesting palladium acts as an electrophile toward the allylic C—H bond[29]:

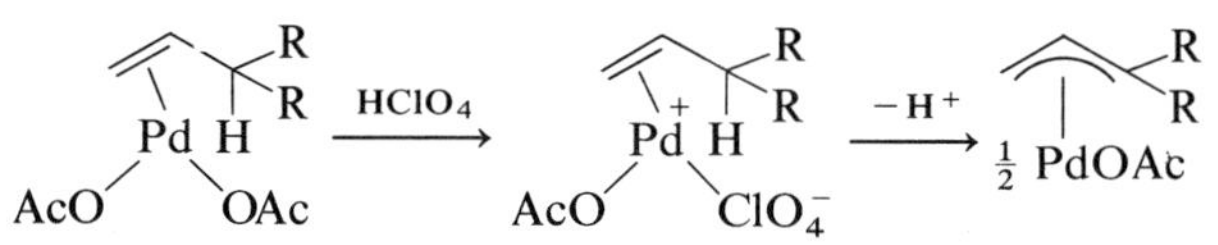

(j)

In agreement, palladium trifluoroacetate anion reacts rapidly with olefins under mild conditions to yield π-allylpalladium complexes[30]:

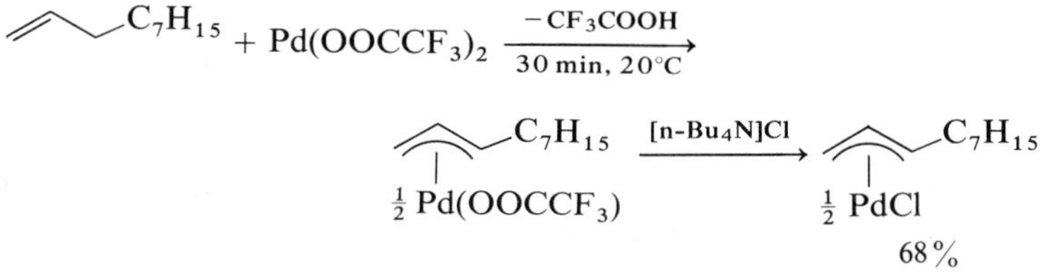

(k)

The latter procedure is a preparative method for the synthesis of π-allylpalladium complexes from olefins.

The cleavage of the allylic C—H bond by palladium is stereospecific[31]; e.g., specifically deuterated 4-cholestene (**IV**) loses a hydrogen or a deuterium syn to the incoming metal atom (Scheme 2):

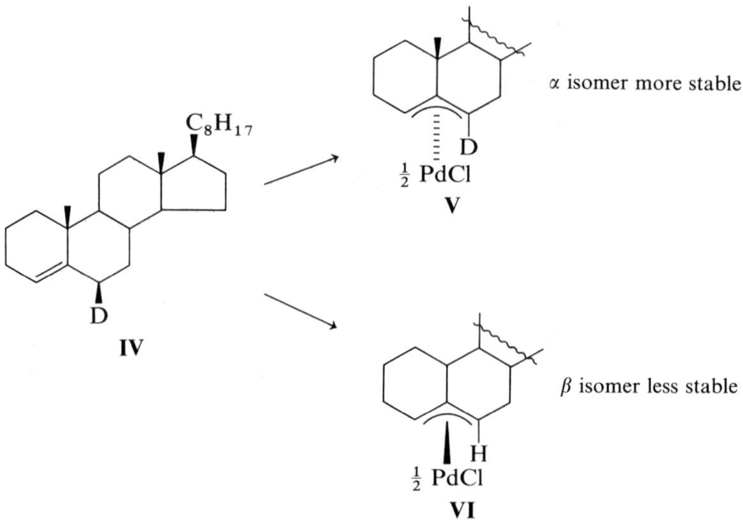

Scheme 2

Undeuterated 4-cholestene gives the more stable α isomer over the β isomer in a ratio of 3.3 : 1. The observed ratio of **V** : **VI**, obtained from deuterated cholestene **IV**, is ~11.5, indicating an isotope effect of ~3.5 in the formation of the π-allyl derivative. In contrast, the formation of π-allylpalladium complexes from α,β-unsaturated systems is nonstereospecific[32–36]. Thus, formation of π-allylpalladium complexes from enones **VII** and **VIII** takes place via a palladium-assisted enolization. Formation of π-allylcomplex **IX** from **VII** takes place with the loss of the β proton, which is anti to the incoming metal atom[33,34]:

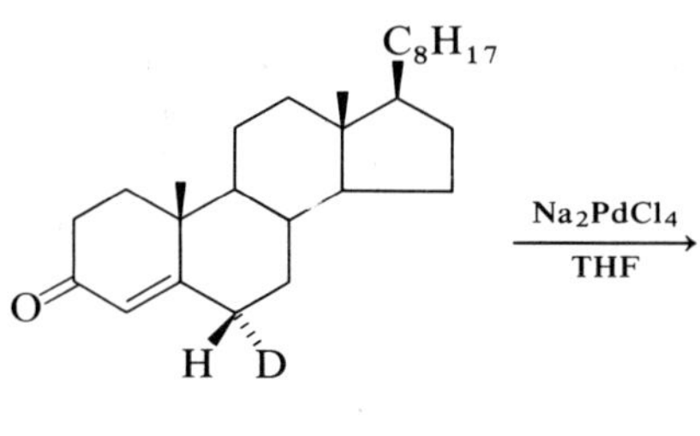

VII **IX** (1)

but elimination of hydrogen (deuterium) in **VIII** is nonstereospecific, giving π-allylpalladium complex **X**, containing both H and D in the allyl group[35,36].

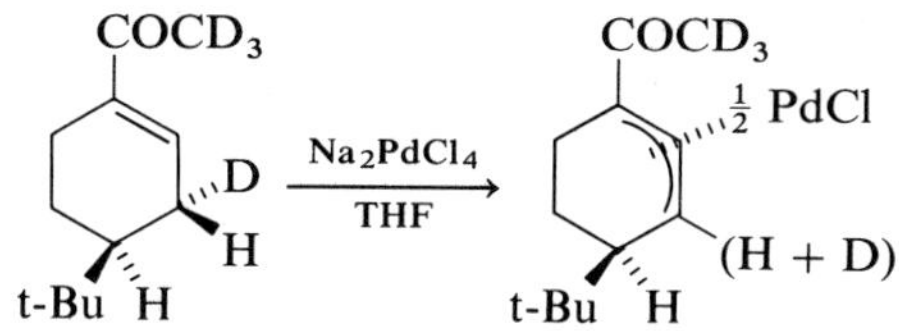

VIII X (m)

A reversible equilibrium between nickel–olefin complexes and π-allylnickel hydride complexes exists at low temperature[37,38]:

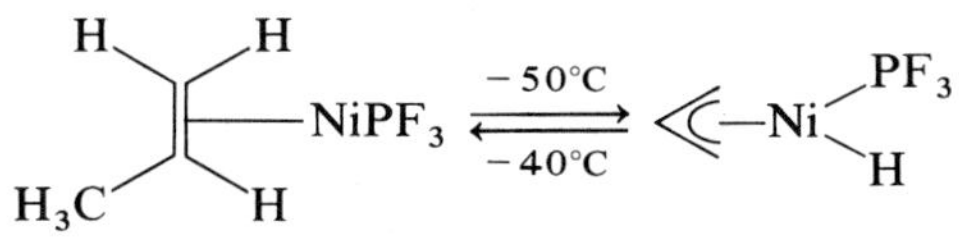

XI XII (n)

When the solution of the olefin complex **XI** is cooled below −50°C, the hydrido-π-allylnickel complex, **XII**, forms. The formation of a π-allylnickel complex from allylic C—H bond cleavage is demonstrated by olefin **XIII** which with a dichloro complex of nickel affords the π-allylnickel complex, **XIV** [39]:

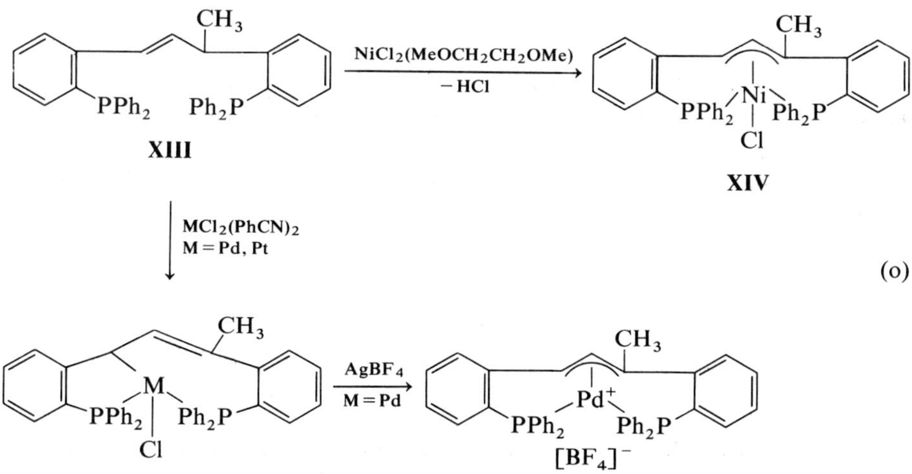

XIII XIV XV XVI (o)

Reaction of **XIII** with Pd(II) and Pt(II) complexes gives only the σ-allyl complex, **XV**. The σ-allyl complex **XV** (M = Pd) can be transformed into the π-allyl complex, **XVI**, by treatment with $AgBF_4$. The formation of the π-allyl-metal chloride complex (i.e., **XIV**) from **XIII** and Ni(II), but not from Pd(II) and Pt(II) is consistent with the ability of Ni(II) to form 18-electron complexes, which are less favored for Pd(II) and Pt(II).

(ii) Rhodium and Iridium. Reaction of **XIII** with Rh(III) chloride in equimolar amounts results in C—H bond cleavage and formation of π-allyl complex **XVII** in 60% yield[40]:

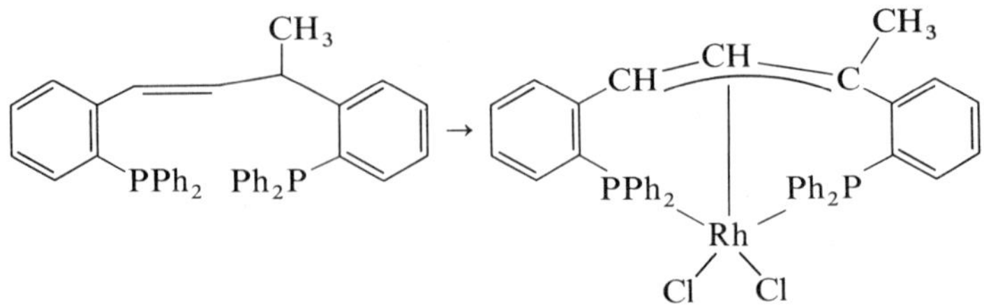

(p)

XIII **XVII**

An allylhydrido–metal complex is initially formed when the allylic C—H bond is cleaved. For Rh such allylhydrido–metal complexes are not isolated during the formation of π-allyl complexes from olefins, but the allylhydridorhodium complex, **XIX**, prepared by oxidative addition of HCl to the allylRh(I) complex, **XVIII**:

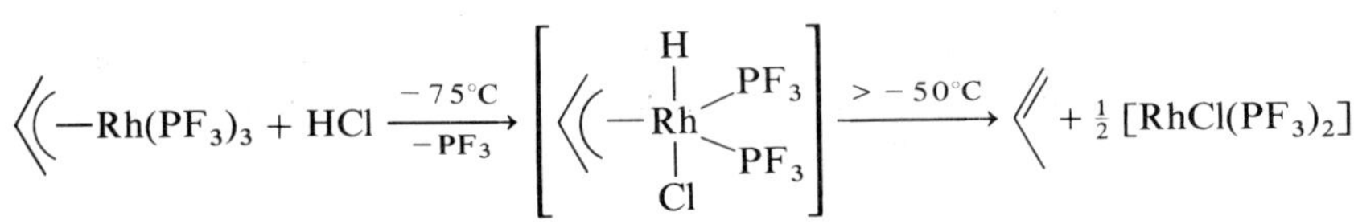

(q)

XVIII **XIX**

is characterized by ^{1}H and ^{19}F NMR spectroscopy at low temperature (−75°C). When the temperature is raised above −50°C, hydride addition to the allyl ligand takes place to release propene.

Reaction of allylbenzene or trans-β-methylstyrene with $IrCl(N_2)(PPh_3)_2$ produces the π-allyl complex **XX** in 71–78% yield[42]:

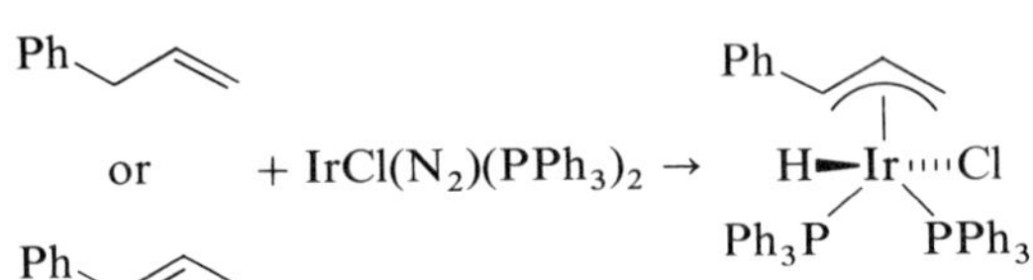

(r)

XX

The metal inserts into the C—H bond and the reaction may be considered as a concerted oxidative addition. This is a rare example of an allyl–hydrido–metal complex being isolated from allylic C—H bond cleavage by the metal. Other examples are the reaction of Mo(0) and Ru(0) complexes with olefins where isolable hydrido-π-allyl-metal(II) complexes are characterized (vide infra).

(iii) Iron and Ruthenium. π-Allyliron complexes are not readily prepared directly from olefins by cleavage of an allylic C—H bond, but π-allylhydridoiron complexes are intermediates in olefin isomerization catalyzed by Fe(0) carbonyl complexes[4,45–47]. The intermediate π-allyliron hydride complexes formed are too reactive to be isolated and undergo a fast reductive elimination to give the isomerized or the starting olefin:

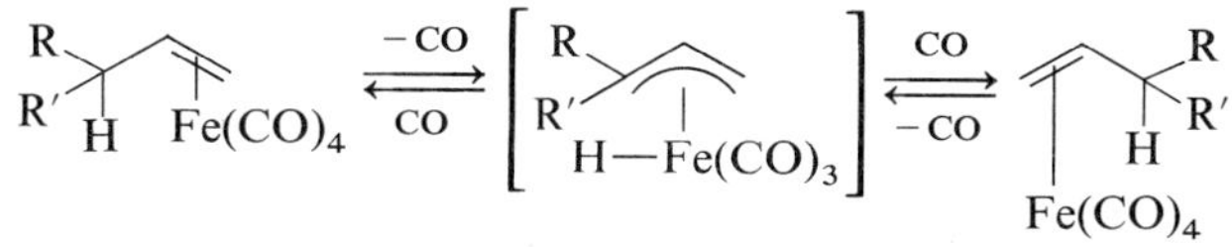

(s)

but π-allylic complexes of iron and ruthenium are prepared from the corresponding π-olefin complex by allylic hydride ion abstraction with $[Ph_3C][BF_4]$[48]. Thus treatment of complexes **XXIa** and **XXIb** with $[Ph_3C][BF_4]$ give the π-allyl complexes **XXIIa** and **XXIIb** in 90 and 62% yield, respectively:

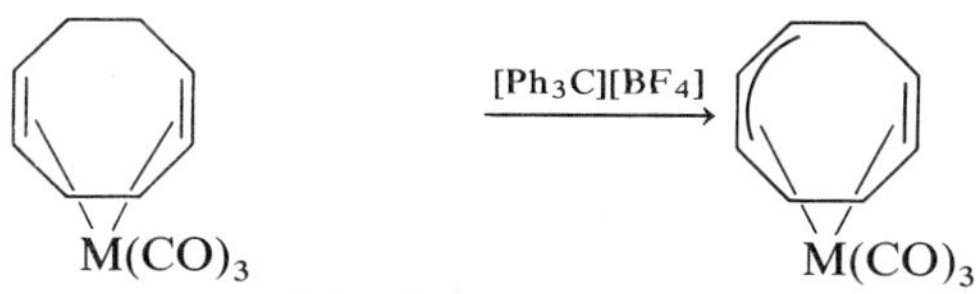

(t)

XXI **a** M = Fe
b M = Ru

XXII

Stable hydrido-π-allylruthenium(II) complexes **XXIV** are obtained from the reaction of Ru(0) complexes **XXIII** with olefins[49]:

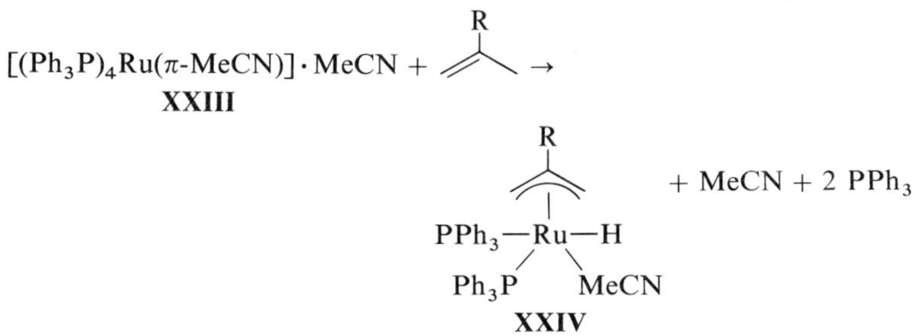

(u)

Although the coordinated olefin intermediate is not detected, π-olefin complex formation preceeds the rupture of the allylic C—H bond.

Complex **XXIII** also catalyzes the isomerization of allylbenzene to cis- and trans-β-methylstyrene at 35–60°C[50]:

$$PhCH_2CH{=}CH_2 \xrightarrow{\textbf{XXIII}} \underset{\text{cis and trans}}{PhCH{=}CH{-}CH_3} \qquad (v)$$

A π-allylruthenium related to **XXIV** is most likely an intermediate in this isomerization:

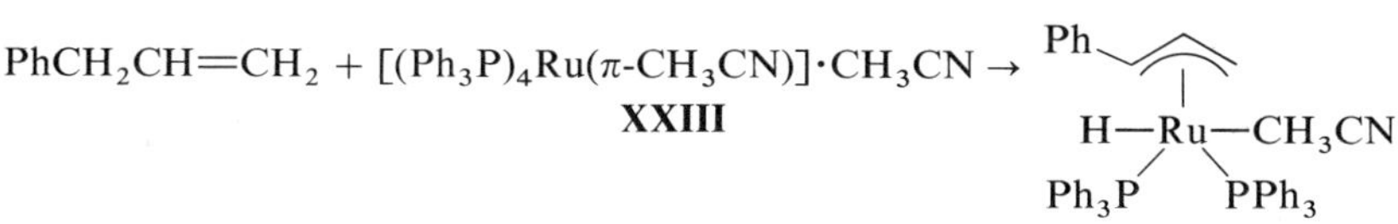

(w)

Reaction of the dinuclear ruthenium complexes, **XXV**, with propene gives π-allylruthenium complexes, **XXVI**[51]. A mechanism is shown in Scheme 3.

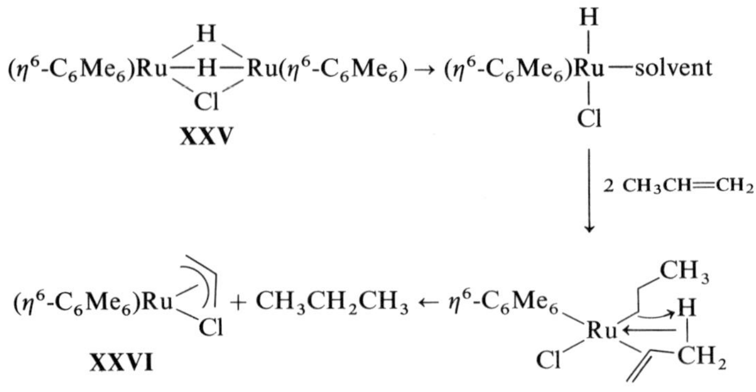

Scheme 3

Reaction of Ru(0) complex, **XXVII**, with 1-hexene gives a syn–anti mixture of η^3-hexenylruthenium hydride complexes, **XXVIIIa** and **XXVIIIb**[52]:

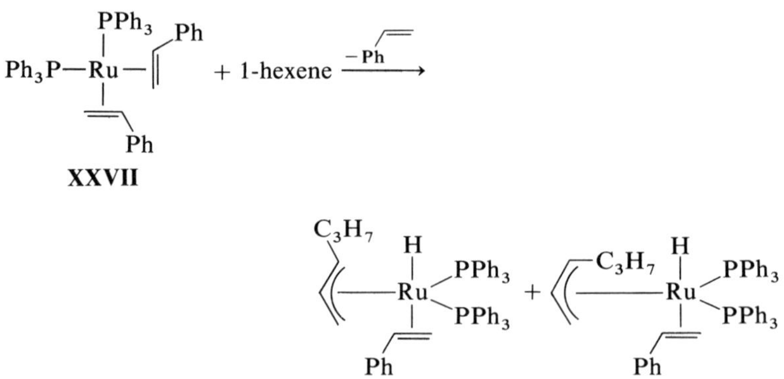

(x)

2-Hexene gives only diene and no π-allyl complex.

(iv) Molybdenum and Tungsten. The interaction of Mo(0) and W(0) complexes with olefins gives π-allyl–metal complexes via allylic C—H bond cleavage[53,54]. Thus, reaction of a Mo(0) species generated in situ from Na–Hg reduction of $MoCl_3(THF)_3$, with propene in the presence of bis-1,2-(diphenylphosphino)ethane (biphos) gives the π-allyl–hydride complex, **XXIX**[53,54]:

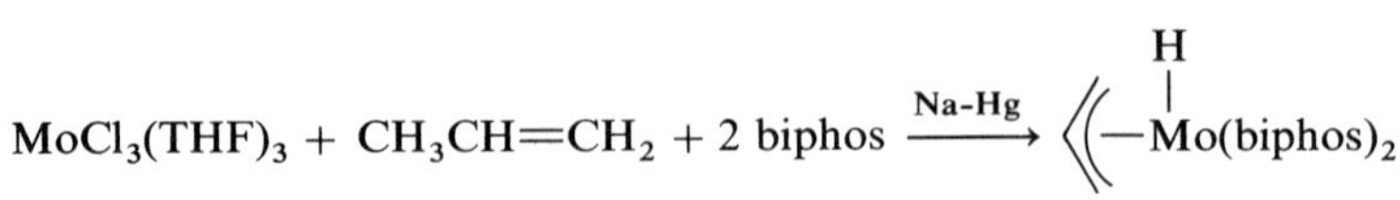

(y)

XXIX

which is stable and decomposes only at $>110°C$ to yield propene. An exchange of metal hydride with the terminal hydrogens of the π-allyl group, observed at 101°C by ^1H NMR spectroscopy, is explained by a reversible hydride addition–elimination:

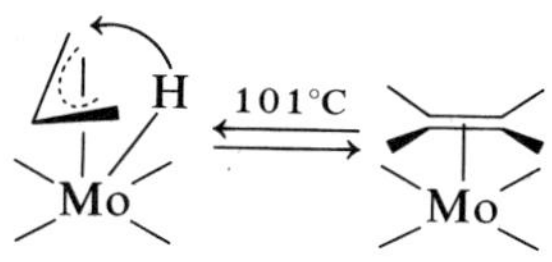

(z)

π-Allyl–metal hydride complexes of W are prepared analogously by reduction of W(IV) complexes in the presence of the appropriate olefin and ligand[54]:

WCl_4(diphos) + $CH_3CH{=}CH_2$ + biphos $\xrightarrow{\text{Na sand}}$ 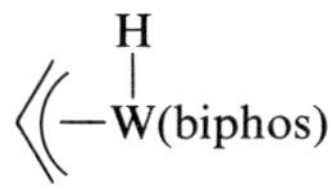 (aa)

WCl_4(dppen) + $RCH_2CH{=}CH_2$ + dppen $\xrightarrow{\text{Na sand}}$ 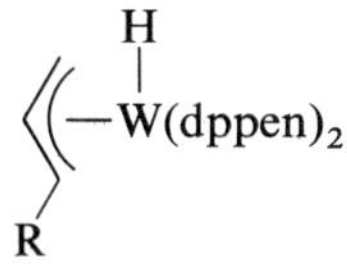 (ab)

where R = H, CH_3, C_2H_5, C_3H_7, C_5H_{11}, C_6H_5.

(J-E. BÄCKVALL, K. ZETTERBERG, B. ÅKERMARK)

1. G. W. Parshall, *J. Mol. Catal.*, *4*, 243 (1978).
2. R. Baker, *Chem. Rev.*, *73*, 487 (1973).
3. B. M. Trost, *Acc. Chem. Res.* *13*, 385 (1980).
4. C. P. Casey, C. R. Cyr, *J. Am. Chem. Soc.*, *95*, 2248 (1973).
5. E. O. Sherman, M. Olson, *J. Organomet. Chem.*, *172*, C13 (1979).
6. C. A. Tolman, in *Transition Metal Hydrides*, E. L. Muetterties, ed., Marcel Decker, New York, 1971, p. 271.
7. J. Tsuji, K. Sakai, H. Nagashima, I. Shimizu, *Tetrahedron Lett.*, *22*, 131 (1981).
8. A. Herrmann, B. Åkermark, *Angew. Chem., Int. Ed. Engl.*, *23*, 453 (1984).
9. J. E. McMurry, P. Kogovský, *Tetrahedron Lett.*, *25*, 4187 (1984).
10. J. E. Bäckvall, R. E. Nordberg, *J. Am. Chem. Soc.*, *103*, 4959 (1981).
11. J. E. Bäckvall, S. E. Ryström, R. E. Nordberg, *J. Org. Chem.*, *49*, 4619 (1984).
12. J. E. Bäckvall, *Acc. Chem. Res.*, *16*, 335 (1983).
13. J. E. Bäckvall, J. E. Nyström, R. E. Nordberg, *J. Am. Chem. Soc.*, *107*, 3676 (1985).
14. P. W. Jolly, G. Wilke, *The Organic Chemistry of Nickel*, Vol. 2, Academic Press, New York, 1975.
15. J. Tsuji, *Organic Synthesis with Palladium Compounds*, Springer-Verlag, Berlin, 1980, p. 88.
16. R. Hüttel, J. Kratzer, *Angew. Chem.*, *71*, 456 (1959).
17. R. Hüttel, J. Kratzer, M. Bechter, *Chem. Ber.*, *94*, 766 (1961).
18. For a review see R. Hüttel, *Synthesis*, 225 (1970).
19. R. Hüttel, H. Christ, *Chem. Ber.*, *96*, 3101 (1963).
20. H. C. Vogler, *Recl. Trav. Chim. Pays-Bas*, *88*, 225 (1969).
21. R. G. Schultz, U.S. Patent 3,446,825 (1966); *Chem. Abstr.* *71*, 61,553 (1969).
22. B. M. Trost, P. E. Strege, *Tetrahedron Lett.*, 2603 (1974).
23. B. M. Trost, P. E. Strege, L. Weber, T. Fullerton, T. J. Dietsche, *J. Am. Chem. Soc.*, *100*, 3407 (1978).
24. A. D. Kelly, J. Braatz, *J. Chem. Soc., Chem. Commun.*, 169 (1968).
25. F. Conti, M. Donati, G. F. Pregaglia, R. Ugo, *J. Organomet. Chem.*, *30*, 421 (1971).
26. D. Morelli, R. Ugo, F. Conti, M. Donati, *J. Chem. Soc.; Chem. Commun.*, 801 (1967).
27. G. W. Parshall, G. Wilkinson, *Inorg. Chem.*, *1*, 896 (1962).
28. J. Tsuji, S. Imamura, J. Kiji, *J. Am. Chem. Soc.*, *86*, 4491 (1964).
29. R. G. Brown, R. V. Chaudhari, J. M. Davidson, *J. Chem. Soc., Dalton Trans.*, 176 (1977).
30. B. M. Trost, P. J. Metzner, *J. Am. Chem. Soc.*, *102*, 3572 (1980).
31. I. J. Harvie, F. J. McQuillin, *J. Chem. Soc., Chem. Commun.*, 747 (1978).
32. K. Henderson, F. J. McQuillin, *J. Chem. Soc., Chem. Commun.*, 15 (1978).

33. D. J. Collins, B. M. K. Gatehouse, W. R. Jackson, G. A. Kakos, R. N. Timms, *J. Chem. Soc., Chem. Commun.*, 138 (1980).
34. D. J. Collins, W. R. Jackson, R. N. Timms, *Aust. J. Chem.*, *33*, 2663 (1980).
35. W. R. Jackson, A. Stragalinou, J. U. G. Strauss, *J. Chem. Soc., Chem. Commun.*, 139 (1980).
36. A. W. Coller, W. R. Jackson, A. Stragalinou, J. U. G. Strauss, *Aust. J. Chem.*, *34*, 171 (1981).
37. H. Bönnemann, *Angew. Chem.*, *82*, 699 (1970).
38. G. Curturan, A. Scrivanti, F. Morandini, *Angew. Chem., Int. Ed. Engl. 20*, 112 (1981).
39. M. A. Bennet, R. N. Johnson, G. B. Robertson, I. B. Tomkins, P. O. Whimp, *J. Am. Chem. Soc.*, *98*, 3514 (1976).
40. M. A. Bennet, R. N. Johnson, G. B. Robertson, I. B. Tomkins, P. O. Whimp, *J. Organomet. Chem. 77*, C43 (1974).
41. J. F. Nixon, B. J. Wilkins, *J. Organomet. Chem.*, *80*, 129 (1974).
42. T. H. Tulip, J. A. Ibers, *J. Am. Chem. Soc.*, *100*, 3252 (1978).
43. T. H. Tulip, J. A. Ibers, *J. Am. Chem. Soc.*, *101*, 4201 (1979).
44. G. F. Emerson, R. Pettit, *J. Am. Chem. Soc.*, *84*, 4591 (1962).
45. W. T. Hendrix, F. G. Cowherd, J. L. von Rosenberg, *J. Chem. Soc., Chem. Commun.*, 97 (1978).
46. F. G. Cowherd, J. L. von Rosenberg, *J. Am. Chem. Soc.*, *91*, 2157 (1969).
47. J. V. Strauss, P. W. Ford, *Tetrahedron Lett.*, 2917 (1975).
48. F. A. Cotton, A. J. Deeming, P. L. Josty, S. S. Ullah, A. J. P. Domingos, B. F. G. Johnson, J. Lewis, *J. Chem. Soc., Dalton Trans.*, 2093 (1974).
49. E. O. Sherman, Jr., P. R. Schreiner, *J. Chem. Soc., Chem. Commun.*, 223 (1978).
50. E. O. Sherman, Jr., M. Olson, *J. Organomet. Chem.*, *172*, C13 (1979).
51. M. A. Bennet, T. N. Huang, T. W. Turney, *J. Chem. Soc., Chem. Commun.*, 312 (1979).
52. B. N. Chaudret, D. J. Cole-Hamilton, G. Wilkinson, *J. Chem. Soc., Dalton Trans.*, 1739 (1978).
53. J. W. Byrne, H. V. Blaser, J. A. Osborn, *J. Am. Chem. Soc.*, *97*, 3871 (1975).
54. J. W. Byrne, Ph.D. Thesis, Harvard University, 1976.

5.8.2.3.9. Metallacyclobutanes from Nucleophilic Attack on π-Allyl–Metal Complexes.

The π-allyl ligand bonded to transition-metal cations can be susceptible to nucleophilic addition reactions either on a terminal or on the central carbon atom. Regioselective addition to a terminal carbon of an allyl ligand produces a metal–π-olefin complex (see §5.8.2.3.10), whereas regioselective addition to the central carbon leads to metallacyclobutane formation:

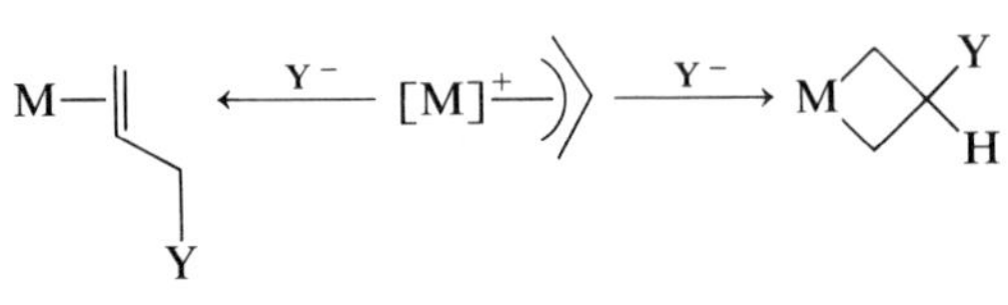

(a)

The regioselectivity of nucleophilic additions to allyl ligands bonded to transition-metal cations varies according to the electron richness of the transition-metal species[1]. When coordinated to an electron-rich transition-metal center, regioselective nucleophilic attack onto the central carbon is observed. This is interpreted in terms of charge control with the allyl ligand resembling the allyl anion in these complexes; i.e., with the central carbon carrying a higher positive charge than the terminal ones[1].

Allyl ligands bonded to the electron-rich species, [bis-h^5-cyclopentadienyl tungsten]$^+$ undergo nucleophilic attack exclusively on the central carbon to generate tungstenacyclobutanes; e.g., $[(\eta^5\text{-}C_5H_5)_2W(\pi\text{-}C_3H_5)]^+$, **I**, is reduced by $NaBH_4$ to bis-η^5-cyclopentadienyltungstenacyclobutane, **2**[2]:

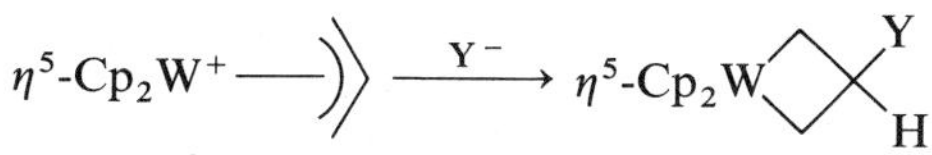

I

II Y = H
III Y = D
IV Y = $CH_2CH{=}CH_2$ (b)

Cation **I** reacts similarly with $NaBD_4$ and allyllithium to give the metallacycles **III** and **IV**, respectively[2]. The analogous Mo cation, **V**, likewise undergoes hydride reduction to **VI**:

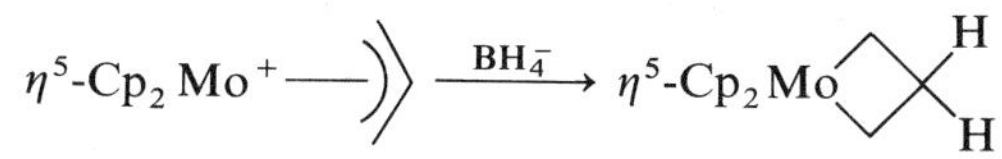

(c)

V **VI**

Reaction of $(\eta^5\text{-}C_5H_5)_2MoCl_2$ with allylmagnesium bromide produces the 3-substituted metallacyclobutane derivative, **VII**, presumably via nucleophilic addition to the initially formed allyl cation, **V**[3]:

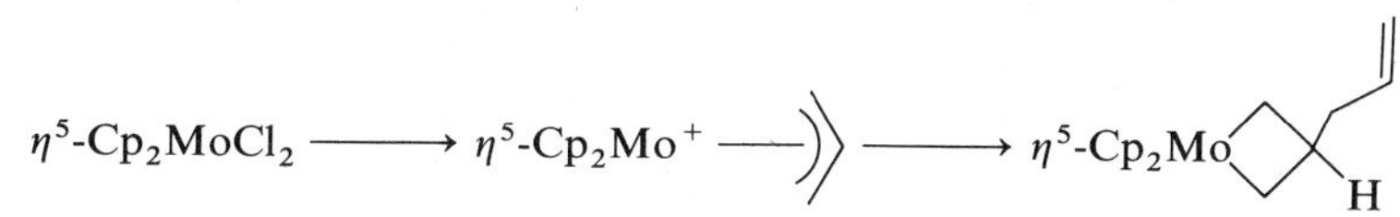

(d)

V **VII**

The 3-methylmetallacyclobutane **VIII**, may be prepared either by the reaction of cation **I** with methyllithium or by hydride reduction of the 2-methylallyl cation **IX**[4]:

(e)

I **VIII** **IX**

Metallacyclobutane derivatives with substituents in the 2 position, e.g., **XI**, may be prepared by hydride reduction of 1-substituted allyl cations, **X**[4]:

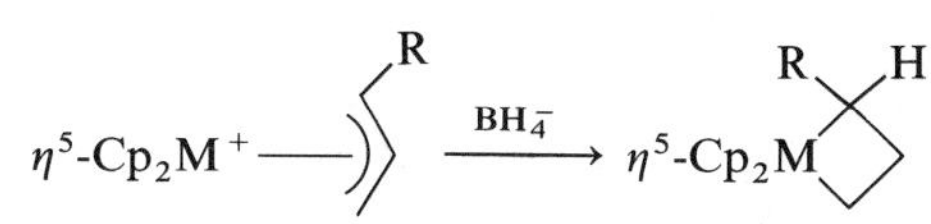

(f)

X M = Mo, W
R = Me, Ph

XI

Reduction of the cations **XII** (M = Mo, W) with $NaBD_4$ occurs stereoselectively to produce single isomers **XIII** which are trans[4]. The trans-configuration is that expected for exo attack on the syn complex, **XII**:

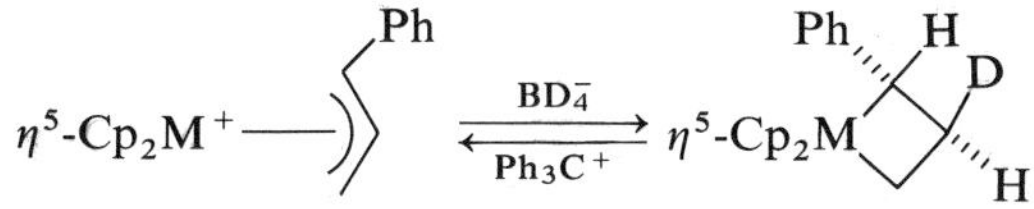

(g)

XII **XIII**

(S. G. DAVIES)

1. S. G. Davies, M. L. H. Green, D. M. P. Mingos, *Tetrahedron, 34,* 3047 (1978); *Reactions of Coordinated Ligands,* Vol. 1, P. S. Brateman, ed., Plenum Press, New York, 1986, p. 897.
2. F. W. Benfield, B. R. Francis, M. L. H. Green, N-T. Luong-Thi, G. Moser, J. S. Poland, D. M. Roe, *J. Less-Common Met., 36,* 187 1974.
3. M. Ephjritikhine, B. R. Francis, M. L. H. Green, R. E. Mackenzie, M. J. Smith, *J. Chem. Soc., Dalton Trans.,* 1131 (1977).
4. G. J. A. Adam, S. G. Davies, K. A. Ford, M. Ephritikhine, P. F. Todd, M. L. H. Green, *J. Mol. Catal., 8,* 15 (1980).

5.8.2.3.10. π-Olefin Complexes from Nucleophilic Attack on π-Allyl–Metal Complexes.

The addition of nucleophiles to the terminal carbon atom of π-allyl ligands bonded to transition-metal complexes produces π-olefin complexes that frequently are not sufficiently stable to be isolated. The overall result is release of the olefin. π-Allylpalladium complexes react with nucleophiles to generate free olefins, and this reaction finds application in organic synthesis[1–4]. π-Olefin complexes sufficiently stable to be isolated result from nucleophilic addition to π-allyl ligands bonded to electron-poor 18-electron transition-metal cations[5]. As with nucleophilic addition reactions to other unsaturated hydrocarbon ligands bonded to 18-electron cations, the nucleophile attacks the uncoordinated face of the π-allyl ligand. This is consistent with an S_N2 reaction with displacement of the M—C bond and inversion of configuration at the carbon under attack.

The cation $[(\pi\text{-allyl})Fe(CO)_4]^+$, **I**, reacts with py or Ph_3P to give the relatively stable monosubstituted olefin complexes **II** (85%) and **III**, respectively.[6]

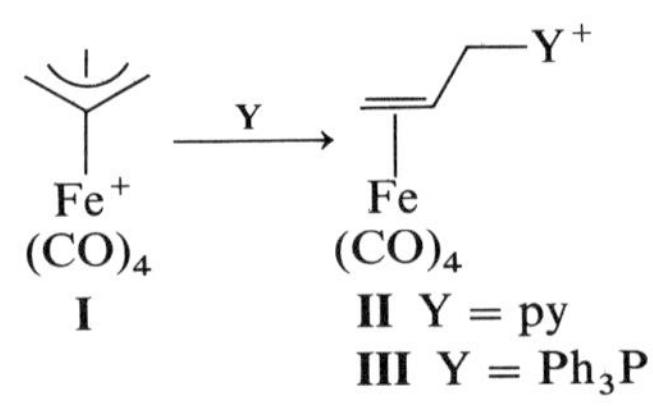

(a)

Decomplexation of the olefin is observed, however, when **I** is treated with carbon nucleophiles[6]. Iron tetracarbonyl complexes of polysubstituted olefins are less stable than those of monosubstituted olefins, and the introduction of substituents onto the allyl ligand does not permit the isolation of the corresponding π-olefin complexes[6].

The 1-methylallyliron tetracarbonyl cation, **IV**, reacts with dialkylcadmium reagents to generate the olefin complexes **V** and **VI** which are stable for a short period in solution[7]:

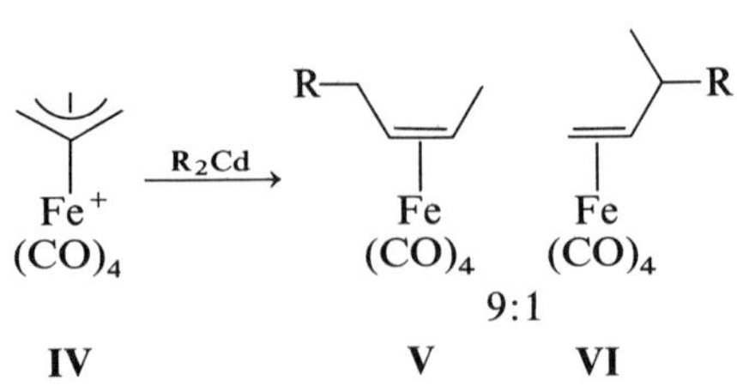

(b)

The cation $[(C_5H_5)Mo(CO)(NO)(allyl)]^+$ reacts with nucleophiles (H^-, $[RS]^-$, $[MeO]^-$, etc.) to give the corresponding neutral $[(C_5H_5)Mo(CO)(NO)(olefin)]$ complexes[8]. The two ends of the allyl ligand are not equivalent here; one is trans to CO, while the other is trans to NO.

The cation $[(C_5H_5)Mo(CO)(NO)(allyl)]^+$ exists as an equilibrating mixture of isomers **VII** and **VIII**. Nucleophilic attack on a terminal carbon of cation **VII** would lead to the olefin complex **IX** if attack occurred onto the carbon trans to the NO ligand, or to complex **XI** if attack occurred onto the carbon trans to the CO ligand. Cation **VIII**, however, would give complex **X** by attack on the carbon atom trans to CO and complex **XII** by attack onto the carbon trans to NO. The olefin complexes **IX** and **X** rapidly equilibrate by rotation about the Mo–olefin bond as do the olefin complexes **XI** and **XII**. Nucleophilic attack by hydride or cyanide shows little regioselectivity, and both diastereoisomeric pairs **IX** $\rightleftharpoons$ **X** and **XI** $\rightleftharpoons$ **XII** are formed[8].

With certain nucleophiles (e.g., 1-pyrrolidino-2-methylpropene), nucleophilic addition occurs stereospecifically to yield only isomers **IX** $\rightleftharpoons$ **X**[7]. These results are interpreted in terms of preferential attack trans to the NO ligand in cation **VII** and trans to the CO ligand in cation **VIII**[9].

The electronic structure of the $[(C_5H_5)Mo(CO)(NO)(allyl)]^+$ isomers **VII** and **VIII** suggests that in cation **VII** the terminal carbon of the allyl ligand trans to NO is the most positive, whereas for cation **VIII** the terminal carbon trans to CO is the most positive:[10]

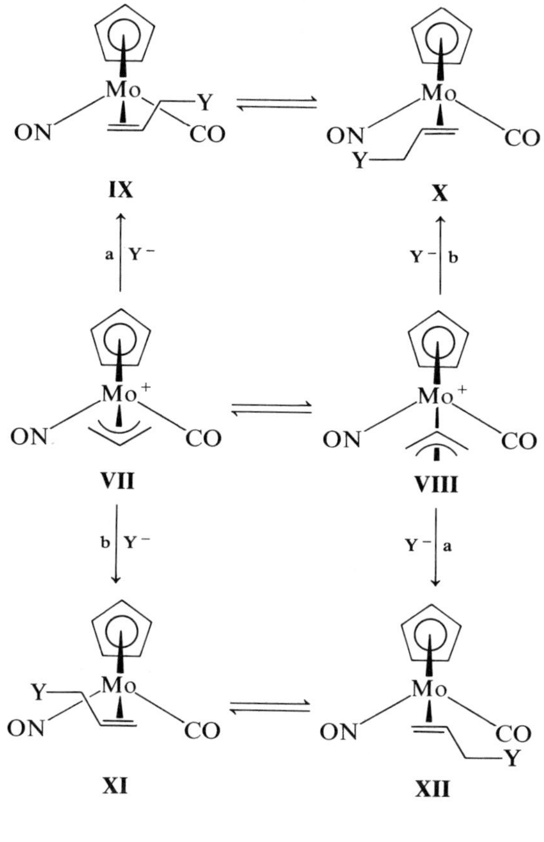

(c)

Thus the observed regioselectivity of nucleophilic addition is consistent with the reaction being subject to charge control.

Either separately or as a mixture, isomers **XIII** and **XIV** undergo nucleophilic attack by 1-pyrrolidino-2-methylpropene on the unsubstituted terminal carbon of the allyl ligand to yield, after hydrolysis, a single product, **XV**:

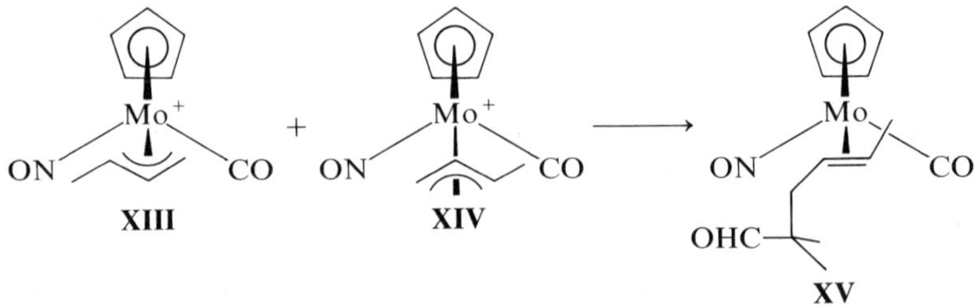

(d)

A mixture of isomers **XVI** and **XVII** containing the (syn,syn-1,3-dimethyl)allyl ligand likewise only yields a single olefin complex, **XVIII**[9]:

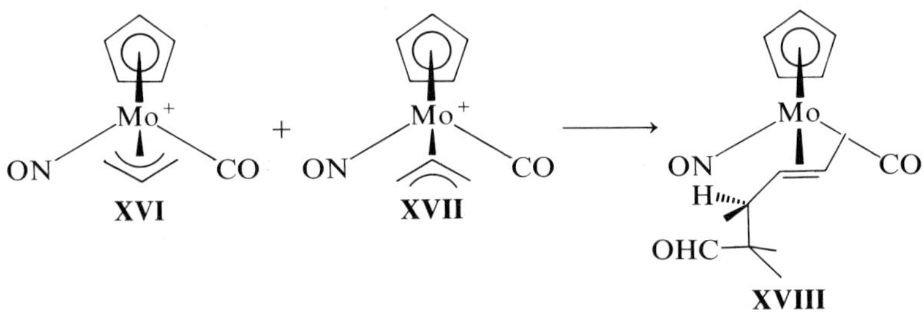

(e)

Nucleophilic addition to the (syn,anti-1,3-dimethyl)allyl ligand of cation **XIX** yields the olefin complex **XX**, a diastereoisomer of complex **XVIII**, demonstrating that the configuration in the cation of the allyl carbon that undergoes nucleophilic attack determines the configuration of that carbon in the product[9].

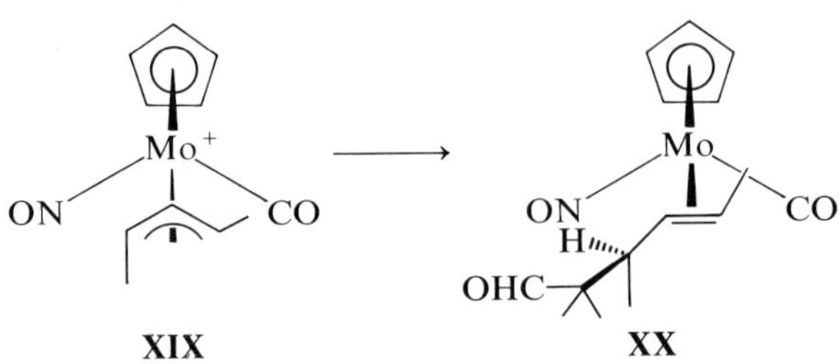

(f)

The relative configurations of the chiral centers in complexes **XVIII** and **XX** confirm that as expected nucleophilic attack occurs on the uncoordinated face of the π-allyl ligand[7].

Cations that contain both π-olefin and π-allyl ligands react with nucleophiles preferentially on the olefin[4], but a significant proportion of attack on the π-allyl ligand can occur; e.g., cation **XXI** yields complex **XXII** from attack onto the olefin and complex **XXIII** from nucleophilic addition to the π-allyl ligand[11]:

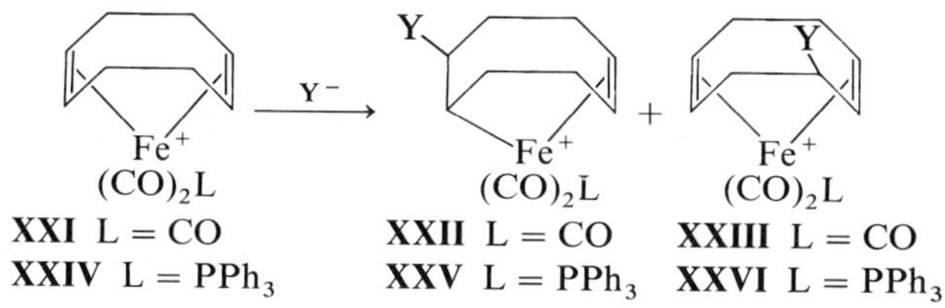

(g)

where $Y^- = H^-$, $[CN]^-$, $[acac]^-$, $[CH(CO_2Et)_2]^-$, $[CPh(CO_2Et)_2]^-$. Complex **XXIII** never predominates in these reactions.

Nucleophilic addition to the more electron-rich cation **XXIV** leads as expected to a higher proportion of attack onto the olefin than for complex **XXI**. The cation **XXVII** reacts with nucleophiles such as D^- and $[CN]^-$ on the olefin ligand as expected[12]. Nucleophilic addition by $[i\text{-PrS}]^-$, however, occurs onto the π-allyl ligand[12]:

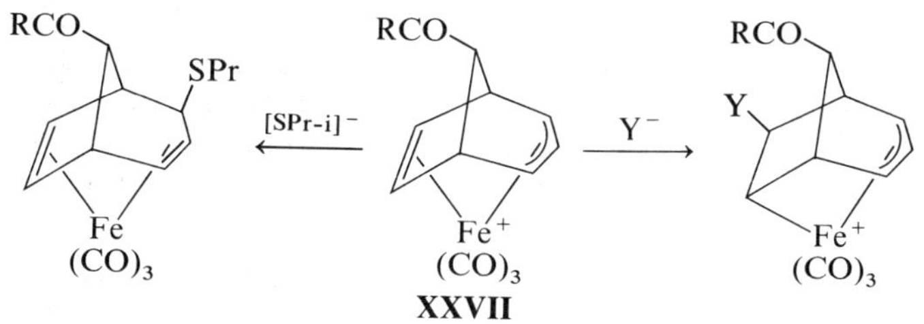

(h)

where R = Me, Ph. For the large and polarizable nucleophile, $[i\text{-PrS}]^-$, the regioselectivity may be sterically or orbitally rather than charge controlled.

(S. G. DAVIES)

1. B. M. Trost, Tetrahedron Report No. 32, *Tetrahedron, 33*, 2615 (1977).
2. J. Tsuji, *Organic Synthesis with Palladium Compounds*, Springer-Verlag, New York, 1980.
3. R. F. Heck, *Palladium Reagent in Organic Synthesis*, Academic Press, New York, 1985.
4. S. G. Davies, M. L. H. Green, D. M. P. Mingos, Tetrahedron Report No. 57, *Tetrahedron, 34*, 3047 (1978).
5. T. H. Whitesides, R. W. Arhart, R. W. Slaven, *J. Am. Chem. Soc., 95*, 5792 (1973).
6. A. J. Pearson, *Tetrahedron Lett.*, 3617 (1975).
7. J. W. Faller, A. M. Rosan, *J. Am. Chem. Soc., 98*, 3388 (1976).
8. R. D. Adams, D. F. Ghodosh, J. W. Faller, A. M. Rosan, *J. Am. Chem. Soc., 101*, 2570 (1979).
9. B. E. R. Schilling, R. Hoffmann, J. W. Faller, *J. Am. Chem. Soc., 101*, 592 (1979).
10. A. J. Deeming, S. S. Ullah, A. J. P. Domingos, B. F. G. Johnson, J. Lewis, *J. Chem. Soc., Dalton Trans.*, 2093 (1974).
11. A. D. Charles, P. Divers, B. F. G. Johnson, K. D. Karlin, J. Lewis, A. V. Rivera, G. M. Sheldrick, *J. Organomet. Chem., 128*, C31 (1977).

5.8.2.3.11. Metal–Carbene Complexes from Olefin Metathesis Reactions.

It is now generally accepted that the chain propagation steps in transition-metal-catalyzed metathesis of olefins involve reaction of an olefin with an intermediate carbene complex, probably via a carbene–olefin complex, to form a metallocyclobutane. This can cleave to give a new carbene complex together with the metathesized olefin[1,2]:

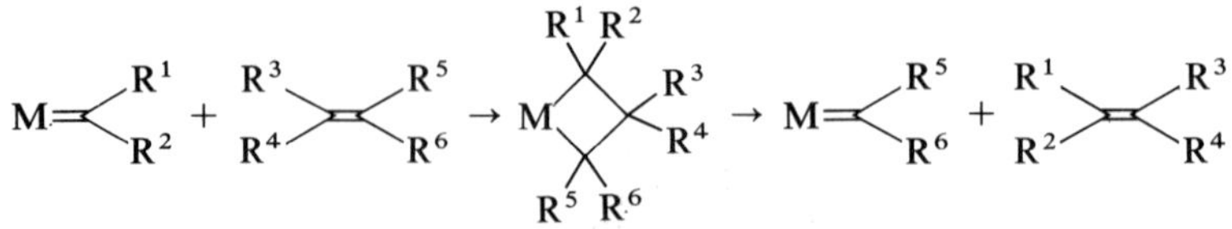
(a)

A stoichiometric carbene complex olefin reaction that models these steps was discovered during studies of the reactions of electron-rich olefins with transition-metal complexes[3,4]. Such Rh(I) complexes as $Rh(PPh_3)_2(L)Cl$ (L = CO, PPh_3) catalyze metathesis of $[CH_2N(Ph)]_2C{=}C[N(Ph)CH_2]_2$ and $[CH_2N(p\text{-}C_6H_4CH_3)]_2C{=}C\text{-}[N(p\text{-}C_6H_4CH_3)CH_2]_2$, and evidence in favor of an alkylidene–metallocyclobutane mechanism for this reaction includes the occurence of the alkylidene exchange:

$$RhCl(PPh_3)C[N(Ph)CH_2]_2 \xrightarrow{[CH_2N(p\text{-}C_6H_4CH_3)]_2C{=}C[N(p\text{-}C_6H_4CH_3)CH_2]_2} RhCl(PPh_3)_2C[N(p\text{-}C_6H_4CH_3)CH_2]_2 \quad \text{(b)}$$

The diphenylalkylidene[5–7], $W(CO)_5CPh_2$, also undergoes alkylidene exchange reactions when treated with some olefins, including ethylvinyl ether, trans-2-butene, isobutylene and 1-phenyl-1-methoxyethylene, as demonstrated by the organic products of the reaction[5]. The product carbene complex is only isolable when it contains a stabilizing alkoxy substituent:

$$W(CO)_5{=}CPh_2 + CH_2{=}CPh(OCH_3) \rightarrow \underset{24\%}{W(CO)_5{=}CPh(OCH_3)} + \underset{26\%}{Ph_2C{=}CH_2} \quad \text{(c)}$$

The stability of the alkoxycarbene complexes $W(CO)_5C(OR)R'$ suggests that the action of $W(CO)_5CPh_2$ on vinyl ethers should provide a general route to alkoxy carbene complexes, but these reactions are sensitive to the nature of the vinyl ether. In, e.g., the ring-opening polymerization of cyclic vinyl ethers, the reactivity with $W(CO)_5CPh_2$ is a function of ring size[8]. At 60°C 1-ethoxycyclopentene reacts to give, inter alia, an ethoxycarbene complex:

$$W(CO)_5CPh_2 + \overline{CH_2CH_2CH_2CHC}OEt \rightarrow W(CO)_5C(OEt)(CH_2)_3CHCPh_2 \quad \text{(d)}$$

In contrast, 1-ethoxycylohexene and 1-ethoxycyclooctene do not react with the W complex. The structure of the carbene complex formed by reacting of $W(CO)_5CPh_2$ with 2-ethoxynorbornene shows no interaction between the metal center and the double bond formed in the modified ligand during such reactions[9].

In one example a carbene ligand in $W(CO)_5CPh_2$ exchanges metathetically for another carbene lacking a stabilizing heteroatomic substituent; the reaction is photochemically promoted[10]:

$$W(CO)_5CPh_2 \xrightarrow[CH_2CPh(p\text{-}C_6H_4OCH_3)]{h\nu} W(CO)_5C(Ph)(p\text{-}C_6H_4OCH_3) + CH_2CPh_2 \quad \text{(e)}$$

Cyclopropanation is always a possible alternative to metathesis when carbene complexes are treated with alkenes[11–13], and differences in either the olefin substrate or the complex can influence the outcome. The reactions of $W(CO)_5C(p\text{-}C_6H_4CH_3)_2$ with internal olefins, e.g., give primarily metathetical products, while those with terminal olefins give both cyclopropanes and metathetical products, probably because cyclopropanation is favored when methathesis would lead to an unstable methylene complex[14,15]. The reaction of $W(CO)_5CPhH$ with olefins, which is facile and gives exclusively cyclopropanation, has been suggested to occur without CO loss through a 7-coordinated metallocyclobutane instead of a 6-coordinated metallocyclobutane analogous to that proposed for reactions of $W(CO)_5CPh_2$ and $W(CO)_5C(p\text{-}C_6H_4CH_3)_2$ [16].

A detailed study of intramolecular cyclopropanation and olefin metathesis reactions of $W(CO)_5C(OCH_2CH_2CH{=}CHOCH_3)C_6H_4\text{-p-}CH_3$, which contains a trans-vinyl-ether side chain, supports these arguments[17].

The utility of carbene complex synthesis via olefin metathesis is restricted by the need to begin with a reactive alkylidene complex prepared by a different route, and except in special circumstances (see below), modification of the primary route provides easier access.

Carbene complexes for which metathesis is the preferred synthetic method include the W alkylidene complexes, $W(O)(PEt_3)_2Cl_2CHR$ (R = CH_3, Ph, C_2H_5, H), prepared from the corresponding neopentylidene complex[18,19]:

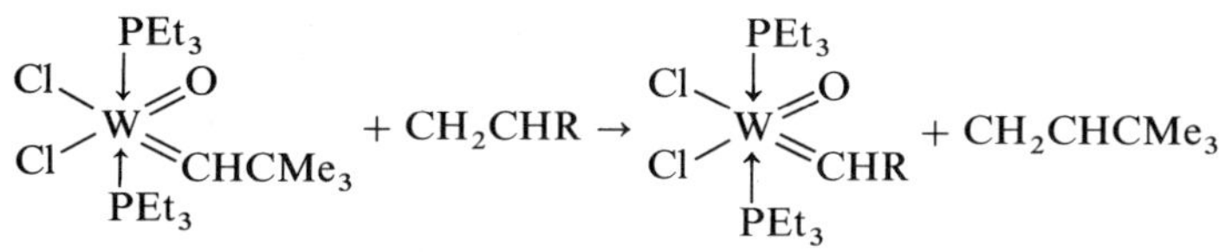 (f)

These reactions are slow in the absence of catalysts ($\approx$1 d), but rapid ($\approx$1 h) in the presence of a trace of $AlCl_3$, which generates a vacant site for alkene coordination by complexing PEt_3.

Metathesis is the best route to these complexes. The preparation of W alkylidenes by α-hydride abstraction has had only limited success[19], and the neopentylidene substrate is prepared by an unusual heterometallic exchange reaction[19,20]:

$$Ta(CHCMe_3)(PEt_3)_2Cl_3 + W(O)(OCMe_3)_4 \rightarrow Ta(OCMe_3)_4Cl + W(O)(PEt_3)_2Cl_2(CHCMe_3) \qquad (g)$$

Analogous reactions can occur with high-valent Nb and Ta complexes.[18,20]. Reaction of $Ta(CHCMe_3)(OCMe_3)_2(PMe_3)Cl$ with phenyl-substituted alkenes forms the corresponding benzylidene complexes:

$$M(CHCMe_3)(OCMe_3)_2(PMe_3)Cl \xrightarrow{PhCHCHR} M(CHPh)(OCMe_3)_2(PMe_3)Cl \qquad (h)$$

where R = H, CH_3, Ph; M = Nb, Ta. The related haloalkylidene complex, $Ta(CHCMe_3)(PMe_3)_2Cl_3$, gives only products derived from an intermediate metallocyclobutane by nonmetathetical rearrangements; the Nb complexes behave similarly.

The discovery of a route to W alkylidene complexes based on hydrogen transfer from amido to alkylidyne ligands[21] has allowed access to imidoneopentylidene complexes, $W(CHCMe_3)(NAr)(OR)_2$ [Ar = 2,6-C_6H_3(Pr-i)$_2$; R = t-Bu, $CMe_2(CF_3)$, $CMe(CF_3)_2$. $C(CF_3)_2CF_2CF_2CF_3$][22]. Their reactivity with alkenes depends on the alkoxide and the alkene, with a fine balance between metathesis and metallocyclobutane formation. The complex with R = $CMe_2(CF_3)$ gives the alkylidene complexes $W(CHR')(NAr)(OR)_2$ (R' = CHEt, $CHSiMe_3$) directly from 3-hexene and $CH_2CHSiMe_3$, whereas the complex with R = $CMe(CF_3)_2$ gives $W(CHEt)(NAr)(OR)_2$ directly from 3-hexene and $W(CHSiMe_3)(NAr)(OR)_2$ from $CH_2CHSiMe_3$ via a stable metallocyclobutane. No stable metathesis product is formed if R = t-Bu or $C(CF_3)_2CF_2CF_2CF_3$, and ethylene, 1-pentene and cis-2-pentene give metallocycles or complex products with all substrates.

Bulky alkoxides or thiolates also control the reactions of alkenes with the Ta neopentylidene complexes $Ta(CHCMe_3)(L)_3(THF)$ [L = O-2,6-C_6H_3-i-Pr_2 (DIPP), O-2,6-$C_6H_3Me_2$ (DMP), S-2,4,6-C_6H_2-i-Pr_3 (TIPT)][23]. The DIPP complex reacts with $CH_2CHSiMe_3$ to give $Ta(CHSiMe_3)(DIPP)_3(THF)$; other olefins give metallocyclobutanes. The TIPT complex is inert to common olefins but reacts with norbornene to form a living polymer shown to be an alkylidene complex.

Cationic W alkylidene complexes formed by halide abstraction from $W(CHCMe_3)(OCH_2CMe_3)_2X_2$ (X = Cl, Br) give observable (NMR) alkylidene complexes with acyclic olefins[24] or norbornene[25].

(N. J. COOPER, R. L. THOMPSON)

1. J. J. Rooney, A. Stewart, *Catal., 1*, 277 (1977).
2. R. H. Grubbs, *Comp. Organomet. Chem., 8*, Ch 54 (1982).
3. D. J. Cardin, M. J. Doyle, M. F. Lappert, *J. Chem. Soc.; Chem. Commun.*, 927 (1972).
4. D. J. Cardin, B. Cetinkaya, M. F. Lappert, *Chem. Rev., 72*, 545 (1972).
5. C. P. Casey, T. J. Burkhardt, *J. Am. Chem. Soc., 96*, 7808 (1974).
6. C. P. Casey, T. J. Burkhardt, C. A. Bunnell, J. C. Calabrese, *J. Am. Chem. Soc., 99*, 2127 (1977).
7. E. O. Fischer, W. Held, F. R. Kreissl, *Chem. Ber., 110*, 656 (1977).
8. J. Levisalles, H. Rudler, D. Villemin, *J. Organomet. Chem., 146*, 259 (1978).
9. J. Levisalles, H. Rudler, D. Villemin, J. Daran, Y. Jeannin, L. Martin, *J. Organomet. Chem., 155*, Cl (1978).
10. L. Fong, N. J. Cooper, *J. Am. Chem. Soc., 106*, 2595 (1984).
11. K. H. Dotz, E. O. Fischer, *Chem. Ber., 105*, 1356, 3966 (1972).
12. K. A. Kremer, P. Helquist, R. C. Kerber, *J. Am. Chem. Soc., 103*, 1862 (1981).
13. F. J. Brown, *Prog. Inorg. Chem., 27*, 1 (1980) and refs. therein.
14. C. P. Casey, H. E. Tuinstra, M. C. Saeman, *J. Am. Chem. Soc., 98*, 608 (1976).
15. C. P. Casey, *J. Organomet. Chem. Library, 1*, 397 (1976).
16. C. P. Casey, S. W. Polichnowski, *J. Am. Chem. Soc., 99*, 6098 (1977).
17. C. P. Casey, N. L. Hornung, W. P. Kosar, *J. Am. Chem. Soc., 109*, 4908 (1987).
18. R. R. Schrock, S. Rocklage, J. Wengrovius, G. Rupprecht, J. Fellmann, *J. Mol. Catal., 8*, 73 (1980).
19. J. H. Wengrovius, R. R. Schrock, M. R. Churchill, J. R. Missert, W. J. Youngs, *J. Am. Chem. Soc., 102*, 4515 (1980).
20. S. M. Rocklage, J. D. Fellman,, G. A. Rupprecht, L. W. Messere, R. R. Schrock, *J. Am. Chem. Soc., 103*, 1440 (1981).
21. S. M. Rocklage, R. R. Schrock, M. R. Churchill, H. Wasserman, *J. Organomet. Chem., 1*, 1332 (1982).
22. R. R. Schrock, R. T. DePue, J. Feldman, C. J. Schaverien, J. C. Dewan, A. H. Liu, *J. Am. Chem. Soc., 110*, 1423 (1988).
23. K. C. Wallace, A. H. Liu, J. C. Dewan, R. R. Schrock, *J. Am. Chem. Soc., 110*, 4964 (1988).
24. J. Kress, M. Wesolek, J. A. Osborn, *J. Chem. Soc., Chem. Commun.*, 514 (1982).
25. J. Kress, J. A. Osborn, R. M. E. Greene, K. J. Ivin, J. J. Rooney, *J. Am. Chem. Soc., 109*, 899 (1987).

5.8.2.4. from Conjugated Dienes

5.8.2.4.1. Conjugated Diene Complexes by Reaction with Metal Complexes by Ligand Displacement.

Diene and polyene complexes with η^4-coordination are known for nearly all the transition metals from group IVA to VIII. Table I offers examples for Cr—Rh. Ligand

TABLE 1. SYNTHESES OF η^4-DIENE COMPLEXES BY LIGAND DISPLACEMENT BY CONJUGATED DIENES.

Metal	Ligand displaced[a]	Typical diene	Conditions[b]	Ref.
Ta	C_6H_5	Cyclooctatetraene		6
Cr	CO	1,3-Butadiene	hν	7
	CH_3CN	Fulvene		8, 9
	Phosphole	1,3-Butadiene	hν	10
Mo	CO	E,E-1,4-Diphenyl-1,3-butadiene		11
	CH_3CN	Fulvene		8, 9
	Phosphole	1,3-Butadiene	hν	10
	CO	1,3-Butadiene	hν	12
W	CO	E,E-1,4-Diphenyl-1,3-butadiene		11
	CH_3CN	Fulvene		8, 9
	Phosphole	1,3-Butadiene	hν	10
Mn	CO and Mn [$Mn_2(CO)_{12}$]	1-Phenylphosphole	hν	13
	CO	1,3-Butadiene	hν	14
	CO	Cyclooctatetraene		15
	NO and CO	1,3-Butadiene	hν	14
Re	CO	1,3,5-Cycloheptatriene	hν	16
	CO	1,3-Butadiene	hν	17
Fe	CO	3,5-Heptadiene-2-one	Thermal or hν	3, 18, 19
	CO	Tropone	$(CH_3)_3NO$	20
	CO and Fe[$Fe_2(CO)_9$]	1,3-Butadiene	Ultrasound	21
	CO and Fe [$Fe_2(CO)_9$]	3,5-Heptadien-2-ol		3, 18
	CO and Fe [$Fe_3(CO)_{12}$]	Cyclooctatetraene		3, 22
	Benzylidineacetone	1,3-Cyclohexadiene		1
	Diene	Isoprene		23
	Hexamethylbenzene	1,3-Cyclohexadiene		24
	PF_3	1,3-Butadiene	Thermal or hν	25, 26
Ru	CO and Ru [$Ru_3(CO)_{12}$]	1,3-Cyclohexadiene		27
	1,5-cyclooctadiene	1,3-Butadiene		28
Os	CO and Os [$Os_3(CO)_{12}$]	Ethyl sorbate; 2,3-dimethyl-1,3-butadiene		29, 30
Co	CO	1,3-Cyclohexadiene		4
	CO and Co [$Co_2(CO)_8$]	1,3-Butadiene	Thermal or hν	31
Co	N_2	2,3-Dimethyl-1,3-butadiene		32
	Diene	1,3-Butadiene		32
	2,2′-Bipyridyl	Isoprene		34
Rh	Cyclooctene	2,3-Dimethyl-1,3-butadiene		35
Ir	μ-Cl	2-Methyl-1,3-cyclohexadiene		36
Ni	CO	Tetracyclone		37
	$P(OC_6H_4\text{-o-}C_6H_5)_3$	Cyclopentadiene		38
Pd, Pt	C_6H_5CN or Cl^-	Pentamethylcyclopentadiene		32

[a] Multinuclear reactants are given in brackets.
[b] No entry for thermal reactions, hν for photoreactions.

displacement by diene nucleophiles is versatile. Complexes of heterodienes such as benzylidine acetone[1] and 2-benzolypyridine anil[2] are also prepared by nucleophilic substitution. The method is general only for Cr—Co (see Table 1), other metals preferring alternative bonding; e.g., with fulvenes $V(CO)_6$ gives only η^5-cyclopentadienyl complexes, while Pd and Pt complexes cause rearrangement of 1,3-cyclooctadiene to give 1,5-cyclooctadiene complexes.

η^4-Diene complexes of Fe form from precursors such as unsaturated alcohols, halides and strained compounds, e.g., cyclopropenes, especially in lieu of unstable dienes[3]. Cis–trans isomerization, hydrogen shifts, skeletal rearrangement and condensation occur, but they can be avoided or minimized by use of reactive metal derivatives, e.g., $Fe_2(CO)_9$ or benzylidine acetone $Fe(CO)_3$, and mild conditions.

Syntheses can involve redox reactions at the metal in addition to nucleophilic displacement:

$$RhCl_3(H_2O)_3 + \text{cyclohexadiene} \xrightarrow{\text{EtOH}} [(\eta^4\text{-cyclohexadiene})RhCl]_2 \qquad (a)^4$$

The hν-induced reductive elimination of biphenyl from diphenylyirconocene yields (η^4-diene)zirconocene complexes[5].

(C. P. LILLYA)

1. G. Evans, B. F. G. Johnson, J. Lewis, *J. Organomet. Chem., 102*, 507 (1975).
2. H. W. Frühauf, F.-W. Grevels, A. Lander, *J. Organomet. Chem., 178*, 349 (1979).
3. The following reviews and compendia on diene Fe complexes are especially valuable: (a) A. J. Birch, I. D. Jenkins, in *Transition Metal Organometallics in Organic Synthesis*, Vol 1, H. Alper, ed., Academic Press, New York, 1976, p. 1; (b) R. B. King, in *The Organic Chemistry of Iron*, Vol 1, E. A. Koerner von Gustorf, F.-W. Grevels, I. Fischler, eds., Academic Press, New York, 1981, p. 525; (c) C. R. Bock, E. A. Koerner von Gustorf, *Adv. Photochem., 10*, 221 (1977); (d) R. C. Kerber, in ref. 7b, Vol 2, 1981, p. 1; (e) *Gmelin Handbuch der Anorganischen Chemie*, 8th ed., *Organoiron Compounds*, Vols. B6, B7, Springer Verlag, Berlin, 1981.
4. B. F. G. Johnson, J. Lewis, D. J. Yarrow, *J. Chem. Soc., Dalton Trans.*, 2084 (1972).
5. G. Erker, J. Wichen, K. Engel, C. Krüger, *Chem. Ber., 115*, 3300 (1982).
6. L. J. Guggenberger, R. R. Schrock, *J. Am. Chem. Soc., 97*, 6693 (1975).
7. I. Fischler, M. Budzwait, E. A. Koerner von Gustorf, *J. Organomet. Chem., 105*, 325 (1976).
8. F. Edelmann, U. Behrens, *J. Organomet. Chem., 134*, 31 (1977).
9. Ph. Narbel, T. Boschi, R. Roulet, P. Vogel, A. A. Pinkerton, D. S. Schuranzenbach, *Inorg. Chim. Acta, 36*, 161 (1979).
10. S. Oskan, Z. Ozen, *Doga*: *Kim. Ser., 10*, 53 (1986); *Chem. Abstr., 107*, 23,441m (1987).
11. A. N. Nesmeyanov, V. V. Krivyhh, V. S. Kaganovich, M. I. Rybinskaya, *J. Organomet. Chem., 102*, 285 (1975).
12. C. G. Kreitzen, G. Wendt, J. Kaub, *Chem. Ber., 122*, 215 (1989).
13. F. Mathey, *J. Organomet. Chem., 93*, 377 (1975).
14. M. Heberhold, A. Razavi, *Angew. Chem., Int. Ed. Engl., 14*, 351 (1975).
15. R. B. King, M. N. Ackermann, *Inorg. Chem., 13*, 637 (1974).
16. R. Davis, J. A. O. Ojo, *J. Organomet. Chem., 110*, C39 (1976).
17. E. Koerner von Gustorf, Z. Pfaffer, R.-W. Grevels, *Z. Naturforsch., Teil B, 208*, 66 (1971).
18. N. A. Clinton, C. P. Lillya, *J. Am. Chem. Soc., 92*, 3058 (1970).
19. M. Cais, N. Maoz, *J. Organomet. Chem., 5*, 370 (1966).
20. Y. Shvo, E. Hazum, *J. Chem. Soc., Chem. Commun.*, 829 (1975).
21. S. V. Ley, C. M. R. Low A. D. White, *J. Organomet. Chem., 302*, C13 (1986).
22. W. McFarlane, G. Wilkinson, *Inorg. Synth., 8*, 178 (1966).
23. F.-W. Grevels, U. Feldhoff, K. Schnieder, *Abstr. VIIth Int. Conf. Organomet. Chem.*, Venice, 1975; see King, ref. 7b.
24. S. R. Weber, H. H. Brintzniger, *J. Organomet. Chem., 127*, 45 (1977).
25. J. D. Warren, R. J. Clark, *Inorg. Chem., 9*, 373 (1970).

26. Th. Kruck, L. Knoll, J. Laufenberg, *Chem. Ber., 106*, 697 (1973).
27. T. H. Whiteside, R. A. Budnik, *Inorg. Chem., 14*, 664 (1975).
28. R. Ruh, W. von Phillipsborn, *J. Organomet. Chem., 127*, C59 (1977).
29. A. Oudeman, Ph.D. Thesis, Univ. of Calgary, Alberta, Canada, 1977.
30. E. O. Fischer, K. Bittler, H. P. Fritz, *Z. Naturforsch., Teil B, 18*, 83 (1963).
31. P. McArdle, A. R. Manning, *J. Chem. Soc., A*, 2123 (1970).
32. P. V. Rinze, *Angew. Chem., Int. Ed. Engl., 13*, 336 (1974).
33. G. Mestroni, A. Camus, *Inorg. Nucl. Chem. Lett., 5*, 215 (1967).
34. G. Mestroni, A. Camus, G. Cocevar, *J. Organomet. Chem., 29*, C17 (1971).
35. L. Porri, A. Lionetti, *J. Organomet. Chem., 6*, 442 (1966).
36. J. Muller, H. Menig, P. V. Rinze, *J. Organomet. Chem., 181*, 387 (1979).
37. E. Weiss, W. Hübel, *J. Inorg. Nucl. Chem., 11*, 42 (1959).
38. M. Englert, P. W. Jolly, G. Wilke, *Angew. Chem., Int. Ed. Engl., 10*, 77 (1971).
39. P. V. Balakrishnan, P. M. Maitlis, *J. Chem. Soc., A*, 1715, 1721 (1971).

5.8.2.4.2. Diene Complexes by Nucleophilic Attack on Metal Cationic Complexes.

Cations derived from Fe, Ru, Mo, Cr, Mn, Co, Rh, W, Re and Os [1] complexed with olefinic or other donor systems such as CO or PPh_3 normally have 18-electron configurations. Although most of the charge is on the metal, the carbon system behaves as a stabilized carbocation and is capable of reacting on carbon with nucleophiles. If a strongly electron-attracting group such as $Fe(CO)_3$ is attached, it reacts at the **I** or **V** positions in the dienyl case, like a classical carbonium ion. Cations such as **I** react stereospecifically, α-exo, to form[2] (**II**) if products are kinetically controlled; thermodynamic contol leads to a mixture of α-exo and β-endo isomers, e.g., with OH [2] or OMe [3] anions. Whether this is due to a genuine equilibrium of products is not certain, since a β substituent is more stable to acid-catalyzed reversal than an α substituent. The absolute configuration of **I**, (R = Me, OMe) as shown is known to be that of the + isomer[4].

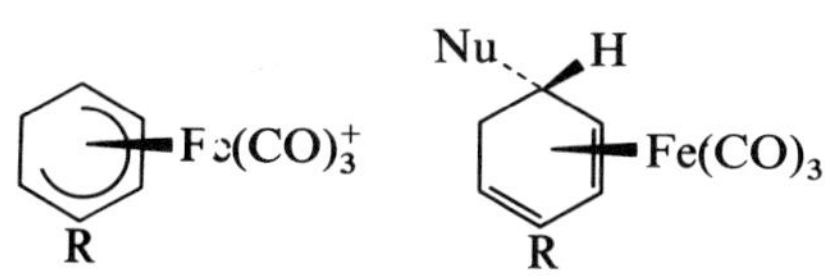

I **II**

With a laterally attached, more electron-donating metal group, reaction occurs at the **II** position, which then has the major charge. This is possible, in contrast to classical carbonium ion behavior, because of the possibility of simultaneous σ and π bonding, exemplified in the products of reaction of the (cycloheptadienyl)$Fe(CO)_3$ cation[5] and the conversion of **III** into **IV**[6]. The 16-electron $MoC_7H_9(CO)_3$ cation reacts at position **III** with PPh_3 with loss of the metal, probably initiated by reaction with the Mo [7].

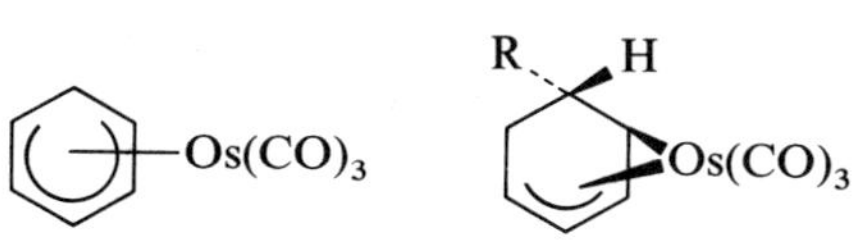

III **IV**

where R = H, CN.

The necessarily cisoid (η^5-diene)$Fe(CO)_3$ cations in the open-chain series retain this carbon configuration with reactive nucleophiles[8], but with weak bases, including water, the carbon becomes transoid to the complexed diene owing to a π-allylic intermediate. In all cases attack is α-exo.

Complexes of unsymmetrical dienes are chiral[2], and the reaction stereospecificity results in formation of a new asymmetric center, fully resolved and of known absolute configuration relative to the precursor[4,9]. This is useful in organic synthesis, since cations can be resolved through related neutral complexes such as bornyl ethers[10] or carboxylic acids[11].

Complexed cations are the synthetic equivalents of often unknown organic cations: e.g., **I**, (R = OMe) is equivalent either to **V** or to **VI** according to the subsequent treatment of the intermediate (**II**, R = OMe)[2].

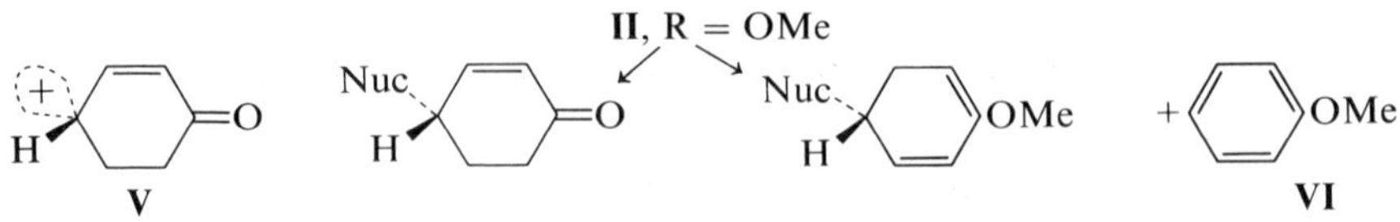

Rules[1] indicate which complexed olefinic structure becomes subject to attack. Some results are surprising to intuition, e.g., reaction of a benzenoid rather than a tropenoid system. Complications are avoided by using rather noncompetitive ligands such as CO, PPh_3.

Kinetics of reacting **I** R = H, OMe with nucleophiles such as the enol of pentan-2,4-dione[12] aromatic amines[13], phosphorus derivatives[14] and some reactive aromatic compounds[15], and relative rates with substituted (cyclohexadienyl)$Fe(CO)_3$ cation[16] have been examined. These behave as classically expected, but in contrast to 1- or 2-OMe, a 3-OMe increases rate through its inductive effect. The kinetics agree with electrophilic substitution with the possible intermediacy of π complexes[12–15]. Because aryl (N-diene)$Fe(CO)_3$ complexes can rearrange by dissociation into C-aryl derivatives[17], intermediates could also involve reaction with an N of an indole or a MeO (oxonium cation) of MeO–aromatics.

Rates of reaction also depend on the metal and the other ligands; e.g., one series of rates decreases: $MnC_7H_8(CO)_3$, $FeC_4H_4(CO)_2NO$, $FeC_6H_7(CO)_3$, $Fe(C_6H_6OMe\text{-}2)(CO)_3$, $OsC_6H_7(CO)_3$, $RuC_6H_7(CO)_3$, $CrC_7H_7(CO)_3$ [18]. With the same complexing structures the rates Fe > Os > Ru differ over several orders of magnitude; Fe is frequently advantageous as well as cheap.

Among nucleophiles used are enols[2,12] (including mono- and bis-TMS ethers, which are particularly favorable[19]); derivatives of P [14,20], N [2,13] and S in hydrogen sulfite; dithionate and sulfide[20] Si, Ge, Sn [21] and alkyl or aryl in derivatives of Cd, Zn [22] or Li [23]. The Li reagents in particular require special experimental conditions, including dichloromethane as solvent and radical scavengers. Many nucleophiles cause electron addition to the cations, with formation of radicals. Aromatic substitutions displace not only H but $SnMe_3$ or $SiMe_3$ [15,24]. Allylsilanes[25] and trialkyl alkynylborates[26] provide synthetically useful products.

Theoretical calculations on unsubstituted $FeC_6H_7(CO)_3$ indicate that the 1,5 positions of reactivity are chosen not on the basis of charge but of free valency[27]. Despite this, with small charged anions the distinction between 1 and 5 in unsymmetrically substituted (cyclohexadienyl)$Fe(CO)_3$ cations may be decided by charge distribution[1,2].

Carbon-13 NMR shifts correlate with reactivities[28]. Transition states may vary with the nucleophile[23]. Steric effects of substituents are predictable: reaction at an occupied position is slow[2], and the seven-membered ring series is slower than the six.

Some aspects of the temperature dependency of the regioselectivity of nucleophilic additions have been studied[29].

(A. J. BIRCH)

1. S. G. Davies, M. L. H. Green, D. M. P. Mingos, *Tetrahedron Rep.*, 57 (1977).
2. A. J. Birch, *Ann. N.Y. Acad. Sci.*, *333*, 107 (1980); A. J. Birch, B. M. R. Bandara, K. Chamberlain, B. Chauncy, P. Dahler, A. I. Day, I. D. Jenkins, L. F. Kelly, T. C. Khor, G. Kretschmer, A. J. Liepa, A. S. Narula, W. D. Raverty, E. Rizzardo, C. Sell, G. R. Stephenson, D. J. Thompson, D. J. Williamson, *Tetrahedron*, (Woodward Issue), *37*(Suppl.), 289 (1981).
3. K. E. Hine, B. F. G. Johnson, J. Lewis, *J. Chem. Soc., Chem. Commun.*, 81 (1975).
4. A. J. Birch, W. D. Raverty, G. R. Stephenson, *J. Chem. Soc., Chem. Commun.*, 857 (1980).
5. R. Edwards, J. A. S. Howell, B. F. G. Johnson, J. Lewis, *J. Chem. Soc., Dalton Trans.*, 2105 (1974).
6. A. L. Burrows, B. F. G. Johnson, J. Lewis, D. G. Parker, *J. Organomet. Chem.*, *194*, C11 (1980).
7. A. Salzer, H. Werner, *J. Organomet. Chem.*, *81*, 101 (1975).
8. J. G. Mahler, R. Pettit, *J. Am. Chem. Soc.*, *85*, 3955, 3959 (1963); see also G. Maglio, R. Palumbo, *J. Organomet. Chem.*, *76*, 367 (1974); C. R. Jablonski, T. S. Soernson, *Can. J. Chem.*, *52*, 2085 (1974); D. G. Gresham, D. J. Kowalski, C. P. Lillya, *J. Organomet. Chem.*, *144*, 71 (1978).
9. A. J. Birch, G. R. Stephenson, *Tetrahedron Lett.*, 779 (1980).
10. A. J. Birch, T. C. Khor, B. M. R. Bandara, to be published.
11. A. J. Birch, B. M. R. Bandara, *Tetrahedron Lett.* 2981 (1980).
12. C. A. Mansfield, L. A. P. Kane-Maguire, *J. Chem. Soc., Dalton Trans.*, 2187 (1976).
13. G. R. John, L. A. P. Kane-Maguire, *J. Chem. Soc., Dalton Trans.*, 873, 1196 (1979); see also T. I. Odiaka, L. A. P. Kane-Maguire, *Inorg. Chim. Acta*, *37* 85 (1979).
14. M. Gower, G. R. John, L. A. P. Kane-Maguire, T. I. Odiaka, *J. Chem. Soc., Dalton Trans.*, 2003 (1979).
15. G. R. John, L. A. P. Kane-Maguire, *Inorg. Chim. Acta*, in press (1981); G. R. John, C. A. Mansfield, L. A. P. Kane-Maguire, *J. Chem. Soc., Dalton Trans.*, 574 (1977).
16. A. J. Birch, D. Bogsanyi, L. F. Kelly, *J. Organomet. Chem.*, *214*, C39 (1981).
17. A. J. Birch, A. J. Liepa, G. R. Stephenson, *Tetrahedron Lett.*, 1723 (1979).
18. R. John, L. A. P. Kane-Maguire, *J. Chem. Soc., Dalton Trans.*, 1196 (1979).
19. A. J. Birch, A. S. Narula, P. Dahler, G. R. Stephenson, L. F. Kelly, *Tetrahedron Lett.*, 979, 2455 (1980).
20. A. J. Birch, I. D. Jenkins, A. J. Liepa, *Tetrahedron Lett.*, 1723 (1975).
21. N. S. Nametkin, V. D. Tyurin, V. I. Ivanov, *Izv. Akad. Nauk. SSSR, Ser. Khim.*, 1911 (1975).
22. A. J. Birch, A. J. Pearson, *J. Chem. Soc., Perkin Trans. I*, 954 (1976).
23. B. M. R. Bandara, A. J. Birch, T. C. Khor, *Tetrahedron Lett.*, 3625 (1980).
24. G. R. John, L. A. P. Kane-Maguire, C. Eaborn, *J. Chem. Soc., Chem. Commun.*, 481 (1975).
25. L. F. Kelly, A. S. Narula, A. J. Birch, *Tetrahedron Lett.*, 871 (1980); *Tetrahedron*, 38, 1813 (1982).
26. A. Pelter, K. J. Gould, L. A. P. Kane-Maguire, *J. Chem. Soc., Chem. Commun.*, 1029 (1974).
27. D. W. Clack, M. Monshi, L. A. P. Kane-Maguire, *J. Organomet. Chem.*, *107*, C40 (1976).
28. A. J. Birch, P. W. Westerman, A. J. Pearson, *Aust. J. Chem.*, *29*, 1671 (1976).
29. A. J. Birch, G. R. Stephenson, *J. Organomet. Chem.*, *218*, 91 (1981).

5.8.2.4.3. Cationic Dienyl Complexes from Metal Diene Complexes.

These methods involve removal of an anion from carbon: the anions may be of classical type or involve hydride or organic groups such as α-exoCN or α-CO_2H (but not β-CO_2H removed by triphenylmethylium ion. Sulfuric acid also can reverse the addition of nucleophiles with C—O, C—N, C—S or C—C cleavage[1]. Obtaining a desired substituted cation relates to obtaining a diene–metal precursor amenable to conversion reaction. Hydride removal is useful, since the precursors are more readily obtained than those with the appropriately positioned leaving group.

In the open-chain (η^4-diene)$Fe(CO)_3$ series, the trans-Me complex (**I**, R = H) does not react with trityl cation, in contrast to the cis-Me (**II**), which yields **III**. However, **I** (R = OH) is readily converted by HBF_4 into the salt of **III**[2]. Product configurations show that the group departs from the α-exo configuration.

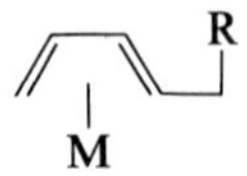

I

II

III

where M = $Fe(CO)_3$.

In the substituted (cyclohexadiene)$Fe(CO)_3$ series, which is the most intensively investigated, the bulk of the trityl group confers some advantages of specificity but also drawbacks due to steric hindrance. The 1-Me, 4-$CHMe_2$ derivative, e.g., reacts only adjacent to the Me [3], but hydride is not removed from a ring position carrying only β-endot H, or from an α-exo position adjacent to an α substituent[4]. In all cases the abstraction is wholly α-exo. When reaction is slow the product may correspond not to the precursor, but to an acid-catalyzed isomerization product; this can be avoided by treatment with a solid base such as $CaCO_3$.

Hydride abstraction from unsymmetrically substituted (cyclohexadiene) $Fe(CO)_3$ derivatives may lead to mixtures of salts[4]. The 1-OMe(**IV**), for example, yields **V** (R = OMe) with some **VI**; the former is hydrolyzed by water to the neutral separable complex 7 from which the salt can be regenerated by reagents such as trimethyloxonium fluoroborate. The 2-OMe (diene) yields **VI** with a little of the 3-OMe cation[4,5]. The transition state of the reaction may resemble the cationic product. The 1-CO_2Me (diene) also reacts in the **V** position[4] through initial attachment of the trityl to the carbonyl oxygen. Considerable experimental information is available on the course of abstractions[2,6].

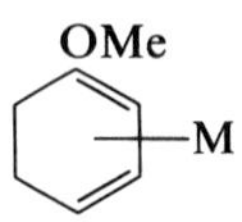

IV

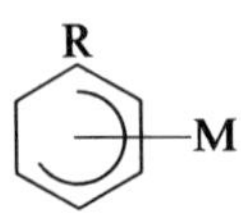

V

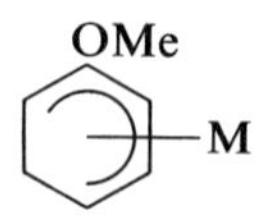

VI

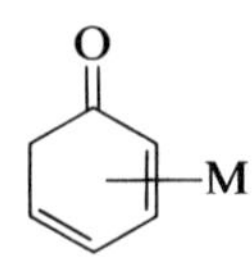

VII

where M = $Fe(CO)_3$.

Carbocations can be generated in the direct conversion of a cisoid system, by reaction of borohydride or a reagent such as LiBu on **VII** and treatment of the resulting carbinol with HBF_4 to yield **V** (R = H or Bu)[7]. The adjacent transoid carbocations 8 or 9 can isomerize to the more stable **5**. The precursors can either be appropriate carbinols[2,4], or olefins, reacted with acid[8]. The cation (**VIII**) loses a 5 B-D stereospecifically as shown, on the way to **V** (R = $CHMe_2$)[9]. Triene or π-allyl complexes may be intermediates[8]. A Wagner-Meerwein rearrangement of the tosylate of a (cyclo-pentadiene)$Fe(CO)_3$ carbinol leads by ring expansion to a cyclohexadienyl cation[10].

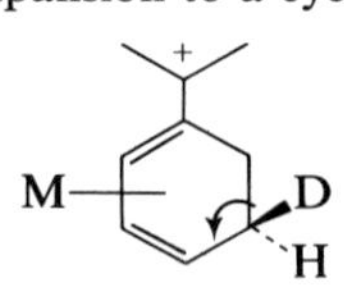

VIII

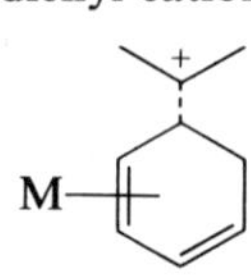

IX

where M = $Fe(CO)_3$.

An olefinic bonds conjugated with the 1 position of a diene complex is highly reactive toward electrophilic reagents because of the stability of the resulting complexed cation. These reagents include not only protons[11] but the trityl cation[11] and bromine[12].

Because of the availability of many required precursors[4], a useful method that is free from the steric demands of the trityl procedure is to react derivatives of (MeO-cyclohexadiene) $Fe(CO)_3$ or cognate complexes with acids under vigorous conditions to cause proton-induced rearrangement of the initial complex structure, with loss of OMe [13], e.g., in the 1,3-dimethoxy complex **X** which by route b yields the synthetically important **XI**[14]. The alternative route a occurs to about 10% with sulfuric acid, but not with TFA. The mechanisms have been confirmed with deuterated acid[4,5]. With optically active complexes some racemization may occur[16].

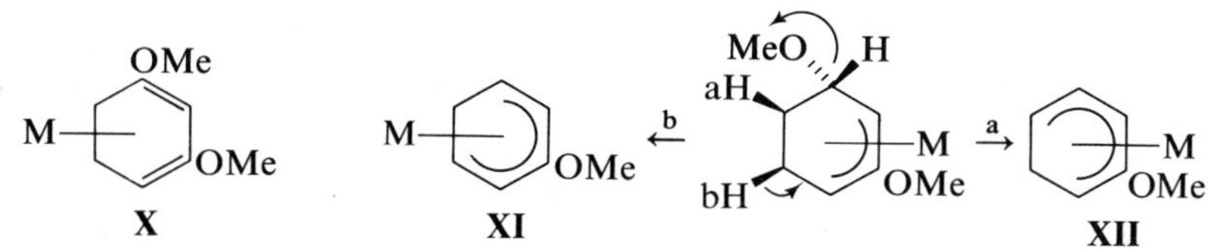

where M = $Fe(CO)_3$;

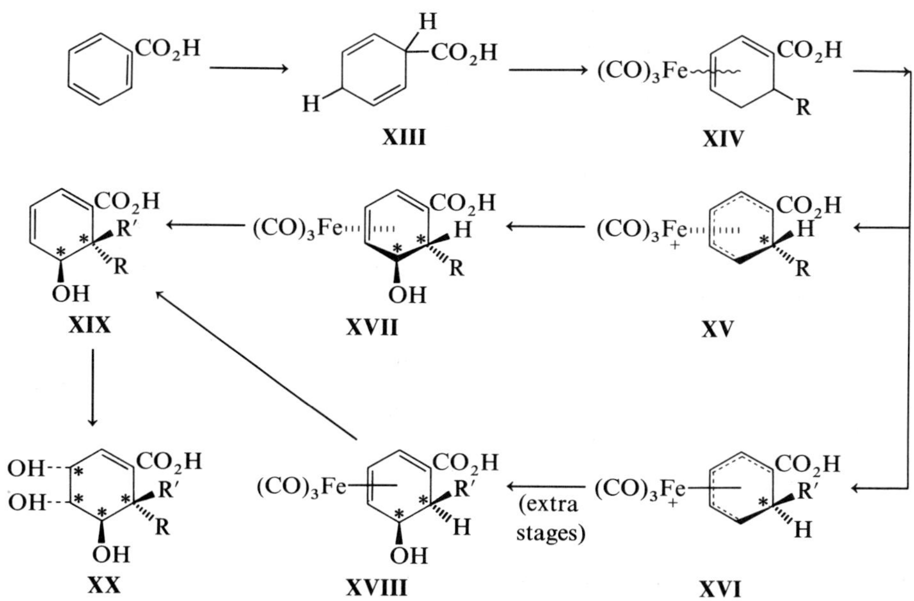

where in both **XV** and **XVI** R = R′ = D; from **XV** on, R = D, R′ = H, and from **XVI** on, R = H, R′ = D; * marks a fully resolved asymmetric center or known configuration, all directed by the original **XV** and **XVI**.

While complexes of Fe have been the most extensively studied, many of the reactions are general, e.g., hydride abstractions occur with complexes of Ru, Os, Co, Cr, Mo and W [17].

Cations resulting from hydride abstraction are byproducts in reactions that involve carbonium ion intermediates, such as strong acids (with disproportionation), or electron-pair acid-catalyzed acylation[18].

(A. J. BIRCH)

1. M. A. Hashmi, J. D. Munro, P. L. Pauson, J. M. Williamson, *J. Chem. Soc.*, A 240 (1967); see also A. J. Birch, H. Fitton, *Aust. J. Chem.*, *22*, 971 (1967); B. M. R. Bandara, A. J. Birch, *Tetrahedron Lett.*, 3499 (1980).
2. J. E. Mahler, R. Pettit, *J. Am. Chem. Soc.*, *85*, 3955 (1963); see also C. R. Jablonski, T. S. Sorenson, *Can. J. Chem.*, *52*, 2085 (1974); D. G. Gresham, D. J. Kowalski, C. P. Lillya, *J. Organomet. Chem.*, *144*, 71 (1978).
3. A. J. Birch, A. I. Day, W. D. Raverty, to be published, 1989.
4. A. J. Birch, K. B. Chamberlain, M. A. Haas, D. J. Thompson, *J. Chem. Soc.*, *Perkin Trans. I*, 1882 (1973); see also A. J. Birch, *Ann. N.Y. Acad. Sci.*, *333*, 107 (1980). A. J. Birch, I. D. Jenkins, in *Transition Metal Complexes of Olefinic Compounds*, H. Alper, ed., Academic Press, New York, 1 (1976).
5. T. H. Whitesides, J. P. Neilan, *J. Am. Chem. Soc.*, *98*, 63 (1976).
6. A. J. Pearson, *J. Chem. Soc.*, *Perkin Trans. I*, 2069 (1977); 400 (1980); *Tetrahedron Lett.*, 3933, 3937 (1980); see also A. J. Birch, A. J. Pearson, *J. Chem. Soc.*, *Perkin Trans. I*, 638 (1978).
7. A. J. Birch, L. F. Kelly, T. C. Khor, to be published (1981).
8. B. F. G. Johnson, J. Lewis, D. G. Parker, G. R. Stephenson, *J. Organomet. Chem.*, *194*, C14 (1980).
9. B. M. R. Bandara, A. J. Birch, *J. Organomet. Chem.*, in press (1981).
10. G. E. Heberich, H. Muller, *Chem. Ber.*, *104*, 2781 (1971).
11. H. J. Dauben, D. J. Bertelli, *J. Am. Chem. Soc.*, *83*, 497 (1961); D. Cunningham, P. McArdle, H. Sherlock, B. F. G. Johnson, J. Lewis, *J. Chem. Soc.*, *Dalton Trans.*, 2340 (1977); M. Brookhart, K. J. Karel, L. E. Nance, *J. Organomet. Chem.*, *140*, 203 (1977); B. F. G. Johnson, J. Lewis, D. G. Parker, S. R. Porte, *J. Chem. Soc.*, *Dalton Trans.*, 794 (1977).
12. A. J. Birch, A. I. Day, to be published.
13. A. J. Birch, M. A. Haas, *J. Chem. Soc.*, C, 2465 (1971).
14. A. J. Birch, L. F. Kelly, A. S. Narula, *Tetrahedron Lett.*, *22*, 1433 (1981); *Org. Synth.* submitted (1981); see also A. J. Birch, L. F. Kelly, D. J. Thompson, *J. Chem. Soc.*, *Perkin Trans. 1*, 1006 (1981); A. J. Birch, P. Dahlert, L. F. Kelly, A. S. Narula, *Tetrahedron Lett.*, *22*, 1433 (1981); A. J. Birch, B. Chauncy, L. F. Kelly, D. J. Thompson, *J. Organomet. Chem.*, *286*, 37 (1985); A. J. Birch, L. F. Kelly, *J. Organomet. Chem.*, *285*, 267 (1985).
15. A. J. Birch, B. Chauncy, L. F. Kelly. D. J. Thompson, *J. Chem. Soc.*, *Perkin Trans. I*, submitted (1981)
16. A. J. Birch, G. R. Stephenson, *Tetrahedron Lett.* 779 (1981).
17. J. Lewis, A. Parkins, *J. Chem. Soc.*, A 1150 (1967); G. R. Johns, L. A. P. Kane-Maguire, *J. Chem. Soc.*, *Dalton Trans.*, 1196 (1979); M. A. Hasmi, S. D. Munro, P. L. Pauson *J. Chem. Soc.*, A, 240 (1967); *J. Chem. Soc.*, 3475 (1961).
18. N. S. Nametkin, A. I. Nekhaev, V. D. Tyurin, S. P. Gubin, L. P. Kolesnikova, *Izv. Akad. Nauk SSSR, Ser. Khim.*, *3*, 676 (1975).

5.8.2.4.4. π-Allyl Complexes by Nucleophilic Attack.

Transition metals promote formation of π-allyl complexes from dienes by nucleophilic attack. Different nucleophiles and dienes participate in the reaction, which may take two routes where the nucleophile and the metal add either from the same (path a) or the opposite face of the diene (path b). Nucleophiles may be divided into two classes: one, which includes alcohols, amines, carboxylates and malonates, reacts preferentially according to (a):

(a)

The other, which includes hydride and alkyl- and phenylcarbanions, reacts preferentially according to (b):

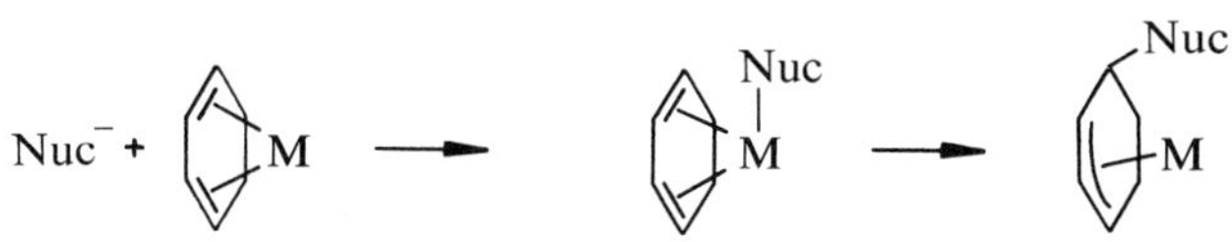

(b)

(For discussion and leading references, see refs. 1–6.) The classification is not completely rigorous, and there are exceptions.

(i) by Direct Nucleophilic Attack on Coordinated Dienes. Most examples of external attack (Eq. a) on coordinated 1,3-dienes involve palladium complexes[7,8]. The formation of π-allyl complex **Ia** in the reaction between palladium chloride and butadiene was first reported in 1957[9], although the true structure of the products was not realized until 1962[10].

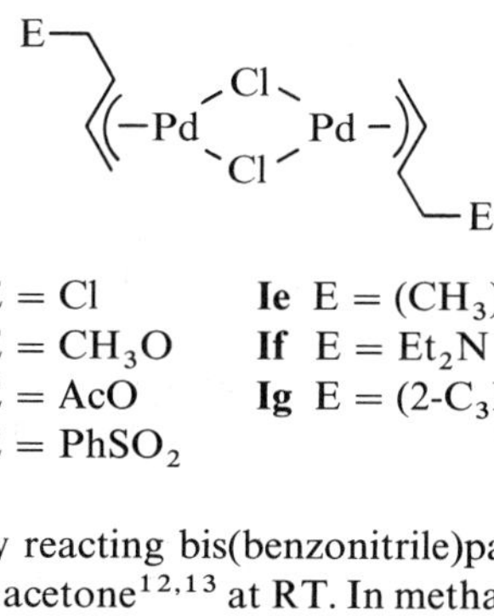

Ia E = Cl	**Ie** E = $(CH_3)_2N$
Ib E = CH_3O	**If** E = Et_2N
Ic E = AcO	**Ig** E = $(2\text{-}C_3H_7)_2N$
Id E = $PhSO_2$	

Compound **Ia** is prepared by reacting bis(benzonitrile)palladium chloride[11] with butadiene in benzene[7–10] THF or acetone[12,13] at RT. In methanol, the methoxy complex **Ib** is formed instead[14–18], and in acetic acid the acetate **Ic** is produced[18,19]. Additional examples of nucleophiles are sulfinates[20] and amines[21]. The E groups are readily displaced; e.g., CH_3O, acetate and $(CH_3)_2N$ can be rapidly displaced in the presence of catalytic amounts of acid.

The alcoxy complexes **Ib** are best prepared by reacting the diene and $[PdCl_4]^{2-}$ at RT in alcohol[15,18]. Different alcohols and dienes may be used, e.g., isoprene, 4-methyl-1,3-pentadiene, 2,3-dimethylbutadiene and various cyclic dienes[14–17].

The acetoxy complexes **Ic** have been prepared either using $[PdCl]^{2-}$ in refluxing acetic acid[18] or with palladium acetate at RT[19]. Amino complexes are preferably prepared in THF or acetone and are labile[12,13]. However, using diisopropylamine, a complex **Ig** may be isolated[21].

In the stable methoxy- and acetoxy complexes **Ib** and **Ic** the bridging chloride may be exchanged for other groups, e.g., acetate, thiocyanate, acetoacetate and cyclopentadienyl[14,18].

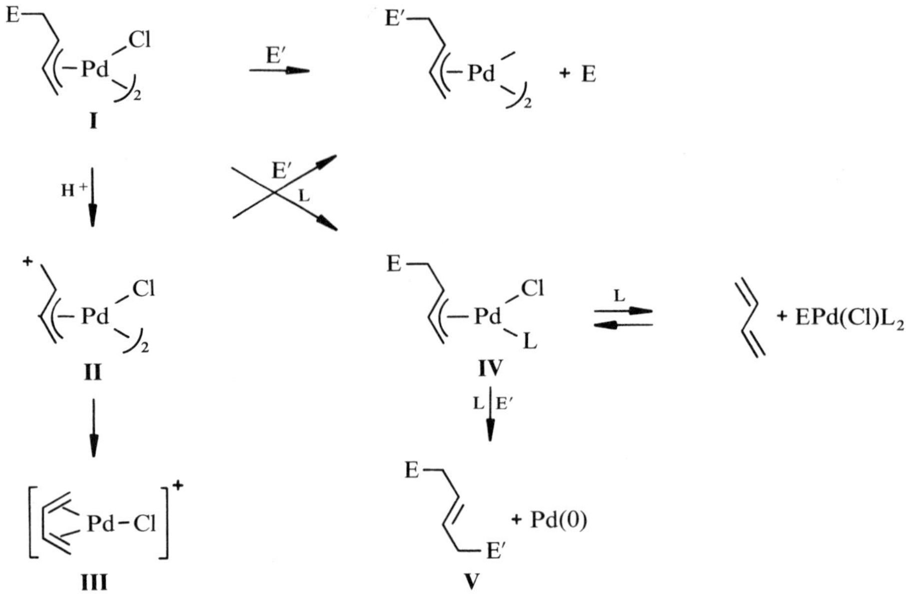

Scheme 1

Scheme 1 summarizes the reaction pathways for π-allylpalladium complexes of type **I**. The facile exchange of the E group can proceed either via direct nucleophilic displacement or via a carbonium ion, **II**. Spectroscopic study suggests[22] that an η^4-diene complex **III** may also be involved. Such ligands as phosphines give mononuclear complexes[23] and also induce reversal of the addition of X^- to the diene. Addition of a second nucleophile to yield Pd(0) and the disubstituted olefin **V** also proceeds readily under the appropriate conditions[4,12,13,24–29].

External nucleophiles add to dienes complexed to metals other than palladium; e.g., in the molybdenum complex **VI** either of the three π-ligands, the arene, the π-allyl group or the diene, can be attacked by the nucleophile:

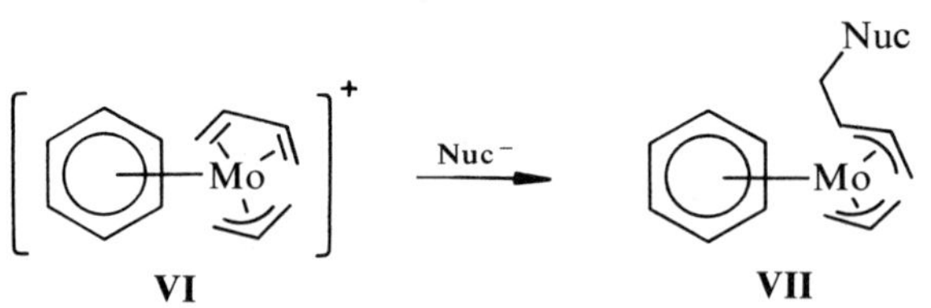

(c)

The methoxide, methylsulfide, cyanide and hydride nucleophiles can be reacted with **VI** to yield **VII**, and that is the only attack on the diene that is observed[30]. Emprical rules that predict the relative reactivity of different types of π-ligands are available[31]. Although the stereochemistry of the addition to **VI** has not been determined, all the nucleophiles, including hydride, may add as external nucleophiles since complex **VI** is coordinatively saturated. The positive charge on the complex is also important. Both these properties are possessed by the other reported examples; e.g., the iron–cyclobutadiene complex **VIII** reacts with triphenylphosphine in acetone at RT to yield **IX**[32]:

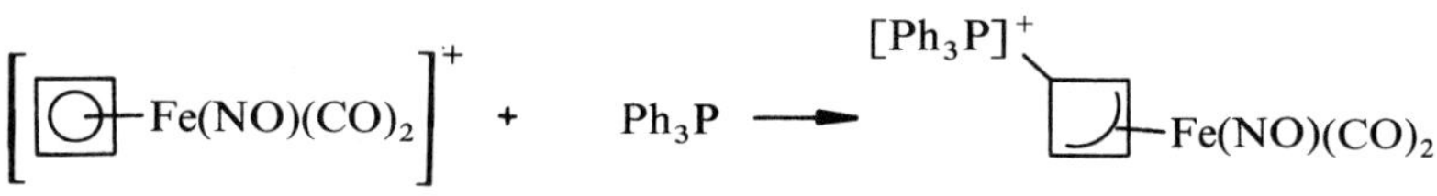

(d)

VIII **IX**

Likewise, the nickel complex **X** adds methoxide to yield **XI**[33]:

(e)

X **XI**

It is not clear how general these reactions are, since an iron complex **XII** of butadiene gives a homoallyl adduct **XIII** (Nuc = malonate) instead of a π-allyl complex[34]:

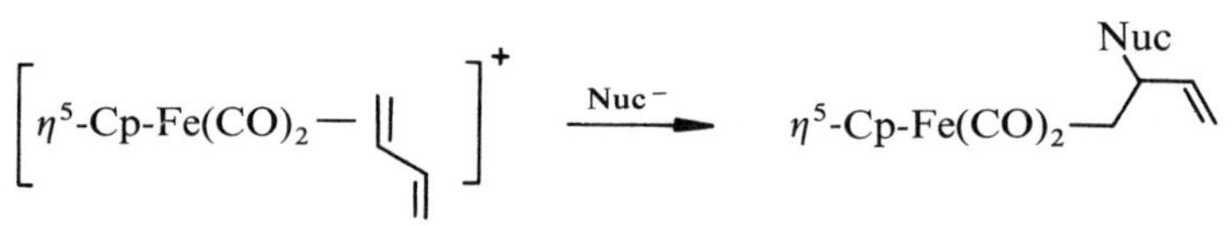

(f)

XII **XIII**

However, a related cationic diene–Co complex gives π-allyl intermediates with nucleophiles such as py, trimethylphosphine, and phenylmagnesium chloride[15].

(ii) by Reaction Between Coordinated Nucleophiles and Dienes. While few metals promote nucleophilic attack of heteroatom nucleophiles, the reaction between hydride and dienes occurs for most of the transition elements. Most may be regarded as insertions of the diene into a metal–hydrogen bond according to path (b) but there are also examples of external addition according to path (a). Stable π-allyl complexes can result from the addition of unstabilized carbanions.

Hydrides and alkyls of the titanium triad metals are polymerization catalysts, so it is understandable that there are few stable π-allyl intermediates. The only example we can locate is the reaction between the zirconium hydride **XIV** and butadiene, which yields the homoallyl complex **XV** in place of the expected π-allyl complex[35]:

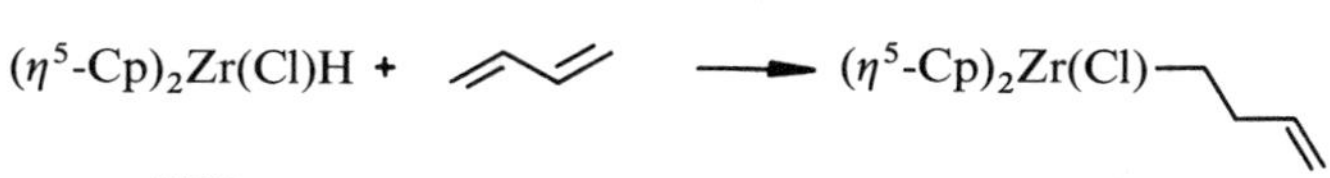

(g)

XIV **XV**

Vanadium and chromium triad metals are also polymerization catalysts. By the use of strongly coordinating ligands such as CO, phosphines and cyclopentadienyl, it is possible to isolate complexes of the vanadium triad metals. By means of photochemical activation, hydrides of the type **XVI** (L = phosphine, arsine, CO) react with butadiene to give the π-allyl complexes **XVII**:

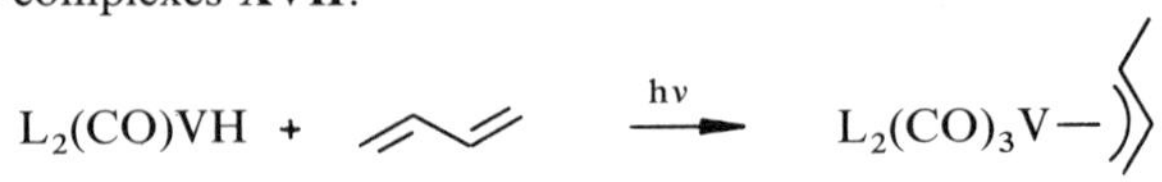
(h)

XVI **XVII**

Complexes **XVII** are labile, in particular in halogenated solvents, and must be prepared at −10 to −40°C[36]. Butadiene, isoprene, cyclohexadiene, etc., may be used. In contrast, the niobium hydride **XVIII** reacts smoothly with butadiene to give **XIX**[37]:

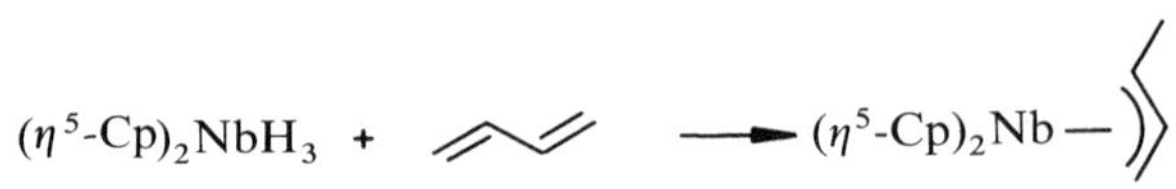
(i)

XVIII **XIX**

For chromium only one example is available[38]:

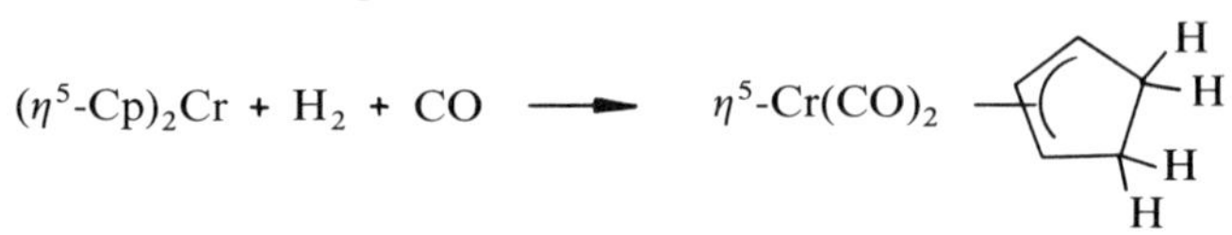
(j)

XX **XXI**

The reactions proceed at high pressure and ~65°C. For the heavier elements in the triad, especially for molybdenum, there are many examples; however, e.g., the hydrides **XXII** react with butadiene to give complexes **XXIII**[39,40]:

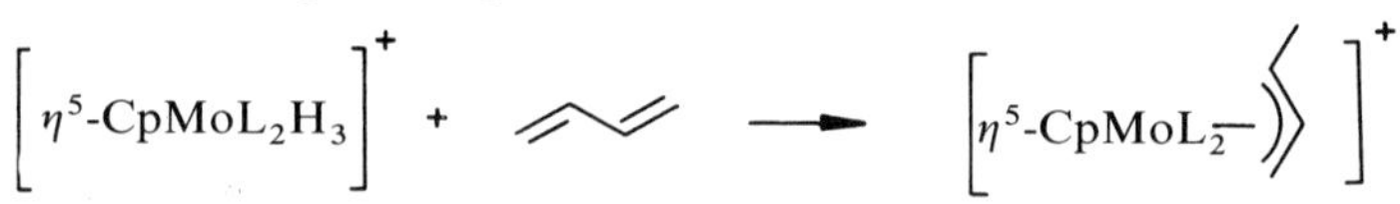
(k)

XXIIa L_2 = biphosphine **XXIIIa**
XXIIb $L_2 = \eta^5$-Cp **XXIIIb**

As in the related reaction **XVIII** → **XIX**, the intermediate is a monohydride. The reactions are sluggish and **XXIIa** → **XXIIIa** requires 12 h at 70°C in toluene for completion. Complex **XXIIIb** may also be prepared from **XXIV** by reaction with hydride[40]:

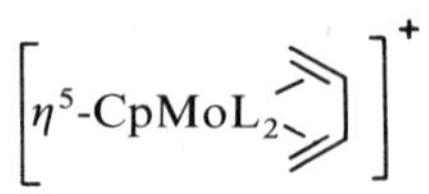

XXIV L_2 = biphosphine

The hydride may coordinate to the metal first or add as an external nucleophile.

In the manganese triad there is both hydride and alkyl addition. The hydride **XXV** and butadiene gives **XXVI** at RT[41]:

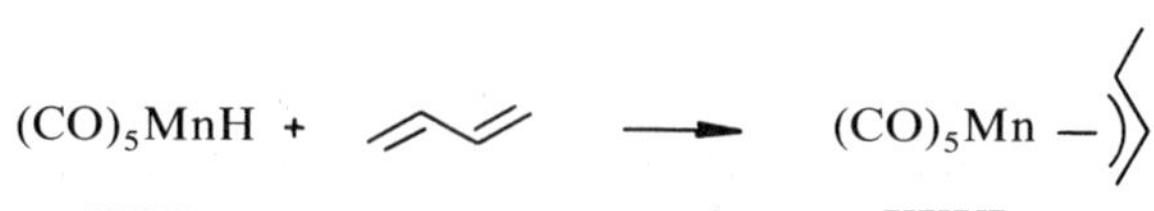
(l)

XXV **XXVI**

The methyl compound **XXVII** adds to butadiene, but only after CO insertion, to give **XXVIII**[42]:

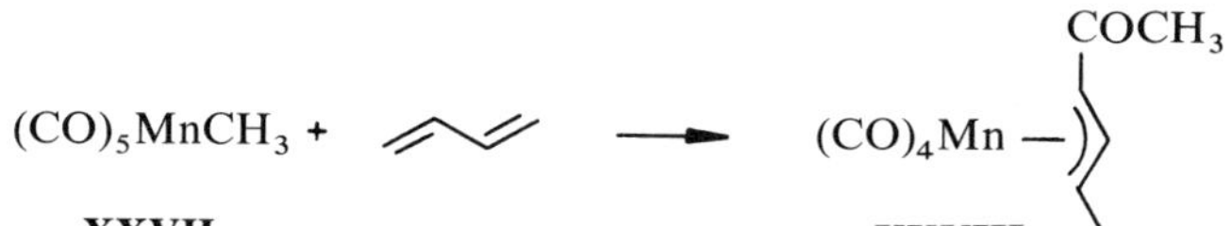

(m)

XXVII **XXVIII**

Among the iron triad metals most complexes are iron π-allyl adducts from dienes. This may be fortuitous or owing to an electronic property of the heavier elements. The ruthenium diene complex **XXIXa** shows no tendency to undergo insertion[43], but the origin of this may be steric, as indicated by the protonation of $(CO)_3$Ru(butadiene), which goes via complex **XXIXb** and yields **XXXa**[44]:

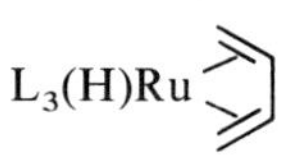

XXIX L = phosphine

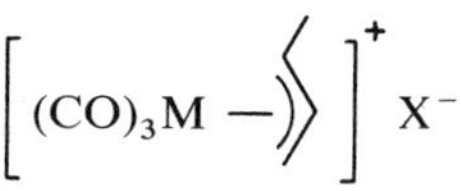

XXX **a** M = Ru
b M = Fe

Similarly, iron complexes **XXXb** are formed when butadiene and $Fe(CO)_5$ react in the presence of a strong, anhydr acid, e.g., HCl, HBF_4, $HClO_4$ or $HSbCl_6$, neat (for HCl) or in nitromethane[45,46]. The X = Cl^- is coordinated to iron while the others are ionized. The mechanism is by oxidative addition of HX to $Fe(CO)_5$ to give an iron hydride, which subsequently reacts with butadiene, but direct protonation of an iron diene complex cannot be excluded.

The addition of excess strong acid, e.g., HSO_3F[47,48] to **XXX**, produces a η^3-,η^1-allyl iron hydride, **XXXI**[47,48].

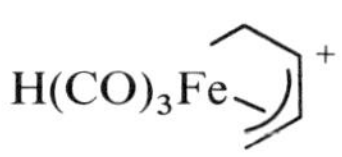

XXXI

$(NO)(CO)_3FeH$

XXXII

The intermediacy of an iron hydride in the formation of **XXX** is supported by the facile reactions between butadiene and the hydrides **XXXII**[49] and **XXXIII**[50]. Owing to coordinative saturation, **XXXIII** gives a σ-allyl complex **XXXIV**, which may be transformed photochemically into a π-allyl complex[50]:

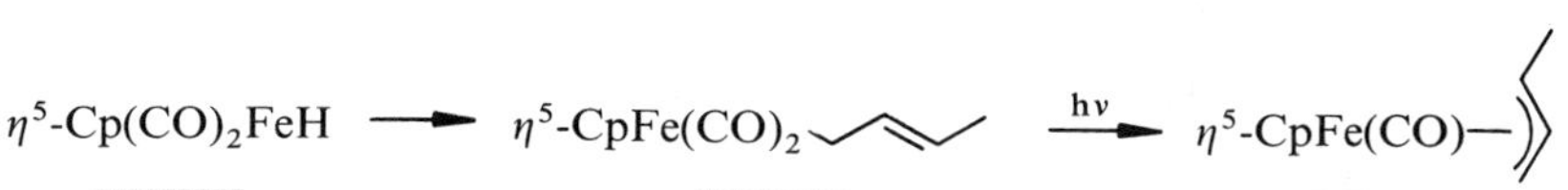

(n)

XXXIII **XXXIV** **XXXV**

Analogous σ-allyl complexes **XXXVI** are also formed from the butadiene complex **XII**[34]:

(o)

XII **XXXVI** R = CH_3, Ph

The use of malonate results in a different reaction, the product being **XIII**.

Cobalt, rhodium and iridium give π-allyl complexes from dienes, e.g., the π-allyl complex **XXXVIII** from the cobalt hydride **XXXVII**[51]:

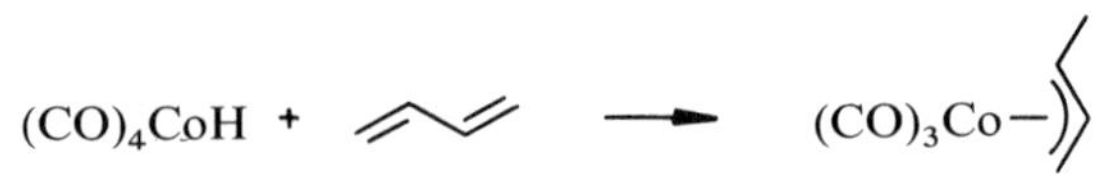

(p)

XXXVII **XXXVIII**

although the true structure of the complex was not realized until later[52,53]. The hydride **XXXVII** is prepared from $K[Co(CO)_4]$ and an acid, such as acetic or phosphoric, and reacted in situ with dienes such as isoprene, 1,3-pentadiene or 2,3-dimethyl-1,3-butadiene[53].

$(PF_3)_4CoH$

XXXIX

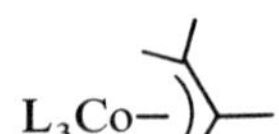

XL

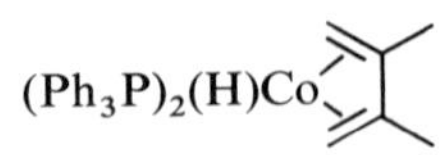

XLI

Like complex **XXXVII**, the phosphine complex **XXXIX** reacts with 2,3-dimethyl-1,3-butadiene[54] to give the expected complex **XL**. Hydridodiene complexes are intermediates. Using an unreactive diene, 2,3-dimethylbutadiene and $(Ph_3P)_2Co(N_2)H$, one such complex can be isolated, **XLI**. Rhodium and iridium hydrides also react with dienes.

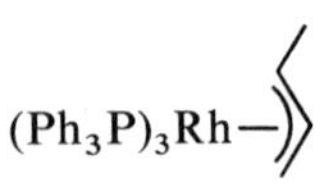

XLII

$(Ph_3P)_2(CO)M{-}H$

XLIII a M = Rh
b M = Ir

XLIV a M = Rh
b M = Ir

Representative products are **XLII** and **XLIV**[55,56]. With butadiene in benzene the rhodium complex **XLIIIa** reacts fast at RT, whereas the iridium complex **XLIIIb** requires ~2 h at reflux. Other dienes give complex products with **XLIIIa**. The complex **XLIIIb** reacts with isoprene (12 h at reflux) but fails to react with cyclic dienes.

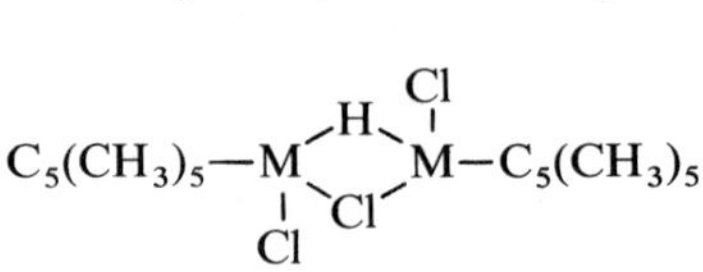

XLV a M = Rh
b M = Ir

$[(COD)MCl]_2$

XLVI a M = Rh
b M = Ir

XLVII

In contrast, the iridium complex **XLVb** is more reactive than the corresponding rhodium complex **XLVa**. Furthermore, the reaction rates in this reaction are independent of the diene (cyclooctadiene ~isoprene ~2,3-dimethylbutadiene)[57].

Cyclooctadiene complexes **XLVI** yield π-allyl complexes **XLVII** when reacted with an isopropyl organomagnesium halide reagent in the presence of 1,3-pentadiene. Rapid β elimination gives an intermediate hydride, which adds to the diene. With isoprene, an intermediate alkyl complex **XLVIII** may be isolated[58], demonstrating a high potential stability of coordinatively saturated species.

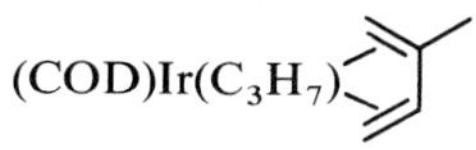

XLVIII

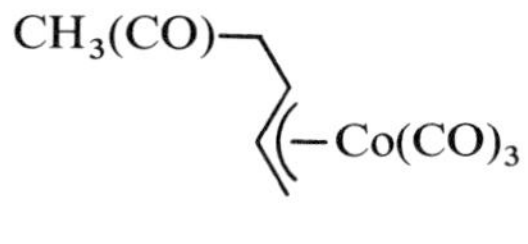

XLIX

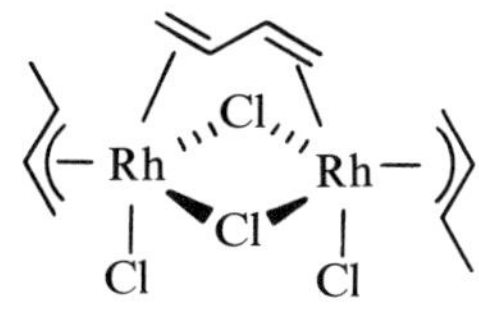

L

With alkyls not capable of β elimination, addition of the alkyl group is possible, e.g.,with methylcobalt carbonyl, which undergoes CO insertion prior to reaction with the diene[59]. The product thus is the acyl complex **XLIX**.

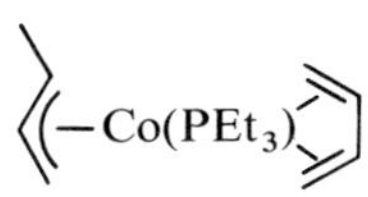

LI

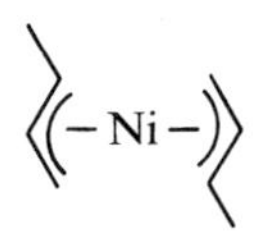

LII

Rhodium complexes catalyze the oligomerization of dienes. The isolation of the complex **L** from, e.g., $RhCl_3$, ethanol and butadiene, under conditions that yield an intermediate Rh(III) hydride, is therefore of mechanistic interest[60].

The stable cobalt complex **LI** is prepared by cocondensing cobalt atoms and butadiene at liquid nitrogen temperature[61]. This technique is also used to prepare π-allyl complexes of the nickel triad metals, e.g., **LII**[62]. Metal hydrides are intermediates in these reactions[62], as shown by the reaction of nickel hydrides with dienes to yield π-allyl complexes[63]. The π-allyl complex **LIII** may be prepared by adding HCl and butadiene to $(PPh_3)_4Ni$ in toluene or THF at $-78°C$, then warming to RT[64].

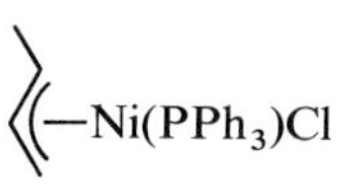

LIII

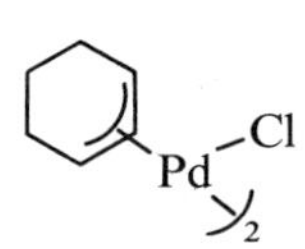

LIV

Palladium hydrides are also reactive; e.g., $Pd(CO)Cl_2$ or $PdCl_2$ react with cyclic dienes via hydrides to yield π-allyl complexes such as **LIV**[65,66].

The less reactive platinum complex **LV** reacts with butadiene similarly at RT to give the π-allyl complex **LVI**[67].

(q)

LV **LVI**

With palladium, high yields of π-allyl complexes may also be obtained from palladium aryls and palladium alkyls lacking β hydrogens[68]. Complexes **LVIII** are prepared via the unstable compounds **LVII** by reacting tin or mercury alkyls or aryls at RT with $LiPdCl_3$ in acetonitrile in the presence of the diene:

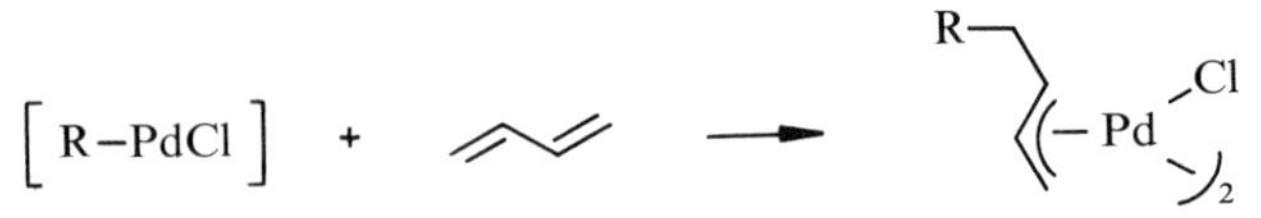

(r)

LVII **LVIII**

where R = Ph, $PhCH_2$, CH_3. The π-allyl complexes of nickel and palladium are reactive[69,70]; e.g., the π-allyl complex **LIX** reacts with butadiene to give a π-allyl complex **LX**:

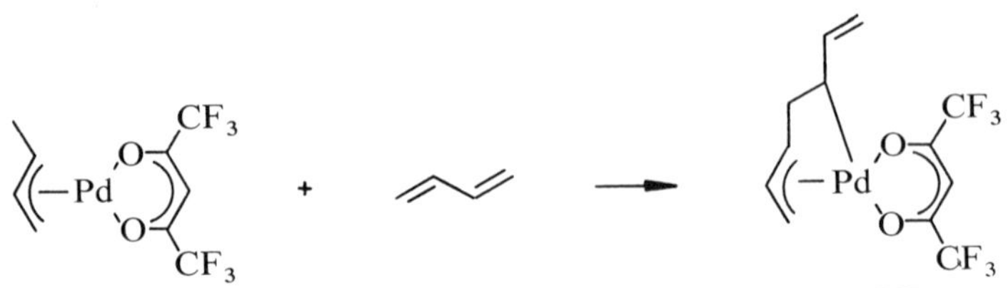

(s)

LIX **LX**

This reaction type may be important in the formation of oligomers and telomers from 1,3-dienes.

Allyl copper compounds are formed in the reaction between magnesium cuprates and dienes[71]. These complexes may be π-allyl or σ-allyl.

(iii) Mechanistic Aspects. The division of nucleophilic addition to dienes into two classes of reactions, e.g., (a) and (b), is ambiguous because such nucleophiles as hydride and carbanions add according to either (a) or (b) depending on reaction conditions. However, palladium-promoted addition to simple olefins indicates that heteroatom nucleophiles add according to (a) even when partly coordinated to the metal, whereas carbanions react according to (b). (For leading references, see Refs. 1–6.) The preference for anti addition to 1,3-dienes is maintained for heteroatom nucleophiles such as dimethylamine, methanol and acetate[12,25–28]. Although the stereochemistry of the addition of simple carbanions to dienes is unknown for the palladium-promoted reaction, syn addition according to (b) is expected. However, with coordinatively saturated systems, where coordination to the metal prior to addition is precluded, simple carbanions may also add as external nucleophiles according to path (a). Coordinative saturation is important in the comparison of the reaction of the phenyl complex **LXI** with cycloheptatriene[72]:

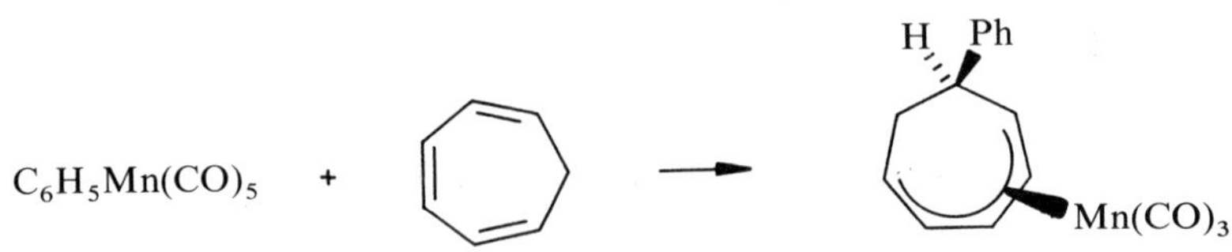

LXI **LXII**

and the addition of phenyl carbanion to the manganese-cycloheptatriene complex **LXIII**[73]:

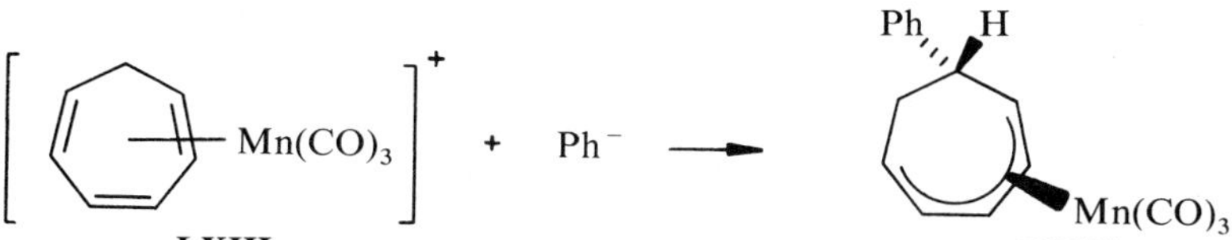

LXIII LXIV

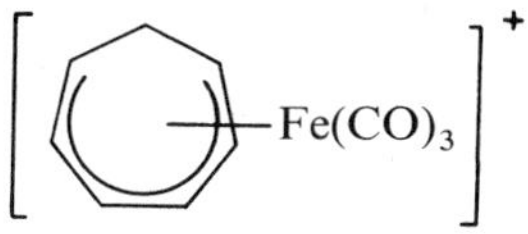

LXV

In the former insertion according to (b) yields the endo complex **LXII**, whereas in the latter direct attack of phenyl carbanion according to (a) produces the exo complex **LXIV**. However, palladium–diene complexes[12,13,24–29] and the iron complex **LXV**[74] react with heteroatom nucleophiles according to path (a) even when the nucleophile is partly coordinated to the metal. Thus there is an intrinsic difference between heteroatom nucleophiles and simple carbanions in accordance with the division made in this section.

(B. ÅKERMARK, K. ZETTERBERG)

1. J. E. Bäckvall, B. Åkermark, S. O. Ljunggren, *J. Am. Chem. Soc., 101*, 2411 (1979).
2. B. Åkermark, J. E. Bäckvall, K. Zetterberg, *Acta Chem. Scand., 836*, 577 (1982).
3. J. E. Bäckvall, *Pure Appl. Chem., 55*, 1669 (1983).
4. J. E. Bäckvall, J. E. Nyström, R. E. Nordberg, *J. Am. Chem. Soc., 107*, 3676 (1985).
5. P. Sautet, O. Eisenstein, K. M. Nicolas, *Organometallics, 6*, 1845 (1987).
6. J. E. Bächvall, in *Advances in Metal-Organic Chemistry*, L. S. Liebeskind, ed., Jai Press, Greenwich, Conn., 1989.
7. P. M. Maitlis, *The Organic Chemistry of Palladium*, Academic Press, New York, 1971.
8. F. R. Hartley, *The Chemistry of Platinum and Palladium*, Applied Science Publishers, London, 1973.
9. P. E. Slade Jr., H. B. Jonassen, *J. Am. Chem. Soc., 79*, 1277 (1957).
10. B. L. Shaw, *Chem. Ind.* (London), 1190 (1962).
11. M. S. Kharasch, R. C. Seyler, F. R. Mayo, *J. Am. Chem. Soc., 60*, 882 (1938).
12. B. Åkermark, J. E. Bäckvall, A. Löwenborg, K. Zetterberg, *J. Organomet. Chem., 166*, C33 (1979).
13. B. Åkermark, K. Zetterberg, J.-E. Bäckvall, unpublished work.
14. S. D. Robinson, B. L. Shaw, *J. Chem. Soc.*, 4806 (1963).
15. S. D. Robinson, B. L. Shaw, *J. Chem. Soc.*, 5002 (1964).
16. G. Agami, J. Levisalles, *J. Organomet. Chem., 64*, 281 (1974).
17. R. Pietropaolo, F. Faraone, D. Pietropaolo, P. Pirano, *J. Organomet. Chem., 64*, 403 (1974).
18. J. M. Rowe, D. A. White, *J. Chem. Soc., A*, 1451 (1967).
19. R. G. Brown, R. V. Chandhari, J. M. Davidson, *J. Chem. Soc., Dalton Trans.*, 176 (1977).
20. Y. Tamaru, M. Kagotani, Z.-I. Yoshida, *J. Chem. Soc., Chem. Commun.*, 367 (1978).
21. K. Zetterberg, J.-E. Bäckvall, B. Åkermark, unpublished work.
22. J. Lukas, P. A. Kramer, *J. Organomet. Chem., 31*, 111 (1971).
23. K. A. Holder, P. Powell, *J. Organomet. Chem., 122*, 275 (1976).
24. R. G. Brown, J. M. Davidson, *J. Chem. Soc., A*, 1321 (1971).
25. J.-E. Bäckvall, R. Nordberg, K. Zetterberg, B. Åkermark, *Organometallics, 2*, 1625 (1983).
26. J.-E. Bäckvall, R. E. Nordberg, *J. Am. Chem. Soc., 103*, 4959 (1981).
27. J.-E. Bäckvall, R. E. Nordberg, E. E. Björkman, C. Moberg, *J. Chem. Soc., Chem. Commun.*, 943 (1980).
28. J.-E. Bäckvall, J. E. Nyström, *J. Chem. Soc., Chem. Commun.*, 59 (1981).
29. B. Åkermark, A. Ljungquist, M. Pannunzio, *Tetrahedron Lett., 22*, 1055 (1981).

30. M. L. H. Green, L. C. Mitchard, W. E. Silverthorn, *J. Chem. Soc., Dalton Trans.*, 1952 (1973).
31. S. G. Davies, M. L. H. Green, D. M. R. Mingos, *Tetrahedron, 34*, 3047 (1978).
32. A. Efraty, J. Potenza, S. S. Sandku Jr., R. H. Hehr, *J. Organomet. Chem., 70*, C24 (1974).
33. P. M. Maitlis, A. Efraty, M. L. Games, *J. Am. Chem. Soc., 87*, 719 (1965).
34. P. Lennon, A. M. Rosan, M. Rosenblum, *J. Am. Chem. Soc., 99*, 8426 (1977).
35. C. A. Bertelo, J. Schwartz, *J. Am. Chem. Soc., 98*, 262 (1976).
36. U. Franke, E. Weiss, *J. Organomet. Chem., 193*, 329 (1980) and earlier references.
37. F. N. Tebbe, G. Parshall, *J. Am. Chem. Soc., 93*, 3793 (1971).
38. E. O. Fischer, K. Ulm, *Chem. Ber., 94*, 2413 (1961).
39. M. Ephritikhine, M. L. H. Green, R. E. MacKenzie, *J. Chem. Soc., Chem. Commun.*, 619 (1976).
40. T. Aviles, M. L. H. Green, A. R. Diaz, C. Romao, *J. Chem. Soc., Dalton Trans.*, 1367 (1979).
41. W. R. McClellan, H. H. Hoehn, H. N. Cripps, E. L. Muetterties, B. W. Howk, *J. Am. Chem. Soc., 83*, 1601 (1961).
42. W. D. Bannister, M. L. H. Green, R. N. Haszeldine, *Proc. Chem. Soc.*, 370 (1964).
43. T. A. Ashworth, E. Singleton, M. Laing, L. Pope, *J. Chem. Soc., Dalton Trans.*, 1032 (1978).
44. K. G. Ihrman, T. H. Coffield, U.S. Patent 3,117,148 (Jan. 7, 1964); *Chem. Abstr., 60*, 15,914e (1964).
45. F. J. Impastato, K. C. Ihrman, *J. Am. Chem. Soc., 83*, 3726 (1961).
46. G. F. Emerson, R. Pettit, *J. Am. Chem. Soc., 84*, 4591 (1962).
47. D. A. T. Young, J. R. Holmes, H. D. Kaesz, *J. Am. Chem. Soc., 91*, 6968 (1969).
48. M. Brookhart, D. L. Harris, *Inorg. Chem., 13*, 1540 (1974).
49. R. Bruce, F. M. Chandhary, G. R. Knox, P. L. Pauson, *Z. Naturforsch., Teil B, 20*, 73 (1965).
50. M. L. H. Green, P. L. I. Nagy, *J. Chem. Soc.*, 189 (1963).
51. W. W. Prichard, U.S. Patent 2,600,571 (June 17, 1952); *Chem. Abstr.*
52. H. B. Jonassen, R. I. Stearns, J. Kenttämaa, *J. Am. Chem. Soc., 80*, 2586 (1958).
53. J. A. Bertrand, H. B. Jonassen, D. W. Moore, *Inorg. Chem., 2*, 601 (1963).
54. H. Eshtiagh-Hosseini, J. F. Nixon, *J. Organomet. Chem., 192*, C9 (1980).
55. C. K. Brown, W. Mowat, G. Yagupsky, G. Wilkinson, *J. Chem. Soc., A*, 850 (1971).
56. C. A. Reilly, H. Thyret, *J. Am. Chem. Soc., 89*, 5144 (1967).
57. H-B. Lee, K. Moseley, C. White, P. M. Maitlis, *J. Chem. Soc., Dalton Trans.*, 2322 (1975).
58. J. Müller, W. Hähnlein, H. Menig, H. Pickardt, *J. Organomet. Chem., 197*, 95 (1980).
59. R. F. Heck, *J. Am. Chem. Soc., 85*, 3381 (1963).
60. R. Cramer, *J. Am. Chem. Soc., 89*, 1633 (1967).
61. H. Nishihara, M. Fujita, H. Takahashi, T. Saito, Y. Sasaki, *Transition Met. Chem., 5*, 98 (1980).
62. P. S. Skell, M. J. McGlinchey, *Angew. Chem., Int. Ed. Engl., 11*, 532 (1972).
63. C. Tolman, *J. Am. Chem. Soc., 92*, 6777, 6785 (1970).
64. M. J. D'Aniello Jr., E. K. Barefield, *J. Am. Chem. Soc., 100*, 1474 (1978).
65. E. O. Fischer, H. Werner, *Chem. Ber., 95*, 695 (1962).
66. R. Hüttel, H. Dietl, H. Christ, *Chem. Ber., 97*, 2037 (1964).
67. A. J. Deeming, B. F. G. Johnson, J. Lewis, *J. Chem. Soc., Chem. Commun.*, 598 (1970).
68. R. F. Heck, *J. Am. Chem. Soc., 90*, 5542 (1968).
69. D. Medema, R. van Helden, *Recl. Trav. Chim. Pays-Bas, 90*, 304 (1971).
70. R. P. Hughes, J. Powell, *J. Am. Chem. Soc., 94*, 7723 (1972).
71. J. Normant, G. Cahiez, J. Villieras, *J. Organomet. Chem., 92*, C28 (1975).
72. J. C. Bart, S. A. R. Knox, J. McKinney, F. G. A. Stone, *J. Chem. Soc., Dalton Trans.*, 1 (1977).
73. F. Haque, J. Miller, P. L. Pauson, J. B. Pd Tripathi, *J. Chem. Soc., C*, 743 (1971).
74. D. A. Brown, W. K. Gloss, F. M. Hussein, *J. Organomet. Chem., 186*, C58 (1980).
75. L. S. Barinelli, K. Tao, K. M. Nicholas, *Organometallics, 5*, 588 (1986).

5.8.2.4.5. π-Allyl Metal Complexes by Oligomerization of Conjugated Dienes.

The oligomerization and cyclooligomerization of dienes is a synthetically useful and commercially significant reaction catalyzed by transition-metal complexes from different triads. In nearly all examples, η^3-allyl complexes are intermediates, formed by either coupling of dienes about a metal template or by insertion of the diene into a metal–H bond. Allyl complexes are isolated from stoichiometric and catalytic diene oligomerizations and are active catalysts themselves. The organometallic compounds isolated are important and useful because of their lability (and attendant extensive chemistry) and catalytic utility. These reactions are of preparative value for η^3-allyl complexes. This development is not only significant for synthetic purposes but promises to shed light on the complex ligand and substrate effects observed in catalytic diene oligomerizations. The chemistry of the metal complexes present in diene oligomerizations is complex and unpredictable. Many species are present in such solutions, and their highly reactive nature makes isolation difficult.

(i) Complexes from Oligomerization of 1,3-Dienes. In the Ni-catalyzed diene cyclooligomerization of butadiene to predominantly trans,trans,trans-1,5,9-cyclo dodecatriene[1] a labile olefin complex, (1,5,9-cyclododecatriene)Ni, can be isolated[2]. Reaction of this complex or bis(1,5-cyclooctadiene)Ni with butadiene at −40°C results in olefin displacement and formation of an unstable bis(η^3-allyl) complex, (2,6,10-dodeca-triene-1,12-diyl)Ni. Warming to RT results in reductive elimination, regenerating (1,5,9-cyclododecatriene)Ni [3]:

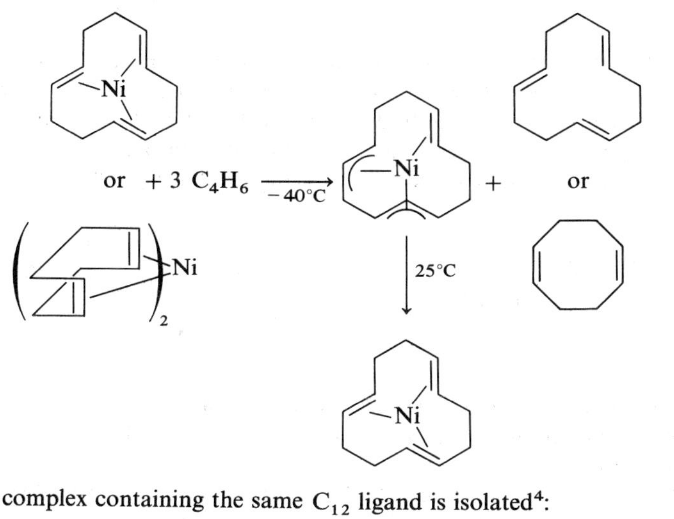

(a)

A stable Ru complex containing the same C_{12} ligand is isolated[4]:

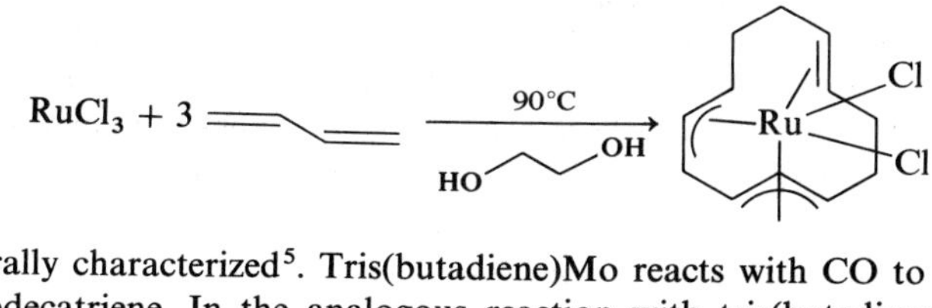

(b)

and structurally characterized[5]. Tris(butadiene)Mo reacts with CO to form $Mo(CO)_6$ and cyclododecatriene. In the analogous reaction with tris(butadiene)W, a tungsten dicarbonyl complex of the 2,6,10-dodecatriene-1,12-diyl ligand can be detected[6]:

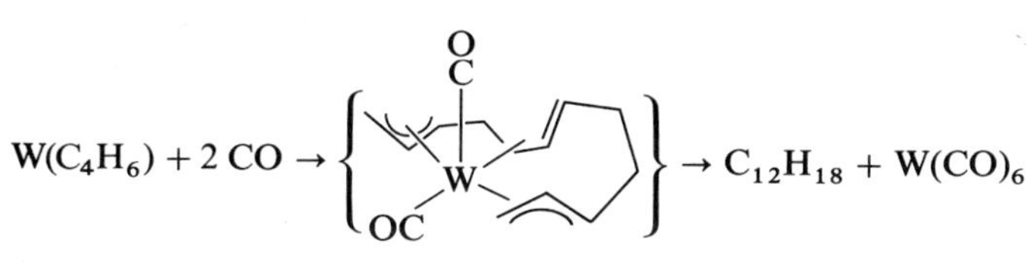

(c)

Substitution of butadiene in Eq. (a) with isoprene affords an analogous cis-Ni complex with the regiochemistry shown:

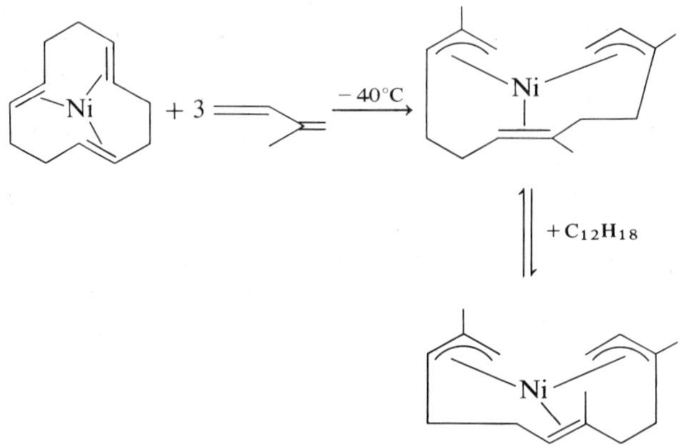

(d)

The complexes derived from both butadiene and isoprene exist as a mixture of equilibrating isomers, related by rotation of the double bond about the Ni–olefin vector from NMR[7]. The isomers of (2,6,10-dodecatriene-1,12-diyl)Ni react at different rates with trimethylphosphine to form an observable η^1,η^3-bis(allyl) intermediate before forming the trimethylphosphine adduct of (cyclododecatriene)Ni [8]:

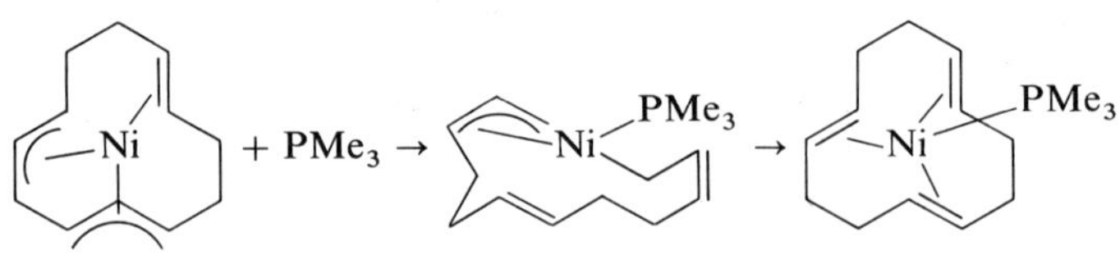

(e)

In contrast to butadiene and isoprene, 2,3-dimethylbutadiene reacts with (cyclododecatriene)Ni to form an olefin complex, bis(2,3-dimethylbutadiene)Ni, rather than a bis(allyl) complex[9].

In the presence of tricyclohexylphosphine, Ni–diene condensations form C_8-containing η^1,η^3-bis(allyl) complexes. Treatment of the tricyclohexylphosphine adduct of (cyclododecatriene)Ni with 1,3-dienes affords thermally stable complexes having the stoichiometry $Ni(diene)_2P(C_6H_{11})_3$ [10]. From IR[12] and NMR studies[8,13] and the structure of the isoprene derivative[9b], these complexes contain the η^1,η^3-bis(allyl) structure, both in the solid state and in solution:

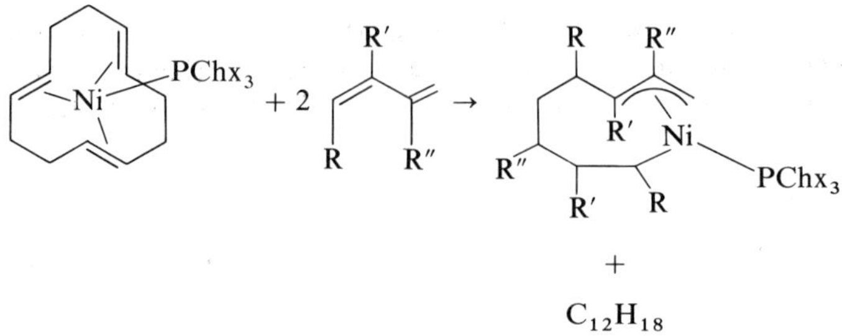

(f)

where the following possible combinations occur: R, R′, R″ = H; R, R′ = H, R″ = CH_3; R = H, R′, R″ = CH_3; R = CH_3, R′, R″ = H. The derivative formed from 2,3-dimethylbutadiene is also formed when bis(2,3-dimethylbutadiene)Ni is treated with tricyclohexylphosphine[9].

An analogous reaction with less bulky trimethylphosphine results in the formation of a mixture of complexes containing (η^1,η^3-octadienediyl)Ni(PMe_3) and (η^3,η^3-octadienediyl)Ni(PMe_3) in 85% + 15%, respectively[4]. This mixture does not isomerize on warming from −30°C to its decomposition point at 60°C.

The η^1,η^3-complexes containing triphenylphosphine and tri-o-biphenylphosphite groups rearrange irreversibly above −10°C to give complexes containing the symmetrical η^3,η^3-octadienediyl group[4]:

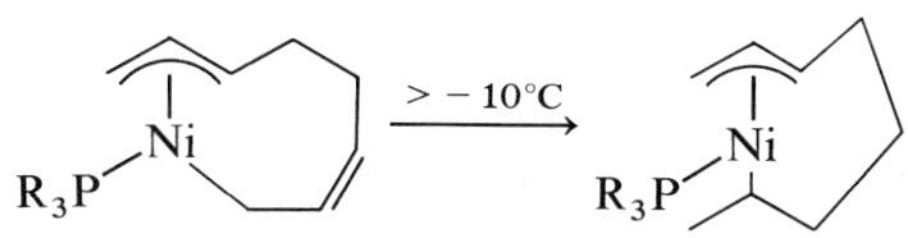

(g)

where R = Ph, OC_6H_4-O-Ph. The allyl groups are found to be syn substituted. Treatment of isoprene with ligand-modified zero-valent Ni compounds leads to a (η^1,η^3-dimethyloctadienediyl)Ni(PR_3) complex irrespective of the ligand. However, the product is a mixture of two isomers, both 2,6- and 2,7-dimethyloctadienediyl groups. The ratio of these isomers is ligand dependent[4]. Reaction of cis-piperylene with Ni complex leads to a mixture of four compounds containing the η^1,η^3-dimethyloctadienediyl group. One of the products contains the methyl groups in a trans orientation[4].

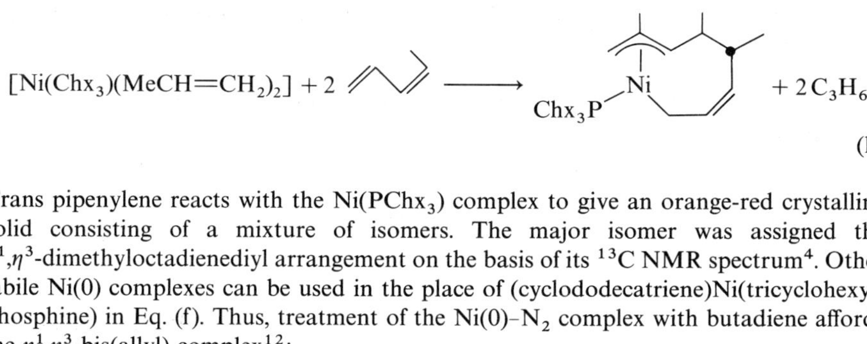

(h)

Trans pipenylene reacts with the Ni($PChx_3$) complex to give an orange-red crystalline solid consisting of a mixture of isomers. The major isomer was assigned the η^1,η^3-dimethyloctadienediyl arrangement on the basis of its ^{13}C NMR spectrum[4]. Other labile Ni(0) complexes can be used in the place of (cyclododecatriene)Ni(tricyclohexylphosphine) in Eq. (f). Thus, treatment of the Ni(0)–N_2 complex with butadiene affords the η^1,η^3-bis(allyl) complex[12]:

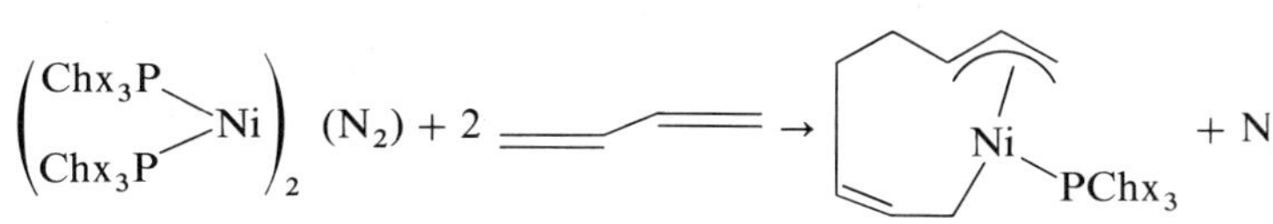

(i)

The C—C bond formation in these complexes is reversible. Treatment of the butadiene or isoprene derivatives with molten triphenylphosphine leads to diene evolution, in moderate yields[11,13], but reductive coupling of the M—C bonds occurs when the complexes are reacted with CO at low T since 4-vinylcyclohexene is formed[11,13]. An unstable olefin complex can be formed from divinylcyclobutane and (cyclododecatriene)Ni(tricyclohexylphosphine) that liberates divinylcyclobutane when reacted with CO at low T. However, if the compound is warmed to RT before cooling and reacting with CO, 4-vinylcyclohexene is liberated:

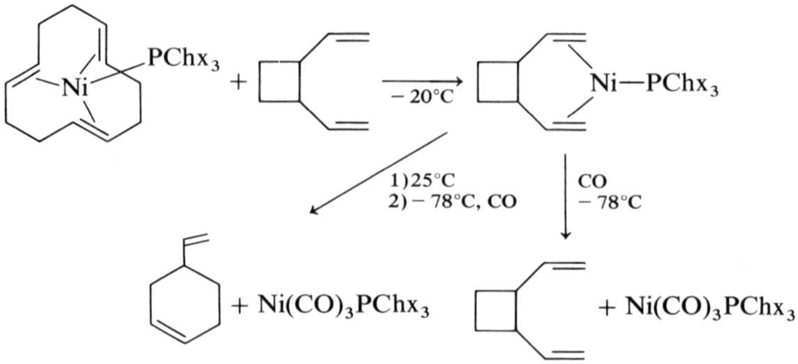

(j)

The divinylcyclobutane complex is transformed into an η^1,η^3-bis(allyl) complex on warming[13]:

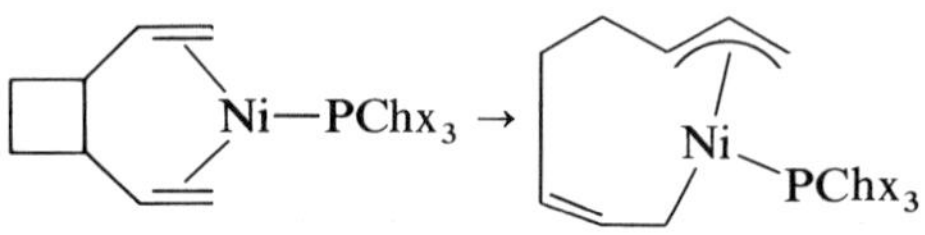

(k)

In the reaction of Ni(0)–phosphine complexes with butadiene C_8 allyl complexes are formed, whereas C_{12} allyl complexes are the products in the absence of phosphines. The Ni-catalyzed cyclotrimerization of butadiene to cyclododecatriene can be made to produce 1,5-cyclooctadiene, vinylcyclohexene or divinylcyclobutane instead by addition of phosphines[1,15].

A related bis(η^3-allyl) complex of Zr is known. The dimeric Zr(III) complex disproportionates when mixed with butadiene in toluene, forming a $ZrCl_4$–phosphine complex and a derivative containing the C_8H_{12} ligand:

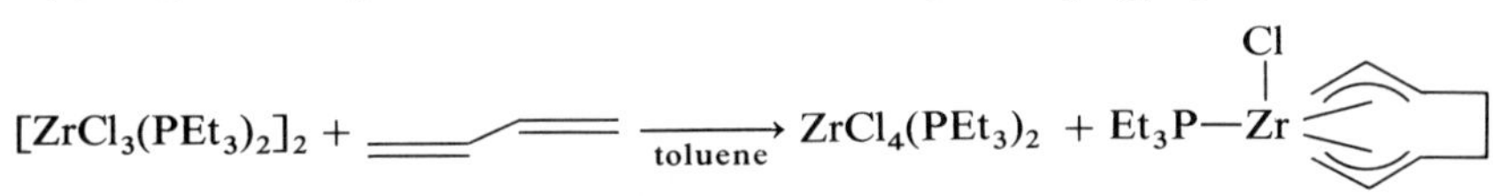

(l)

Both allyl groups are η^3 on the basis of ^{13}C NMR [16]. Similar Pt compounds are known[17].

An analogous cyclooctatetraene complex can be prepared by displacement of hexadiene from a bis(η^3-allyl) complex with butadiene. The resulting allylic, C_8H_{12}-containing complex is a catalyst for the dimerization of butadiene to 1,3,6-trans,cis-octatriene[18]:

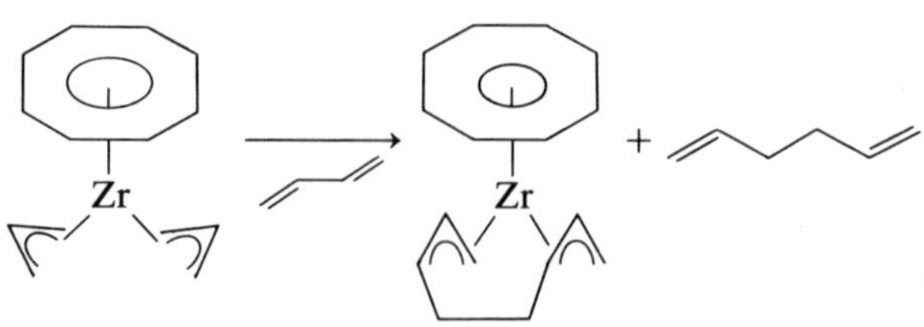

(m)

Alternatively, the same compound may be prepared by condensation of butadiene with

(η^8-cyclooctatetraene) (η^4-butadiene)Zr, a procedure also effective for Ti and Hf analogs[19]:

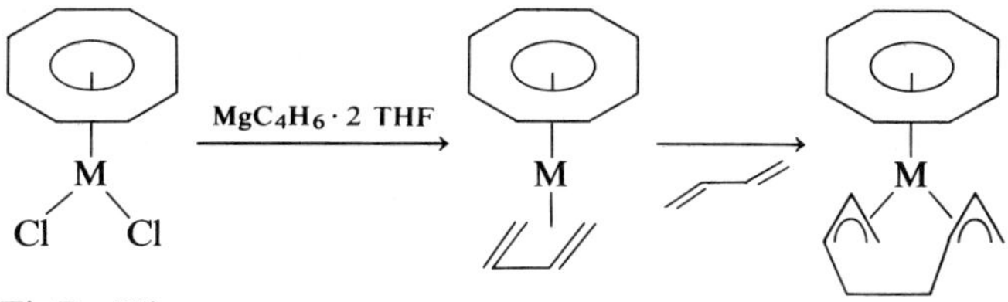

(n)

where M = Ti, Zr, Hf.

Cobalt(II) chloride reacts with butadiene in the presence of ethanolic Na borohydride to yield a complex having the stoichiometry $CoC_{12}H_{19}$[20]:

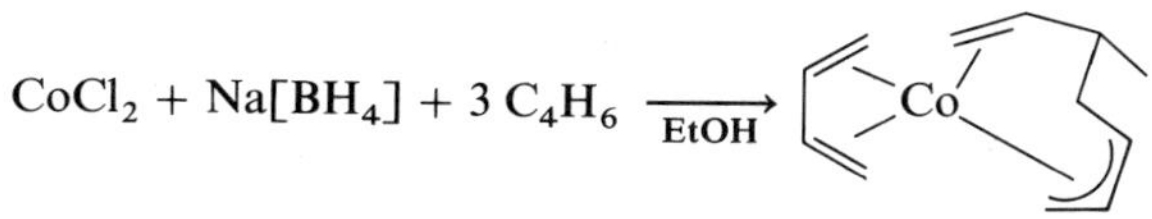

(o)

i.e., it contains the 3-methyl-1,6-heptadiene-8-yl ligand[21]. The compound catalyzes the oligomerization of butadiene to 3-methyl-1,4,6-heptatriene[20].

Several binuclear allyl complexes are obtained by condensation of butadiene, as in:

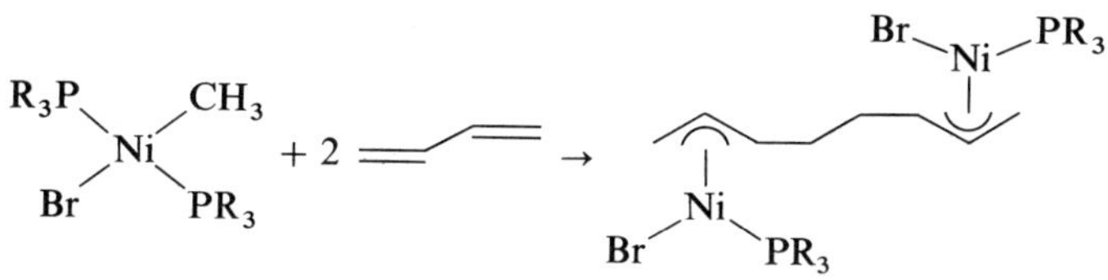

(p)[22]

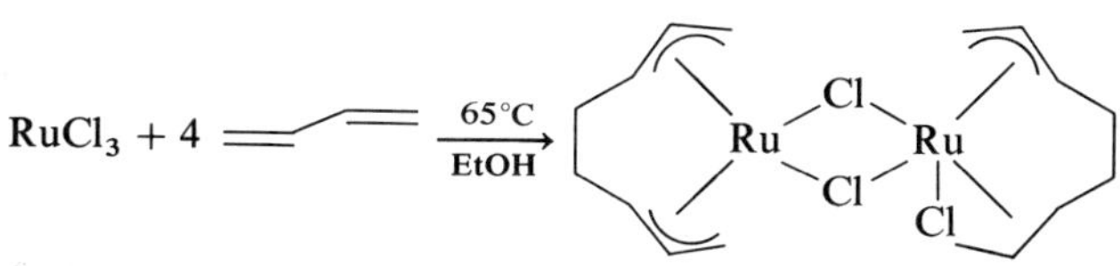

(q)[23]

One binuclear complex may be involved in the catalytic cycle for butadiene oligomerization. Allylpalladium acetate reacts with butadiene to form an acetate-bridged allyl complex. Heptadiene is displaced from this intermediate when it is treated with additional butadiene, and a binuclear, acetate-bridged complex of the 2,6,10-dodecatriene-1,12-diyl ligand is claimed to be formed:

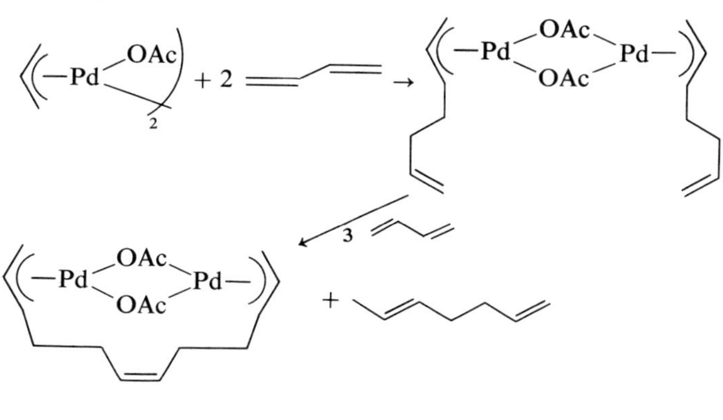

(r)

This complex catalyzes the trimerization of C_4H_6 to 1,3,6,10-dodecatetraene[24].

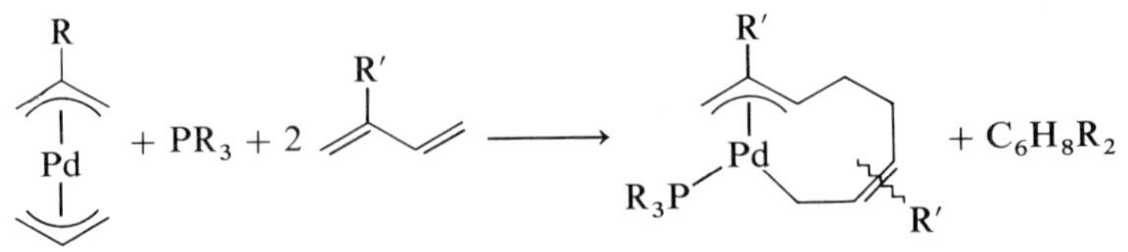

(s)

where R = Me, Chx. η^1,η^3-Octadienediylpalladium–trialkylphosphine complexes are synthesized by treating bis(η^3-allyl)- or bis(η^3-2-MeC_3H_4)Pd with trialkyl phosphine and butadiene at or near RT in ether or THF [25,26]. The products are of variable thermal stability depending on the phosphine, and are air-sensitive, yellow crystalline solids. They can be characterized by ^{1}H and ^{13}C NMR and x-ray crystallography.

The reaction of isoprene with bis(η^3-allyl)Pd [25] or bis(η^3-2-MeC_3H_4)Pd [26] in the presence of trialkylphosphine leads to a mixture of isomers containing [Pd(PR_3)(η^1,η^3-2,6-$Me_2C_8H_{10}$)] and [Pd(PR_3)(η^1,η^3-2,7-$Me_2C_8H_{10}$)] complexes. The relative ratio of the two isomers can be determined by ^{31}P NMR and is found to be ligand dependent[26].

Treatment of [Pd(dba)$_2$] (dba = dibenzylidene acetone) with isoprene in the presence of P(NMe_2)$_3$ leads to [η^1,η^3-dimethyloctadienediyl]Pd[P(NMe_2)$_3$] complex[26].

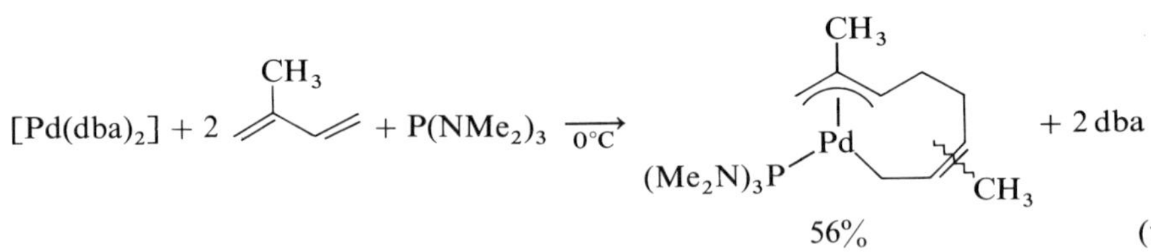

(t)

η^1,η^3-Dimethyloctadienediyl)Pd(PChx$_3$) is also synthesized by displacing the diolefin in [Pd(PChx$_3$)(η^2,η^2-1,6-C_8H_{13}OMe)] with isoprene[26]. The (η^1,η^3-octadienediyl)Pd–trialkyl phosphine complexes can also be synthesized by reducing palladium salts with magnesium anthracenes in the presence of a ligand and butadiene[27,28].

Treatment of $PdCl_2$ with Mg(anthracene) in the presence of isoprene and PR_3 leads to the (η^1,η^3-dimethyloctadienediyl)Pd(PR_3) complexes[26].

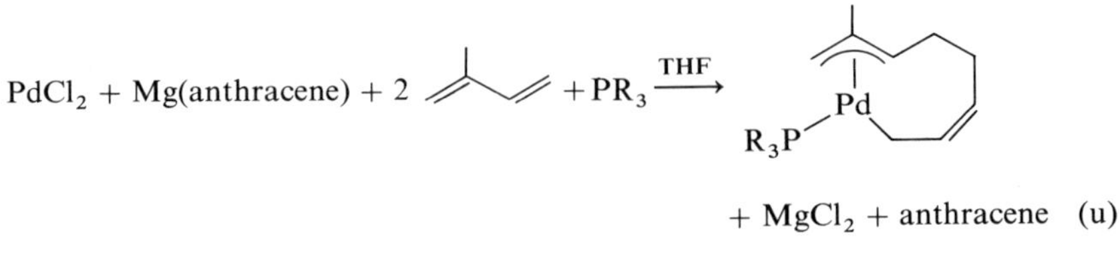

(u)

where R = i-Pr, Ph.

(ii) Complexes from Oligomerization of 1,2-Dienes. Binuclear bis(allyl) complexes are formed from the reactions of Fe carbonyls with allene[29] and substituted allenes[30]:

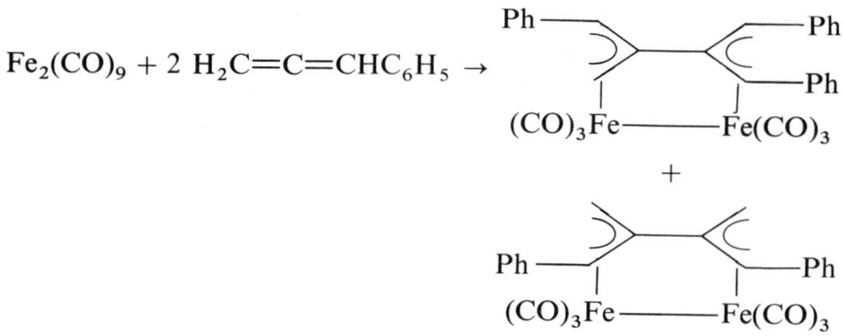
(v)

Similar complexes are formed in the reaction of allylpalladium acetylacetonate (acac) and allene[31]: Palladium acetate, however, leads to a binuclear complex containing a trimer of allene[32]:

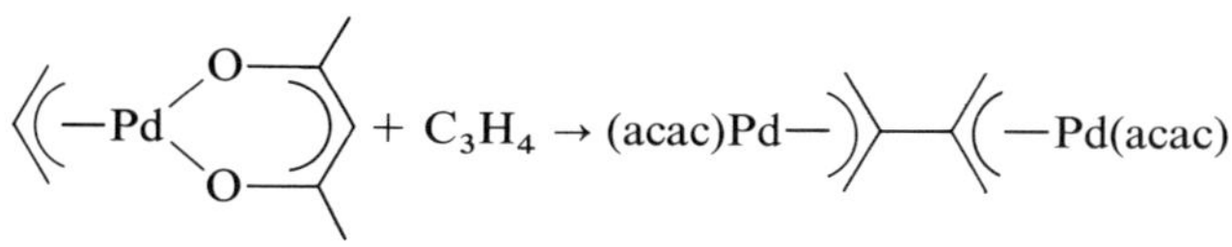
(w)

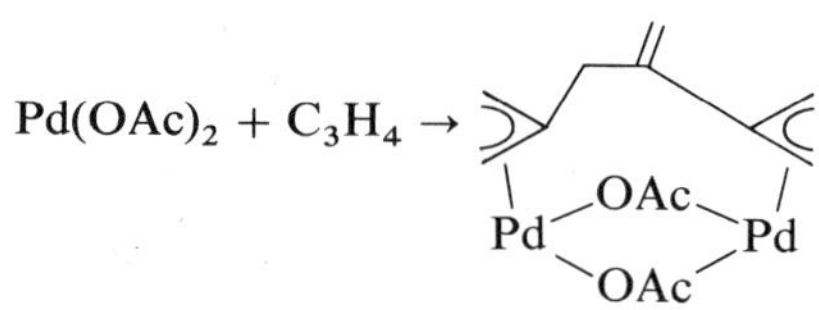
(x)

Reaction of allene[33] with dichlorobis(benzonitrile)Pd(II) in benzonitrile produces a high yield of a (π-allyl)Pd complex.

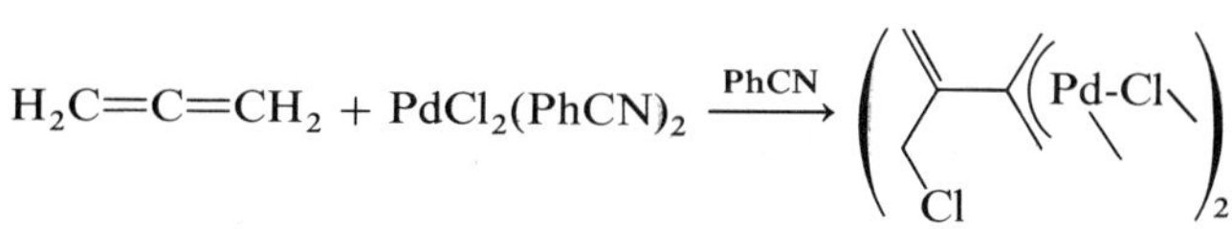
(y)

The bridging halide can be exchanged for bromide, iodide, thiocyanate or acetylacetone.

Nickel(0) complexes rapidly oligomerize allene; selectivity depends on the detailed ligand environment[34,35]. Allylnickel complexes are isolable from these reactions and are involved in the catalytic cycle. Treatment of bis(cyclooctadiene)Ni at −30°C with, sequentially, allene and triphenylphosphine affords:

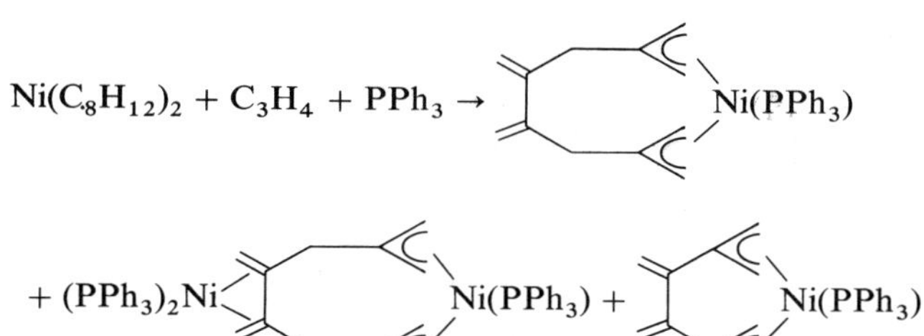
(z)

Heating the $(C_9H_{12})Ni(PPh_3)$ product with allene leads to a cyclic allene tetramer, catalytically. The other products in Eq. (z) also catalyze this oligomerization but are converted to the more stable $(C_9H_{12})Ni(PPh_3)$ during the reaction[36–38]. Other phosphine derivatives of the bis(allylic) C_9H_{12} ligand are prepared by reaction of $Ni(PR_3)_2$ complexes with allene[39]; the structure[40] of $(C_9H_{12})NiP(C_6H_{11})_3$ is known.

Rhodium(I) acetylacetonate derivatives form allyl complexes when reacted with allene[41,42]:

$(C_2H_4)_2Rh(acac) + 4\ C_3H_4 \rightarrow 2\ C_2H_4 +$ 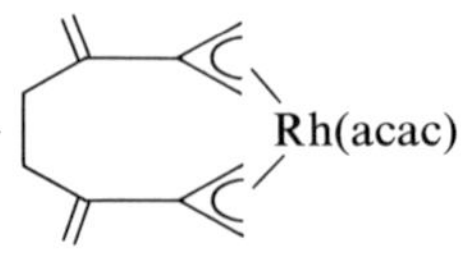 (aa)

Allyl formation is dependent on the substituent ligands, since the analogous reaction with $[RhCl(C_2H_4)_2]_2$ affords only olefin complexes of a cyclic allene pentamer[43].

(S. S. WREFORD)

(R. PANKAYATSELVAN, K. NICHOLAS)

1. G. Wilke, *Angew. Chem., Int. Ed. Engl., 2,* 105 (1963).
2. G. Wilke, *Angew, Chem., 72,* 581 (1960).
3. B. Bogdanovic, P. Heimback, M. Kröner, G. Wilke, E. G. Hoffman, J. Brandt, *Justus Liebigs Ann. Chem., 727,* 143 (1969).
4. J. K. Nicholson, B. L. Shaw, *J. Chem. Soc. (A),* 807 (1966).
5. J. E. Lydon, J. K. Nicholson, B. L. Shaw, M. R. Truter, *Proc. Chem. Soc.,* 421 (1964).
6. W. Gausing, G. Wilke, *Angew. Chem., Int. Ed. Engl., 20,* 186 (1981).
7. B. Henc, P. W. Jolly, R. Salz, G. Wilke, R. Benn, E. G. Hoffman, R. Mynott, G. Schroth, K. Seevogel, J. C. Sekutowski, C. Krüger, *J. Organomet. Chem., 191,* 425 (1980).
8. B. Henc, P. W. Jolly, R. Salz, S. Stubbe, G. Wilke, R. Benn, R. Mynott, K. Seevogel, R. Goddard, C. Krüger, *J. Organomet. Chem., 191,* 449 (1980).
9. (a) P. W. Jolly, R. Mynott, R. Salz, *J. Organomet. Chem., 184,* C49 (1980); (b) B. Here, P. W. Jolly, R. Salg, G. Wilke, R. Benn, E. G. Hoffmann, R. Mynott, G. Schroth, K. Seevogel, J. C. Sebutowski, C. Kruger, *J. Organomet. Chem., 191*(2), 425 (1980).
10. P. W. Jolly, I. Tkatchenko, G. Wilke, *Angew. Chem. Int. Ed. Engl., 10,* 328 (1971).
11. B. Barnett, B. Büssemeier, P. Heimbach, P. W. Jolly, C. Krüger, I. Tkatchenko, G. Wilke, *Tetrahedron Letters,* 1457 (1972).
12. J. M. Brown, B. T. Golding, M. J. Smith, *J. Chem. Soc., Chem. Commun.,* 1240 (1971).
13. P. W. Jolly, I. Tkatchenko, G. Wilke, *Angew. Chem., Int. Ed. Engl., 10,* 329 (1971).
14. R. Benn, B. Bussemeier, S. Holle, P. W. Jolly, R. Mynott, I. Tkatchenko, G. Wilke, *J. Organomet. Chem., 279,* 63 (1985).
15. W. Brenner, P. Heimbach, H. Hey, E. W. Müller, G. Wilke, *Justus Leibigs Ann. Chem., 727,* 161 (1969).
16. J. H. Wengrovius, R. R. Schrock, C. S. Day, *Inorg. Chem., 20,* 1844 (1981).
17. G. K. Barker, M. Green, J. A. K. Howard, J. L. Spencer, F. G. A. Stone, *J. Chem. Soc. Dalton,* 1839 (1978).
18. H. J. Kablitz, G. Wilke, *J. Organomet. Chem., 51,* 241 (1973).
19. G. Wilke, in *Fundamental Research in Homogeneous Catalysis,* Vol. 3, M. Tsutsui, ed., Phenum Press, New York, NY, 1978, p. 1.
20. G. Natta, U. Giannini, P. Pino, A. Cassata, *Chimica e Industria, 47,* 524 (1965).
21. G. Allegra, F. Lo Giudice, G. Natta, U. Giannini, G. Fagherazzi, P. Pino, *J. Chem. Soc., Chem. Commun.,* 1263 (1967).
22. T. S. Cameron, M. L. H. Green, H. Munakata, C. K. Prout, M. J. Smith, *J. Coord. Chem., 2,* 43 (1972).
23. L. Porri, M. C. Gallazzi, A. Colombo, G. Allegra, *Tetrahedron Lett.,* 4187 (1965).
24. D. Medema, R. van Helden, C. F. Kohll, *Inorg. Chim. Acta, 3,* 255 (1969).
25. A. Dohring, P. W. Jolly, R. Mynott, K. P. Schick, G. Wilke, *Z. Naturforsch., Teil B, 36,* 1198 (1981).
26. R. Benn, P. W. Jolly, R. Mynott, B. Raspel, G. Schenker, K. P. Schick, G. Schroth, *Organometallics, 4,* 1945 (1985).
27. B. Bogdanovic, S. T. Liao, R. Mynott, K. Schlichte, U. Westeppe, *Chem. Ber., 117,* 1378 (1984).
28. B. Bogdanovich, *Angew. Chem., 97,* 253 (1985).

29. A. Nakamura, *Bull. Chem. Soc. Jpn., 39*, 543 (1966).
30. S. Otsuka, A. Nakamura, K. Tani, *J. Chem. Soc. A*, 1404 (1969).
31. R. P. Hughes, J. Powell, *J. Organomet. Chem., 20*, P17 (1969).
32. T. Okamoto, *Bull. Chem. Soc. Jpn, 44*, 1353 (1971).
33. L. S. Hegedus, N. Kambre, R. Tamura, P. D. Woodgate, *Organometallics, 2*, 1658 (1983).
34. M. Englert, P. W. Jolly, G. Wilke, *Angew. Chem. Int. Ed. Engl., 83*, 84 (1971).
35. S. Otsuka, A. Nakamura, *Adv. Organomet. Chem., 14*, 245 (1976).
36. S. Otsuka, A. Nakamura, S. Ueda, K. Tani, *J. Chem. Soc., Chem. Commun.*, 863 (1971).
37. S. Otsuka, A. Nukamura, T. Yamagata, K. Tani, *J. Am. Chem. Soc., 94*, 1037 (1972).
38. S. Otsuka, K. Tani, T. Yamagata, *J. Chem. Soc. Dalton Trans.*, 2491 (1973).
39. M. Englert, P. W. Jolly, G. Wilke, *Angew. Chem., Int. Ed. Engl., 84*, 120 (1972).
40. B. L. Barnett, C. Krüger, Y. H. Tsay, *Angew. Chem., Int. Ed. Engl., 84*, 121 (1972).
41. G. Pantini, P. Racanelli, A. Immirzi, L. Porri, *J. Organomet. Chem., 33*, C17 (1971).
42. G. Ingrosso, A. Immirzi, L. Porri, *J. Organomet. Chem., 60*, C35 (1973).
43. S. Otsuka, K. Tani, A. Nakamura, *J. Chem. Soc., A*, 1404 (1969).

5.8.2.4.6. π-Allyl Metal Complexes by Metal-Atom and Related Reactions.

Conjugated dienes codeposited with metal atoms yield unique materials. All the transition metals form complexes with 1,3-butadiene at low temperature[1-6], but only the Mo- and W-butadiene complexes are stable in the homoleptic form[7,8]: The other early transition metals simply yield oligomers or polymers of butadiene upon matrix warming[1,4,5]:

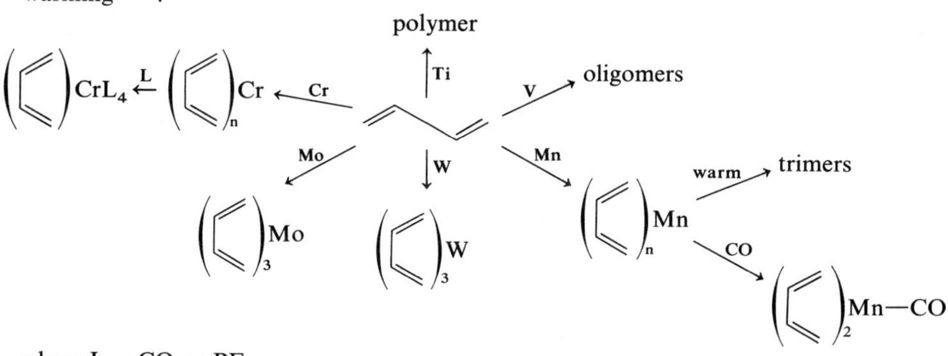

where L = CO or PF_3.

Chromium codeposited with 1,3-cyclohexadiene[1] yields an unstable $(diene)_nCr$ complex, which can form η^4-$C_6H_8)_2Cr(PF_3)_2$ by PF_3 addition.

The later transition metals also form complexes with 1,3-butadiene at low temperatures, but upon warming these decompose. The Ni–butadiene complex (with xs diene present) yields 2,6,10-dodecatrienylNi upon warming[9], while the Fe–butadiene complex decomposes unless trapping ligands are added[10]. Mixtures of phosphines or phosphites with butadiene plus Fe atoms yield $Fe(butadiene)_2[P(OCH_3)_3]$ and $Fe(butadiene)[P(OCH_3)_3]_3$[11]. Similarly, $Fe(1,3\text{-cyclooctadiene})[P(CH_3)_3)_3]_3$ was prepared as well as $Fe(1,3\text{-cyclohexadiene})P[(OCH_3)_3]_3$. The 1,3-cyclohexadiene reaction also yields a bis(diene) adduct[11]:

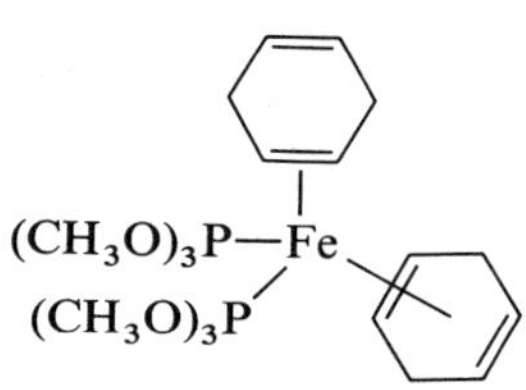

TABLE 1. CYCLOPENTADIENE–METAL ATOM REACTIONS AND PRODUCTS (CpH = CYLCOPENTADIENE)

Reaction	Products	Ref.
CpH + Cr →	$[(\eta^5\text{-Cp})_2CrH_2] \rightarrow (\eta^5\text{-Cp})_2Cr + H_2$	13,14
CpH + Mo →	$(\eta^5\text{-Cp})_2MoH_2$ isolable	15,16
CpH + W →	$(\eta^5\text{-Cp})_2WH_2$	15,16
CpH + C_6H_6 + Mn →	$(\eta^5\text{-Cp})$Mn  $+ \frac{1}{2} H_2$	17
CpH + Fe →	$(\eta^5\text{-Cp})_2Fe + H_2$	13,14
CpH + Co →	$(\eta^5\text{-Cp})$Co 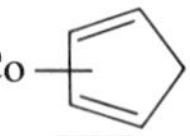 $+ \frac{1}{2} H_2$	13,14
CpH + Ni →	$(\eta^5\text{-Cp})$Ni 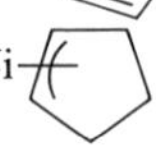	13,14

In addition, hydrogen transfers occur in the 1,3-cyclohexadiene system to yield the π-arene–Fe complexes, $(\eta^6\text{-}C_6H_6)Fe\,[P(OCH_3)_3]_2$ and $(\eta^6\text{-}C_6H_6)Fe(\eta^4\text{-}1,3\text{-}C_6H_8)$. Cobalt atoms with butadiene, in the presence of a hydrogen donor, yield a $(diene)_2$–Co–H compound[9]:

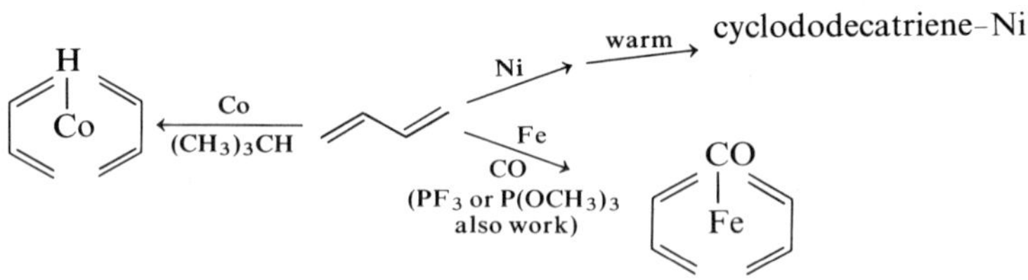

Cyclopentadiene with metal atoms leads to cyclopentadienyl complexes with either the release of hydrogen gas or the simultaneous formation of metal hydrides. Cyclopentadiene can bond to the metal both as a cyclopentadienyl ring and as a conjugated diene. Hydrogen transfer and hydrogen release occurs to yield the most stable complex, usually with 18 valence electrons. Thus, hydrogen transfer and movement is extremely facile at low temperature in these metal atom reactions. A series of cyclopentadiene–metal atom reactions are shown in Table 1. In a few cases additional ligands are added. Spirocycles containing the cyclopentadiene system react with Fe atoms to yield mixtures of unusual ferrocene derivatives[12]:

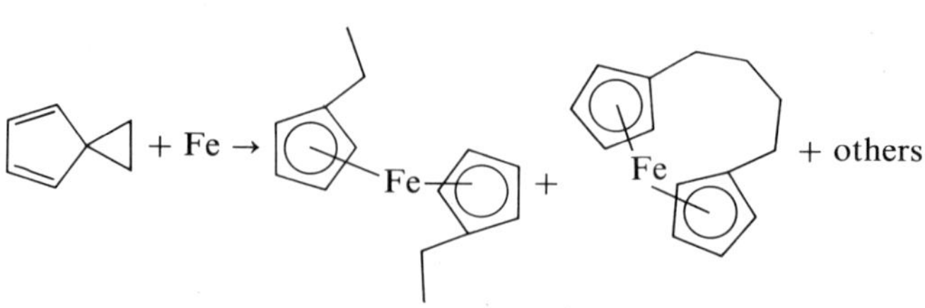

(a)

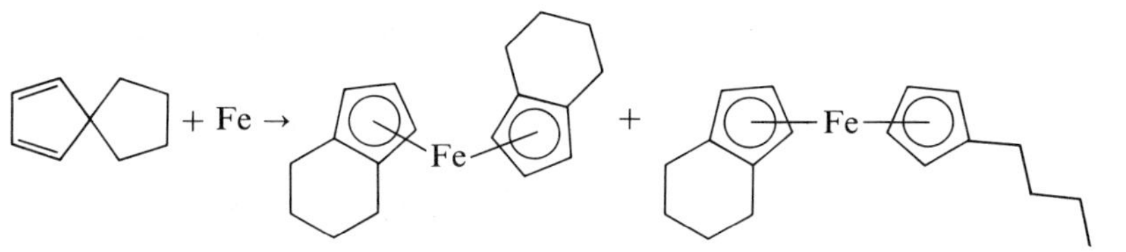

(b)

See §5.8.2.7.7 for a discussion of Co atom reactions with cyclopentadiene–alkyne mixtures.

(K. J. KLABUNDE)

1. J. R. Blackborow, R. H. Grubbs, A. Miyashita, A. Scrivanti, E. A. Koerner von Gustorf, *J. Organomet. Chem., 122* (1), C6 (1976).
2. P. S. Skell, D. L. Williams-Smith, M. J. McGlinchey, *J. Am. Chem. Soc. 95*, 3337 (1973).
3. E. Koerner von Gustorf, O. Jaenicke, O. E. Polansky, *Angew. Chem., Int. Ed. Engl., 11*, 532 (1972).
4. V. M. Akhmedov, M. T. Anthony, M. L. H. Green, D. Young, *J. Chem. Soc., Dalton Trans.*, 1412 (1975).
5. V. M. Akhmedov, M. T. Anthony, M. L. H. Green, D. Young, *J. Chem. Soc., Chem. Commun.*, 777 (1974).
6. M. Heberhold, *Chem. Unserer Zeit, 10*, 120 (1976).
7. P. S. Skell, E. M. Van Dam, M. P. Silvon, *J. Am. Chem. Soc., 96*, 626 (1974).
8. M. Yevitz, P. S. Skell, *Int. Union Crystallography Intercong. Symp. Intra-Intermol. Forces*, 1974 P. State Univ., Ser. 2, Vol. 2, Pap, E9 (1974).
9. P. S. Skell, M. J. McGlinchey, *Angew. Chem., Int. Ed. Engl., 14*, 195 (1975).
10. D. L. Williams-Smith, L. R. Wolf, P. S. Skell, *J. Am. Chem. Soc., 94*, 4042 (1972).
11. S. D. Ittel, F. A. VanCatledge, J. P. Jesson, *J. Am. Chem. Soc., 101*, 3874 (1979).
12. A. J. L. Hanlan, R. C. Ugolick, J. G. Fulcher, S. Togashi, A. B. Bocarsly, J. A. Gladysz, *Inorg. Chem., 19*, 1543 (1980).
13. P. L. Timms, *Adv. Inorg. Chem. Radiochem., 14*, 121 (1972).
14. P. L. Timms, *J. Chem. Soc., Chem. Commun.*, 1033 (1969).
15. M. J. D'Aniello, Jr. E. K. Barefield, *J. Organomet. Chem. 76*, C50 (1974).
16. E. M. Van Dam, W. N. Brent, M. P. Silvon, P. S. Skell, *J. Am. Chem. Soc. 97* 465 (1975).
17. E. A. Koerner von Gustorf, O. Jaenicke, O. Wolfbeis, C. R. Eady, *Angew. Chem., Int. Ed. Engl. 14*, 278 (1975).

5.8.2.5. from Chelating Dienes

5.8.2.5.1. Chelating Diene Complexes by Direct Reaction with Metal Complexes via Ligand Displacement.

In §5.8.2.3.1 it is pointed out that olefins are relatively weak ligands for transition metals[1]. Diolefins that can act as bidentate ligands form stronger complexes than monoolefins owing to the chelate effect. As a consequence most of the methods described in §5.8.2.3.1 have been used to prepare diolefin complexes. Thus, for example, diolefins displace anionic ligands from Pt(II)[2–8], Pd(II)[6,9], Rh(I)[10,11] and Ir(I)[12,13]:

$$Na_2[PtCl_4] + \|\frown\| \rightarrow [PtCl_2(\|\frown\|)] + 2\ NaCl \tag{a}$$

where $\|\frown\|$ = 1,5-COD, Dewar hexamethylbenzene, norbornadiene, hexa-1,5-diene, 5-methylenecycloheptene;

$$2\ RhCl_3 + 2\|\frown\| \rightarrow [Rh(\|\frown\|)Cl]_2 \tag{b}$$

where ‖⌒‖ = 1,5-COD, dicyclopentadiene, cyclooctatetraene, carvone;

$$2\ IrCl_3 + 2\ 1{,}5\text{-COD} \xrightarrow{EtOH} [Ir(1{,}5\text{-COD})Cl]_2 \quad (c)$$

$$H_2[IrCl_6] + 1{,}5\text{-COD} \xrightarrow[(2)\ aq.\ NaOAc]{(1)\ heat\ in\ PrOH} [Ir(1{,}5\text{-COD})Cl]_2 \quad (d)$$

The conditions of reaction of cyclooctadiene with Pd(II) and Pt(II) are critical in determining whether the olefin isomerizes during complex formation. Thus complexes of 1,4- and 1,5-cyclooctadiene with both metals have been prepared without any apparent isomerization[6,7], whereas under different conditions phosphine complexes of Pd(II)[14] and Pt(II)[15] isomerise the nonconjugated cyclooctadienes to the 1,3 isomer. Similarly while 1,4-cyclooctadiene reacts with $K_2[PtCl_4]$ in water to form $Pt(1,4\text{-COD})Cl_2$, it forms $[Pt(1,5\text{-COD})Cl_2]$ in 50% aqueous ethanol[7].

The formation of chelating diolefin complexes by displacement of olefin ligands is rare because complexes containing two monoolefin ligands in suitable mutual positions to be replaced by one diene ligand are themselves rare. Thus diolefins such as 1,3-cyclohexadiene, 1,4-cyclohexadiene and norbornadiene all displace isobutylene from $[Fe(\eta^5\text{-Cp})(CO)_2(isobutylene)][BF_4]$, but only one double-bond of the diolefin coordinates to the iron[16–18]. Diolefins readily displace 1,5-cyclooctadiene from $Ru(1,5\text{-COD})(CO)_3$ to form new diolefin complexes[19]:

$Ru(1,5\text{-COD})(CO)_3$ + 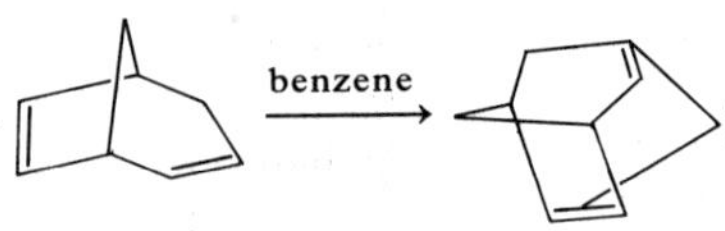

$Ru(CO)_3$ (e)

An exception to the generalization that diolefins rarely displace monoolefins to form chelate complexes is where an anionic chloride ligand is lost as hydrogen chloride from Pt(II) together with ethylene when $H[PtCl_3(C_2H_4)]$ is treated with a diolefin such as norbornadiene[20]:

$H[PtCl_3(C_2H_4)]$ + 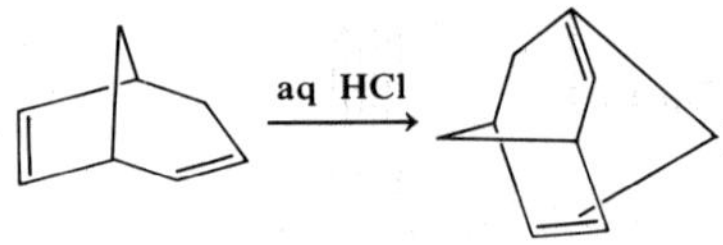

$PtCl_2 + C_2H_4 + HCl$ (f)

A number of chelating diolefin complexes of group-VIA metal carbonyls have been prepared by refluxing the hexacarbonyl in a hydrocarbon solvent[21–23]:

$$M(CO)_6 + NBD \xrightarrow[reflux]{hydrocarbon} M(NBD)(CO)_4 + 2\ CO \quad (g)$$

where M = Cr, Mo. The experimental difficulty with sublimation of the metal hexacarbonyl during this reaction can be overcome using a special apparatus[24,25]. Although few olefin complexes of Co(0) are known, chelating diolefins displace CO from $Co_2(CO)_8$ without cleavage of the $Co_2(CO)_2$ bridge[26]:

$$Co_2(CO)_8 + NBD \xrightarrow[pet.\ ether]{heat\ in} Co_2(NBD)_2(CO)_4 + 4\ CO \quad (h)$$

Similarly, heating $Rh_6(CO)_{16}$ with norbornadiene in methylcyclohexane yields a series of diolefin complexes [27]:

$$Rh_6(CO)_{16} + NBD \xrightarrow[\text{methylcyclohexane}]{\text{heat in}} Rh_6(NBD)_n(CO)_{16-2n} + 2n\,CO \qquad (i)$$

Acetonitrile complexes of the group-VIA metals are useful intermediates in the development of the chemistry of those metals. Three carbonyl ligands of $M(CO)_6$, where M = Cr, Mo or W, are replaced by acetonitrile when the hexacarbonyls are refluxed in that solvent[28,29]. When $[M(CO)_3(CH_3CN)_3]$ is treated with chelating diolefins the three acetonitrile ligands together with one carbonyl ligand are displaced to form[30] $M(diolefin)_2(CO)_2$:

$$W(CO)_3(CH_3CN)_3 + \text{1,3-cyclohexadiene} \rightarrow [W(\text{1,4-cyclohexadiene})_2(CO)_2] + 3\,CH_3CN + CO \qquad (j)$$

Hexamethyl-Dewar-benzene reacts with $M(CO)_3(CH_3CN)_3$ to form products with four carbonyl ligands, although the mechanism of formation of $M(CO)_4$ from the $M(CO)_3$ moiety is not known[31,32]:

$M(CO)_3(CH_3CN)_3$ + HMDB → 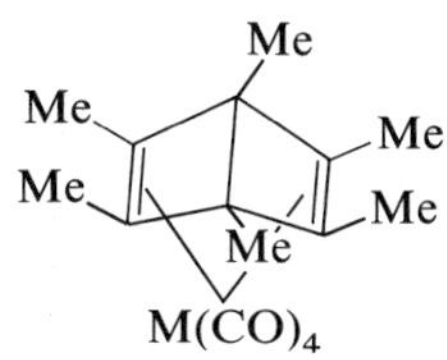 (k)

A similar reaction occurs between $W(CO)_3(CH_3CN)_3$ and either 1,5-cyclooctadiene or bicyclo[2.2.1]-heptadiene[30]:

$$W(CO)_3(CH_3CN)_3 + \text{1,5-COD} \xrightarrow{\text{hexane}} W(\text{1,5-COD})(CO)_4 \qquad (l)$$

The displacement of benzonitrile from $[Pd(PhCN)_2Cl_2]$ is a simple and commonly used route for the preparation of $[Pd(diolefin)Cl_2]$ complexes[20,33–36].

The most facile syntheses of diolefin complexes are those in which the diolefin reacts with a coordinatively unsaturated metal complex without ligand displacement. Thus a unidentate norbornadiene complex of iron, $[Fe(\eta^5\text{-}Cp)(CO)_2(\eta^2\text{-}NBD)][BF_4]$, can be prepared by treating $[Fe(\eta^5\text{-}Cp)(CO)_2]_2$ with xs $Ph_3C[BF_4]$, followed by quenching with cycloheptatriene prior to addition of norbornadiene[37]. Tetracoordinate complexes of Ir(I) undergo diolefin addition[38,39]:

[Ir(N⌒N)(1,5-COD)]X +  → $[Ir(N⌒N)(1,5\text{-}COD)(C_4H_6)]X$ (m)

where N⌒N = bipy, phen;

$[IrN⌒N)(CO)_2]X$ + diolefin → $[Ir(N⌒N)(CO)_2(diolefin)]X$ (n)

where N⌒N = bipy, phen; diolefin = 1,5-COD, NBD. In the dicarbonyl complexes formed in Eq. (n) the Ir is five-coordinate, so that the diolefin acts only as a monodentate ligand.

Diolefin complexes of Ag(I) salts[40] with a range of anions, including nitrate[41] perchlorate[41,42], tetrafluoroborate[43], hexafluoroacetylacetonate[44], β-diketonate[45] and trifluoromethylsulfonate[46], have been prepared by adding the salt(I) salt to the liquid diolefin[40]. Corresponding Cu(I) salts with halide[47], perchlorate[47], acetate[48], trifluoroacetate[49,50], tetrafluoroborate[51] and trifluoromethylsulfonate[52,53] anions have been prepared. The Cu(I) complexes with anions of low coordinating ability often disproportionate to Cu(0) and Cu(II). Generally, the most stable complexes of both Cu(I) and Ag(I) are with the trifluoromethylsulfonate anion[46,52,53]. The stoichiometry of these complexes is dependent upon solvent, diolefin and anion. The most complete series studied has been the 1,5-COD complexes of Cu(I), where poorly coordinating anions such as perchlorate, tetrafluoroborate and trifluoromethylsulfonate form $Cu(1,5\text{-}COD)_2X$, while more strongly coordinating anions form Cu(1,5-COD)X. With acetate and trifluoroacetate $Cu_2(1,5\text{-}COD)X_2$ are formed in which the anions may be bridging[48].

(F. R. HARTLEY)

1. F. R. Hartley, *Chem. Rev.*, *73*, 163 (1973).
2. J. Chatt, L. M. Vallarino, L. M. Venanzi, *J. Chem. Soc.*, 3413 (1957).
3. K. A. Jensen, *Acta Chem. Scand.*, *7*, 866 (1953).
4. C. R. Kistner, *Diss. Abs.*, *24*, 4391 (1964).
5. C. R. Kistner, J. H. Hutchinson, J. R. Doyle, J. C. Storlie, *Inorg. Chem.*, *2*, 1255 (1963).
6. G. B. McCauley, *Diss. Abs.*, *B28*, 86 (1967).
7. H. A. Tayim, A. Bouldoukian, M. Kharbousch, *Inorg. Nucl. Chem. Lett.*, *8*, 231 (1972).
8. C. B. Anderson, J. T. Michalowski, *J. Chem. Soc., Chem. Commun.*, 459 (1972).
9. P. J. Hendra, D. B. Powell, *Spectrochim. Acta*, *17*, 913 (1961).
10. J. Chatt, L. M. Venanzi, *J. Chem. Soc.*, 4735 (1957).
11. V. Schurig, *J. Organomet. Chem.*, *74*, 457 (1974).
12. J. L. Herdé, J. C. Lambert, C. V. Senoff, *Inorg. Synth.*, *15*, 18 (1974).
13. R. H. Crabtree, G. E. Morris, *J. Organomet. Chem.*, *135*, 395 (1977).
14. D. Wittenburg, H. Seibt, German Patent, 1,136,329 (1962); *Chem. Abstr.*, *58*, 4442b (1963).
15. H. A. Tayim, J. C. Bailar, *J. Am. Chem. Soc.*, *89*, 3420 (1967).
16. W. P. Giering, M. Posenblum, *J. Chem. Soc., Chem. Commun.*, 441 (1971).
17. M. Rosenblum, *Acc. Chem. Res.*, *7*, 122 (1974).
18. K. M. Nicholas, *J. Am. Chem. Soc.*, *97*, 3254 (1975).
19. A. J. P. Domingos, B. F. G. Johnson, J. Lewis, *J. Organomet. Chem.*, *49*, C33 (1973).
20. R. A. Alexander, N. C. Baenziger, C. Carpenter, J. R. Doyle, *J. Am. Chem. Soc.*, *82*, 535 (1960).
21. M. A. Bennett, L. Pratt, G. Wilkinson, *J. Chem. Soc.*, 2037 (1961).
22. R. Pettit, *J. Am. Chem. Soc.*, *81*, 1266 (1959).
23. A. R. L. Bursics, E. Bursics-Szekeres, F. G. A. Stone, *J. Fluorine Chem.*, *7*, 619 (1976).
24. W. Strohmeier, *Chem. Ber.*, *94*, 2490 (1961).
25. W. Strohmeier, *Angew. Chem., Int. Ed. Engl.*, *3*, 730 (1964).
26. P. McArdle, A. R. Manning, *J. Chem. Soc.*, *A*, 2123 (1970).
27. J. A. J. Jarvis, R. Whyman, *J. Chem. Soc., Chem. Commun.*, 562 (1975).
28. D. P. Tate, W. R. Knipple, J. M. Augl, *Inorg. Chem.*, *1*, 433 (1962).
29. B. L. Ross, J. G. Grasselli, W. M. Ritchey, H. D. Kaesz, *Inorg. Chem.*, *2*, 1023 (1963).
30. R. B. King, A. F. Fronzaglia, *Inorg. Chem.*, *5*, 1837 (1966).
31. E. O. Fischer, C. G. Kreiter, W. Berngruber, *Angew. Chem., Int. Ed. Engl.*, *6*, 634 (1967).
32. E. O. Fischer, W. Berngruber, C. G. Kreiter, *Chem. Ber.*, *101*, 824 (1968).
33. J. R. Doyle, P. E. Slade, H. B. Jonassen, *Inorg. Synth.*, *6*, 216 (1960).
34. I. A. Zakharova, L. A. Leites, V. A. Aleksanyan, *J. Organomet. Chem.*, *72*, 283 (1974).
35. M. S. Kharasch, R. C. Seyler, F. R. Mayo, *J. Am. Chem. Soc.*, *60*, 882 (1938).
36. H. P. Fritz, H. Keller, *Chem. Ber.*, *95*, 158 (1962).
37. P. F. Boyle, K. M. Nicholas, *J. Organomet. Chem.*, *114*, 307 (1976).
38. G. Mestroni, A. Camus, *Inorg. Nucl. Chem. Lett.*, *9*, 261 (1973).
39. G. Mestroni, A. Camus, G. Zassinovich, *J. Organomet. Chem.*, *73*, 119 (1974).
40. C. D. M. Beverwijk, G. J. M. Van Der Kerk, A. J. Leusink, J. G. Noltes, *Organomet. Chem. Rev.*, *A5*, 215 (1970).
41. A. E. Comyns, H. J. Lucas, *J. Am. Chem. Soc.*, *79*, 4339 (1957).
42. G. Bressan, R. Broggi, M. P. Lachi, A. L. Segre, *J. Organomet. Chem.*, *9*, 355 (1967).

43. A. Albinati, S. V. Meille, G. Carturan, *J. Organomet. Chem.*, *182*, 269 (1979).
44. W. Partenheimer, E. H. Johnson, *Inorg. Chem.*, *11*, 2840 (1972).
45. W. Partenheimer, E. H. Johnson, *Inorg. Synth.*, *16*, 117 (1976).
46. G. S. Lewandos, D. K. Gregston, F. R. Nelson, *J. Organomet. Chem.*, *118*, 363 (1976).
47. S. E. Manahan, *Inorg. Chem.*, *5*, 2063 (1966).
48. D. A. Edwards, R. Richards, *J. Organomet. Chem.*, *86*, 407 (1975).
49. M. B. Dines, *Inorg. Chem.*, *11*, 2949 (1972).
50. D. L. Reger, M. D. Dukes, *J. Organomet. Chem.*, *113*, 173 (1976).
51. S. E. Manahan, *Inorg. Nuc. Chem. Lett.*, *3*, 383 (1967).
52. R. G. Salomon, J. K. Kochi, *J. Am. Chem. Soc.*, *95*, 1889 (1973).
53. R. G. Salomon, J. K. Kochi, *J. Chem. Soc., Chem. Commun.*, 559 (1972).

5.8.2.5.2. η^1 σ-Alkyl Complexes by Nucleophilic Attack on Metal-Chelating Diene Complexes.

Nonconjugated dienes bound to electrophilic transition-metal centers are activated toward nucleophilic attack analogous to olefin activation by transition metals discussed in §5.8.2.3.4. Although secondary reactions or rearrangements can occur, the pattern of the nucleophilic addition step (Nuc = nucleophile) leading to σ, π complexes is shown below:

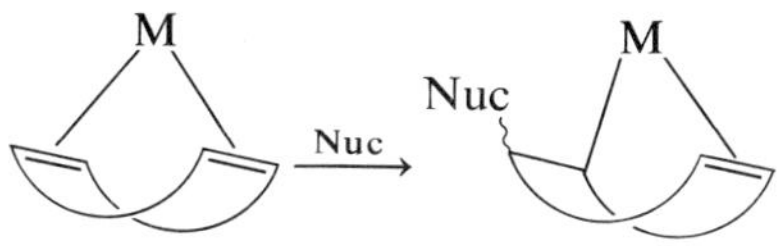

(a)

The majority of cases (Table 1) involve activation of the diene by complexation to Pd(II) and Pt(II) centers. However, other metal–diene complexes have been examined including Ni(II), Ir(I), and Mn(I) complexes. Cationic or neutral complexes are used in the nucleophilic addition reactions. The most common nucleophiles employed are oxygen or nitrogen bases (hydroxide, alkoxides, carboxylates, amines) or stabilized carbon nucleophiles (malonate, β-diketonates). The dienes employed include 1,5-cyclooctadiene, norbornadiene, dicyclopentadiene, 4-vinylcyclohexene, 7-(alkylidene)norbornene and endo-4-vinylnorbornene.

The stereochemistry of addition in these reactions is of interest in that the two general mechanisms for nucleophilic attack lead to products of differing stereochemistries. If nucleophilic attack occurs initially at the metal center followed by transfer of the nucleophile to the olefinic ligand, then endo stereochemistry is expected with the M—C and Nuc—C bonds cis to one another:

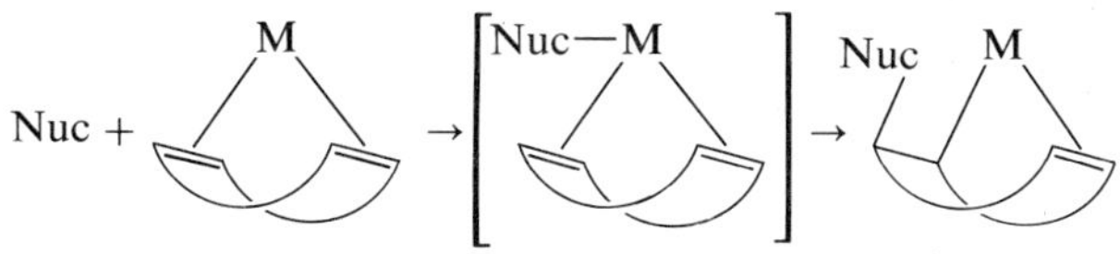

(b)

cis product

Alternatively, direct attack of the nucleophile at the olefinic bond is expected to occur exo and lead to a trans product:

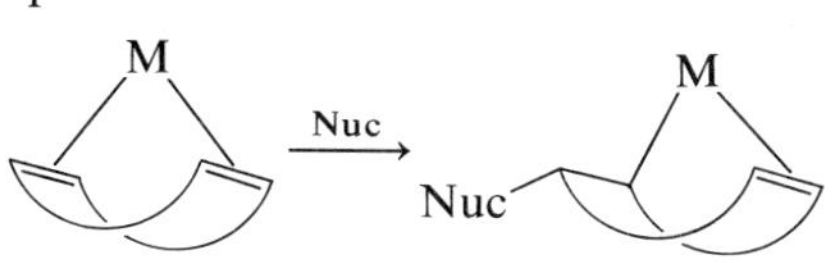

trans product (c)

TABLE 1. NUCLEOPHILIC REACTION PRODUCTS

A. 1,5-Cyclooctadiene complexes

Diene	ML_n	Nucleophile, reaction conditions	Product	Y	Ref.
1,5-Cyclooctadiene 	MX_2 M = Pd,Pt X = Cl,Br, I	MeOH, Na_2CO_3		—OMe	7,8
	MCl_2 M = Pd,Pt X = Cl	2 NH_2R R = (S)-α-phenylthylamine R = C_6H_5, M = Pt R = CH_2-p-$(CH_3)C_6H_4$, M = Pt		—NRH	8–11
	$PtCl_2$	2 NaOH, H_2O, MeOH X = Y = OMe	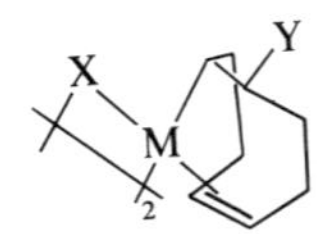	—OMe	12, 13
	PdCl X = Cl	H_2O,Na_2CO_3, CH_2Cl_2		—OH	2, 9, 10
	MX_2 M = Pd,Pt X = Cl,Br	ZCH_2CO_2R, Na_2CO_3, Et_2O Z = CO_2Et, R = Et Z = CO_2Me, R = Me Z = COMe, R = Et		—CH(CO_2R)Z	14
	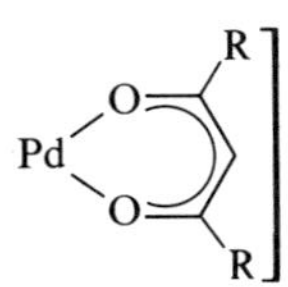 BF_4	H_2O, Na_2CO_3, CH_2Cl_2 R = t-Bu	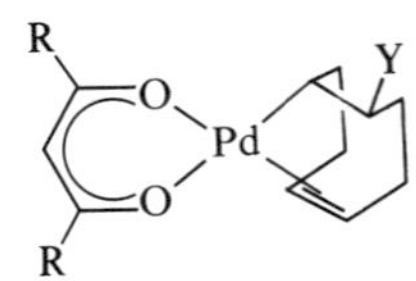	—OH	15

MX_2 M = Pd,Pt X = Cl,Br,I	2 Tl(OC)(R)CH(R′)(CO), $CHCl_3$ R = R′ = CH_3 R = CH_3, R′ = C_6H_5 R = R′=C_6H_5		—CH(COR′)(COR)	16
$PdCl_2$	$Pb(OAc)_2$, HOAc	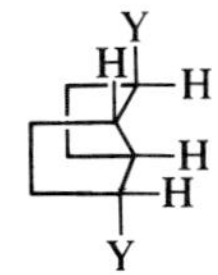	—OAc	17
Ni(η^5-Cp)]BF_4	NaOMe, MeOH	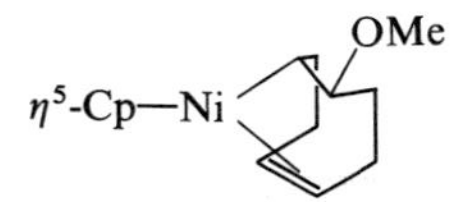		18
(phen)Ir]Cl	MeOH, (NC)CHH(CN)	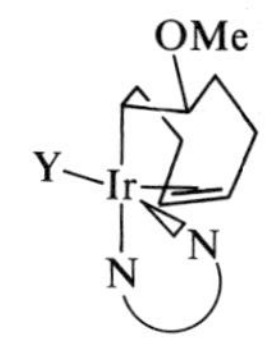	(NC)CHCH(CN)	19, 20
$Mn(CO)_4]BF_4$	NaBH R = H NaOMe R = OMe NaN_3 R = N_3	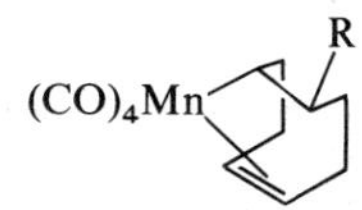		21

Continues over

TABLE 1. (*Continued*)

B. endo-Dicyclopentadiene complexes

Diene	ML_n	Nucleophile, reaction conditions	Product	Ref.
endo-Dicyclopentadiene 	$PdCl_2$	ROH R = Me, Et, n-Pr, i-Pr MeOH, Na_2CO_3	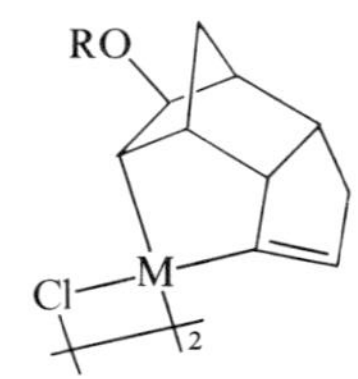	7, 22
	$PtCl_2$	ROH R = Me NaOR, ROH or NaOR, Na_2CO_3 R = Me, Et, n-Pr		23, 24
	$PtCl_2$	2 $PhNH_2$, CH_2Cl_2 or 1 $PhNH_2$ + NEt_3	[(PhNHdi-η^5-Cp)PtCl]$_2$ [a]	9, 11
	$PtCl_2$	2 Tl[RCOCHCOR′] R = CH_3, R′ = Ph R = R′ = Ph	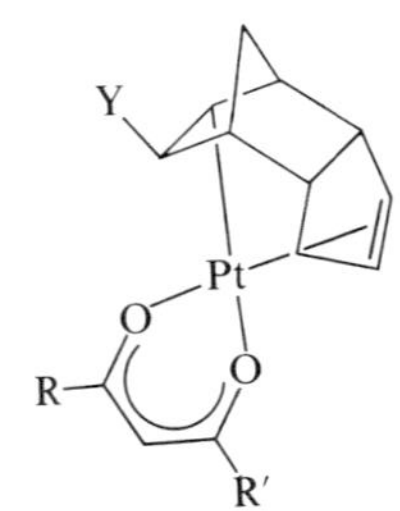 Y = RCOCHCOR′	16
	$PtCl_2$	acac, Na_2CO_3 R = R′=CH_3		25

C. Norbornadiene complexes

Diene	ML_n	Nucleophile, reaction conditions	Product	Y	Ref.
Norbornadiene 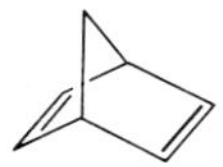	MCl_2 M = Pd, Pt	2 NH_2R R = $PhCH_2$ R = $Ph(CH_3)CH$ R = Ph	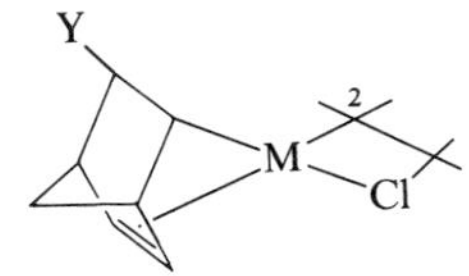	—NHR	9, 10
	$PdCl_2$	NaOMe, MeOH or Na_2CO_3, MeOH		—OMe	22
	$PdCl_2$	Ph_2Hg or $NaBPh_4$	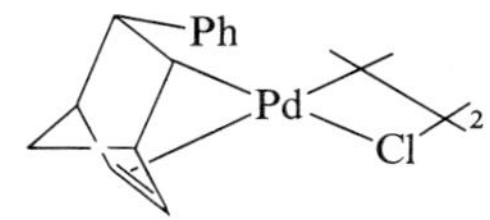		1
	$Ni(\eta^5\text{-}Cp)]BF_4$	NaOMe, MeOH	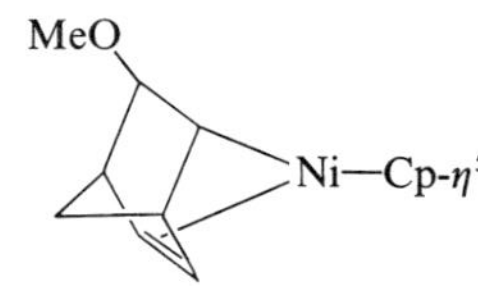		18

Continues Over

D. Miscellaneous dienes (*Continued*)

Diene	ML_n	Nucleophile, reaction conditions	Product	Ref.
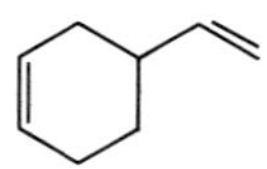	MCl_2 M = Pd, Pt	2 NH_2R R = CH_2Ph R = $Ph(CH_3)CH$ R = Et	[(diene NHR)$MCl]_2$	8
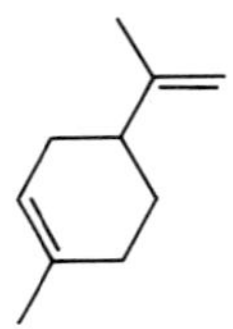	$PtCl_2$	MeOH, Na_2CO_3	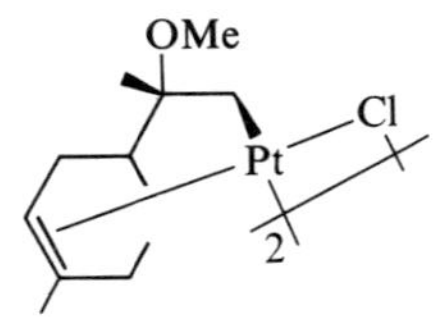	7
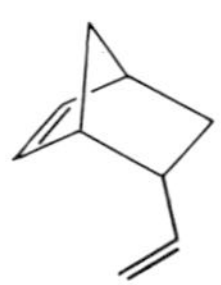	$PdCl_2$	Cl^-	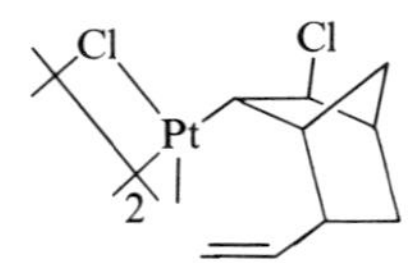	22
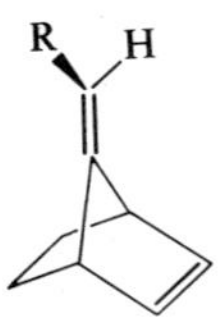	PdX_2 X = Cl, Br	X^-, CH_2Cl_2 or $CHCl_3$	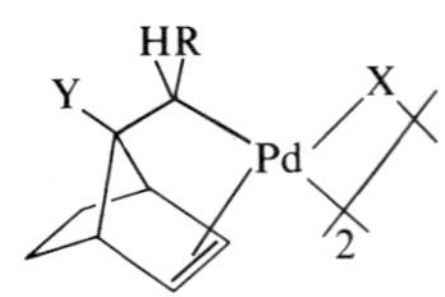 X = Y = Cl R = H, CO_2Et, CO_2Me X = Y = Br R = H, CO_2Et	23

Diene	Metal salt	Reagent / conditions	Product	Ref.
	MCl_2 M = Pd, Pt	2 RNH_2 R = $PhCH_2$ R = $Ph(CH_3)CH$ R = Et	$[(diene\ NMR)MCl]_2$	8
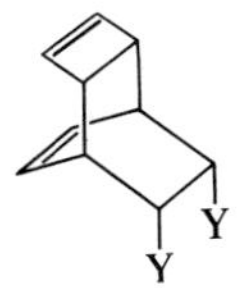	$PdCl_2$	CH_3COOH, reflux	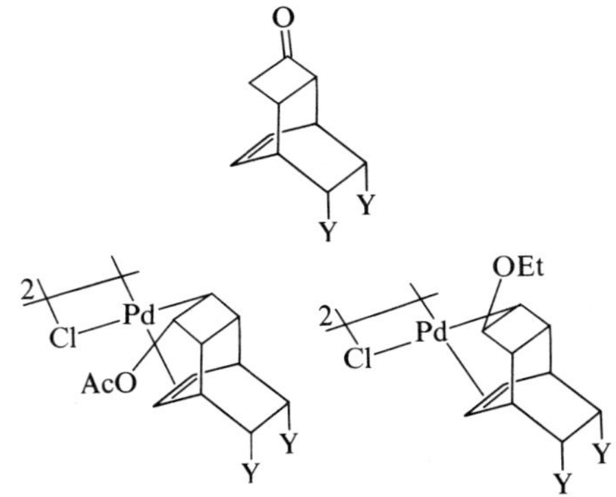 Y = CO_2Me	24

[a] di-η^5-Cp = endo-dicyclopentadiene.

Experimentally it has been found, generally through X-ray analysis of the product complexes, that nucleophilic addition is almost always exo leading to trans products (see Table 1). An interesting case[1] of endo addition involves the phenylation of norbornadiene–$PdCl_2$. The phenylation reaction was postulated to proceed via a Pd–aryl intermediate, **I**, which could not be isolated or detected:

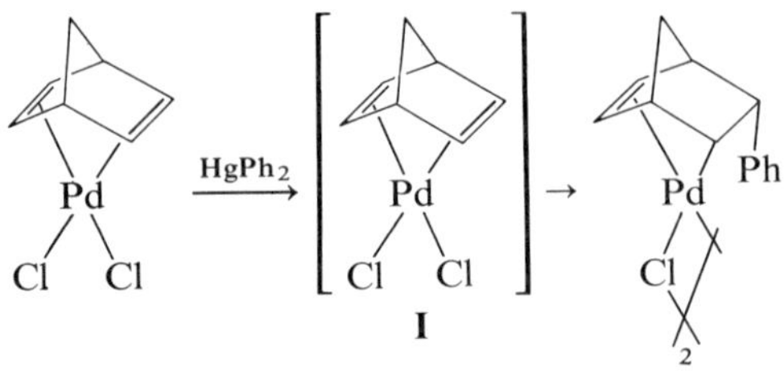

(d)

In support of this contention, phenylation of the analogous Pt complex leads to the isolable Pt–aryl species **II**.

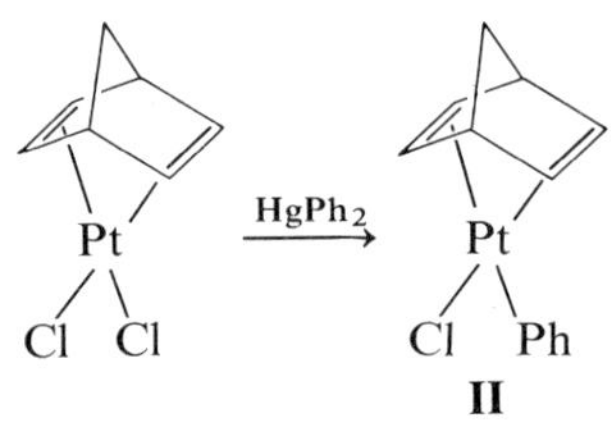

(e)

No aryl migration form Pt to the norbornadiene ligand was observed.

The stereochemistry of nucleophilic addition is especially relevant to the mechanism of the Pd(II)-catalyzed conversion of ethylene and water to acetalydehyde. The mechanism originally proposed involved hydroxide attack at Pd followed by migration to coordinated ethylene:

$$[C_2H_4PdCl_3]^- \xrightarrow{OH^-} [C_2H_4PdCl_2(OH)]^- \rightarrow$$

$$HOC_2H_4PdCl \rightarrow CH_3CHO + HCl + Pd(0) \qquad (f)$$

In view of the many documented cases of exo nucleophilic attack of Pd(II)–diene complexes, exo hydroxide attack at ethylene appeared an attractive alternative. Indeed, it has been shown[2] that hydroxide attack on 1,5-cyclooctadiene–$PdCl_2$ in aqueous base results in exo attack of hydroxide, a result which clearly suggests exo attack in the Pd(II)-catalyzed conversion of ethylene and H_2O to acetaldehyde. More recent mechanistic investigations also support exo hydroxide attack[3,4].

Recently, the regioselectivity of nucleophilic attack on (endo-dicyclopentadiene–$PdCl_2$) has been used to confirm theoretical predictions concerning the mode of olefin activation by transition metals[5]. Calculations suggest that olefin activation can occur by a lateral slipping of the olefin away from the symmetrical h^2 geometry. The carbon most distant from the metal develops a partial positive charge and is predicted to be the site of nucleophilic attack[6].

X-Ray analysis of the ridgid tricyclic Pd complex **III** indicates that the C_5-C_6 double bond is symmetrically coordinated, but that the C_1-C_2 double bond is slipped and tilted with C_1—Pd = 228 pm and C_2—Pd = 219 pm[5].

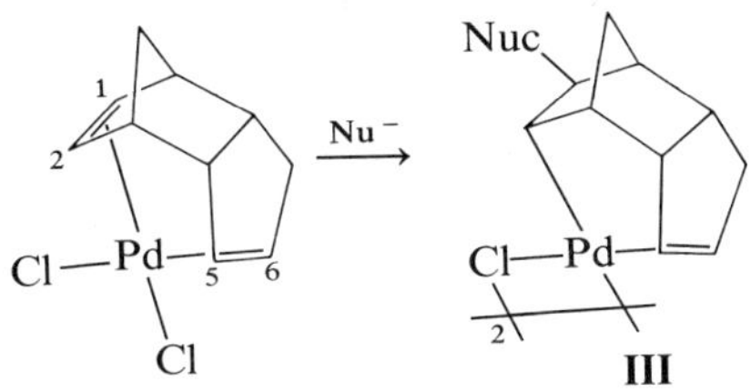

III

In agreement with the theoretical predictions and the observed distortion, attack by a variety of nucleophiles on complex **III** occurs exclusively at C_1.

(G. F. SCHMIDT, M. BROOKHART)

1. A. Segnitz, P. M. Bailey, P. M. Maitlis, *J. Chem. Soc., Chem. Commun.*, 698 (1973).
2. J. K. Stille, D. E. James, *J. Organomet. Chem.*, *108*, 401 (1976).
3. J. K. Stille, D. E. James, *J. Am. Chem. Soc.*, *97*, 674 (1975).
4. J. E. Bäckvall, B. Åkermark, S. O. Ljunggren, *J. Am. Chem. Soc.*, *101*, 2411 (1979).
5. L. L. Wright, R. M. Wing, M. F. Rettig, *J. Am. Chem. Soc.*, *104*, 610 (1982).
6. O. Eisenstein, R. Hoffmann, *J. Am. Chem. Soc.*, *103*, 4308 (1981).
7. J. Chatt, L. M. Vallarino, L. M. Venanzi, *J. Chem. Soc.*, 3413 (1957).
8. J. Chatt, L. M. Vallarino, L. M. Venanzi, *J. Chem. Soc.*, 2496 (1957).
9. G. Paiaro, A. DeRenzi, R. Palumbo, *J. Chem. Soc., Chem. Commun.*, 1150 (1967).
10. R. Palumbo, A. DeRenzi, A. Panunzi, G. Paiaro, *J. Am. Chem. Soc.*, *91*, 3874 (1969).
11. R. N. Haszeldine, R. V. Parish, D. W. Robbins, *J. Organomet. Chem,*, *23*, C33 (1970).
12. A. B. Goel, S. Goel, D. G. Vanderveer, *Inorg. Chim. Acta*, *54*, L169 (1981).
13. A. B. Goel, S. Goel, *Inorg. Chim. Acta*, *77*, L5 (1983).
14. H. Takahashi, J. Tsuji, *J. Am. Chem. Soc.*, *90*, 2387 (1968).
15. D. A. White, *J. Chem. Soc.* (*A*), 145 (1971).
16. B. F. G. Johnson, J. Lewis, M. S. Subramanian, *J. Chem. Soc.*, *A*, 1993 (1968).
17. P. M. Henry, M. Davies, G. Ferguson, S. Phillips, R. Restivo, *J. Chem. Soc., Chem. Commun.*, 112 (1974).
18. G. Parker, A. Salzer, H. Werner, *J. Organomet. Chem.*, *67*, 131 (1974).
19. N. Bresciana-Pahor, M. Calligaris, G. Nardin, P. Delise, *J. Chem. Soc., Dalton Trans.*, 762 (1976).
20. G. Mestroni, G. Zassinovich, A. Camus, *Inorg. Nucl. Chem. Lett.*, *11*, 359 (1975).
21. P. J. Harris, S. A. R. Knox, F. G. A. Stone, *J. Organomet. Chem.*, *148*, 327 (1978).
22. M. Green, R. I. Hancock, *J. Chem. Soc.*, *A*, 2054 (1967).
23. Whitla, H. M. Powell, L. M. Venanzi, *J. Chem. Soc., Chem. Commun.*, 310 (1966).
24. K. A. Hofmann, J. v. Narbutt, *Chem. Ber.*, *41*, 1625 (1908).
25. J. K. Stille, D. B. Fox, *Inorg. Nucl. Chem. Lett.*, *5*, 157 (1969).
26. P. M. Henry, *Adv. Organomet. Chem.*, *13*, 363 (1975).
27. G. Wiger, G. Albelo, M. F. Rettig, *J. Chem. Soc., Dalton Trans.*, 2242 (1974).
28. T. Sasaki, K. Kanematsu, A. Kondo, *J. Chem. Soc., Perkin Trans. I*, 2516 (1976).

5.8.2.5.3. by Metal-Atom and Related Reactions.

Cyclooctadiene yields stable complexes when codeposited with a variety of transition metal atoms. With Cr both 1,3- and 1,5-cyclooctadiene yield an intermediate complex that upon treatment with PF_3 yields a 1,5-cyclooctadiene complex[1]. Hydrogen migration to give isomerization to the 1,5-complex is facile. Hydrogen transfer is also evident from the formation of a hydride byproduct[1], $C_8H_{11}Cr(PF_3)_3H$. Similarly, hydrogen transfers are facile with 1,3- or 1,4-cyclohexadiene in the presence of Cr atoms (in a warming matrix)[2]. Disproportionation is induced catalytically to yield benzene and cyclohexane[3]. Similar results are obtained with Fe, Co and Ni atoms[3]:

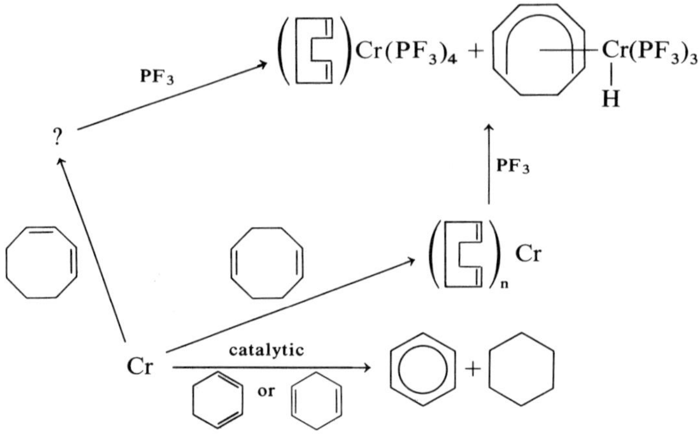

Nickel atoms codeposited with 1,5-cyclooctadiene yield the known bis(diene)Ni(0), and the analogous Pd and Pt complexes can be prepared by metal atom reactions[4–7]. However, these complexes can be more conveniently prepared on large scale by conventional methods[8,9,10].

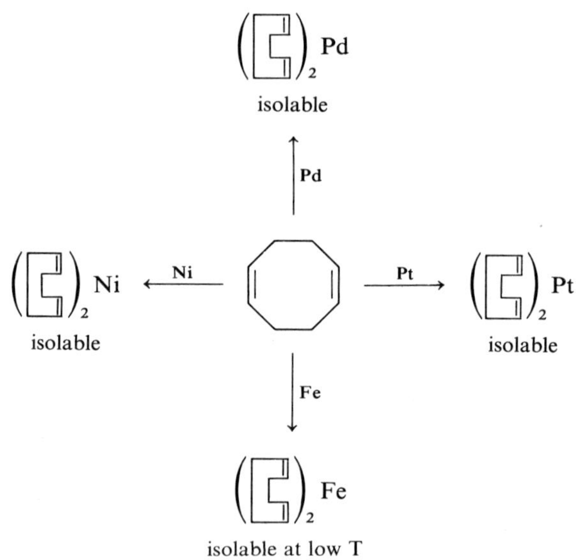

Bis(1,5-cyclooctadiene)Fe(0), is prepared using Fe atoms impinging into a diene-methycyclohexane solution. This paramagnetic complex is stable to $-30°C$ in hexane and 0°C in 1,5-cyclooctadiene solution and is extremely reactive with added ligands.

Norbornadiene is cyclodimerized by Ni atoms via a nickelacyclopentane intermediate[11]. No stable (norbornadiene)M(0) complexes are known, although see §5.8.2.9.3 for a discussion of metal atom generated (η^6-$C_6H_5CH_3$)Ni(C_6F_5)$_2$ which reacts with norbornadiene to yield[12] norbornadiene polymer plus (norbornadiene)Ni-(C_6F_5)$_2$.

(K. J. KLABUNDE)

1. E. A. Koerner von Gustorf, O. Jaenicke, O. Wolfbeis, C. R. Eady, *Angew. Chem., Int. Ed. Engl., 14*, 278 (1975).
2. J. R. Blackborow, R. H. Grubbs, A. Miyashita, A. Scrivanti, E. A. Koerner von Gustorf, *J. Organomet. Chem., 122* (1), C6 (1976).
3. P. S. Skell, M. J. McGlinchey, *Angew. Chem., Int. ed. Engl., 14*, 195 (1975).
4. R. M. Atkins, R. MacKenzie, P. L. Timms, T. W. Turney, *J. Chem. Soc., Chem. Commun.*, p. 764 (1975).
5. P. S. Skell, J. J. Havel, *J. Am. Chem. Soc., 93*, 6687 (1971).
6. P. L. Timms, private communication (1974).
7. G. Wilke, private communication (1974).
8. G. Wilke, *Angew. Chem., Int. Ed. Engl., 2*, 105 (1963).
9. M. Green, J. A. K. Howard, J. L. Spencer, F. G. A. Stone, *J. Chem. Soc., Chem. Commun.*, p. 449 (1975).
10. B. Bogdanovic, M. Kroner, G. Wilke, *Justus Liebigs Ann. Chem., 699*, 1 (1966).
11. J. R. Blackborow, U. Feldhoff, F. W. Grevels, R. H. Grubbs, A. Miyashita, *J. Organomet. Chem., 173*, 253 (1979).
12. R. G. Gastinger, B. B. Anderson, K. J. Klabunde, *J. Am. Chem. Soc., 102*, 4959 (1980).

5.8.2.6. from Arenes

5.8.2.6.1. Formation of (η^6-Arene)Metal Complexes by Ligand Displacement

The most common method of formation of η^6-arene–metal complexes is by direct coordination of the arene to the metal, usually by replacement of one or more ligands from the coordination sphere. This section begins with the reactions involving bis-η^6-arene complexes, followed by the methods leading to mono-η^6-arene complexes. The examples illustrate the scope, and the list is not comprehensive[1–3]. Formation of new arene complexes by exchange of non-arene ligands is not included[1–3].

(i) Formation of Bis-(η^6-Arene) Metal Complexes. General methods are developed systematically: (i) from metal halides, (ii) by arene exchange, (iii) by reductions of metal halides with arene radical anions and (iv) from reaction of metal halides with arylmagnesium halide reagents.

The most general way to prepare bis(η^6-arene) complexes[4] uses the arene, a metal halide, an electron-pair acceptor acid such as AlX_3, and Al metal as a reducing agent:

$$6 \text{ arene} + 3\ MX_3 + AlX_3 + 2\ Al \rightarrow 3\ [(\text{arene})_2M]AlX_4 \qquad (a)$$

where if M = Tc, C_6H_6 is ligand[5]; if M = Ru, naphthalene[6]; if M = V, mesitylene[7]: if M = Co and Rh, hexamethylbenzene[8]; if M = Re, mesitylene and benzene[9]; and if M = Mo, toluene; X = Cl, Br[10]. The yields in some cases such as the bis(η^6-mesitylene)Re(I) complex[9] are low, but typical yields are in the range 25–75%.

Excluding Al from the reaction mixture leads to formation of an arene complex without change in the oxidation state of the metal:

$$2 \text{ arene} + MX_2 + AlX_3 \rightarrow [(\text{arene})_2M][AlX_4]_2 \qquad (b)$$

e.g., for M = Fe and benzene (79% yield)[11] and for M = Ni with hexamethylbenzene as ligand[12]. It is one of the few methods of formation of arene–Ni complexes other than employing metal atom reactors[13]. The cationic complexes can be reduced to the neutral arene complexes by reaction with alkali metals[14]:

$$6\,[W(C_6H_6)_2]^+ + 8\,[OH]^- \rightarrow 5\,W(C_6H_6\text{-}\eta^6)_2 + [WO_4]^{2-} + 4\,H_2O + 2\,C_6H_6 \quad \text{(c)}$$

The major advantages of this synthesis are the convenient starting materials (metal halide) and the generality with respect to the transition metal employed. A serious limitation is the relatively harsh reaction environment, which restricts the substituents that can be carried on the arene.

Exchange of one arene ligand for another is potentially a powerful method for preparing of arene–metal complexes. With bis(η^6-arene)metal complexes, only a few examples are known with no systematic development. Equilibrium of bis(η^6-benzene)Cr(0) with biphenyl at 115°C in the presence of $AlCl_3$ allows isolation of bis(η^6-biphenyl)Cr in 20% yield[15]:

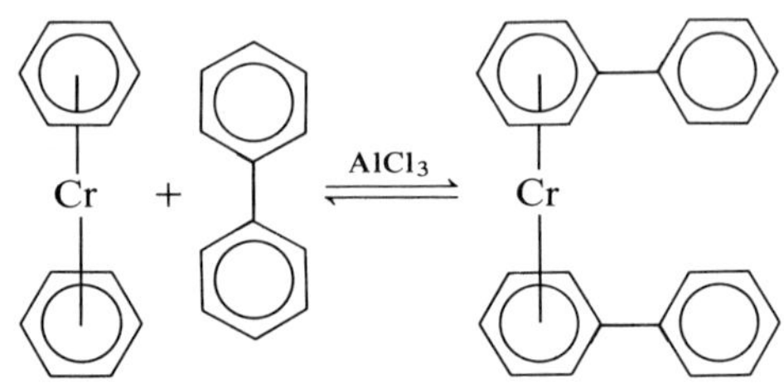

(d)

The corresponding Cr(I) cationic complexes react similarly[16]. Using perdeutero-hexamethylbenzene, exchange with bis(η^6-hexamethylbenzene)Ru is observed[17] at 80°C in the presence of $AlCl_3$.

A serendipitous preparation of arene–metal complexes is the reaction of phenylmagnesium bromide with chromium trichloride:

$$\text{arylMgX} + MX_n \xrightarrow{\text{low T}} \text{arene}_2M \quad \text{(e)}$$

With $CrCl_3$ and arylmagnesium halide reagents, the yields of Ar_2Cr are not high[18,19]. Examples of conversions of manganese dichloride to bis(η^6-benzene)Mn(I) cation[20] and of vanadium tetrachloride to bis(η^6-benzene)Va(0) are also known[21,22], where different degrees of reduction of the metal occur (no change for Mn, −1 for Cr, −2 for V), with PhMgBr serving as the reducing agent.

The process involves initial formation of Ph_nM by halide replacement, followed by hydrogen transfer to allow arene π-coordination[23]. The reaction conditions complement those of the metal halide–aluminum halide–aluminum metal method and may be useful with specifically substituted arenes.

The technique of forming π complexes in a metal atom generator is useful for arene–metal complexes[3], and is outlined in §5.8.2.9.3. Related in mechanism is a solution-phase technique for reducing metal halides to lower valent states with alkali metals in the presence of arenes. By this method bis(η^6-biphenyl)Cr(I) cation is obtained in 60% yield[24]:

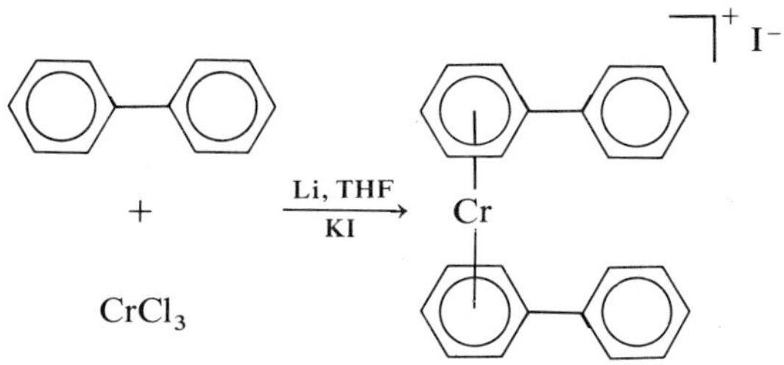

(f)

The mechanism involving an intermediate biphenyl radical anion suggests limitations in the nature of the arene.

(ii) Formation of Mono(η^6-Arene)Metal Complexes. The three general methods outlined above (metal halide–aluminum halide–aluminum metal, arene exhange and addition of arylmagnesium halide reagents) also are used to prepare mono-η^6-arene complexes. In addition, thermal displacement of neutral donor ligands by arenes and electron-pair acceptor acid-promoted replacement of cyclopentadienyl ligands are often employed.

The first of the above methods is less widely used for preparing mono-η^6-arene complexes than the bis-η^6-arene complexes, e.g., in the reaction of chlorotetra-carbonyl-Mn(I) with arenes and $AlCl_3$ at 60–80°C to give 40–76% yields of η^6-arene-Mn(I) tricarbonyl cations[25]. Several η^6-arene complexes of early transition elements are also prepared by this method[26–29], e.g., the chloro-bridged dimer[28]:

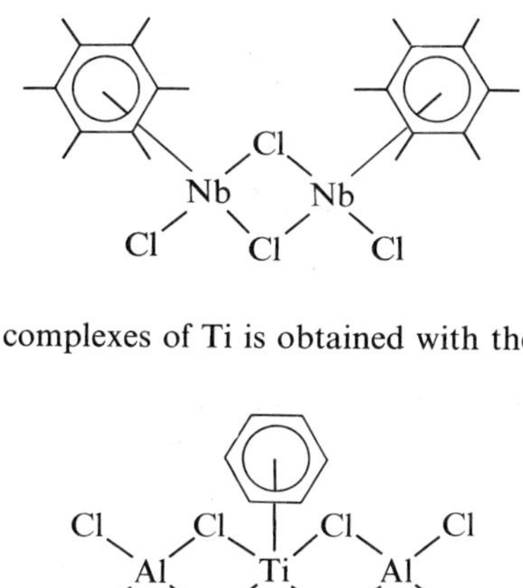

(g)

One of the rare η^6-arene complexes of Ti is obtained with the Al included[29–31]:

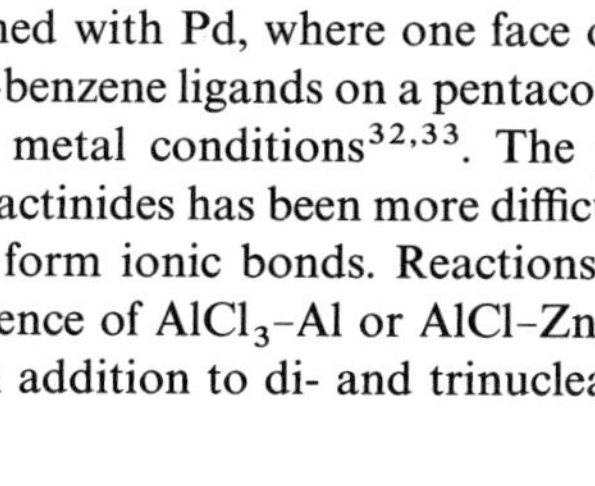

(h)

Related complexes are obtained with Pd, where one face of a benzene ring is shared by two Pd atoms[26,27]. Mono-η^6-benzene ligands on a pentacobalt cluster also form from the aluminum halide–aluminum metal conditions[32,33]. The preparation of η-arene complexes of the lanthanides and actinides has been more difficult, presumably because of the proclivity of these metals to form ionic bonds. Reactions between hexamethylbenzene and metal halides in the presence of $AlCl_3$–Al or AlCl–Zn reagents in the isolation of a mononuclear Sm complex, in addition to di- and trinuclear U complexes.[34,35]

I

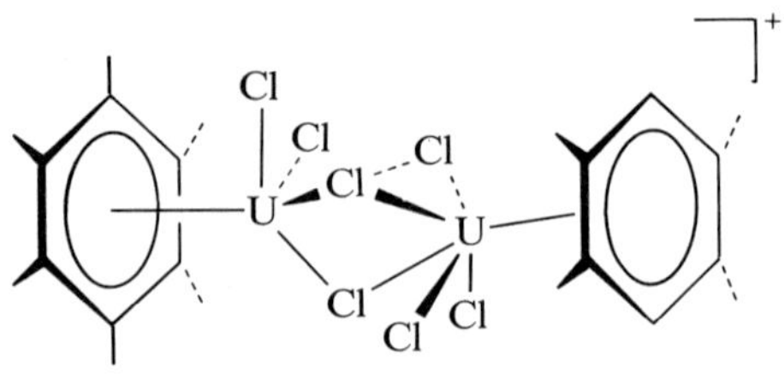

II

Addition of phenylmagnesium bromide to organometallic complexes can lead to mono(η^6-benzene)derivatives, e.g., with bis-(η^5-cyclopentadienyl)manganese, η^6-benzene-η^5-cyclopentadienyl–Mn is obtained in low yield[36,37]. Reacting aryllithium reagents with (1,5-COT)Fe(CO)$_3$ and quenching the intermediate anions with Et$_3$OBF$_4$ or Me$_3$SiCl results in a formal phenylalkoxycarbene insertion into the 1,5-COT ligand to form a [4.2.1]non-2-ene, with concomitant migration of the Fe(CO)$_2$ fragment to the aromatic ring[38]:

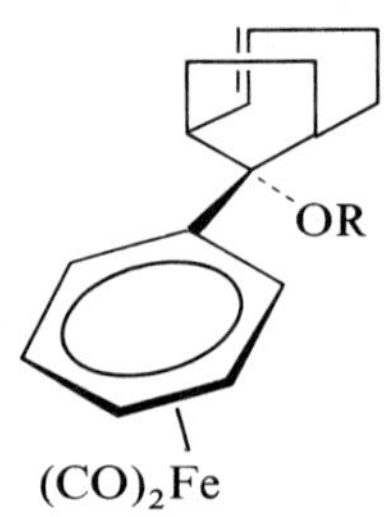

V

However, the method is not systematically developed.

The reduction of Mo$_2$Cl$_{10}$ with Mg in the presence of PPhR$_2$ (R = Me or Ph) produces[39] dinuclear and mononuclear η^6-phosphinoaryl complexes of Mo:

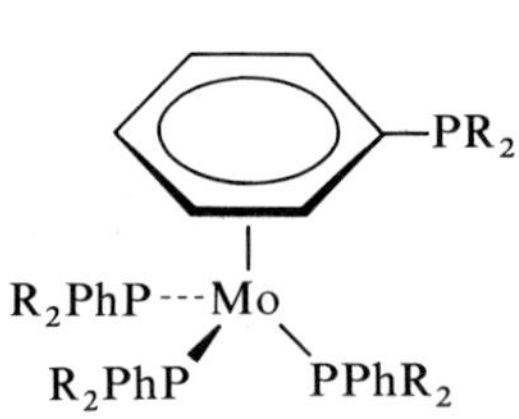

III, R = Ph

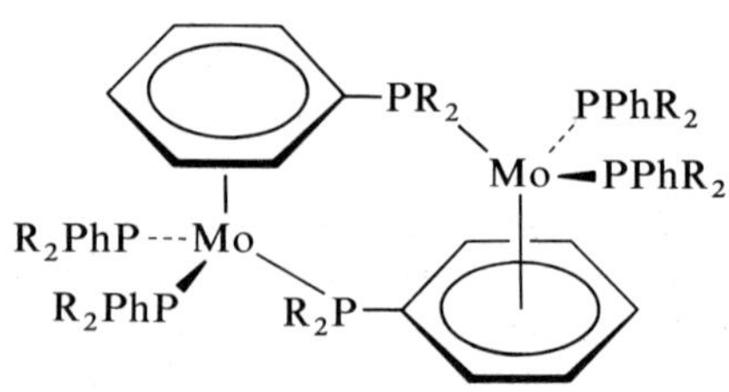

IV, R = Me

This method has not yet been generalized for the synthesis of dialkylphosphino-substituted complexes.

Direct exchange of arene ligands in mono-η^6-arene complexes is utilized in a limited way as a preparative method; e.g., if (η^6-arene)M(CO)$_3$ (where M = Cr, Mo, W) is treated with ^{14}C-labeled arene to measure the rate of incorporation of labeled arene into the complex, a complex mechanism is revealed. Donor ligands such as THF show an accelerating effect on the exchange of arene ligands in (η^6-arene)Co(CO)$_3$ complexes[40,41] but this is not yet a general preparative method. Arene exchange with (η^6-naphthalene)Cr(CO)$_3$ has gained popularity as a method of preparing (η^6-arene)Cr(CO)$_3$ complexes under relatively mild thermal conditions (refluxing dioxane or THF)[42]. Detailed mechanistic studies have more fully characterized the mechanisms of intramolecular Cr(CO)$_3$ fragment migration and intermolecular arene exchange reactions[43]. Arene exchange in coordinating solvents is apparently initiated via coordination of a solvent molecule and slippage of the arene to an "η^4" bonding mode[43]. Arene exchange in noncoordinating solvents relies on a preassociation step to form, η^4-slipped Cr dimers prior to a rate-determining reaction with free arene[43]:

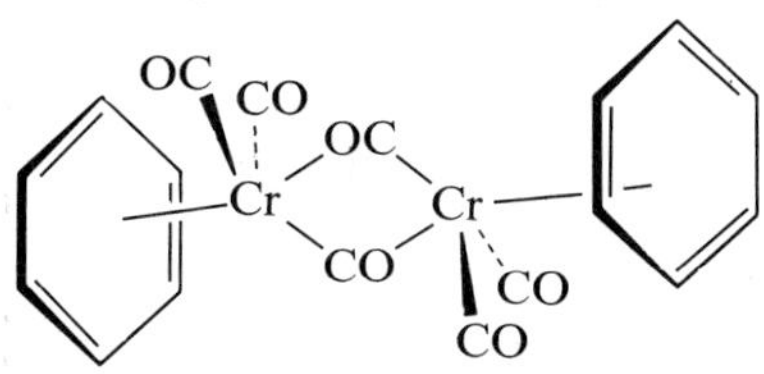

VI

The temperature for exchange with (η^6-arene)Cr(CO)$_3$ ranges from 80°C (in the presence of THF) to almost 180°C (in noncoordinating solvents). The corresponding (η^6-arene)Mo(CO)$_3$ complexes undergo exchange at temperatures as low as 25°C using a donor solvent[17–33,36,37,44–49]. The isoelectronic η^6-arene-η^4-(1,5-cyclooctadiene)-Ir(I) and Ru(0) complexes equilibrate with free arenes at 25°C [50]. The iridium complex is particularly efficient with polymethylbenzene ligands[50].

Arene exchange can be induced by irradiation, but give few results of preparative significance. Irradiation of η^6-(p-xylene)-η^5-cyclopentadienyl-Fe(I) in the presence of hexamethylbenzene gives the corresponding hexamethylbenzene complex in high yield[51]:

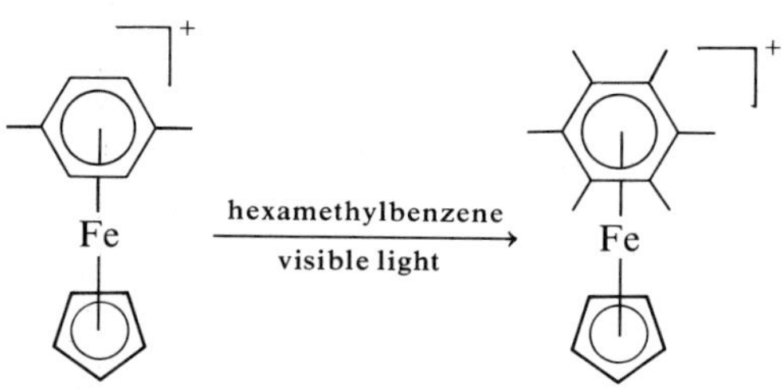

(i)

The generality of this process is not clear.

Replacement of donor ligands by an arene is a thermal reaction, although photochemical processes are possible[52]. It is by far the most useful method for preparing

η^6-arene complexes. The widely studied (η^6-arene)$Cr(CO)_3$ complexes and related species with Mo and W are readily prepared:

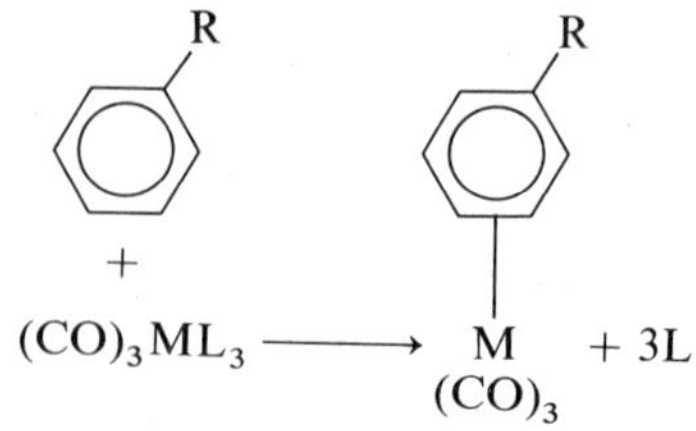

(j)

where L = CO, CH_3CN, py, NH_3, diglyme. While $Cr(CO)_6$ is the simplest source of $Cr(CO)_3$, lower reaction T are possible if the more reactive $Cr(CO)_3L_3$ species listed in Eq. (j) are preformed[53–55]. With reactive arenes such as phenylacetylene, direct reaction with $Cr(CO)_6$ fails, whereas $Cr(CO)_3(CH_3CN)_3$ is efficient in forming (η^6-phenylacetylene)$Cr(CO)_3$ [55]. Similarly, small amounts of THF enhance the rate of complexation.[56] Compatibility of the method with polar functional groups on the arene is obvious from the long list of complexes prepared by this method[1]. The η^6-arene complexes involving a tetracobalt cluster are formed in the same way[57]:

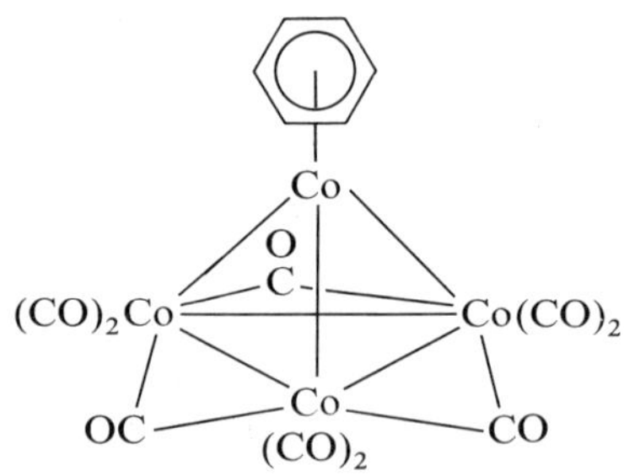

(k)

and a hexaruthenium mono arene complex is also known[58,59]. The displacement of acetonitrile ligands from $[CpRu(CH_3CN)_3]^+$ and $[Cp^*Ru(CH_3CN)_3]^+$ (Cp^* = C_5Me_5) by arene ligands in THF or CH_2Cl_2 has become the preferred route for the synthesis of [η^5-cyclopentadienyl-η^6-arene Ru]$^+$ complexes[60,61]. Iron pentacarbonyl reacts with hexamethylbenzene to give[62] a carbonyl-bridged dimer of η^6-arene-$Fe(CO)_2$. For vanadium hexacarbonyl, internal electron transfer and ligand displacement produce[63] a salt combining an η^6-mesitylene–vanadium tetracarbonyl cation and the vanadium hexacarbonyl anion.

η^6-Arene manganese tricarbonyl cations are characteristically prepared by a pathway involving both halogen abstraction and CO loss. The most common method for preparing manganese cations is through abstraction of Br^- from $Mn(CO)_5Br$ by $AlCl_3$ in the presence of the arene, followed by metathesis to the $[BF_4]^-$ or $[PF_6]^-$ salt[64,65]. An alternate method that is useful for thermally or acid-sensitive arenes is to treat $Mn(CO)_5Br$ with $AgBF_4$ or $AgClO_4$ in CH_2CL_2, preparing a labile $Mn(CO)_5X$ ($X = [BF_4]^-$ or $[ClO_4]^-$) complex in situ. Reacting this intermediate with arene produces the target π-arene complex[64,65]. Flash vacuum pyrolysis (FVP) of (η^1-benzyl)$Mn(CO)_5$ at 250°C results in the extrusion of two CO ligands from mn and the formation of an intermediate thought to exist as a tautomer of alternate zwiterionic η^6-arene and non-charge-separated η^5-pentadiene-exo-methene structures[66]:

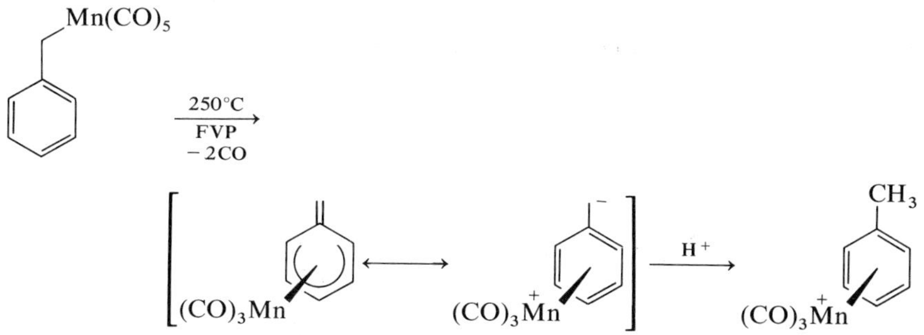

(l)

Upon protonation with trifluoroacetic acid, the intermediate acquires a proton and is converted to $[(\eta^6\text{-toluene})Mn(CO)_3]^+$ in 35% yields.

Ligand-exchange reactions synthesize the chiral ortho-substituted π-arene complexes[67–69]. Coordinating heteroatom substituents are known to accelerate the rate of arene complexation in direct reactions with $Cr(CO)_6$, presumably by ligating and delivering the Cr fragment to the arene. The existence of a preferred conformation of a heteroatom-containing side chain with respect to an ortho-substituted arene group can be used to induce face selectivity in the delivery of the $Cr(CO)_3$ unit. This process effects the diastereospecific synthesis of a particular planar chiral arene enantiomer:

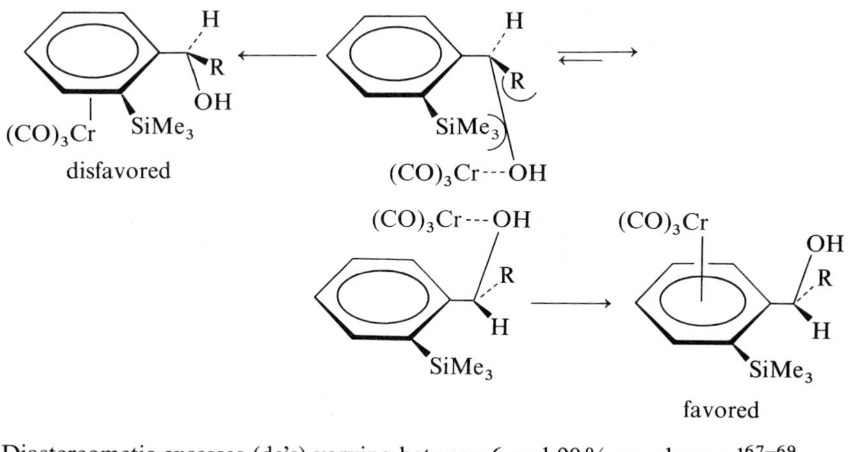

(m)

Diastereometic excesses (de's) varying between 6 and 90% are observed[67–69].

Exchange of cyclopentadienyl for arene is a powerful method in the Fe series because of the easy availability of ferrocene and generally good yields in the electron-pair acceptor acid-promoted sequence[70]:

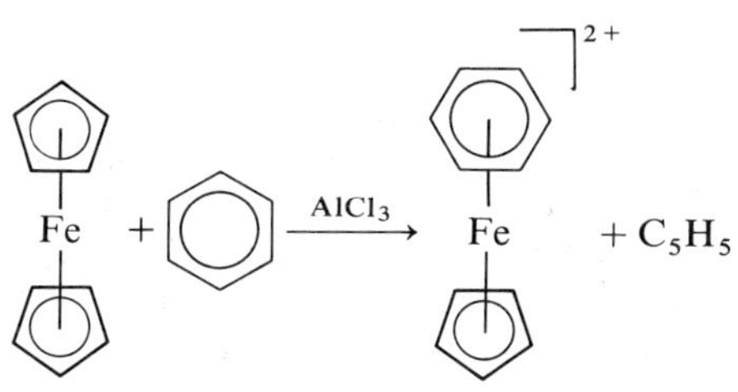

(n)

With bis(carboxylate)-η^5-cycylopentadienyl–Rh and –Ir complexes, strong acid promotes replacement of the anionic ligands by benzene[71]:

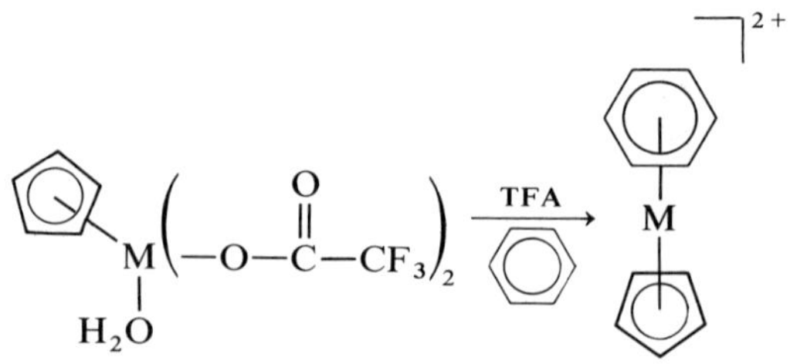

(o)

where M = Rh, Ir. The hydride ligands in $Cp^*_2Ru_2(\mu\text{-}H)_4$ can be replaced with η^6-arene ligands, producing two mononuclear $[Cp^*Ru(\eta^6\text{-arene})]^+$ moieties. These reactions are performed either through protonation with $HOSO_2CF_3$ in benzene or by a direct reaction between the Ru hydride and chlorinated solvents in the presence of an arene[72]:

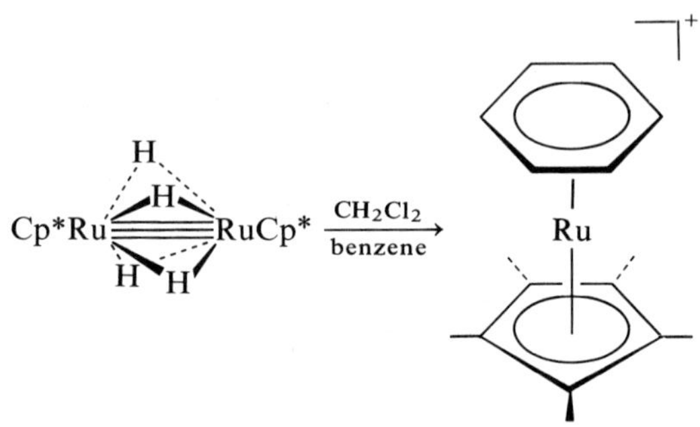

(p)

In a related process, silver ion opens up coordination sites from η^4-norbornadiene–Rh chloro-bridged dimers and allows coordination of arenes at 25°C [73]. Dienes such as norbornadiene can be replaced in a direct thermal reaction at 25°C starting from bis(η^4-diene)–Rh(I) complexes[74]:

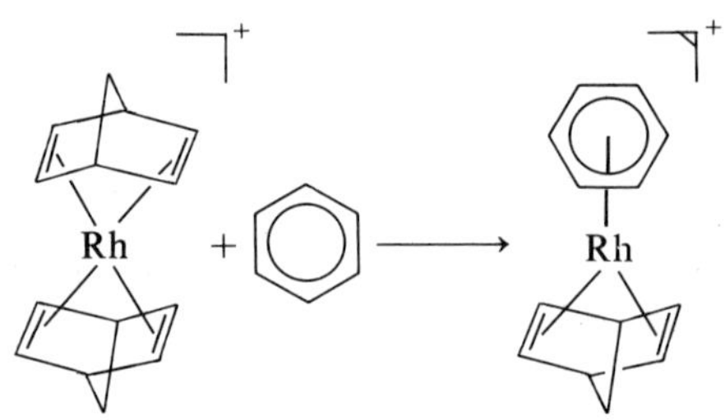

(q)

This process is general although undeveloped as a preparative method.

Ligand-substitution reactions that facilitate η^6-arene coordination are gaining increasing prominence. 2,6-Diphenylphenoxide and 2,6-diphenylthiophenoxide ligands have been found to undergo competitive η^6-bonding and orthometallation with a variety of d^4 Mo and W fragments[75–77], whereas U(III) centers in "U(O-2,6-$C_6H_4Me_2)_3$" dimerize through η^1,η^6-phenoxide bridges in the absence of coordinating solvents[78]:

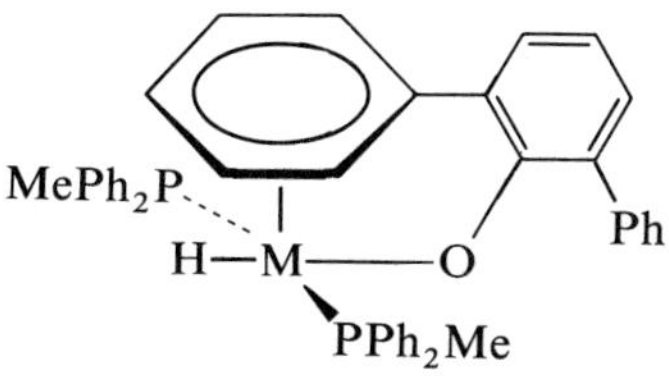

VII

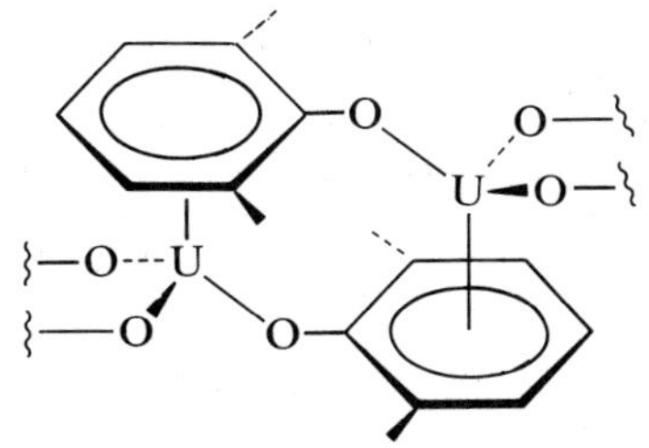

VI

In the absence of arene molecules, $(CF_3CO_2)_2Ni$ reacts with polyhalogenated aryllithium (LiAr) reagents to form $[Ar_2Ni]_n$ polymers[79]. Treatment of these reaction mixtures with arenes can retard polymer formation and produce (η^6-arene)Ni$(Ar)_2$ complexes[79]. Analogous complexes of Co(II) can be prepared similarly[79].

The $Cr_2(O_2CCPh_3)_4$ molecule[80], which has highly labile axial ligands, coordinates C_6H_6 solvent on crystallization:

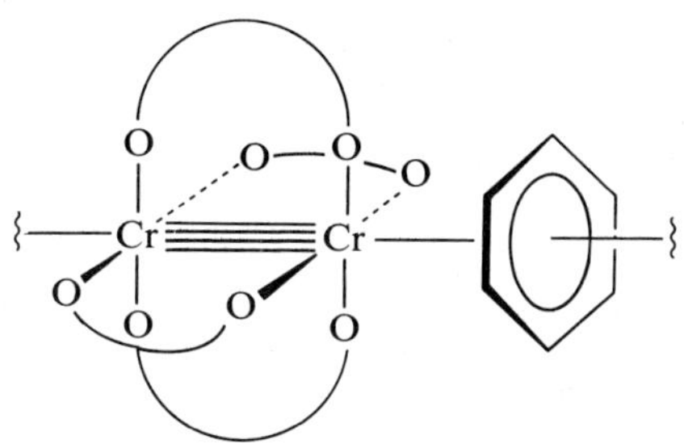

IX

(J. J. GARCIA, M. F. SEMMELHACK)
(J. A. HEPPERT)

1. W. E. Silverthorn, *Adv. Organomet. Chem., 13,* 47 (1975).
2. H. Zeiss, P. J. Wheatley, H. J. S. Winkler, *Benzenoid Metal Complexes,* Ronald Press, New York, 1966.
3. R. G. Gastinger, K. Klalounds, *Trans. Met. Chem., 4,* 1 (1979).
4. E. O. Fischer, W. Hafner, *Z. Naturforsch., Teil B, 10,* 665 (1955).
5. C. Palm, E. O. Fischer, F. Baumgartner, *Tetrahedron Lett., 6,* 253 (1962).
6. E. O. Fischer, C. Elschenbroich, C. G. Kreiter, *J. Organomet. Chem., 7,* 481 (1967).
7. F. Calderazzo, *Inorg. Chem., 3,* 810 (1964).
8. E. O. Fischer, H. H. Lindner, *J. Organomet. Chem., 1,* 307 (1964).
9. E. O. Fischer, A. Wirzmuller, *Chem. Ber., 90,* 1725 (1957).
10. E. O. Fischer, N. Driebitzsch, R. D. Fischer, *Chem. Ber., 92,* 3214 (1959).
11. J. F. Helling, S. L. Rice, D. M. Braitsch, T. Mayer, *J. Chem. Soc., Chem. Commun.,* 931 (1971).
12. H. H. Lindner, E. O. Fischer, *J. Organomet. Chem., 12,* P18 (1968).
13. K. L. Klabunde, B. B. Anderson, M. Bader, L. J. Radonovich, *J. Am. Chem. Soc., 100,* 1313 (1978).
14. E. O. Fischer, F. Scherer, H. O. Stahl, *Chem. Ber., 93,* 2065 (1960).
15. F. Hein, K. Kartte, *Z. Anorg. Allg. Chem., 307,* 22 (1960).
16. E. O. Fischer, J. Seeholzer, *Z. Anorg. Allg. Chem., 312,* 244 (1961).
17. E. L. Mueterties, J. R. Bleeke, A. C. Sievert, *J. Organomet. Chem., 178,* 197 (1979).
18. H. Zeiss, in *Organometallic Chemistry,* H. Zeiss, ed., Reinhold, New York, 1960, p. 380.
19. H. H. Zeiss, M. Tsutsui, *J. Am. Chem. Soc., 79,* 3062 (1957).
20. E. O. Fischer, M. W. Schmidt, *Chem. Ber., 100,* 3782 (1967).

21. P. G. Harrison, J. J. Zuckerman, T. V. Long, T. P. Poeth, B. R. Willeford, *Inorg. Nucl. Chem. Lett.*, *6*, 627 (1970); *Inorg. Chem.*, *10*, 522 (1971).
22. E. Kurras, *Z. Anorg. Allg. Chem.*, *351*, 268 (1967).
23. H. Zeiss, W. Herwig, *Justus Liebigs Ann. Chem.*, *606*, 209 (1957).
24. K. Luhder, *Z. Chem.*, *9*, 31 (1969).
25. G. Winkhaus, L. Pratt, G. Wilkinson, *J. Chem. Soc.*, 3807 (1961).
26. G. Allegra, G. T. Casagrande, A. Immirzi, L. Porri, G. Vitulli, *J. Am. Chem. Soc.*, *92*, 289 (1970).
27. G. Allegra, A. Immirzi, L. Porri, *J. Am. Chem. Soc.*, *87*, 1394 (1965).
28. E. O. Fischer, F. Rohrscheid, *J. Organomet. Chem.*, *6*, 53 (1966).
29. H. Martin, F. Vohwinkel, *Chem. Ber.*, *94*, 2416 (1961).
30. S. Pasynkiewicz, R. Gieznski, S. D. Zierzgowski, *J. Organomet. Chem.*, *54*, 203 (1973).
31. F. Vohwinkel, G. Preusser, U.S. Pat. 3,322, 803; *Chem. Abstr.*, *64*, P55,762 (1967).
32. P. Chini, R. Ercoli, *Gazz. Chim. Ital.*, *88*, 1170 (1958).
33. E. O. Fischer, O. Beckert, *Angew. Chem.*, *70*, 744 (1958).
34. F. A. Cotton, W. Schwotzer, *Organometallics*, *6*, 1275 (1987).
35. G. C. Campbell, F. A. Cotton, J. F. Haw, W. Schwotzer, *Organometallics*, *5*, 274 (1986).
36. R. G. Denning, R. A. Wentworth, *J. Am. Chem. Soc.*, *88*, 4619 (1966).
37. E. O. Fischer, M. W. Schmidt, *Chem. Ber.*, *99*, 2213 (1966).
38. J. Chem, J. Ying, W. Xu, *Huaxue Xuebao*, *45*, 1140 (1987).
39. F. A. Cotton, R. L. Luck, R. H. Morris, *Organometallics*, *8*, 1282 (1989).
40. C. A. L. Mahaffy, P. L. Pauson, *J. Chem. Res.*, *(S)*, 352 (1978).
41. C. A. L. Mahaffy, Ph.D. Thesis, University of Strathclyde, Glasgow, 1978.
42. E. P. Kundig, *Helv. Chim. Acta.*, *64*, 1288 (1981).
43. T. G. Traylor, K. J. Stewart, M. J. Goldberg, *Organometallics*, *5*, 2062 (1986).
44. W. Strohmeier, D. Von Hobe, *Z. Naturforsch.*, *Teil B.*, *18*, 981 (1963).
45. W. Strohmeier, H. Mittnacht, *Z. Phys. Chem.* (*Frankfurt am Main*), *29*, 339 (1961).
46. W. Strohmeier, H. Mittnacht, *Chem. Ber.*, *93*, 2085 (1960).
47. W. Strohmeier, R. Müller, *Z. Phys. Chem.* (*Frankfurt am Main*), *40*, 85 (1964).
48. W. Strohmeier, E. H. Staricco, *Z. Phys. Chem.* (*Frankfurt am Main*), *38*, 315 (1963).
49. T. Bargar, Ph.D. Thesis, Cornell University, 1977.
50. A. Sievert, E. L. Muetterties, *Inorg. Chem.*, *20*, 489 (1981).
51. T. P. Gill, K. R. Mann, *Inorg. Chem.*, *19*, 3007 (1980).
52. K. H. Pannell, D. C. Hambrick, G. S. Lewandos, *J. Organomet. Chem.*, *99*, C21 (1975).
53. For applications of tris(pyridine)tricarbonylchromium, see: K. Ofele, *Chem. Ber.*, *99*, 1972 (1966).
54. For tris(ammine)tricarbonylchromium, see: G. A. Moser, M. D. Rausch, *Syn. React. Inorg. Metal-Org. Chem.*, 37 (1974).
55. For tris(acetonitrile)tricarbonylchromium, see: G. R. Knox, D. G. Leppard, P. L. Pauson, W. E. Watts, *J. Organomet. Chem.*, *34*, 347 (1972).
56. C. A. L. Mahaffy, P. L. Pauson, *Inorg. Syn.*, *19*, 154 (1978).
57. G. Sbrignadello, F. Marcati, *J. Organomet. Chem.*, *46*, 357 (1972).
58. B. F. G. Johnson, R. D. Johnston, J. Lewis, *J. Chem. Soc.*, *A*, 2865 (1968).
59. B. F. G. Johnson, R. D. Johnston, J. Lewis, *J. Chem. Soc.*, *Chem. Commun.*, 1057 (1967).
60. A. M. McNair, J. L. Schrenk, K. R. Mann, *Inorg. Chem.*, *23*, 2633 (1984).
61. P. J. Fagan, M. D. Ward, J. C. Calabrese, *J. Am. Chem. Soc.*, *111*, 1698 (1989).
62. E. O. Fischer, W. Berngruber, C. G. Kreiter, *J. Organomet. Chem.*, *14*, P25 (1968).
63. F. Calderazzo, *Inorg. Chem.*, *3*, 1207 (1964).
64. P. L. Paulson, K. K. Bhasin, W. G. Galkeen, *J. Organomet. Chem.*, *204*, C25 (1981).
65. P. L. Paulson, A. J. Pearson, I. C. Richards, *J. Organomet. Chem.*, *258*, C41 (1981).
66. H. Y. Parker, C. E. Klopfenstein, R. A. Wielesek, T. Koenig, *J. Am. Chem. Soc.*, *107*, 5276 (1985).
67. M. Uemura, T. Minami, T. Hirotsu, Y. Hayashi, *J. Org. Chem.*, *54*, 469 (1989).
68. M. Uemura, T. Kobayashi, K. Isobe, T. Minami, Y. Hayashi, *J. Org. Chem.*, *51*, 2860 (1986).
69. A. Solladie-Cavallo, G. Solladie, E. Tsamo, *J. Org. Chem.*, *44*, 4189 (1979).
70. A. N. Nesmeyanov, N. A. Vol'kenau, I. N. Bolesova.
71. C. White, P. M. Maitlis, *J. Chem. Soc.*, *A*, 3322 (1971).
72. H. Suzuki, H. Omori, D. H. Lee, Y. Yoshida, Y. Moro-oka, *Organometallics*, *7*, 2243 (1988).
73. R. R. Schrock, J. A. Osborn, *J. Am. Chem. Soc.*, *93*, 3089 (1971).

74. M. Green, T. A. Kuc, *J. Chem. Soc., Dalton Trans., 832* (1972).
75. J. Dilworth, *J. Chem. Soc., Chem. Commun.*, 257 (1985).
76. I. P. Rothwell, J. L. Kirschner, P. E. Fanwick, *J. Am. Chem. Soc., 109*, 5840 (1987).
77. J. L. Kercher, I. P. Rothwell, J. C. Huffman, W. E. Streib, *Organometallics, 7*, 1871 (1988).
78. W. G. Van Der Skis, C. J. Burns, J. C. Huffman, A. P. Sattlberger, *J. Am. Chem. Soc., 110*, 5924 (1988).
79. S. T. Lin, R. N. Naiske, K. J. Klabunde, *Organometallics, 4*, 571 (1985).
80. E. A. Cotton, X. Feng, P. A. Kibala, M. Matusz, *J. Am. Chem. Soc., 110*, 2807 (1988).

5.8.2.6.2. Formation of η^5-Cyclohexdienyl Complexes by Nucleophilic Attack on (η^6-Arene)Metal Complexes.

Coordination of an aromatic ring to a transition metal can reverse the usual reactivity of the ring, from nucleophilic to electrophilic. Nucleophilic reagents can add to η^6-arene ligands to produce η^5-cyclohexadienyl complexes:

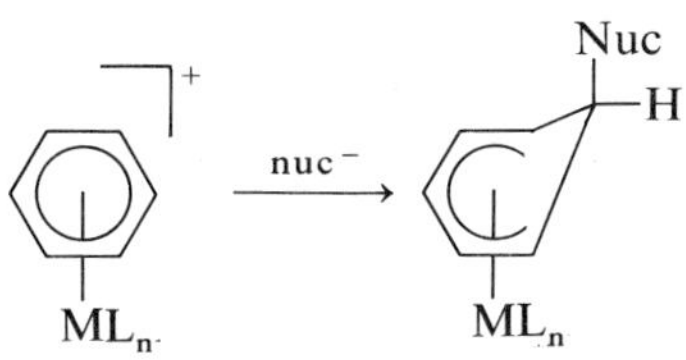

(a)

which are often stable and easily isolated but sometimes are only implicated as transient intermediates. This chapter surveys the formation of η^5-cyclohexadienyl ligands, both as stable products and transient intermediates, through nucleophilic addition to η^6-arene ligands. The reactivity of the arene ligand correlates with the net charge on the η^6-arene complex, and the presentation is organized into the three known charge types, dication, monocation, and neutral. Other factors such as ligand donor ability, formal oxidation state or arene substituents also influence the ease of nucleophilic addition.

(i) Addition to η^6-Arene–Metal Complex Dications. The nucleophilic addition of hydride donors ($LiAlH_4$, $Na[BH_4]$) with the Ru(I) complex **I** gives a mixture of the bis-η^5-cyclohexadienyl species **II** and the diene complex **III**[1]:

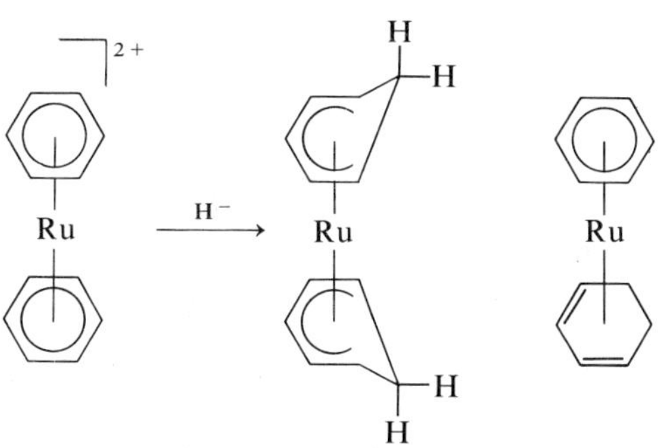

(b)

I **II** **III**

Phenyllithium (2 mol-equiv) also gives an adduct (**IV**)[1]:

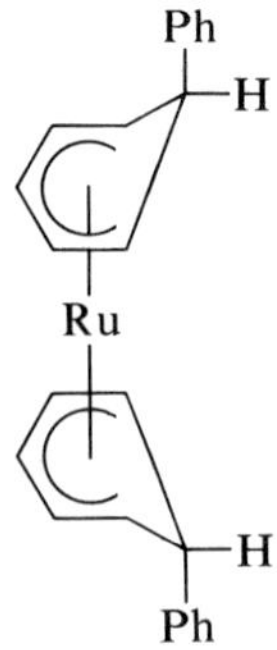

IV

presumably the result of anti-addition. Nucleophilic addition to coordinatively saturated η^6-arene complexes proceeds by anti-addition to give the exo-substituted product[2,3].

The ruthenium complex **V**, with three donor ligands, is also highly electrophilic, reacting with hydride, cyanide and methoxide to give the η^5-cyclohexadienyl-Ru(II) complexes, **VI**[4]:

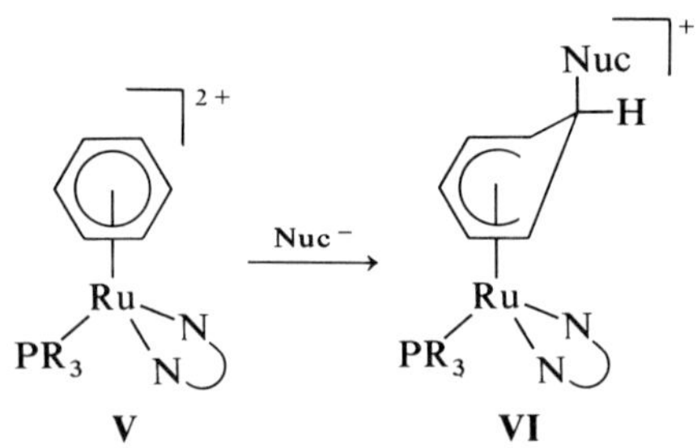

(c)

The corresponding η^6-mesitylene complex **VII** reacts exclusively at an unsubstituted arene position to give **VIII**:

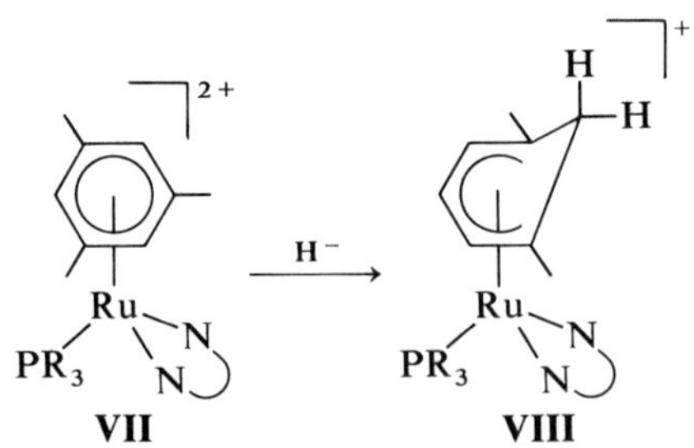

(d)

an example of a general pattern where unsubstituted positions are more reactive than substituted.

Bis-η^6-(arene)-Fe(II) complexes, e.g., **IX** and **X** react with nucleophiles of varying reactivity to give η^5-cyclohexadienyl-η^6-arene complexes (**XI**) or, with excess (and reactive) nucleophile, the bis-η^5-cyclohexadienyl species, **XII**[5].:

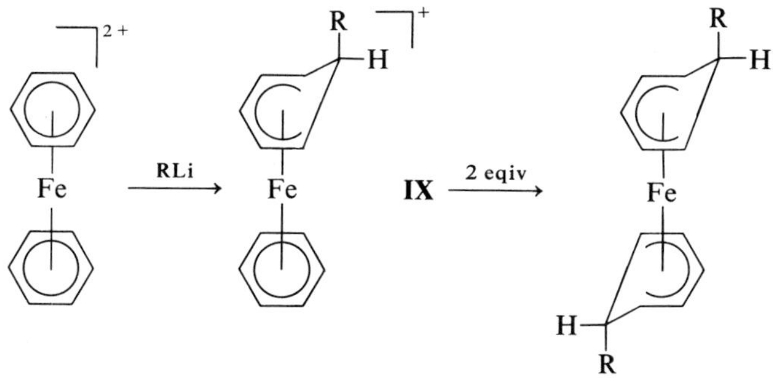

(e)

IX XI XII

Alkyllithium reagents readily give addition to both rings, while stabilized carbanions give single addition. Triethyl aluminum reagents also add a single alkyl group to a mesityl ring in $[(\eta^6\text{-arene})_2Fe]^{2+}$ complexes[6]. Simple phosphines also add, producing phosphonium ion complexes (**XIII**)[7]:

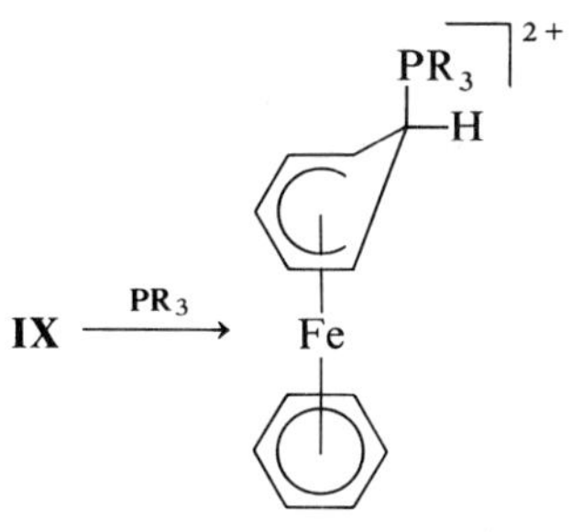

(f)

XIII

With alkyl substituents on the arene ligand, proton abstraction from the benzylic carbon can compete with nucleophilic addition, especially with hard nucleophiles such as amide and alkoxide anions[8]. The result of benzylic proton abstraction is the formation, in low yield, of dimer **XIV** from nucleophilic addition of the benzyl carbon to another η^6-mesitylene complex:

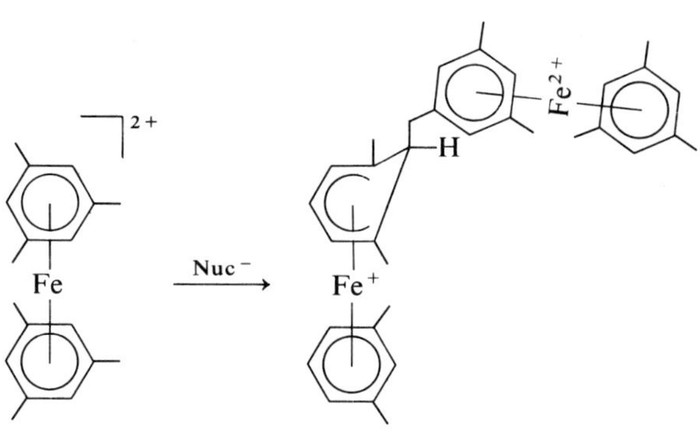

(g)

X XIV

The η^5-(pentamethylcyclopentadienyl)-η^6-arene–Ir(III) and –Ir(III) complexes such as **XV** undergo hydride attack from the exo face to give η^5-cyclohexadienyl ligands **(XVI)** with an endo-methyl substituent[9]. The preference for addition to an unsubstituted position is clear from the corresponding reaction of the η^6-pentamethylbenzene ligand (in **XVII**) where a simple product is formed **(XVIII)**[9]:

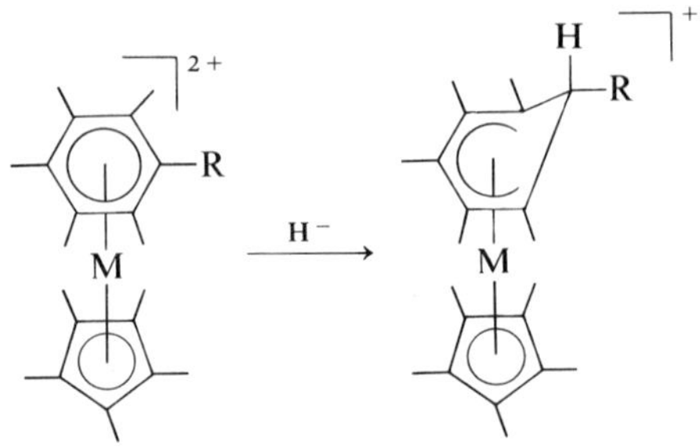

XV **XVI**
XVII **XVIII**

(h)

where when M = Rh and R = Me, **15** → **16**; and when M = Ir and R = H, **17** → **18**. The closely related complex, η^5-(ethyltetramethylcyclopentadienyl)-η^6-benzene–Rh(III) **XVIIIb** reacts with trimethylphosphite to give **XIX**:

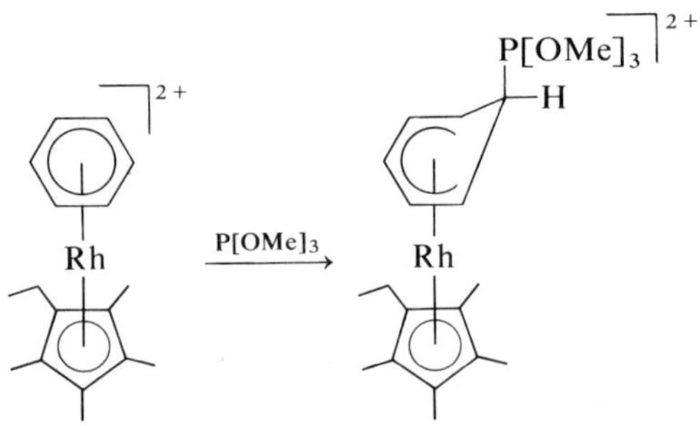

XVIIIb **XIX**

(i)

presumably via a phosphonium ion intermediate[10]. The same complex gives an intractable product mixture with lithium aluminum hydride[11], but other hydride donors give the η^5-cyclohexadienyl product, **XX**, in good yield[11]:

XVIIIb $\xrightarrow{Nuc^-}$

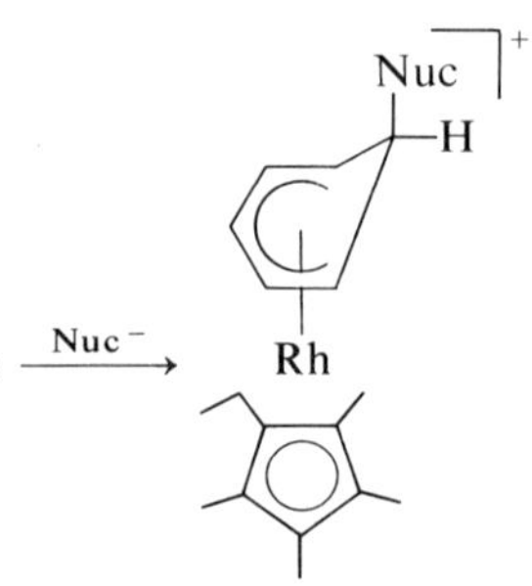

XX, **XXI**

(j)

where if Nuc = LiAl[OBu-t]$_3$H structure **XX** is produced and if Nuc = Tl[acac] structure **XXI** is produced. Thallium acetylacetonate delivers the carbon nucleophile (to give **XXI**) but also causes displacement of the arene ligand[11].

The dimeric Mo dication **XXII** is a special case that shows reactivity similar to the complexes mentioned earlier and adds nucleophiles such as cyanide and methoxide anions to give mono-η^5-cyclohexadienyl complexes (**XXIII**)[12]. Electro oxidation of **XXIII** (Nuc = CN^-) induces loss of the endo hydrogen and generates the η^6-benzonitrile complex, **XXIV**[12]:

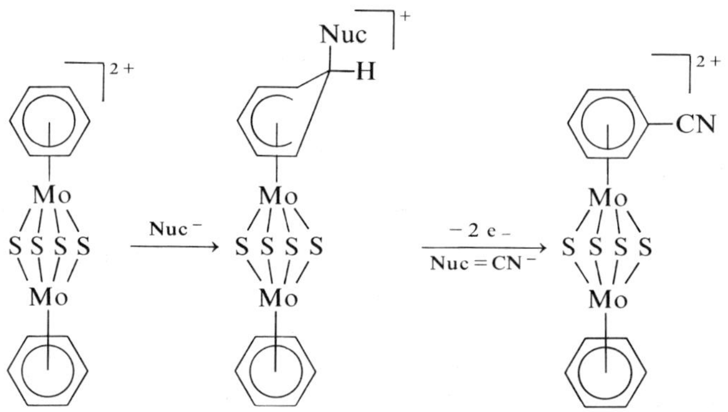

XXII **XXIII** **XXIV** (k)

Sodium borohydride adds hydride to the unsubstituted aromatic positions of an arene-bound [(η^6-benzthiophene)(η^5-Cp*)Ir]$^{2+}$ **XXV** complex, giving all four possible isomers of [(η^5-cyclohexadienylthiophene)(η^5-Cp*)Ir]$^+$ **XXVI**[13]:

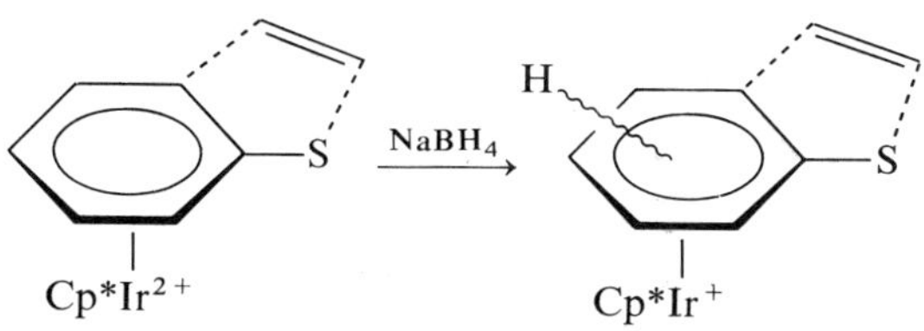

XXV **XXVI** (l)

(ii) Addition to η^6-Arene–Metal Complex Monocations. η^6-Arenetricarbonyl–Mn(I) cations (**XXV**) react with a variety of nucleophiles by the usual anti-addition[14–18], contrary to earlier hypotheses[1]. Exceptions to the usual pattern of simple addition are the halide anions, which preferentially displace the arene ligand[17]. The η^5-cyclohexadienyl complex from addition of malonate anion is stable and readily isolated. However, with other anions (N_3^-, $^-OCH_3$, R_3P, ^-NCS), the η^5-cyclohexadienyl intermediate is only detected and not fully characterized[14–17], except a η^5-cyclohexadienyl complex from addition of tri(n-butyl)phosphine[19]. The substituted complex **XXVI** gives addition primarily ortho and meta with various nucleophiles[20]. With

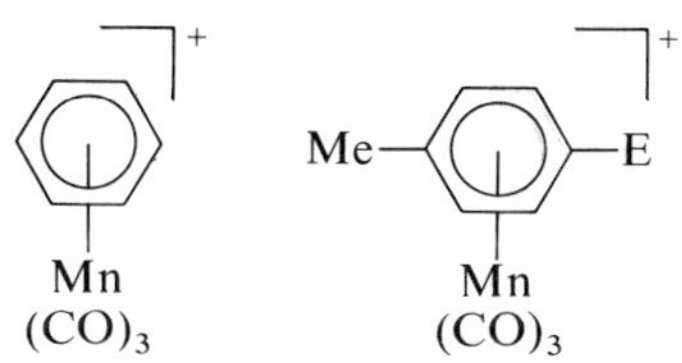

XXVII **XXVIII**

η^6-(N,N-diethylaniline)Mn tricarbonyl, addition of phenyllithium is predominately $[(\eta^6$-octahydrophenanthrene)$Mn(CO)_3]^+$ suffers exo addition by MeLi at $-70°C$, producing only the regioisomer derived from attack at the unsubstituted aromatic carbon[21]. Double nucleophilic addition to π-arene manganese cations can be effected by reacting the neutral η^5-cyclohexadienide manganese addition product (**XXIX**) with $NOPF_6$, forming an $[(\eta^5$-cyclohexadienyl)$Mn(CO)_2(NO)]^+$ complex (**XXX**) that is subject to attack by a second nucleophile[22]. The corresponding η^6-(hexamethylbenzene)Rh tricarbonyl cation readily adds hydride to the area[11].

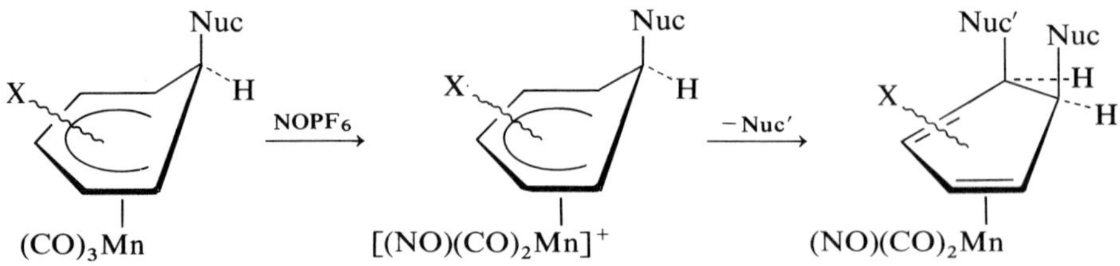

XXIX **XXX**

(m)

where E = Cl, OMe, NMe_2. Nucleophiles will add to η^6-arene-η^5-cyclopentadienyl-Fe cations (**XXXI**), including hydride, alkyl- and aryllithium reagents[23–27]:

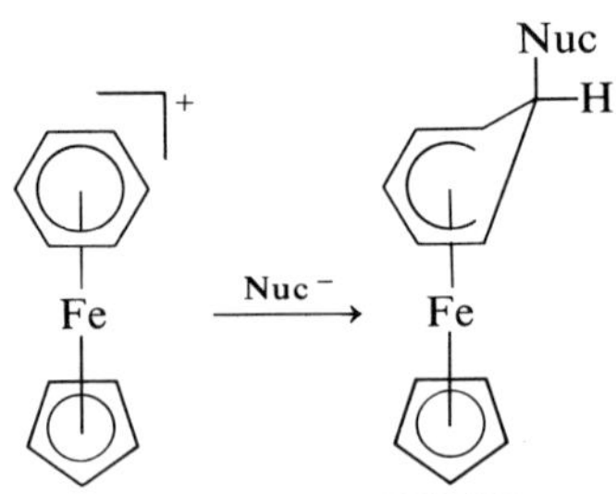

XXXI **XXXII**

(n)

The product η^5-cyclohexadienyl-η^5-cyclopentadienyliron(I) complexes such as **XXXII** are stable, soluble in non polar solvents and readily characterized[24]. The selectivity for the arene ligand (even, open) in preference to the cyclopentadienyl ring is consistent with prediction[28]. The nucleophiles add to an unsubstituted position as before, and the regioselectivity in addition to monosubstituted arene ligands is variable[25], and is sometimes dependent on the reaction conditions[24]. With complex **XXXIII**, the η^5-cyclohexadienyl species are obtained in 91% yield with the ratio o-:m-:p- of 89:7:4[26].

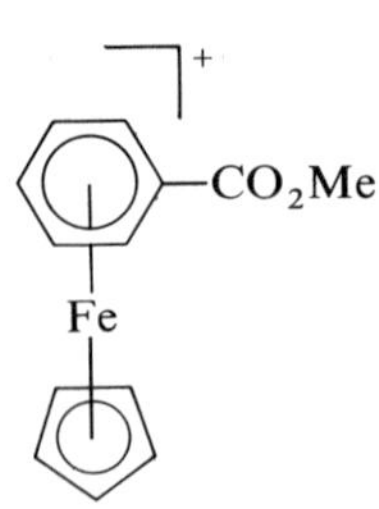

XXXIII

With η^6-chlorobenzene-η^5-cyclopentadienyl–Fe, addition of hydride occurred at the ortho, meta, and para positions (not ipso)[23], but with methoxide and secondary amines, the chloride is replaced via transient η^5-cyclohexadienyl complexes[29,30].

The cationic η^6-(arene)nitrosyldicarbonyl–Cr complexes (**XXXIV**) are more reactive toward nucleophiles than the tricarbonyl analogs and lead to stable η^5-cyclohexadienylnitrosyldicarbonyl–Cr derivatives (**XXXV**)[31]:

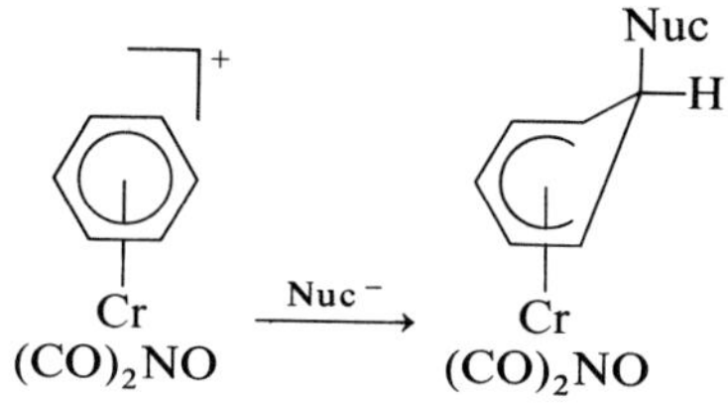

XXXIV **XXXV** (o)

The scope of this reaction is limited by the few analogs of **XXXIV** that can be isolated. Bis-η^6-(chlorobenzene)Cr(I) undergoes reaction with methoxide to form the bis-anisole analog, via η^5-cyclohexadienyl complexes, but none characterized[32].

Extensive studies on the molybdenum complex **XXXVI** with various bidentate phosphorus and sulfur ligands demonstrate that hydride, n-butyllithium and cyanide nucleophiles add to the arene to give **XXXVII**[33]:

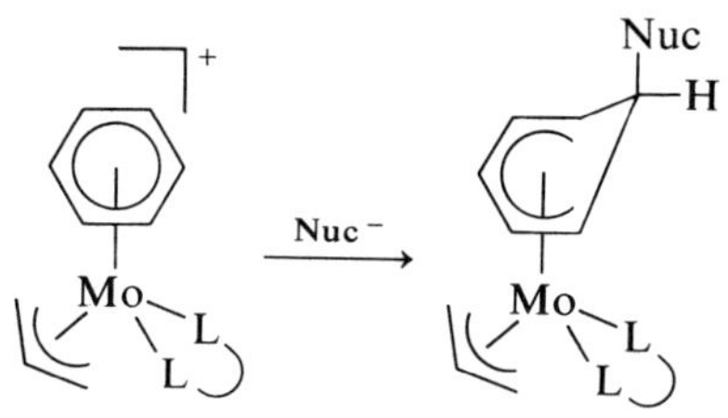

XXXVI **XXXVII** (p)

Borohydride and alkyllithium reagents engage in nucleophilic addition to η^6-benzene or η^6-hexamethylbenzene ligands in various cationic third-row transition-metal complexes, producing neutral η^5-cyclohexadienide products[34–36]. Similarly, the cobalt complex **XXXVIII** adds hydride and n-butyllithium to give the η^4-cyclobutadienyl-η^5-cyclo-hexadienyl–Co complex **XXXIX** in moderate yield[37]:

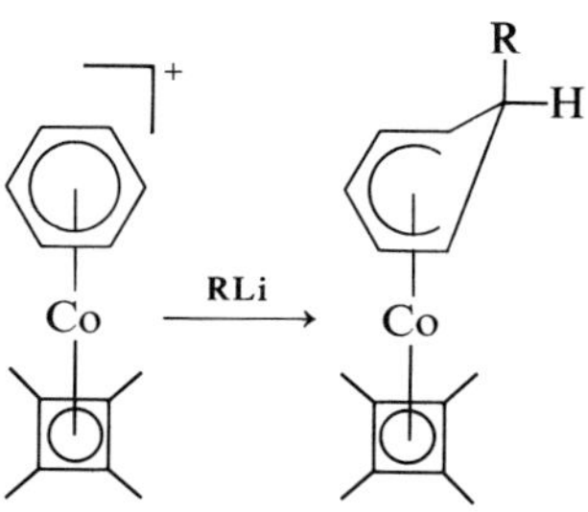

XXXIX **XL** (q)

The vanadium complexes **XLI**, with a variety of methyl substituents, add hydride at the unsubstituted positions[38]:

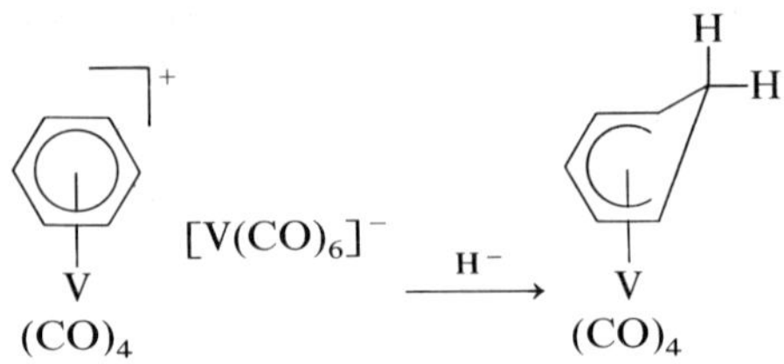

XLI

(r)

The reaction of the rhenium complex **XLII** with lithium at 200°C is a special case, producing the dimeric η^5-cyclohexadienyl species (**XLIII**)[39]:

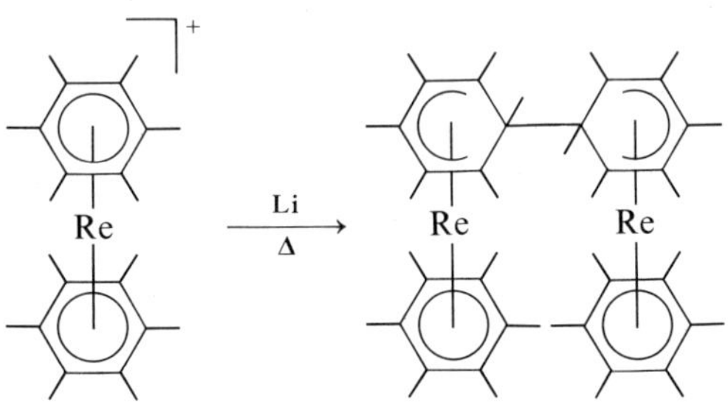

XLII **XLIII**

(s)

The treatment of $[(\eta^6\text{-mesitylene})Mn(CO)_3]^+$ with triethyl aluminum results in a competition between exo ethyl group addition to the mesityl ring and the coupling of two mesitylene ligands at nonmethylated carbon centers, producing a bis-η^5-cyclohexadienyl dimer analogous to **XLII** [40]. The scope of this reaction is not yet defined, but a parallel coupling process occurs upon electrochemical reduction of the iron analog, **XXXI**[41].

(iii) Addition to Neutral η^6-Arene–Metal Complexes. The dimeric complex of Ru(II), **XLIV**, undergoes bridge splitting and addition when treated with cyanide, hydride or hydroxide nucleophiles[37,42,43]. Excess hydride leads to a second addition and produces η^4-cyclohexadiene ligands (in **XLV**)[42,43]:

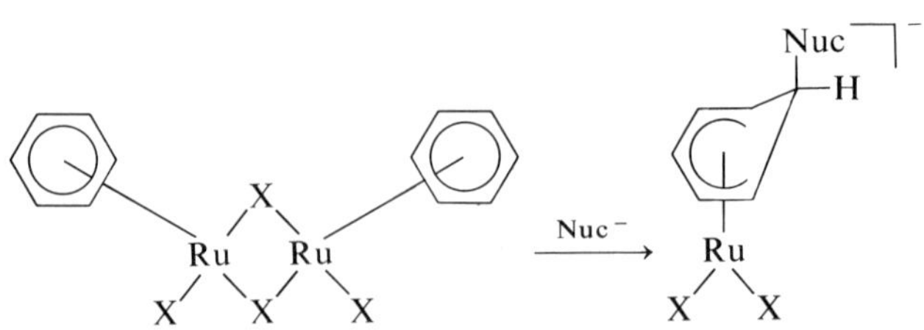

XLIV **XLV**

(t)

The η^5-cyclohexadienyltricarbonyl–Cr anions (**XLVI**) are potentially useful in organic synthesis. The η^6-arenetricarbonyl–Cr complexes (**XLVI**) are not strongly electrophilic but add reactive carbanions to give **XLVII**[43,44]:

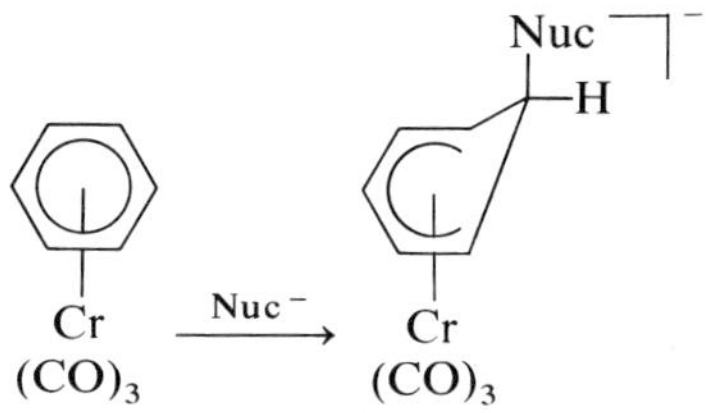

XLVI **XLVII**

(u)

With certain nucleophiles, the η^6-chlorobenzene and η^6-fluorobenzene analogs can proceed through the η^5-cyclohexadienyl species to give substitution for halide[44]. With other substituted-arene ligands, the addition is often highly selective and correlates approximately with a frontier orbital picture of the complex[45]. The anionic η^5-cyclohexadienyl complexes (**XLVII**) are air sensitive and not easily isolated. x-Ray diffraction verifies the exo configuration[46]. Both (η^6-1,4-dimethoxynaphthalene)$Cr(CO)_3$ and (η^6-1-methoxynaphthalene)$Cr(CO)_3$ (**XLVIII**) undergo regioselective reactions with nucleophiles, producing trans-5,6-dihydronaphthalene derivatives (**XLIX**) on quenching with electrophiles under CO pressure[47–49]:

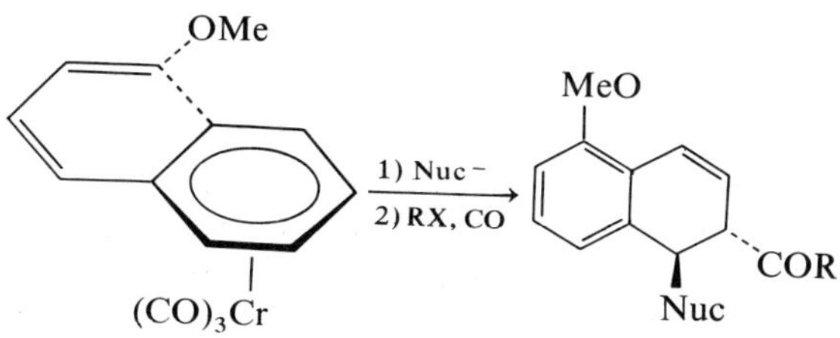

XLVIII **XLIX**

(v)

The addition of nucleophiles to chloro-substituted aromatic derivatives (**L**), followed by protonation at $-78°C$, results in the extrusion of HCl and formation of a simple nucleophile-substituted π-arene complex (**LI**):[50,51]

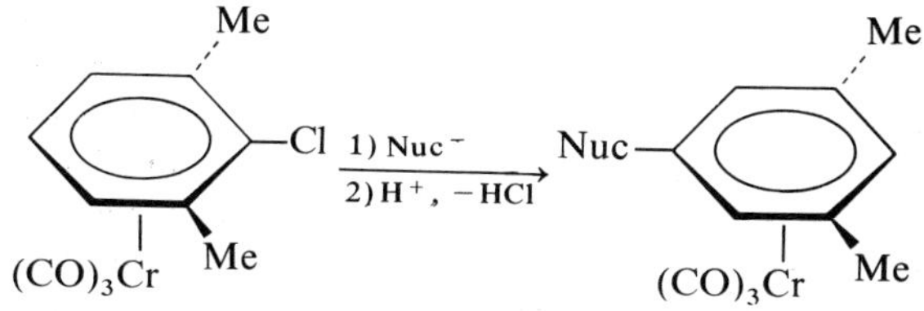

L **LI**

(w)

Treating (η-benzyl alcohol)Cr(Co_3) (**LII**) with 2 equiv of various alkyllithium (RLi) reagents results in the exo addition of an R substituent ortho to the CH_2OLi directing group, forming the proposed cyclohexadienide intermediate (**LIII**). Complex **LIII** rearranges on protonation to form the first class of (η^4, η^2-2-R-cyclohexadiene-exo-1-methene)$Cr(CO)_3$ (**LIV**) complexes[52]:

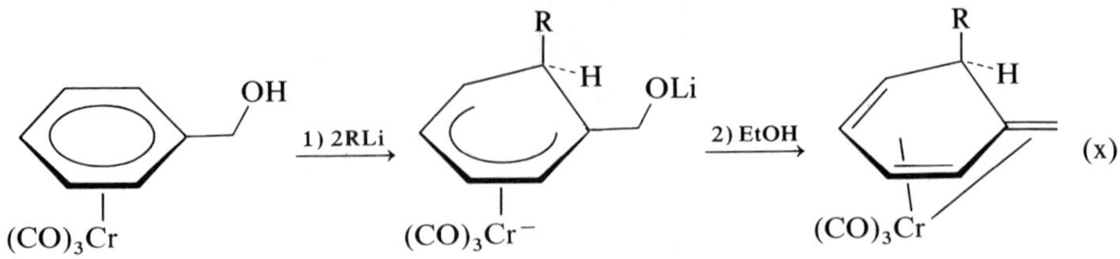

LII **LIII** **LIV**

When a carbon monoxide ligand is replaced by triphenylphosphine (in **LV**), the electrophilic reactivity of the arene ligand is lower, but transient η^5-cyclohexadienyl ligands are again implicated in reactions with nucleophiles[53].

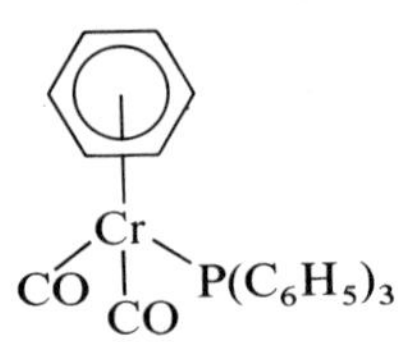

LV

(iv) The Mechanism of Nucleophilic Addition to η^6-arene Complexes. In order to avoid potential pitfalls in the application of this methodology to synthesis, a brief summary of available information concerning the mechanism of nucleophilic addition to η^6-arene complexes is appropriate. Relative rate profiles for nucleophilic attack by phosphines on $[(\eta^6\text{-}C_6H_6)_2M]^{2+}$ (M = Fe, Ru, Os) complexes followed the anticipated trend, Fe > Ru > Os. This behavior is rationalized by the greater ability of the second- and third-row transition metals to donate electron density into π^*-ligand orbitals, which are the site of attack for the phosphine nucleophiles[54]. A similar trend was not observed for $[(\eta^6\text{-arene})M(CO)_3]^+$ (M = Mn or Re), suggesting that π back-donation to the arene ligand is a far less important bonding interaction in these carbonyl-substituted molecules[55]. Chlorobenzene, toluene and anisole substrates bound to a $Cr(CO)_3$ template all appear to engage in regioselective meta nucleophilic addition[44,46], although this picture is somewhat clouded by the demonstrated ability of $[(\eta^5\text{-cyclohexadienide})Cr(CO)_3]^-$ derivatives to undergo rearrangement[44,56]. For example, although (η^6-chlorobenzene)$Cr(CO)_3$ **LVI** reacts over a period of 3 h at 0°C with lithiumisobutyrontrile to produce a mixture of ortho and meta addition products (68% conversion) (**LVII**, **LVIII**), this mixture rearranges over 1 day at 25°C to an apparent gem-disubstituted cyclohexadienide intermediate, subsequently losing chloride ion to produce a phenylisobutyronitrile complex (**LIX**)[44]:

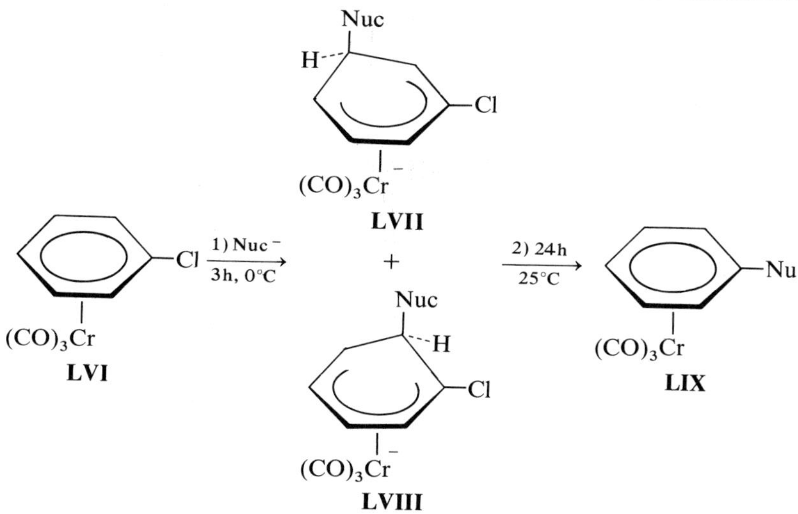

(y)

Similar rearrangement processes involving a chlorobenzene substrate are observed only for highly stabilized carbanions. Nucleophilic addition to a 1,4-dimethoxynaphthalene complex (**LX**) also dramatically demonstrates these rearrangement effects[56]. Reactions employing kinetic conditions (−60°C/0.5 h) produce approximate 1:4 mixtures of the cyclohexadienide products derived from .addition to the 5 and 6 positions of the naphthalene ring (**LXI**, **LXII**), respectively, whereas reactions run for a longer period (0°C/46 h) produce only **LXI**:

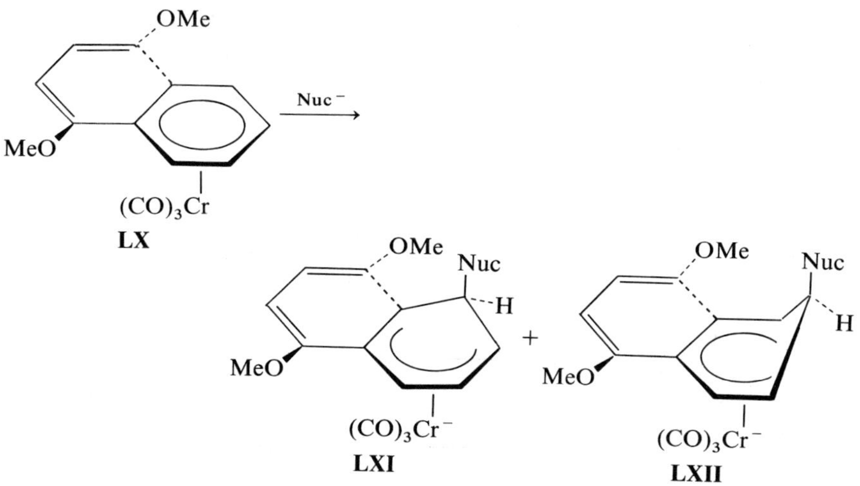

(z)

Temp. (°C)	Time (h)	**LXI:LXII**
−60	0.5	19.81
0	46	98:2

where Nuc$^-$ = $[CMe_2CN]^-$. Double-crossover experiments employing an isobutyronitrile-substituted cyclohexadienide complex (**LXIII** and **LXIII**-d_9) unequivocally established the complete dissociation of the nucleophile from the π-naphthalene chromium substrate with a half life of as 2 h at $-30°C$:

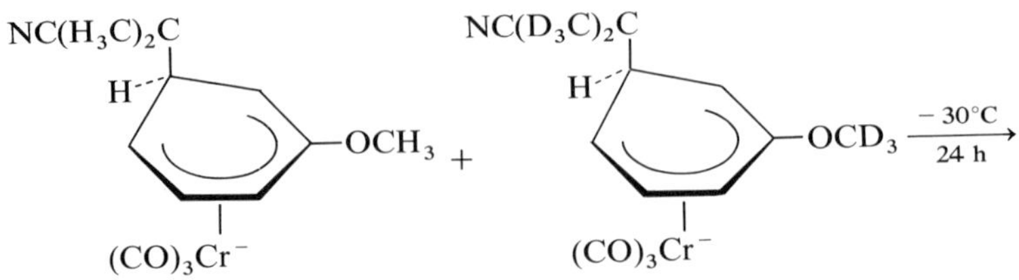

LXIII

LXIII + **LXIII** − d_3
LXIII − d_6 +
LXIII − d_9 (aa)
in essentially
equimolar quantities

(J. J. GARCIA, M. F. SEMMELHACK)
(J. A. HEPPERT)

1. D. Jones, L. Pratt, G. Wilkinson, *J. Chem. Soc., 44*, 58 (1962).
2. J. W. Faller, *Inorg. Chem., 19*, 2857 (1980).
3. P. H. Bird, M. R. Churchill, *J. Chem. Soc., C*, 777 (1967); for similar structures, see ref. 30.
4. D. A. Robertson, T. A. Stephenson, *J. Organomet. Chem., 162*, C31 (1977).
5. J. F. Helling, D. M. Braitsch, *J. Am. Chem. Soc., 92*, 7207 (1970).
6. T. S. Cameron, M. D. Clerk, A. Linden, K. C. Sturge, M. J. Zaworothko, *Organometallics, 7*, 2571 (1988).
7. S. G. Davis, L. S. Gelford, P. A. Sweigart, *J. Chem. Soc., C*, 762 (1979).
8. J. F. Helling, G. G. Cash, *J. Organomet. Chem., 73*, C10 (1974).
9. C. White,m P. M. Maitlis, *J. Chem. Soc., A*, 3322 (1971).
10. G. Fairhurst, C. White, *J. Organomet. Chem., 166*, C23, (1979).
11. N. A. Bailey, E. H. Blunt, G. Fairhurst, C. White, *J. Chem. Soc., Dalton, Trans.*, 829 (1980).
12. W. E. Silverthorn, *J. Am. Chem. Soc., 102*, 842 (1980).
13. S. C. Huckett, R. J. Angelici, *Organometallics, 7*, 1491 (1988).
14. A. Mawby, P. J. C. Walker, R. J. Mawby, *J. Organomet. Chem., 55*, C39 (1973).
15. M. R. Churchill, F. R. Scholer, *Inorg. Chem., 8*, 1950 (1969).
16. P. J. C. Walker, R. J. Mawby, *J. Chem. Soc., Dalton Trans.*, 622 (1973).
17. P. J. C. Walker, R. J. Mawby, *Inorg. Chem. Acta, 7*, 621 (1973).
18. M. Brookhart, W. Lamanna, A. R. Pinhas, *Organometallics, 2*, 638 (1983).
19. D. A. Sweigart, L. A. P. Kane-Maguire, *J. Chem. Soc., C*, 13 (1976).
20. P. L. Pauson, J. A. Segal, *J. Chem. Soc., Dalton Trans.*, 1683 (1975).
21. S. R. Stobart, M. J. Zaworothko, *J. Chem. Soc., Chem. Commun.*, 1700 (1984).
22. Y. K. Chung, E. D. Honig, W. T. Robinson, D. A. Sweigart, N. G. Connelly, S. D. Ittel, *Organometallics, 2*, 1476 (1983).
23. I. V. Khand, P. L. Pauson, W. E. Watts, *J. Chem. Soc., C*, 2024 (1969).
24. A. N. Nesmeyanov, N. A. Volkerau, L. S. Shiloutseva, V. A. Petrokova, *J. Organomet. Chem., 85*, 365 (1975).
25. I. V. Khand, P. L. Pauson, W. E. Watts, *J. Chem. Soc.*, 2257, 2261 (1968).
26. J. F. McGreer, W. E. Watts, *J. Organomet. Chem., 110*, 103 (1976).
27. J. C. Boutonnet, E. Rose, *J. Organomet. Chem., 221*, 157 (1981).
28. S. G. Davies, M. L. H. Green, D. M. P. Mingos, *Tetrahedron, 34*, 3067 (1978).
29. V. V. Litvak, L. S. Filatova, G. A. Selivanova, V. D. Shteingarts, *Zh. Organ. Khim., 16*, 2, 342 (1980).

30. V. V. Litvak, L. S. Filatova, K. Yo Khalikova, V. D. Shteingarts, *Zh. Organ. Khim.*, *16*, 336 (1980).
31. N. G. Connelly, R. L. Kelly, *J. Chem. Soc., Dalton Trans.*, 2334 (1976).
32. V. V. Litvak, P. P. Kun, V. D. Shteingarts, *Zh. Organ. Khim.*, *10*, 5, 1009 (1974).
33. M. L. H. Green, L. C. Mitchard, W. E. Silverthorn, *J. Chem. Soc., Dalton Trans.*, 2177 (1973).
34. A. C. Sienert, E. L. Muetterties, *Inorg. Chem.*, *20*, 2276. (1981).
35. H. Werner, R. Werner, C. Burschka, *Chem. Ber.*, *117*, 152 (1984).
36. K. Zeukert, H. Werner, *Chem. Ber.*, *121*, 811 (1988).
37. A. Efraty, P. M. Maitlis, *Tetrahedron Lett.* 4025 (1966).
38. A. C. Knipe, S. J. McGuinness, W. E. Watts, *J. Chem. Soc., C*, 842 (1979).
39. I. V. Khand, P. L. Pauson, W. E. Watts, *J. Chem. Soc., C*, 116 (1969).
40. M. V. Gandet, A. W. Hanson, P. S. White, M. J. Zaworothko, *Organometallics*, *8*, 286 (1989).
41. C. Moinet, E. Roman, D. Astruc, *J. Organomet. Chem.*, *128*, C45 (1977).
42. R. A. Zelanka, M. C. Baird, *J. Organomet. Chem.*, *35*, C43 (1972).
43. M. F. Semmelhack, *Ann. N.Y. Acad. Sci.*, *295*, 36 (1977).
44. M. F. Semmelhack, H. T. Hall, Jr., *J. Am. Chem. Soc.*, *96*, 7091 (1974).
45. M. F. Semmelhack, G. Clark, R. Farina, M. Saeman, *J. Am. Chem. Soc.*, *101*, 217 (1979).
46. M. F. Semmelhack, H. T. Hall, Jr., R. Farina, M. Yoshifuji, G. Clark, T. Barget, K. Hirotsu, J. Clardy, *J. Am. Chem. Soc.*, *101*, 3535 (1979).
47. M. F. Semmelhack, A. Yamashita, *J. Am. Chem. Soc.*, *102*, 5926 (1980).
48. E. P. Kündig, V. Desobry, D. P. Simmons, *J. Am. Chem. Soc.*, *105*, 6962 (1983).
49. E. P. Kündig, D. P. Simmons, *J. Am. Chem. Soc., Chem. Commum.*, 1320 (1983).
50. F. Rose-Munch, E. Rose, A. Semra, *J. Chem. Soc., Chem. Commum.*, 1551 (1986).
51. F. Rose-Munch, E. Rose, A. Semra, *J. Chem. Soc., Chem. Commun.*, 942 (1987).
52. S. G. Davies, J. Blagg, C. L. Goodfellow, K. H. Sutton, *J. Chem. Soc., Chem. Commun.* 1283 (1986).
53. I. Ogata, R. Iweta, Y. Ikeda, *Reports of the Government Chemical Industrial Institute*, Tokyo., *65*, #3, 96 (1970).
54. P. S. Domaille, S. D. Ittel, J. P. Jesson, D. A. Sweigart, *J. Organomet. Chem.*, *202*, 191 (1980).
55. Y. K. Chung, E. D. Honig, D. A. Sweigart, *J. Organomet. Chem.*, *256*, 277 (1983).
56. E. P. Kündig, V. Desobry, D. P. Simmons, E. Wenger, *J. Am. Chem. Soc.*, *111*, 1804 (1989).

5.8.2.6.3. Formation of η^5-Cyclohexadienide Complexes via Electrophilic Addition to Metal Arene Complexes.

Reduced arene–metal carbonyl complexes of chromium and manganese, in which the additional electrons formally borne by the metal center either complete the valence configuration of an otherwise electron-deficient metal center or force an arene ligand to slip an η^4 or η^5 bonding mode, have become available via convenient syntheses[1–6]. The Cr and Mn complexes among the former class of highly reduced molecules have been synthesized by the chemical reduction of $(\eta^6\text{-}C_6H_6)Cr(CO)_2(NMe_3)$ and $(\eta^6\text{-}C_6H_6)$-$Mn(CO)_2I$, producing $[(\eta^6\text{-}C_6H_6)Cr(CO)_2]^{2-}$ (**I**) and $(\eta^6\text{-}C_6H_6)Mn(CO)_2]^-$ (**II**), respectively[1,2]. The second class of highly reduced complexes is known only for

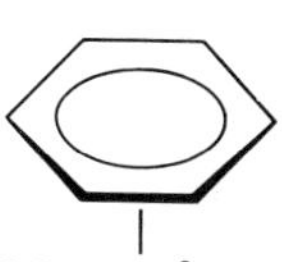
$[(CO)_2Cr]^{2-}$
I

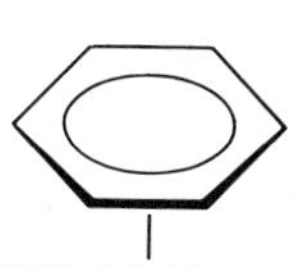
$[(CO)_2Mn]^-$
II

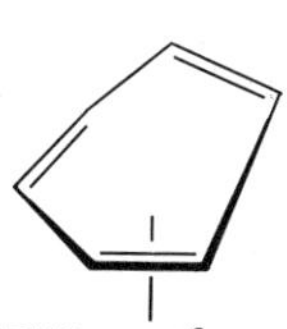
$[(CO)_3Cr]^{2-}$
III

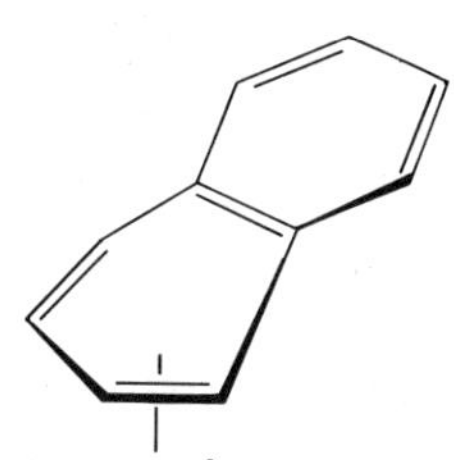
$[(CO)_3Cr]^{2-}$
IV

chromium. These molecules are prepared by a potassium naphthelide or electrochemical reduction of π-benzene or π-naphthalene chromium tricarbonyl complexes to form the η^4-slipped dianions **III** and **IV**, respectively[3–5], or via two-electron reductions of conjugated bis-π-arene chromium complexes, such as $[(CO)_3Cr]_2(\eta^6, \eta^{6'}$-trans-stilbene), to produce bis-η^5-cyclohexadiene-exo-ene complexes (**V**)[6]:

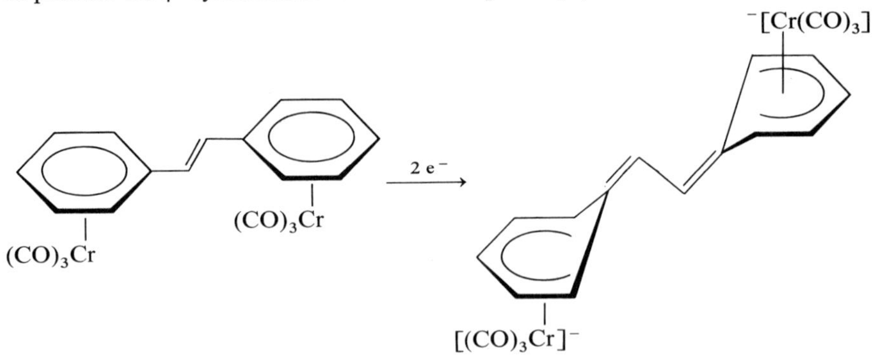

(a)

V

The $[(\eta^4\text{-arene})Cr(CO)_2]^{2-}$ complexes are protonated by water (and benzyl bromide in the case of the benzene complex), forming η^5-cyclohexadienide chromium comlexes. Compound **V** reacts with electrophiles to produce mono-η^5-cyclohexadienide dichromium derivatives (**VI**)[7]:

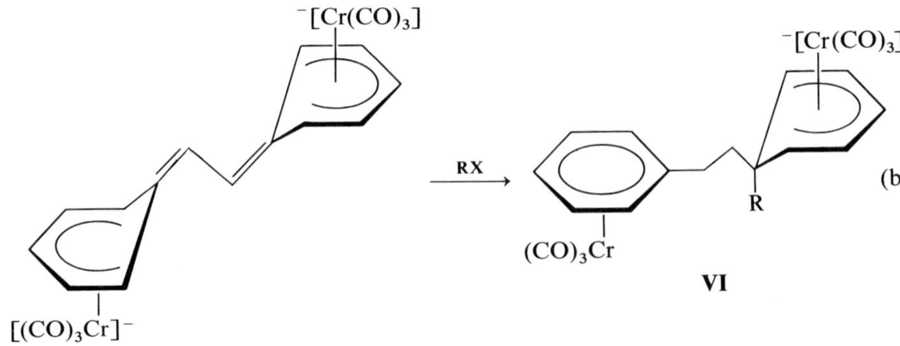

(b)

VI

(J. A. HEPPERT, M. L. MILLIGAN)

1. P. K. Rush, S. K. Noh, M. Brookhart, *Organometallics, 5*, 1745 (1986).
2. V. S. Leong, N. J. Cooper, *Organometallics, 6*, 2000 (1987).
3. R. D. Reike, W. P. Henry, J. S. Arney, *Inorg. Chem., 26*, 420 (1987).
4. V. S. Leong, J. N. Cooper, *J. Am. Chem. Soc., 110*, 2644 (1988).
5. V. S. Leong, J. N. Cooper, *Organometallics, 7*, 2058 (1988).
6. R. D. Reike, S. N. Milligan, L. D. Schulte, *Organometallics, 6*, 699 (1987).
7. L. D. Schulte, R. D. Reike, *J. Org. Chem., 52*, 4829 (1987).

5.8.2.6.5. η^1 σ-Aryl–Metal Complexes by Oxidative Addition of Arenes.

Aromatic compounds with side chains containing coordinating atoms or groups of atoms react with transition-metal compounds to form chelated, ortho-substituted, σ-aryl metal complexes in a reaction known as ortho-metallation. Most examples have employed benzene derivatives, but other aromatic systems have also been used. Four-, five- or six-membered chelate rings plus a few three-membered rings have been reported. The majority of reactions have been with the group-VII and -VIII elements. Examples of orthometallation reactions of the earlier transition elements are much rarer. The usual variation of the reaction is:

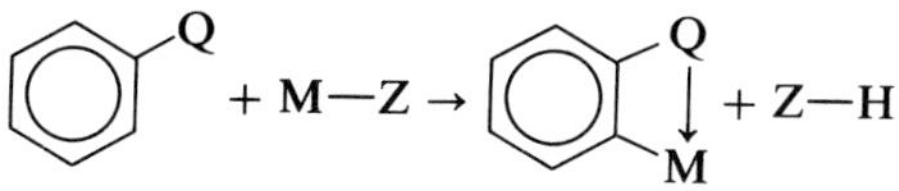

(a)

where Q = a substitiuent containing a coordinating atom or group of atoms and M—Z = a metal complex with anionic ligand Z.

The reaction is closely related to cyclometallation of neopentyl complexes (§5.8.2.2.2) and to cyclometallation of donor ligands (§5.8.2.2.3).

Selective orthometallation occurs only if the side chain contains a good coordinating atom or group of atoms. Metallation of aromatic compounds without the coordinating side chain occurs more slowly and not selectively at the ortho position to any substituents which may be present. The σ-aryl metal derivatives are presumed to be produced on the basis of products formed because the σ-aryl complexes without the stabilizing, chelating ligands are normally unstable under the conditions for metallation. The function of the coordinating side chain in orthometallation reactions is to complex initially with the metal and reduce the entropy of the metallation. Electronic effects of the side chain are unimportant since the coordinating sites may be insulated from the aromatic ring by intervening methylene or similar groups.

The mechanism of metallation varies with the reactants. Both electrophilic and nucleophilic metallations are known[1], and charge-transfer intermediates are likely present in some metallations.

A variety of functional groups in the aromatic side chain facilitate orthometallation, most commonly containing nitrogen or phosphorus, but oxygen, sulfur, arsenic and carbon groups also work. Examples of the nitrogen donor groups are: tertiary amine[2], azo[3–7], azine[8], hydrazone[8,9], 2-pyridyl[10], nitroso[11], imine[12–15] and oxime[17–19], e.g.:

$[Ru(CO)_3Cl_2]_2 + C_6H_5N{=}NC_6H_5 \rightarrow$

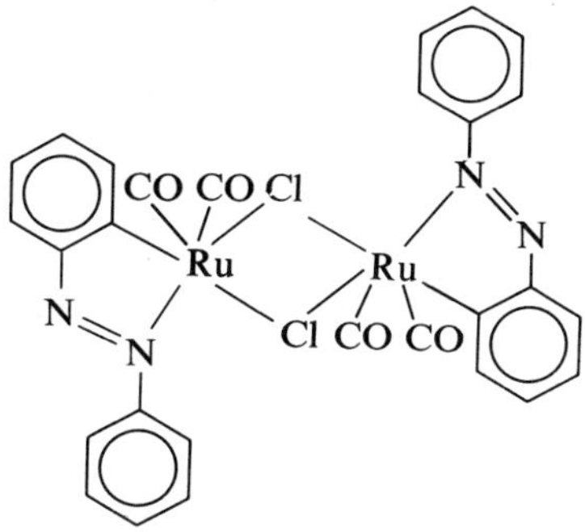

[+HCl] (b)[20]

$(\eta^5\text{-Cp})_2Ni + C_6H_5N{=}NC_6H_5 \rightarrow$

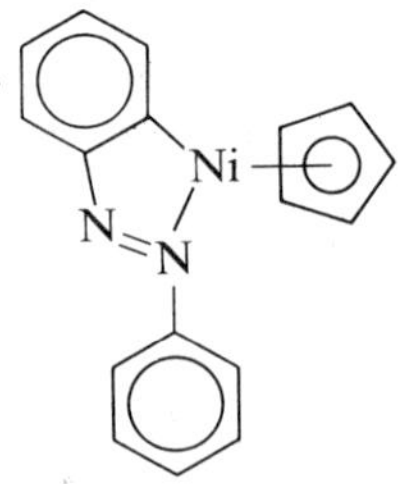

$[+C_5H_6]$ (c)[21]

$C_6H_5N(CH_3)N{=}CH_2 + Pd(OAc)_2 \rightarrow$

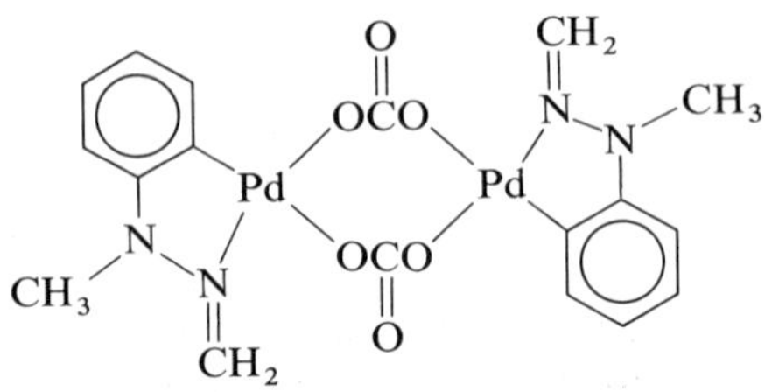

$[+HOAc]$ (d)[8]

Bis-orthometallations have been achieved in rare instances:

$RhCl_3 + C_6H_5N{=}NC_6H_5 \rightarrow$

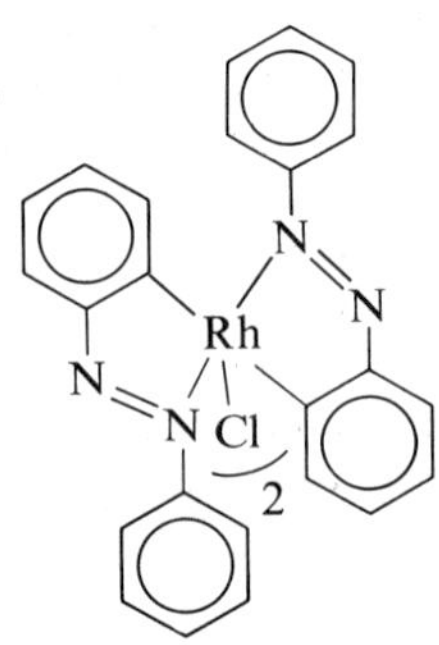

$[+HCl]$ (e)[4]

$K_2PtCl_4 + C_6H_5CH_2N(CH_3)(CH_2)_3N(CH_3)CH_2C_6H_5 \rightarrow$

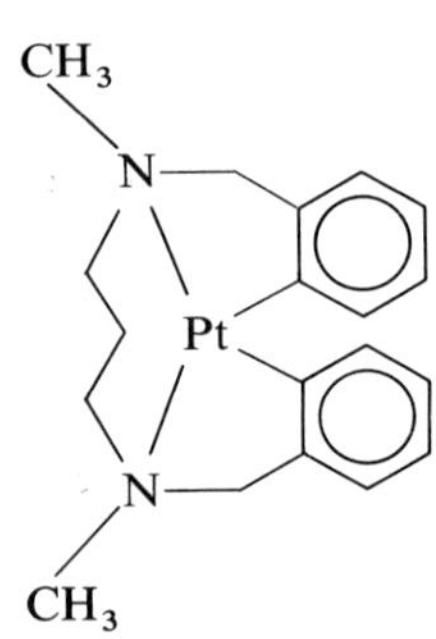

$[+KCl + HCl]$ (f)[22]

In one instance an aromatic ring has been dimetallated[23]:

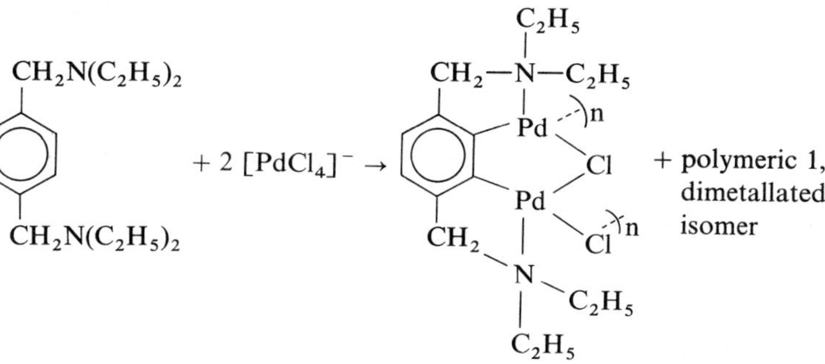

(g)[23]

[+HCl]

The orthometallation reaction with N donor atoms in the side chains usually forms five-membered chelated rings. There are exceptions, such as in the orthopalladation of N-arylamidines where six-membered rings are formed[24]:

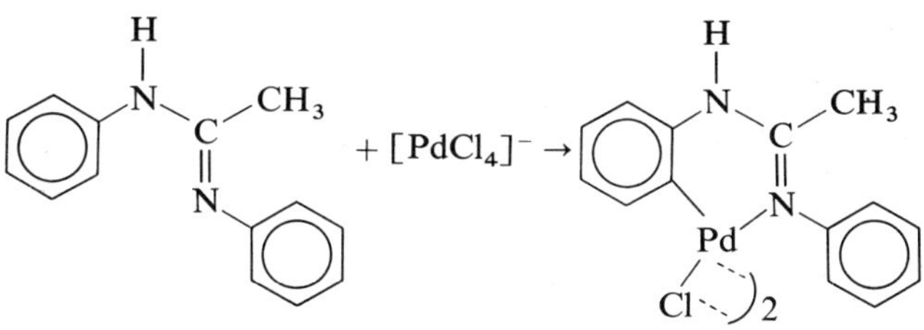

$[+HCl + 2\,Cl^-]$ (h)[24]

and in the titanation of pyridine where the pyridine nitrogen functions as the donor atom and a three-membered ring is formed[25]:

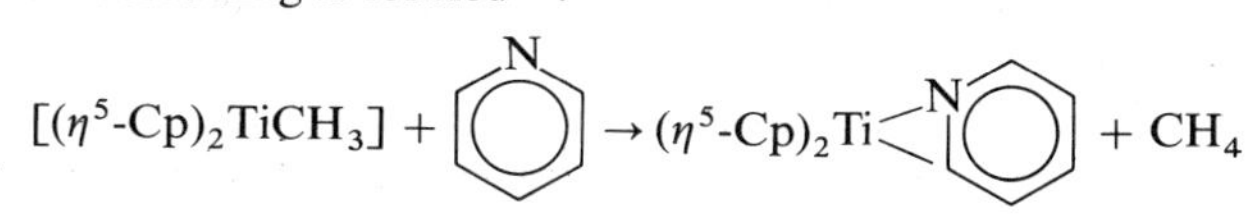

(i)[25]

Stable orthometallated complexes with nitrogen donors in four-membered rings are made by exchange reactions from ortholithium derivatives,[26] but not by direct orthometallation.

Orthometallated complexes are intermediates in numerous cyclic carbonylation reactions involving aromatics with nitrogen donor groups in their side chains. For example, aromatic ketoximes[27] and azobenzene[28] are carbonylated at elevated temperatures and pressures with a cobalt carbonyl catalyst to form five-membered ring heterocycles in which a carbonyl group has replaced the metal in the presumed orthometallated intermediate[8]:

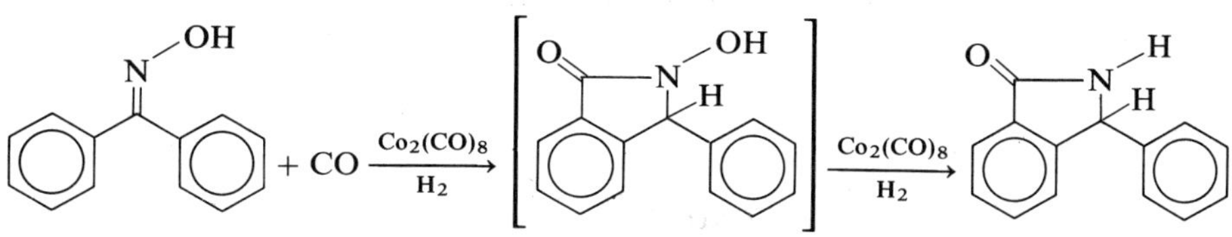

(j)[27]

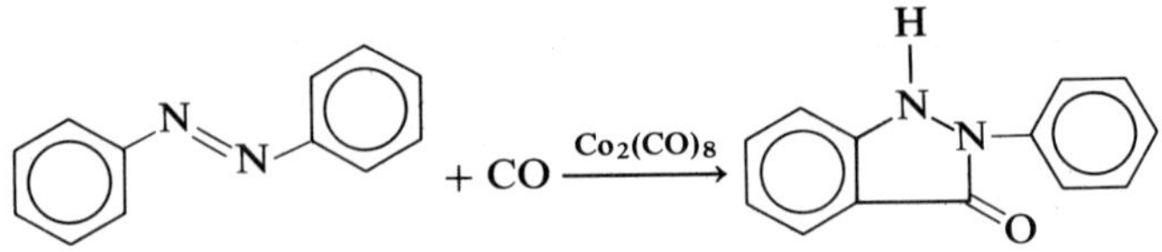

(k)[28]

Many examples of orthometallation of aryl phosphines and phosphites are known. The phosphines produce four-membered rings when phosphorus is attached directly to the ring being metallated[29]:

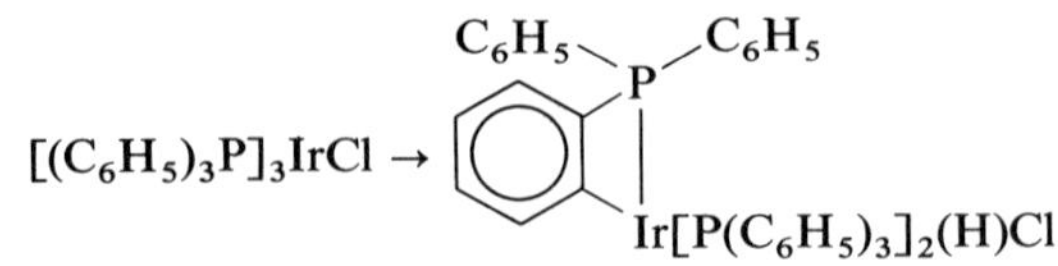

(l)[29]

Orthometallations of arylphosphines involve oxidative additions of the ortho-C—H group, while the ortho-hydrogen atom remains with the metal.

The orthometallation of arylphosphines is accelerated by bulky substituents on the phosphorus[30]:

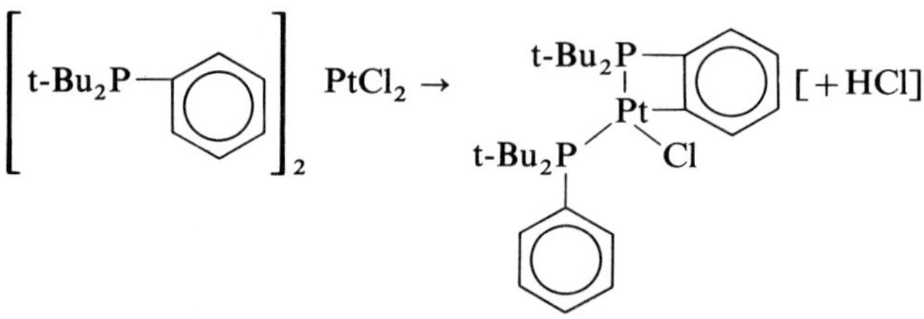

(m)[30]

Similar steric accelerations will probably be found in the orthometallations of aromatics with other donor atoms.

The arylphosphite orthometallations form five-membered chelated rings with the phosphite oxygen in the ring. In the following a metal hydride undergoes orthometallation on heating with loss of hydrogen[31]:

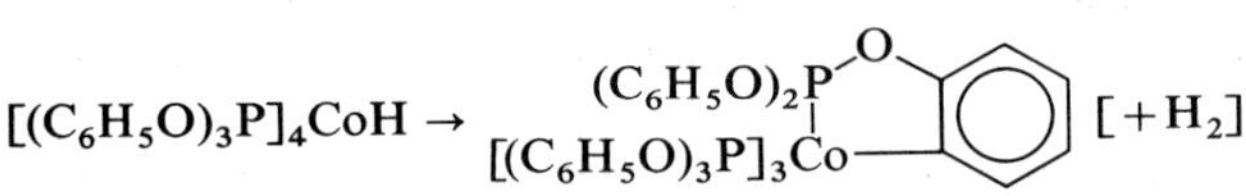

(n)[31]

These reversible reactions explain metal-catalyzed hydrogen–deuterium exchanges at the ortho positions of phosphites[32] and phosphines.[33]

Examples of the promotion of orthometallation by oxygen donors are limited to side chains containing ketonic and amide carbonyl groups. The use of ketone oxygen as the donor atom to promote orthometallation has been reported only for manganese and rhenium carbonyl derivatives[34,35]:

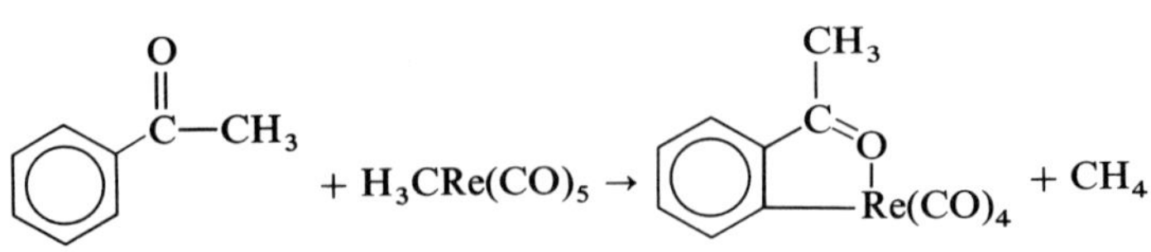

(o)[34]

The orthopalladation of N-arylamides gives six-membered ring chelates[24,36]:

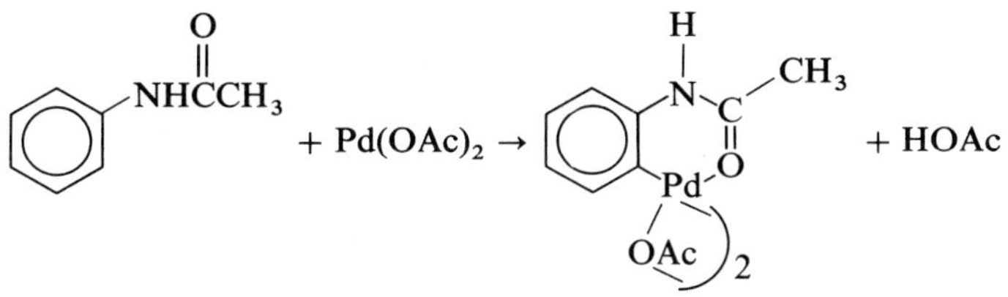

(p)[36]

Aromatic compounds with side-chain sulfur donors orthometallate when the sulfur is in a thioketone[37], thiobenzoate ester[38] or a thioether group[39,40]. In the thioethers an additional coordinating group is required to achieve orthometallation:

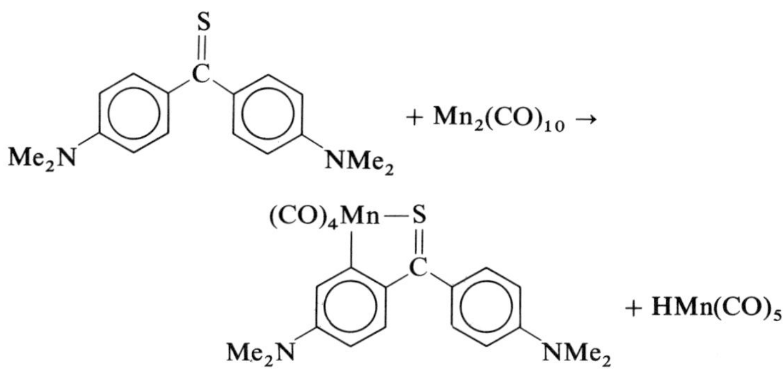

(q)[37]

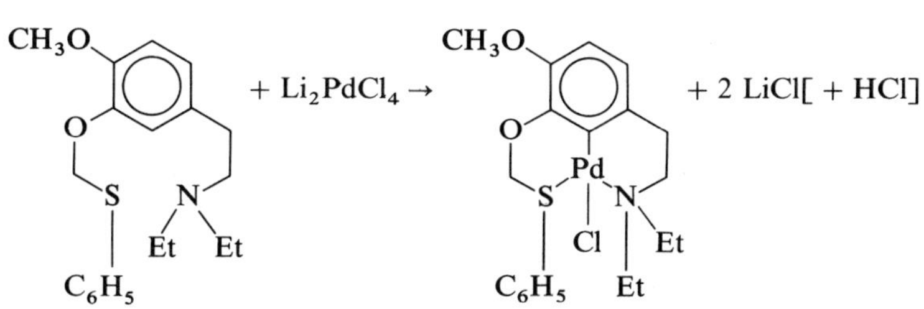

(r)[40]

Neither the compound with the sulfur side chain alone nor the one with the nitrogen side chain alone undergo orthometallation.[40]

Arsenic has received little attention as a donor atom in cyclometallations. For example, 1-naphthylarsines have been metallated and react at the 8-position, showing a preference for the formation of five-membered rings[41]:

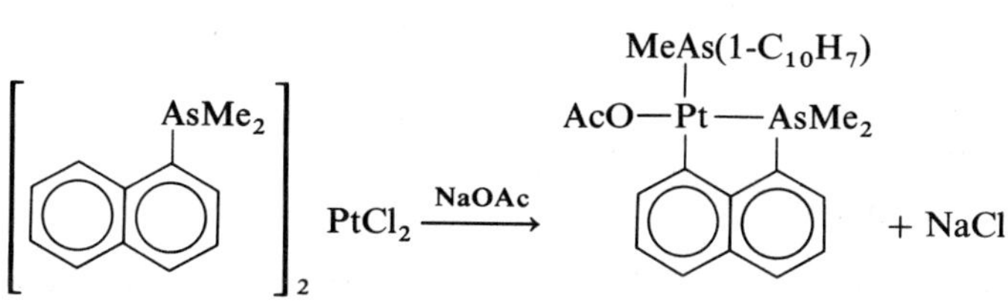

(s)[41]

Orthometallation usually occurs by replacement of an ortho-hydrogen atom by a metal. It is even easier to replace ortho-halogen substituents by transition-metal groups in a similar fashion[42]:

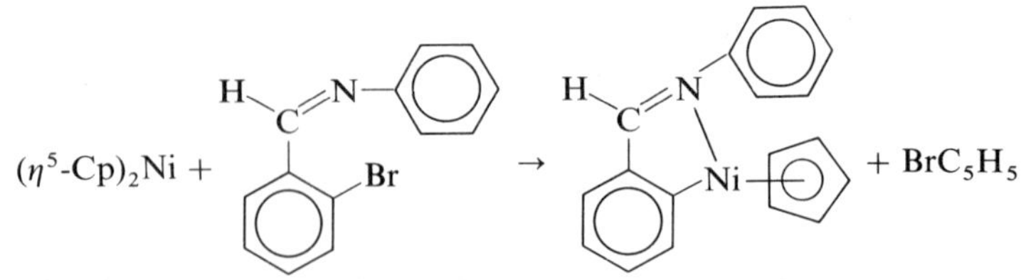

(t)[43]

Even a carbon–carbon double bond is capable of facilitating orthometallation when a halogen group is being replaced.

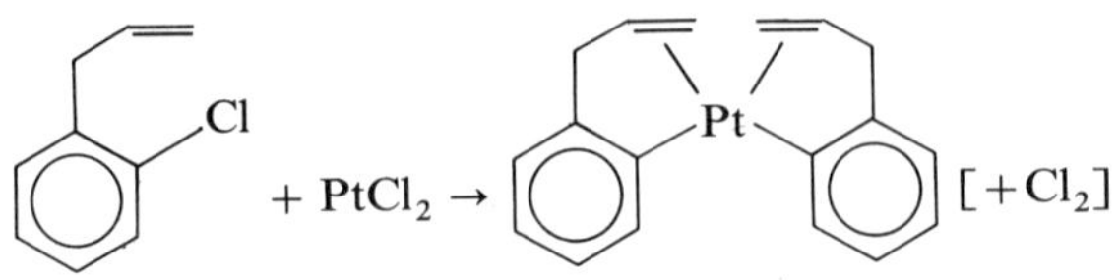

(u)[44]

Three useful reviews have been published on orthometallation[45–47].

(R. F. HECK)

1. M. I. Bruce, B. L. Goodall, F. G. A. Stone, *J. Chem. Soc., Dalton Trans.*, 687 (1978).
2. A. C. Cope, E. C. Friedrich, *J. Am. Chem. Soc.*, *90*, 909 (1968).
3. A. C. Cope, R. W. Siekman, *J. Am. Chem. Soc.*, *87*, 3272 (1965).
4. M. I. Bruce, B. L. Goodall, M. Z. Iqbal, F. G. A. Stone, *J. Chem. Soc., Chem. Commun.*, 661 (1971).
5. M. I. Bruce, M. Z. Iqbal, F. G. A. Stone, *J. Chem. Soc., A*, 3204 (1970).
6. M. M. Bagga, W. T. Flannigan, G. R. Knox, P. L. Pauson, *J. Chem. Soc., C*, 1534 (1969).
7. R. J. Cross, N. H. Tennent, *J. Organomet. Chem.*, *72*, 21 (1974).
8. J. M. Thompson, R. F. Heck, *J. Org. Chem.*, *40*, 2667 (1975).
9. A. G. Constable, W. S. McDonald, L. C. Sawkins, B. L. Shaw, *J. Chem. Soc., Dalton Trans.*, 1992 (1980).
10. A. Kasahara, *Bull. Chem. Soc. Jpn.*, *41*, 1272 (1968).
11. A. G. Constable, W. S. McDonald, B. L. Shaw, *J. Chem. Soc., Dalton Trans.*, 2282 (1980).
12. M. M. Bagga, W. T. Flannigan, G. R. Knox, P. L. Pauson, F. J. Preston, R. I. Reed, *J. Chem. Soc., C*, 36 (1968).
13. R. L. Bennett, M. I. Bruce, B. L. Goodall, M. Z. Iqbal, F. G. A. Stone, *J. Chem. Soc., Dalton Trans.*, 1787 (1972).
14. I. Jardine, F. J. McQuillin, *Tetrahedron Lett.*, 459 (1972).
15. H. Onoue, I. Moritani, *J. Organomet. Chem.*, *43*, 431 (1972).
16. N. D. Cameron, M. Kilner, *J. Chem. Soc., Chem. Commun.*, 687 (1975).
17. A. G. Constable, W. S. McDonald, L. C. Sawkins, B. L. Shaw, *J. Chem. Soc., Chem. Commun.*, 1061 (1978).
18. H. Onoue, I. Moritani, *J. Organomet. Chem.*, *44*, 189 (1972).
19. N. Onoue, K. Nakagawa, *J. Organomet. Chem.*, *35*, 217 (1972).
20. M. I. Bruce, M. Z. Iqbal, F. G. A. Stone, *J. Chem. Soc., A*, 2820 (1971).
21. I. P. Kleiman, M. Dubeck, *J. Am. Chem. Soc.*, *85*, 1544 (1963).
22. M. G. Clerici, B. L. Shaw, B. Weeks, *J. Chem. Soc., Chem. Commun.*, 516 (1973).
23. S. Trofimenko, *J. Am. Chem. Soc.*, *93*, 1808 (1971).
24. N. D. Cameron, M. Kilner, *J. Chem. Soc., Chem. Commun.*, 687 (1975).
25. B. Klei, J. H. Teuben, *J. Chem. Soc., Chem. Commun.*, 659 (1978).
26. G. Longoni, P. Fantucci, P. Chini, F. Canziani, *J. Organomet. Chem.*, *39*, 413 (1972).

27. A. Rosenthal, R. Fastbury, J. Hubscher, *J. Org. Chem.*, *23*, 1037 (1958).
28. S. Horiie, S. Murahashi, *Bull. Chem. Soc. Jpn*, *33*, 247 (1960).
29. M. A. Bennett, D. L. Milner, *J. Am. Chem. Soc.*, *91*, 6983 (1969).
30. A. J. Cheney, B. E. Mann, B. L. Shaw, R. M. Stade, *J. Chem. Soc., Chem. Commun.*, 1176 (1970).
31. L. W. Gosser, *Inorg. Chem.*, *14*, 1453 (1975).
32. G. W. Parshall, W. H. Knoth, R. A. Schunn, *J. Am. Chem. Soc.*, *91*, 4990 (1969).
33. G. W. Parshall, *J. Am. Chem. Soc.*, *90*, 1669 (1968).
34. R. J. McKinney, G. Firestein, H. D. Kaesz, *Inorg. Chem.*, *14*, 2057 (1975).
35. S. S. Crawford, H. D. Kaesz, *Inorg. Chem.*, *16*, 3193 (1977).
36. H. Horino, N. Inoue, *Tetrahedron Lett.*, 2403 (1979).
37. H. Alper, *J. Organomet. Chem.*, *73*, 359 (1974).
38. H. Alper, A. S. K. Chan, *J. Chem. Soc., Chem. Commun.*, 724 (1973).
39. J. Errington, W. S. McDonald, B. L. Shaw, *J. Chem. Soc., Dalton Trans.*, 2312 (1980).
40. R. A. Holton, R. V. Nelson, *J. Organomet. Chem.*, *201*, C35 (1980).
41. J. M. Duff, B. E. Mann, B. L. Shaw, B. Turtle, *J. Chem. Soc., Dalton Trans.*, 139 (1974).
42. I. V. Barinov, T. I. Vohevodskaya, Y. A. Ustynyuk, *J. Organomet. Chem.*, *30*, C28 (1971).
43. Y. A. Ustynyuk, V. A. Chertkov, I. V. Barinov, *J. Organomet. Chem.*, *29*, C53 (1971).
44. M. Aresta, R. S. Nyholm, *J. Chem. Soc., Chem. Commun.*, 1459 (1971).
45. G. W. Parshall, *Acc. Chem. Res.*, *3*, 139 (1970).
46. J. Dehand, M. Pfeffer, *Coord. Chem. Rev.*, *18*, 327 (1976).
47. E. C. Constable, *Polyhedron*, *3*, 1037 (1984).

5.8.2.6.5. η^1 σ-Aryl–Metal Complexes by Oxidative Addition of Arenes.

Direct formation of isolable σ-aryl complexes by intermolecular oxidative addition of an aromatic C—H bond to a transition-metal complex was first observed in an alkyl phosphine complex of Ru and has now been extended to most of the transition metals. The relevant literature is summarized in Table 1. There are reviews of C—H activation by homogeneous metal complexes and its relevance to homogeneous and heterogeneous catalysis[48–58].

There are several general features common to all systems in which aromatic C—H bonds undergo oxidative addition across a metal center. The high strength of aromatic C—H bonds requires that metal complexes capable of carrying out the reaction be relatively energetic. This normally means coordinative unsaturation and a high electron density on the metal center. For this reason, the complexes are very reactive and capable of undergoing many other reactions in addition to the desired C—H cleavage. The more common reactions, intramolecular C—H activation and ortho-metallation, are covered in §5.8.2.6.3. Electrophilic attack of transition-metal complexes on aromatic substrates to yield σ-aryl complexes is covered in §5.8.2.6.5. Another reaction commonly observed in methyl-substituted aromatics is activation of the methyl group to give benzyl-type derivatives. The formation of η^1 σ-alkyls by direct activation of sp^3 C—H bonds is covered in §5.8.2.2.1.

Oxidative addition of aromatic C—H bonds is favored by electron-withdrawing groups on the aromatic ring. There are several examples of systems that do not cleave the C—H bonds of benzene but that react with aromatic compounds bearing electron-withdrawing groups. In substituted aromatics, electronic factors can play a significant role in determining the selectivity of the reaction. More often than not, however, the steric constraints imposed by the substituents play a deciding role in determining the nature of the product.

Generation of the electron-rich, coordinatively unsaturated species, $Ru(dmpe)_2$, by sodium naphthalenide reduction of $Ru(dmpe)_2Cl_2$ leads to insertion into a C—H bond

TABLE 1. SYNTHESES OF σ-ARYL COMPLEXES BY OXIDATIVE ADDITION OF ARENES

Starting complex[a]	Substrate	Product[a]	Ref.
$Zr(Cp)_2Cl_2 + Na(C_{10}H_8)$	$C_{10}H_8$	$[Zr(Cp)_2]_2(\mu\text{-}H)(\mu\text{-}C_{10}H_7)$	1
$Zr(Cp)_2$(NHBu-t)Me	C_6H_6	$Zr(Cp)_2$(NHBu-t)C_6H_5	2
ZrMe(t-Bu_3SiNH)$_3$	C_6H_6	Zr(C_6H_5)(t-Bu_3SiNH)$_3$	3
$ZrCp_2(PMePh_2)_2 + Me_2C{=}O$	C_6H_6	$ZrCp_2(OCHMe_2)(C_6H_5)$	4
$ZrCp_2(PMePh_2)_2 + Me_2C{=}O$	C_6H_5Me	$ZrCp_2(OCHMe_2)(C_6H_4Me)$	4
$W(Cp)_2H_2 + h\nu$	C_6H_6	$W(Cp)_2H(C_6H_5)$	5, 6
	C_6H_5F	$W(Cp)_2H$(m, p-C_6H_4F)	6
	C_6H_5Me	$W(Cp)_2H$(p-C_6H_4Me)	6
	C_6H_5OMe	$W(Cp)_2H$(p-C_6H_4OMe)	6
	$C_6H_5CO_2Me$	$W(Cp)_2H$(m, p-$C_6H_4CO_2Me$)	6
$W(Cp)_2H_2$ + diene	C_6H_6	$W(Cp)_2H(C_6H_5)$	7, 8
	C_6H_5Me	$W(Cp)_2H$(p-C_6H_4Me)	7, 8
$W(Cp)_2H_2$ + Li(n-Bu)	C_6H_5COPh	$W(Cp)_2H$(p-C_6H_4COPh)	9
$W(Cp)_2HR$, R = $CH(CO_2Me)CH_2CO_2Me$	C_6H_6	$W(Cp)_2H(C_6H_5)$	9
$W(Cp)_2HR$, R = Me	C_6H_6	$W(Cp)_2H(C_6H_5)$	10
	C_6H_5F	$W(Cp)_2H$(m, p-C_6H_4F)	10
$W(Cp)_2CO + h\nu$	C_6H_6	$W(Cp)_2H(C_6H_5)$	11
$W(Cp)_2CO + h\nu$	C_6H_5Me	$W(Cp)_2H(C_6H_4Me)$	11
Fe(dmpe)$_2H_2 + h\nu$	C_6H_6	Fe(dmpe)$_2$H(C_6H_5)	12
	C_6H_5Me	Fe(dmpe)$_2$H(m, p-C_6H_4Me)	12
Fe(dmpe)$_2Cl_2$ + Na[$C_{10}H_8$]	$C_{10}H_8$	Fe(dmpe)$_2$H(2-naphthyl)	13
Fe(dmpe)$_2$H(2-naphthyl)	C_6H_6	Fe(dmpe)$_2$H(C_6H_5)	13, 14
	C_6H_5F	Fe(dmpe)$_2$H(o-C_6H_4F)	13, 14
	p-$C_6H_4F_2$	Fe(dmpe)$_2$H(2,4-$C_6H_3F_2$)	14
	C_5H_5N	Fe(dmpe)$_2$H(o, m, p-C_5H_4N)	14
	C_6H_5Me	Fe(dmpe)$_2$H(m, p-C_6H_4Me)	14
	$C_6H_5CF_3$	Fe(dmpe)$_2$H(m, p-$C_6H_4CF_3$)	14
	o-$C_6H_5(CF_3)_2$	Fe(dmpe)$_2$H[$C_6H_4(CF_3)_2$]	14
	C_6H_5CN	Fe(dmpe)$_2$H(o, m, p-C_6H_4CN)	14
	p-MeC_6H_4CN	Fe(dmpe)$_2$H(2-NCC_6H_3-5-Me)	14

$Ru(dmpe)_2Cl_2 + Na(C_{10}H_8)$	$C_{10}H_8$	$Ru(dmpe)_2H$(2-naphthyl)	15
$RuCl_2(P(OMe)_3)_2(PMe_3)_2$ + reduction	C_6H_6	$Ru[P(OMe)_3]_2(PMe_3)_2H(C_6H_5)$	16
$Ru(C_6Me_6)H_2(PChx_2H) + h\nu$	C_6H_6	$Ru(C_6Me_6)H(PChx_2H)(C_6H_5)$	17
$Ru(C_6H_6)H_2(PPr\text{-}i_3) + h\nu$	C_6H_6	$Ru(C_6H_6)H_2(PPr\text{-}i_3)(C_6H_5)$	18
	C_6H_5Me	$Ru(C_6H_6)H_2(PPr\text{-}i_3)(C_6H_4Me)$	18
$Ru(C_6Me_6)H_2(PMe_3) + h\nu$	C_6H_6	$Ru(C_6Me_6)H_2(PMe_3)(C_6H_5)$	18
	C_6H_5Me	$Ru(C_6Me_6)H_2(PMe_3)(C_6H_4Me)$	18
$Os(C_6H_6)I_2(PR_3) + NaC_{10}H_8$ R = Me, i-Pr	C_6H_6	$Os(C_6H_6)(PR_3)H(C_6H_5)$	18
$OsCl_2[P(OMe)_3]_4$ + reduction	C_6H_6	$Os[P(OMe)_3]_4H(C_6H_5)$	16
$OsCl_2[P(OMe)_3]_2(PMe_3)_2$ + reduction	C_6H_6	$Os[P(OMe)_3]_2(PMe_3)_2H(C_6H_5)$	16
	C_6H_5Me	$Os[P(OMe)_3]_2(PMe_3)_2H(C_6H_4Me)$	16
$Os(dmpe)_2Cl_2 + Na(C_{10}H_8)$	$C_{10}H_8$	$Os(dmpe)_2H$(2-naphthyl)	19
$Os(PMe_3)_4HR$ R = Me, (CH_2SiMe_3), (CH_2CMe_3)	C_6H_6	$Os(PMe_3)_4H(C_6H_5)$	20, 21
$Os(arene)(PMe_3)(C_2H_4) + PMe_3$ arene = C_6H_6, C_6H_5Me	C_6H_6	$Os(PMe_3)_4H(C_6H_5)$ + $Os(PMe_3)_3(C_2H_4)H(C_6H_5)$	22
$Os(\eta^6\text{-}C_6H_6)(\eta^4\text{-}C_6H_6) + PMe_3$	C_6H_6	$Os(PMe_3)_4H(C_6H_5)$	23
$Os_3(CO)_{12}$	C_6H_6	$Os_3(CO)_9H_2(C_6H_4)$	24
$RhCp^*(PMe_3)H_2 + h\nu$	C_6H_6	$RhCp^*(PMe_3)H(C_6H_5)$	25, 26
	C_6H_5Me	$RhCp^*(PMe_3)H$(m, p-C_6H_4Me)	25, 26
	p-$C_6H_4Me_2$	$RhCp^*(PMe_3)H$(2,5-$C_6H_3Me_2$)	25, 26
	m-$C_6H_4Me_2$	$RhCp^*(PMe_3)H$(3,5-$C_6H_3Me_2$)	25, 26
	o-$C_6H_4Me_2$	$RhCp^*(PMe_3)H$(3,4-$C_6H_3Me_2$)	25, 26
	p-$C_6H_4Pr_2$	$RhCp^*(PMe_3)H$(2,5-$C_6H_3Pr_2$)	25, 26
$Rh(HBPz^*_3)(CO)_2$	C_6H_6	$Rh(HBPz^*_3)(CO)H(C_6H_5)$	27
$Rh(HBPz^*_3)(CO)H(Chx)$	C_6H_6	$Rh(HBPz^*_3)(CO)H(C_6H_5)$	27
$Rh(HBPz^*_3)CO(alkene)$ alkene = C_2H_4, C_3H_6, C_8H_{14}	C_6H_6	$Rh(HBPz^*_3)(CO)H(C_6H_5)$	28
$Rh(NP_3)H + MeSO_3CF_3$	C_6H_6	$Rh(NP_3)H(C_6H_5)(SO_3CF_3)$	29, 30
$Rh(NP_3)H + MeSO_3CF_3$	$C_6H_5CF_3$	$Rh(NP_3)H(C_6H_4CF_3)(SO_3CF_3)$	30
$[Ir(PP_3)H_2](SO_3CF_3) + h\nu$	C_6H_6	$[Ir(PP_3)H(C_6H_5)](SO_3CF_3)$	31

Continues over

TABLE 1. (*Continued*)

Starting complex[a]	Substrate	Product[a]	Ref.
$IrCp^*(PPh_3)H_2 + h\nu$	C_6H_6	$IrCp^*(PPh_3)H(C_6H_5)$	32, 33
$IrCp^*(PMe_3)H_2 + h\nu$	C_6H_6	$IrCp^*(PMe_3)H(C_6H_5)$	32, 33
$IrCp^*(PPh_3)HR + Al_2O_3$	C_6H_6	$IrCp^*(PPh_3)H(C_6H_5)$	32, 33
R = Chx, CH_2CMe_3			
$IrCp^*(PMe_3)H_2 + h\nu$	p-$C_6H_4Me_2$	$IrCp^*(PMe_3)H(C_6H_3Me_2)$	33
$IrCp^*(CO)_2 + h\nu$	C_6H_6	$IrCp^*(CO)H(C_6H_5)$	34, 35
$IrCp(CO)_2 + h\nu$	C_6H_6	$[IrCp(CO)]_2(\mu\text{-}C_6H_4)$	36
		$IrCp(CO)H(C_6H_5)$ intermediate	36
$IrCp^*H(CH_2CHCH_2) + PR_3$	C_6H_6	$IrCp^*H(CH_2CH_2CH_3)(C_6H_5)(PR_3)$	37, 38
R = Me, Ph			
	C_6H_6	$IrCp^*H(C_6H_5)(\mu\text{-}CHCHCH_2)(\mu\text{-}H)IrCp^*$	38
$IrCp^*L(\mu\text{-}CHCHCH_2)(\mu\text{-}H)IrCp^*$	C_6H_6	$IrCp^*H(C_6H_5)(\mu\text{-}CHCHCH_2)(\mu\text{-}H)IrCp^*$	38
$+ h\nu$ L = CO, C_2H_4			
$IrCp^*(PMe_3)H(Chx)$	C_6H_6	$IrCp^*(PMe_3)H(C_6H_5)$	39, 40
$Ir(N(SiMe_2CH_2PPh_2)_2(\eta^2\text{-}C_8H_{14})$	C_6H_5Me	$Ir(N(SiMe_2CH_2PPh_2)_2$(m, p-$C_6H_4Me)$	41
$IrCp^*(CO)[C(p\text{-}ClC_6H_4){=}NOC{=}O]$	C_6H_6	$IrCp^*(CO)H(C_6H_5)$	42
$Pt(PChx_3)_2$	C_6F_5H	$Pt(PChx_3)_2H(C_6F_5)$	43
	$C_6F_3H_3$	$Pt(PChx_3)_2H(C_6F_3H_2)$	43
	m-$C_6F_2H_4$	$Pt(PChx_3)_2H(2,6\text{-}C_6F_2H_4)$	43
$Pt(PEt_3)_3$	C_6F_5H	$Pt(PEt_3)_2H(C_6F_5)$	43
$Pt(DChxPE)H(CH_2CMe_3)$	C_6H_6	$Pt(DChxPE)H(C_6H_5)$	44, 45
	$C_6H_3Me_3$	$Pt(DChxPE)H(C_6H_2Me_3)$	44, 46
$Pt(PMe_3)_2(SO_3CF_3)(CH_2CMe_3)$	C_6H_6	$Pt(PMe_3)_2(SO_3CF_3)C_6H_5$	47
	C_6H_5Me	$Pt(PMe_3)_2(SO_3CF_3)$(m, p-$C_6H_4Me)$	47

[a] Abbreviations: dmpe, bis(dimethylphosphino)ethane; Cp, η^5-cyclopentadienyl; Cp*, η^5-pentamethylcyclopentadienyl; NP_3, $N(CH_2CH_2PPh_2)_3$; PP_3, $P(CH_2CH_2PPh_2)_3$; $HBPz^*_3$, tris(3,5-dimethylpyrazolylborato; DChxPE, bis(dicyclohexylphosphino)ethane; Chx, cyclohexyl; $C_{10}H_8$, naphthalene.

at the 2 position of the naphthalene[15] to give cis-Ru(dmpe)$_2$H(2-naphthyl)[19]. Similar complexes are obtained with Fe[13,14] and Os[19]. In solution, the Ru complex is entirely cis and the Os analog is predominantly cis, but the Fe analog exists as a mixture of 40% cis and 60% trans[14].

In solution Fe(dmpe)$_2$H(2-naphthyl), **I**, is in facile equilibrium with an undetected intermediate, presumably "Fe(dmpe)$_2$"[15], which inserts into a variety of C—H bonds[14,59,60]:

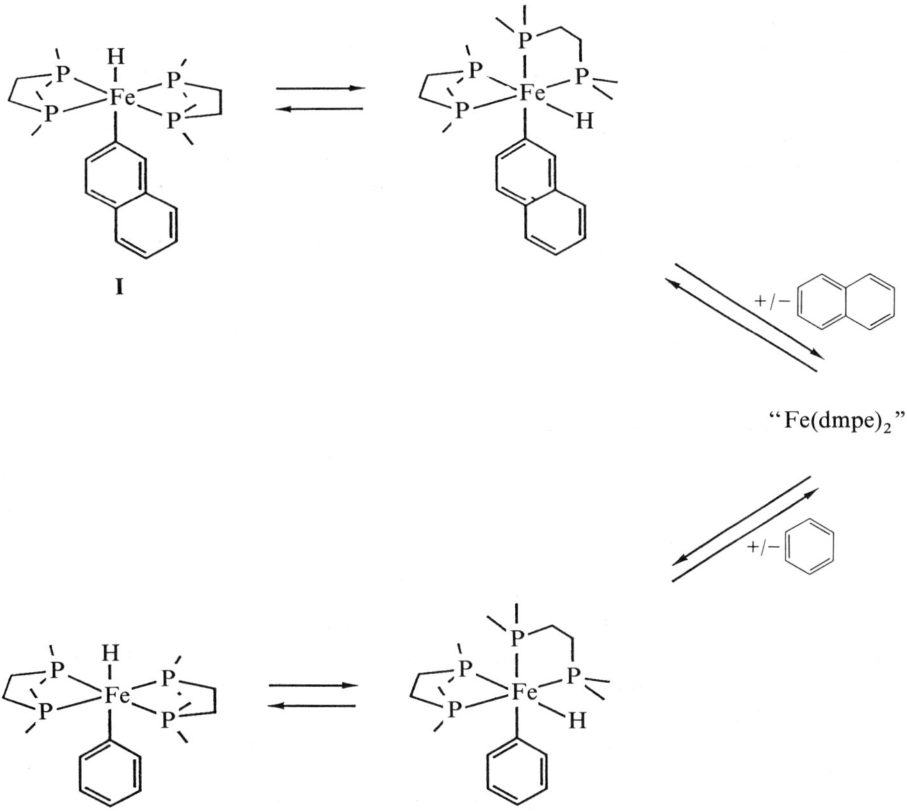

(See §5.8.2.2.1 for insertion into sp^3 C—H bonds, §5.8.2.10.1 for insertion into aldehyde C—H bonds, and §5.8.3.2.5 for insertion into Si—H bonds.) Steric effects play a predominant role in determining the various isomers observed in these reactions, but electronic effects are also important.

When **I** is dissolved in benzene, a mixture of cis- and trans-Fe(dmpe)$_2$H(C$_6$H$_5$) is obtained[13,14]. The rate-limiting step in the reaction is reductive elimination of naphthalene from cis-**I**. Reaction with C$_6$D$_6$ gives the D(C$_6$D$_5$) product, and there is no H–D exchange between aromatic molecules when mixtures of C$_6$H$_6$ and C$_6$D$_6$ are employed[14].

When there is a single substituent on the benzene ring, there are six possible isomeric aryl hydride complexes; the metal center can be cis or trans and the attack on the ring can be ortho, meta, or para. The products are all in slow equilibrium, so their isolation is complicated. Fluoro substituents are ortho directing and cause no steric problems. Therefore, reaction of **I** with fluorobenzene or p-difluorobenzene gives single products, trans-Fe(dmpe)$_2$(o-C_6H_4F) or (2,4-$C_6H_3F_2$), respectively. At equilibrium, no cis product is observed. On the other hand, pyridine shows little selectivity in its reaction with **I**, giving a complete mixture of cis and trans ortho, meta and para products.

Substituents larger than fluoro rule out ortho attack because of steric interactions, thereby limiting the number of products. Reactions of **I** with toluene or $CF_3C_6H_5$ give complex mixtures of cis and trans meta and para isomers[14]. Compound **I** does not react with mesitylene or p-xylene; all aromatic hydrogen atoms are ortho to a substituent. Reaction with o-$C_6H_4(CF_3)_2$ gives only one cis- and one trans-[3,4-$C_6H_4(CF_3)_2$] product.

Benzonitrile reacts with **I** to give a mixture of cis and trans meta and para isomers and the cis-o isomer[16]. In p-toluonitrile, addition of a p-methyl to the ring blocks the meta and para isomers, so the only product observed is cis-o-(p-toluonitrile). The absence of any trans product is attributed to unfavorable steric interaction of the o-cyano group with an equatorial plane of phosphorus atoms; in the cis isomer, the cyano group is directed over the smaller hydride ligand.

The reactive intermediate Fe(dmpe)$_2$ cleaves the C—H bonds of other aromatic substrates, but the reactions are complex[14]. The final products with anisole or methyl benzoate are methyl derivatives derived from cleavage of their respective O—CH_3 bonds. Chlorobenzene gives a transient aryl hydride, but the ultimate product is the phenyliron chloro complex. Both Ru(dmpe)$_2$H($C_{10}H_7$) and Os(dmpe)$_2$H($C_{10}H_7$) are also in equilibrium with their respective M(dmpe)$_2$ intermediates, but the equilibria are very sluggish and of little synthetic utility. An alternative approach to these species is photochemical activation of Fe(dmpe)H_2 [12]; the products can be obtained more cleanly, and the intermediate "Fe(dmpe)$_2$" cleaves even aliphatic C—H bonds.

The reactivity of tungstenocene, "W(Cp-η^5)$_2$", is similar to that observed for zero-valent alkyl phosphine complexes of the iron triad. Generation of tungstenocene in aromatic solvents gives aryl hydrides in moderate to good yields. The various routes to the coordinatively unsaturated intermediate include photolysis of W(Cp-η^5)$_2H_2$ [5,6] or W(Cp-η^5)$_2$CO [11], removal of H_2 from W(Cp-η^5)$_2H_2$ by hydrogenation of a diene[7,8] or thermolysis of alkyl hydrides, W(Cp-η^5)$_2$HR [9,10].

Generation of tungstenocene in benzene gives W(Cp-η^5)$_2$H(C_6H_5) as the product. When C_6D_6 is used as a solvent, the product is W(Cp-η^5)$_2$D(C_6D_5), indicating formation by simple insertion into the C—H or C—D bonds. There is no exchange between W(Cp-η^5)H(C_6H_5) and C_6D_6 even upon irradiation. Photolysis of a solution of W(Cp-η^5)$_2H_2$ (**II**) in fluorobenzene[6] gives a mixture of the meta and para isomers of W(Cp-η^5)$_2$H(C_6H_4F). Photolysis in toluene gives the p-tolyl product exclusively, and o-xylene gives the 3,4-xylyl.

When **II** is photolyzed in p-xylene, attack does not occur at the ortho H atoms; instead, the bis-η^1-benzyl products, W(Cp-η^5)$_2$(η^1-$CH_2C_6H_4$Me-3)$_2$ and W(Cp-η^5)$_2$(η^1-$CH_2C_6H_3Me_2$-2,4)$_2$, are formed[61]. This is unlike reactions observed in the Fe(dmpe)$_2$ system where no reaction is observed in p-xylene. Substitution of methoxy for methyl gives analogous results; anisole yields W(Cp-η^5)$_2$H(C_6H_4OMe-4) and p-methylanisole gives W(Cp-η^5)$_2$(η^1-$CH_2C_6H_4$OMe-4)$_2$ [6]. Photolysis of **II** in p-chlorotoluene gives W(Cp-η^5)$_2$Cl(C_6H_4Me-4) by attack on the C—Cl bond. Photolysis in methyl benzoate gives a mixture of meta- and para-$W(Cp\text{-}\eta^5)_2H(C_6H_4CO_2Me)$, together with $W(Cp\text{-}\eta^5)_2H(OCOC_6H_5)$, which results from attack on the O—CH_3 bond of the ester[10]. Many other reactions not involving formation of a phenyl complex by C—H activation are also observed.

Reactions of aromatic C—H bonds with Pt are limited to the coordinatively unsaturated complexes PtL_2 and PtL_3, where L are basic trialkylphosphine ligands[43]. Reactions of the fluorobenzenes C_6HF_5, 1,3,5-$C_6H_3F_3$ or 1,3-$C_6H_4F_2$ with PtL_2 (L = tricyclohexylphosphine) give the respective trans-$PtL_2H(Ar)$ complexes in moderate isolated yield. The additions occur to give products with two ortho-fluorine substituents. Thus 1,2-$C_6H_4F_2$, 1,4-$C_6H_4F_2$ and C_6H_5F do not add. When fluorobenzenes are added to PtL_3 (L = PEt_3), $PtL_2H(Ar)$ is formed only when Ar is pentafluorophenyl. Thus, with less basic ligands on the metal center, a more electron-deficient arene is required. Another approach to aromatic C—H activation by Pt is to use bulky, chelating alkylphosphine ligands. With the cis geometry imposed upon the metal center, it is more reactive with arenes; benzene and mesitylene are activated[44–46].

The pentamethylcyclopentadienyl Rh and Ir systems have proved capable of activating not only aromatic but also aliphatic sp^3 C—H bonds. Photolysis of the $M(Cp^*\text{-}\eta^5)(PMe_3)H_2$ complexes (M = Rh[25,26], Ir[32,33]) liberate H_2, leaving a very reactive metal center. Reactions with benzene give the hydridophenyl derivatives and reactions with p-xylene give the 2,5-dimethylphenyl derivatives. Other variations of ligands have been found to give similar products. Additional dialkylphenyl derivatives have been synthesized in the Rh system, but there is little additional information on those factors which determine selectivities.

A number of examples of aromatic C—H oxidative addition to transition-metal complexes involve polynuclear compounds. In some cases, mononuclear species may be responsible for the initial attack on the C—H bond, but most mechanisms remain unresolved. Photolysis of $Ir(Cp\text{-}\eta^5)(CO)_2$ in benzene gives a low yield of **III**:

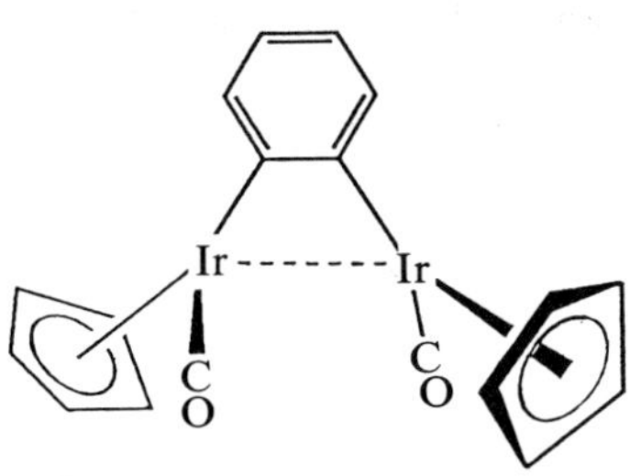

III

The complex, formally derived from a double oxidative addition of benzene to two Ir centers[36], is in contrast to the mononuclear product obtained with the pentamethylcyclopentadienyl derivative[34,35]. In the absence of light there is no reaction; presumably photolysis displaces one carbonyl ligand to give a coordinatively unsaturated precursor. An apparent intermediate in the photolysis has ^1H NMR resonances in the aromatic, cyclopentadienyl and hydride regions and may be $IrCp\text{-}\eta^5(CO)H(C_6H_5)$, but its nature remains uncertain.

The product isolated from the reaction of the allyl complex, η^5-Cp*IrH(CH_2CHCH_2), with benzene has been shown to be dependent on reaction conditions and the concentration of the starting complex[37,38]:

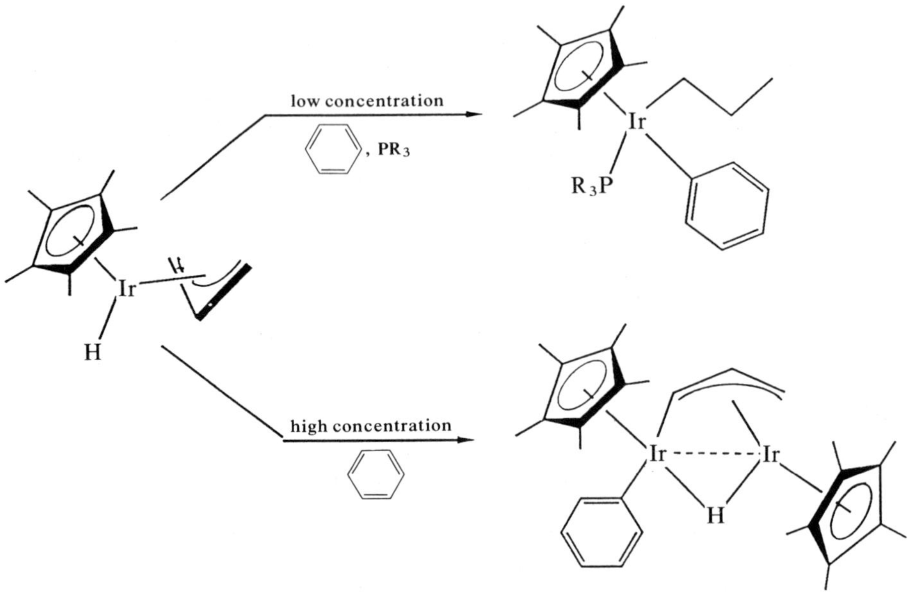

At low metal concentrations, in the presence of phosphine, the product is a monomeric alkylphenyl complex resulting from addition of two hydrogen atoms to the allyl. At higher concentrations, the resulting binuclear complex is bridged by both a hydride and a σ, π-C_3H_4 group. The phenyl and hydrido groups are lost as benzene upon addition of a variety of monodentate ligands, but these reactions can be reversed by photolytic dissociation of π-acid ligands such as CO or olefin[38].

Reduction of Zr(Cp-η^5)$_2$Cl$_2$ with potassium naphthalenide at low T in THF yields **IV**:

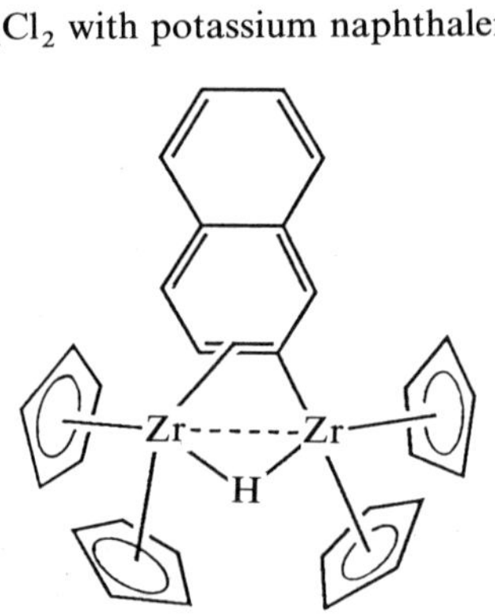

IV

The product is derived by oxidation of a 2-naphthyl C—H bond to a divalent zirconocene species[1]. The μ-naphthyl group is 2-η^1 to one zirconium atom and 1,2-η^2 to the other.

The Os cluster $Os(CO)_{12}$ reacts with benzene (as solvent and reactant) under vigorous conditions[24]. Heating at 194°C in a sealed tube for 2 d yields variable quantities (low to 34%) of *V* derived from oxidative addition of two o-C—H bonds.

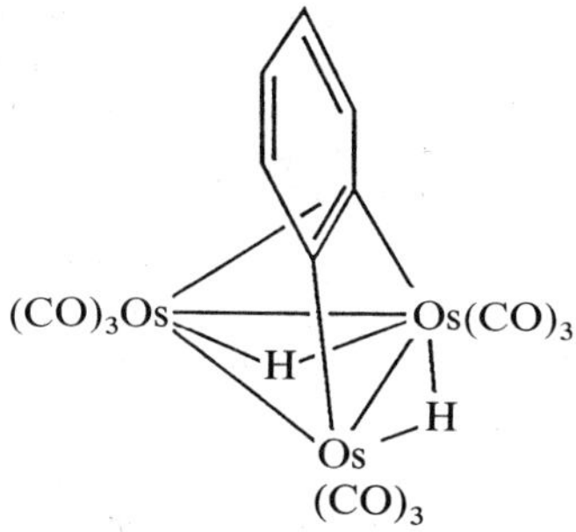

V

The benzene has adjacent η^1-σ bonds to two Os centers and an η^2-π bond to the third Os center. The proposed structure can also be interpreted as a coordinated benzyne.

Homogeneous catalytic activation of C—H bonds occurs by initial cleavage of a C—H bond to give a C—M—H species. The C—M bond is then cleaved with a reagent such as D_2 or Cl_2 to give a substituted product.

(S. D. ITTEL)

1. G. P. Pez, C. F. Putnik, S. L. Suib, G. D. Stucky, *J. Am. Chem. Soc., 101*, 6933 (1979).
2. P. J. Walsh, F. J. Hollander, R. G. Bergman, *J. Am. Chem. Soc., 110*, 8729 (1988).
3. C. C. Cummins, S. M. Baxter, P. T. Wolczanski, *J. Am. Chem. Soc., 110*, 8731 (1988).
4. K. I. Gell, J. Schwartz, *J. Am. Chem. Soc., 103*, 2687 (1981).
5. C. Gianotti, M. L. H. Green, *J. Chem. Soc., Chem. Commun.*, 1114 (1972).
6. M. Berry, K. Elmitt, M. L. H. Green, *J. Chem. Soc., Dalton Trans.*, 1950 (1979).
7. M. L. H. Green, P. J. Knowles, *J. Chem. Soc., Chem. Commun.*, 1677 (1970).
8. M. L. H. Green, P. J. Knowles, *J. Chem. Soc., A*, 1508 (1971).
9. B. R. Francis, M. L. H. Green, G. G. Roberts, *J. Chem. Soc., Chem. Commun.*, 1290 (1971).
10. N. J. Cooper, M. L. H. Green, R. Mahtab, *J. Chem. Soc., Dalton Trans.*, 1557 (1979).
11. K. L. Tang Wong, J. L. Thomas, H. H. Brintzinger, *J. Am. Chem. Soc., 96*, 3694 (1974).
12. M. V. Baker, L. D. Field, *J. Am. Chem. Soc., 108*, 7433 (1986).
13. S. D. Ittel, C. A. Tolman, A. D. English, J. P. Jesson, *J. Am. Chem. Soc., 98*, 6073 (1976).
14. C. A. Tolman, S. D. Ittel, A. D. English, J. P. Jesson, *J. Am. Chem. Soc., 101*, 1742 (1979).
15. J. Chatt, J. M. Davidson, *J. Chem. Soc.*, 843 (1965).
16. H. Werner, J. Gotzig, *J. Organomet. Chem., 284*, 73 (1985).
17. R. H. Morris, M. Shiralian, *J. Organomet. Chem., 260*, C47 (1984).
18. H. Werner, H. Kletzin, K. Roder, *J. Organomet. Chem., 355*, 401 (1988).
19. U. A. Gregory, S. D. Ibekwe, B. T. Kilbourn, D. R. Russell, *J. Chem. Soc., A*, 1118 (1971).
20. P. J. Desrosiers, R. S. Shinomoto, T. C. Flood, *J. Am. Chem. Soc., 108*, 7964 (1986).
21. T. Gregory, P. Harper, R. S. Shinomoto, M. A. Deming, T. C. Flood, *J. Am. Chem. Soc., 110*, 7915 (1988).
22. R. Werner, H. Werner, *Angew. Chem., Int. Ed. Engl., 20*, 793 (1981).
23. J. A. Bandy, M. L. H. Green, D. O'Hare, K. Prout, *J. Chem. Soc., Chem. Commun.*, 1402 (1984).
24. A. J. Deeming, M. Underhill, *J. Chem. Soc., Dalton Trans.*, 1415 (1974).
25. W. D. Jones, F. J. Feher, *J. Am. Chem. Soc., 106*, 1650 (1984).
26. R. A. Periana, R. G. Bergman, *Organometallics, 3*, 508 (1984).

27. C. K. Ghosh, W. A. G. Graham, *J. Am. Chem. Soc.*, *109*, 4726 (1987).
28. C. K. Ghosh, D. P. S. Rodgers, W. A. G. Graham, *J. Chem. Soc., Chem. Commun.*, 1511 (1988).
29. C. Bianchini, A. Meli, M. Peruzzini, F. Zanobini, *J. Chem. Soc., Chem. Commun.*, 971 (1987).
30. C. Bianchini, D. Masi, A. Meli, M. Peruzzini, F. Zanobini, *J. Am. Chem. Soc.*, *110*, 6411 (1988).
31. C. Bianchini, M. Peruzzini, F. Zanobini, *J. Organomet. Chem.*, *326*, C79 (1987).
32. A. H. Janowicz, R. G. Bergman, *J. Am. Chem. Soc.*, *104*, 352 (1982).
33. A. H. Janowicz, R. G. Bergman, *J. Am. Chem. Soc.*, *105*, 3929 (1983).
34. J. K. Hoyano, W. A. G. Graham, *J. Am. Chem. Soc.*, *104*, 3723 (1982).
35. J. K. Hoyano, A. D. McMaster, W. A. G. Graham, *J. Am. Chem. Soc.*, *105*, 7190 (1983).
36. M. D. Rausch, R. G. Gastinger, S. A. Gardner, R. K. Brown, J. S. Wood, *J. Am. Chem. Soc.*, *99*, 7870 (1977).
37. W. D. McGhee, R. G. Bergman, *J. Am. Chem. Soc.*, *110*, 4246 (1988).
38. W. D. McGhee, F. J. Hollander, R. G. Bergman, *J. Am. Chem. Soc.*, *110*, 8428 (1988).
39. M. J. Wax, J. M. Stryker, J. M. Buchanan, C. A. Kovac, R. G. Bergman, *J. Am. Chem. Soc.*, *106*, 1121 (1984).
40. J. M. Buckanan, J. M. Stryker, R. G. Bergman, *J. Am. Chem. Soc.*, *108*, 1537 (1986).
41. M. D. Fryzuk, P. A. MacNeil, N. T. McManus, *Organometallics*, *6*, 882 (1987).
42. P. A. Chetcuti, M. F. Hawthorne, *J. Am. Chem. Soc.*, *109*, 942 (1987).
43. J. Fornies, M. Green, J. L. Spencer, F. G. A. Stone, *J. Chem. Soc., Dalton Trans.*, 1006 (1977).
44. M. Hackett, J. A. Ibers, P. Jernakoff, G. M. Whitesides, *J. Am. Chem. Soc.*, *108*, 8094 (1986).
45. M. Hackett, J. A. Ibers, G. M. Whitesides, *J. Am. Chem. Soc.*, *110*, 1436 (1988).
46. M. Hackett, G. M. Whitesides, *J. Am. Chem. Soc.*, *110*, 1449 (1988).
47. R. L. Brainard, W. R. Nutt, T. R. Lee, G. M. Whitesides, *Organometallics*, *7*, 2379 (1988).
48. G. W. Parshall, *Acc. Chem. Res.*, *8*, 113 (1975).
49. A. E. Shilov, A. A. Shteinman, *Coord. Chem. Rev.*, *24*, 97 (1977).
50. G. W. Parshall, in *Catalysis*, C. Kemball, ed., Specialist Periodical Reports, Vol. 1, The Chemical Society, London, 1977, p. 335.
51. D. E. Webster, *Adv. Organomet. Chem.*, *15*, 147 (1977).
52. G. W. Parshall, *Homogeneous Catalysis*, Wiley Interscience, New York, 1980.
53. E. L. Muetterties, *Chem. Soc. Rev.*, *11*, 283 (1982).
54. R. G. Bergman, *Science*, *223*, 902 (1984).
55. A. E. Shilov, *Activation of Saturated Hydrocarbons by Transition Metal Complexes*, Reidel, Dordrecht, 1985.
56. R. H. Crabtree, *Chem. Rev.*, *85*, 245 (1985).
57. J. Halpern, *Inorg. Chim. Acta*, *100*, 41 (1985).
58. J. Halpern, in *Fundamental Research in Homogeneous Catalysis*, Vol. 1, A. E. Shilov, ed., Gordon and Breach, New York, 1986, p. 383.
59. C. A. Tolman, S. D. Ittel, A. D. English, J. P. Jesson, *J. Am. Chem. Soc.*, *100*, 7577 (1978).
60. C. A. Tolman, S. D. Ittel, A. D. English, J. P. Jesson, *J. Am. Chem. Soc.*, *100*, 4080 (1978).
61. E. Elmitt, M. L. H. Green, R. A. Forder, I. Jefferson, K. Prout, *J. Chem. Soc., Chem. Commun.*, 747 (1974).

5.8.2.6.6. η^1 σ-Aryl–Metal Complexes by Electrophilic Attack on Arenes.

One important mechanism for homogeneous catalytic activation of aromatic C—H bonds[1] is electrophilic attack by transition-metal complexes on the aromatic substrates. It is presumed η^1-aryl complexes are important intermediates in these reactions[2], but they are rarely isolated. Direct electrophilic metallation of aromatic substrates is closely related to reactions observed with nontransition metals (§5.6., auration[3,4]; §5.7.2., mercuration[5,6] and §5.3., thallation[7,8]). References to metal–aryl complexes synthesized by electrophilic attack on arenes by transition metals are summarized in Table 1. Reviews are available[1,2,19–28].

The first observation[14] of η^1-aryl transition-metal species by direct electrophilic metallation in the reaction of benzene with H_2PtCl_6 and $Na_2[PtCl_4]$ in aq CF_3CO_2H

TABLE 1. EXAMPLES OF σ-ARYL COMPLEXES SYNTHESIZED BY ELECTROPHILIC ATTACK ON ARENES

Starting complex[a]	Substrate	Product[a]	Ref.
$ScCp^*_2Me$	C_6H_6	$ScCp^*_2(C_6H_5)$	9
$ScCp^*_2Me(THF)$	C_5H_5N	$ScCp^*_2(C, N\text{-}\eta^2\text{-}C_5H_4N)$	9
$[ScCp^*_2H]_n$	C_5H_5N	$ScCp^*_2(C, N\text{-}\eta^2\text{-}C_5H_4N)$	9
$LuCp^*_2H$	C_6H_6	$LuCp^*_2(C_6H_5)$	10
$LuCp^*_2(C_6H_5)$	C_6H_6	$LuCp^*_2(C_6H_4)LuCp^*_2$	10
$LuCp^*_2H$	C_5H_5N	$LuCp^*_2(C, N\text{-}\eta^2\text{-}C_5H_4N)$	10
$LuCp^*_2Me$	C_6H_6	$LuCp^*_2(C_6H_5)$	10
$RhCp^*Me_2(Me_2SO)$	C_6H_6	$RhCp^*Me(Me_2SO)(C_6H_5)$	11
$IrCp^*Me_2(Me_2SO)$	C_6H_6	$IrCp^*Me(Me_2SO)(C_6H_5)$	12
	C_6H_5R R = I, NO_2, CF_3, Br, OMe, Cl, F, Me	$IrCp^*Me(Me_2SO)(C_6H_4R)$	12
$IrCp^*Me(Me_2SO)(C_6H_5)$	C_6H_6	$IrCp^*(Me_2SO)(C_6H_5)_2$	12
$[(IrCp^*)_2Cl_4] + Al_2Me_6$	C_6H_6	$(IrCp^*(C_6H_5))(\mu\text{-}CH_2)_2(IrCp^*Me)$	13
	C_6H_5Me	$(IrCp^*(m, p\text{-}C_6H_4Me))(\mu\text{-}CH_2)_2(IrCp^*Me)$	13
$Na_2[PtCl_4]$, H_2PtCl_6	C_6H_6	$[H_3O][Pt(H_2O)Cl_4(C_6H_5)]$	14
H_2PtCl_6	C_6H_6	$[H_3O][Pt(H_2O)Cl_4(C_6H_5)]$	15, 16
	$C_{10}H_8$	$[H_3O][Pt(H_2O)Cl_4(2\text{-}C_{10}H_7)]$	15, 16
	$C_{10}H_8$	$[H_3O][Pt(NH_3)Cl_4(2\text{-}C_{10}H_7)]$	16
	$m\text{-}C_6H_4Me_2$	$[H_3O][Pt(NH_3)Cl_4(3,5\text{-}C_6H_4Me_2)]$	16
	$o\text{-}C_6H_4Me_2$	$[H_3O][Pt(NH_3)Cl_4(3,4\text{-}C_6H_4Me_2)]$	16
	$m\text{-}C_6H_4ClMe$	$[H_3O][Pt(NH_3)Cl_4(C_6H_4\text{-}5\text{-}Cl\text{-}3\text{-}Me)]$	16
	C_6H_5R R = Me, Et, Cl, OMe	Inseparable meta–para mixture	15, 16
$H_2PtCl_6 + h\nu$	C_6H_6	$[H_3O][Pt(NH_3)Cl_4(C_6H_5)]$	17
	C_6H_5Me	$[H_3O][Pt(NH_3)Cl_4(p\text{-}C_6H_4Me)]$	17
$ThCp^*_2(CH_2CMe_3)_2$	C_6H_6	$ThCp^*(C_6H_5)_2$	18

[a] Abbreviations: Cp*, pentamethylcyclopentadienyl; C_5H_5N, = py; $C_{10}H_8$, naphthalene.

was followed by isolation and X-ray characterization[15,16] of the η^1-(2-naphthyl) complex, **I**:

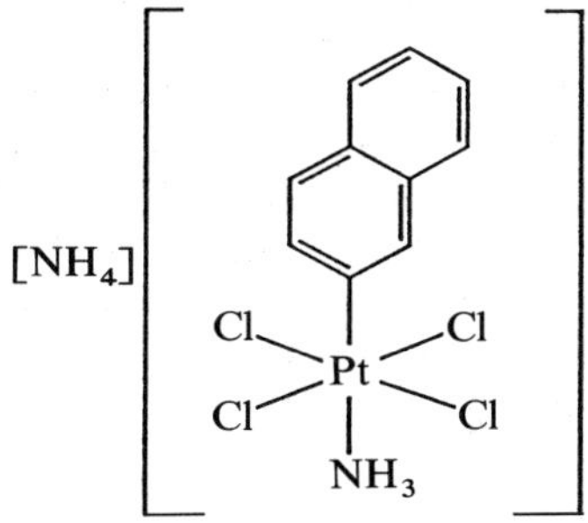

I

The probable product of the reaction is $[H_3O][Pt(H_2O)Cl_4(\eta^1\text{-}2\text{-}C_{10}H_7)]$, but chromatography on silica gel contaminated with NH_3 effected the metathesis of NH_3 for H_2O [16].

Reaction of H_2PtCl_6 with benzene in aq CF_3CO_2H acid gives the phenyl analog of **I** in 40% yield after workup from acetone[16]. Reactions with toluene, ethylbenzene, chlorobenzene and anisole give inseparable mixtures of meta and para products in which the meta/para ratio varies in different runs of the same reaction. As in the C—H oxidative addition reactions in §5.8.2.6.4, no ortho product is observed.

Reaction with m-xylene gives the 3,5-xylyl product, and o-xylene the 3,4-xylyl product, each in 30% yield. Reaction with m-chlorotoluene gives the 5-chloro-3-tolyl product in > 60% yield. As expected, there is no reaction with p-xylene, mesitylene, pentamethylbenzene, p-dimethoxybenzene or p-dichlorobenzene. The products are in good agreement with distributions observed for chlorination and H–D exchange of aromatics in the presence of H_2PtCl_6 [29].

The lanthanides and actinides are also active in electrophilic attack on arenes. Bis(pentamethylcyclopentadienyl)lutetium methyl or hydrido complexes react readily with benzene to give phenyl complexes and methane or dihydrogen[11]. Similar reactions are observed for Sc [9] and Th [19]. In a remarkable reaction[11], benzene can be dimetallated by Lu, yielding **II**:

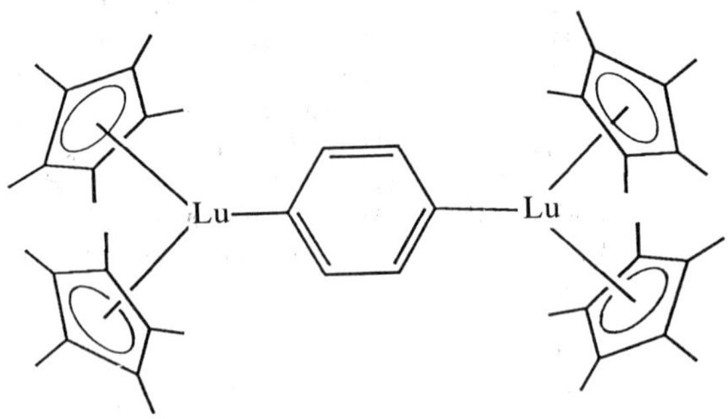

II

The structure of **II** is greatly distorted, probably as a result of electrophilic interaction of the metal with the ortho C—H bonds of the arene. The high reactivity of $LuCp^*_2Me$ and YCp^*_2Me is further demonstrated by the first well-characterized organometallic reaction with saturated hydrocarbons including methane[28,30]. The formation of η^1-σ-alkyls by direct hydrocarbon activation is covered in §5.8.2.2.1.

Electrophilic addition of higher valent transition-metal species to aromatic substrates is more widespread than current syntheses indicate. All six Pt group metals catalyze H–D exchange between naphthalene and D_2O–CH_3COOD [31]. A variety of others, including $TaCl_5$, $MoCl_5$ and WCl_6, catalyze H–D exchange between PhMe and C_6D_6 [32]. Under photochemical conditions, $Re(\eta^6\text{-}C_6H_6)(PPh_3)_2H$ catalyzes H–D exchange between C_6D_6 and other arenes[33]. These and other metals should ultimately yield isolable η^1-aryl species by electrophilic addition.

(S. D. ITTEL)

1. G. W. Parshall, *Acc. Chem. Res.*, *8*, 113 (1975).
2. A. E. Shilov, A. A. Shteinman, *Coord. Chem. Rev.*, *24*, 97 (1977).
3. M. S. Kharasch, H. S. Isbell, *J. Am. Chem. Soc.*, *53*, 3053 (1931).
4. K. S. Liddle, C. Parkin, *J. Chem. Soc.*, *Chem. Commun.*, 26, (1972).
5. O. Dimoth, *Chem. Ber.*, *32*, 758 (1899).
6. F. C. Whitemore, E. B. Middleton, *J. Am. Chem. Soc.*, *43*, 622 (1921).
7. H. Gilman, R. K. Abbott, *J. Am. Chem. Soc.*, *65*, 122 (1943).
8. A. McKillop, J. S. Fowler, M. J. Zelesko, J. D. Hunt, E. C. Taylor, G. McGillivray, *Tetrahedron Lett.*, 2427 (1969).
9. M. E. Thompson, J. E. Bercaw, *Pure Appl. Chem.*, *56*, 1 (1984).
10. P. L. Watson, *J. Chem. Soc.*, *Chem. Commun.*, 276 (1983).
11. M. Gomez, D. J. Robinson, P. M. Maitlis, *J. Chem. Soc.*, *Chem. Commun.*, 825 (1983).
12. M. Gomez, D. J. Robinson, P. M. Maitlis, *J. Chem. Soc.*, *Chem. Commun.*, 825 (1983).
13. K. Isobe, A. Vazquez de Miguel, P. M. Maitlis, *J. Organomet. Chem.*, *250*, C25 (1983).
14. N. F. Gol'dshleger, V. M. Nekipelov, A. T. Nikitaev, K. I. Zamaraev, A. E. Shilov, A. A. Shteinman, *Kinet. Catal.*, *Engl. Transl.*, *20*, 445 (1979).
15. G. P. Shul'pin, L. P. Rosenberg, R. P. Shibaeva, A. E. Shilov, *Kinet. Catal.*, 1570 (1979); *Kinet. Catal.*, *Engl. Transl.*, *20*, 1296 (1979).
16. G. B. Shul'pin, A. E. Shilov, A. N. Kitaigorodskii, J. V. Zeile Krevor, *J. Organomet. Chem.*, *201*, 319 (1980).
17. G. B. Schul'pin, G. V. Nizova, A. E. Shilov, *J. Chem. Soc.*, *Chem. Commun.*, 671 (1983).
18. P. J. Fagan, J. M. Manriquez, E. A. Maatta, A. M. Seyam, T. J. Marks, *J. Am. Chem. Soc.*, *103*, 6650 (1981).
19. G. W. Parshall, in *Catalysis*, C. Kemball, ed., Specialist Periodical Reports, Vol. 1, The Chemical Society, London, 1977, p. 335.
20. D. E. Webster, *Adv. Organomet. Chem.*, *15*, 147 (1977).
21. G. W. Parshall, *Homogeneous Catalysis*, Wiley Interscience, New York, 1980.
22. A. E. Shilov, *Sov. Sci. Rev.*, *B*, *4*, 71 (1982).
23. E. L. Muetterties, *Chem. Soc. Rev.*, *11*, 283 (1982).
24. M. E. Thompson, J. E. Bercaw, *Pure Appl. Chem.*, *56*, 1 (1984).
25. A. E. Shilov, *Activation of Saturated Hydrocarbons by Transition Metal Complexes*, Reidel, Dordrecht, 1985.
26. R. H. Crabtree, *Chem. Rev.*, *85*, 245 (1985).
27. I. P. Rothwell, *Polyhedron*, *4*, 177 (1985).
28. P. L. Watson, G. W. Parshall, *Acc. Chem. Res.*, *18*, 51 (1985).
29. J. L. Garnett, J. C. West, *Aust. J. Chem.*, *27*, 129 (1974).
30. P. L. Watson, *J. Am. Chem. Soc.*, *105*, 6491 (1983).
31. M. Preece, S. D. Robinson, *Inorg. Chim. Acta*, *29*, L199 (1978).
32. J. L. Garnett, M. A. Long, R. F. W. Vining, T. Mole, *J. Chem. Soc.*, *Chem. Commun.*, 1172 (1972).
33. W. D. Jones, M. Fan, *Organometallics*, *5*, 1057 (1986).

5.8.2.6.7. by Metal-Atom and Related Reactions.

The synthesis of stable π-arene complexes of the transition metals has progressed substantially owing to metal atom chemistry. Particularly rewarding has been work on the production of bis(arene)metal(0) complexes of the early transition metals; e.g., V, Cr, Mo and W yield bis(benzene)V(0), Cr(0), Mo(0) and W(0) complexes[1–6]. In addition, metal-atom reactions have yielded new arene-substituted compounds of V, Cr, Mo and W that cannot be prepared by classical methods[1–13]. Hitherto unknown bis(arene)-metal(0) complexes can be prepared for Ti and Nb[7–9]. The metal atom method is the most versatile (and often the only method) for the preparation of such complexes. Table 1 summarizes the bis(arene)metal(0) complexes prepared to date.

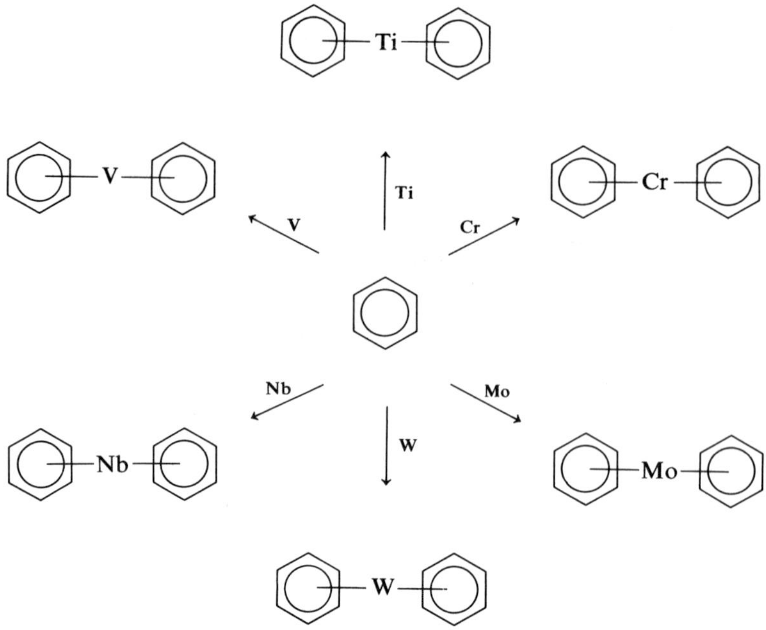

Polynuclear aromatics can also be employed. Napthalene and Cr atoms yield the reactive bis(η^6-napthalene)-chromium(0), which exhibits interesting arene displacement chemistry[14]. Phosphines [PF_3, $P(OCH_3)_3$ and $P(CH_3)_3$] readily displace one π-complexed napthalene, whereas CO and isonitriles are capable of displacing both π-napthalene ligands. 1,4-Dimethylnapthalene yields mainly (95%) complexation of the unsubstituted ring to Cr[14]:

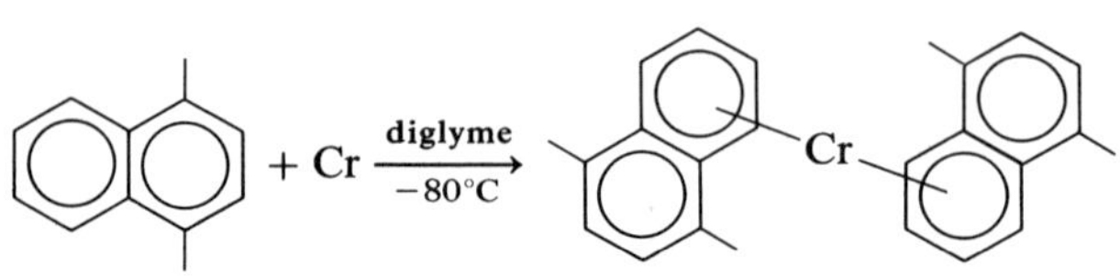

(a)

Polymers containing aryl groups also react efficiently with metal atoms to yield bis(arene)metal(0) complexes within the polymer chain[15]. Thus, poly(methylphenylsiloxanes) react at 0°C as a liquid with Ti, V, Cr, Mo and W atoms to give high yields of colored bis(η^6-arene)metal complexes. In contrast Co, Fe and Ni atoms yield only metal slurries. Similarly, poly(oxyphenylene) and polystyrene (in a solvent) react with V and Cr atoms to yield colored solutions[15]. By adjusting the metal-atom flux, small, polymer-supported Ti and Mo clusters can be prepared[16]. In some cases the complexed metal atoms spontaneously migrate through the polymer fluid to form dimers[16].

Mixed arene–arene′, arene–ligand and pi sandwich compounds of the early transition metals have also been prepared and isolated. Mixtures of C_6F_6–C_6H_6 with Cr atoms yield the predicted products[10,17]. However, the $(C_6F_6)_2Cr$ compound is not thermally

TABLE 1. BIS(ARENE)METAL(0) COMPLEXES PREPARED BY METAL-ATOM REACTIONS

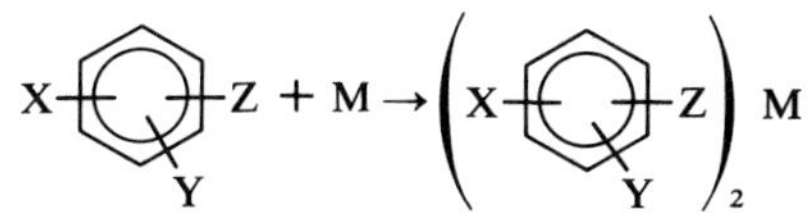

Metal	X	Y	Z	Ref.
Ti	H	H	H	7, 8
	CH_3	H	H	7, 8
	CH_3	CH_3	CH_3 (1,3,5)	7, 8
Nb	H	H	H	9
	CH_3	H	H	9
	CH_3	CH_3	CH_3 (1,3,5)	9
V	H	H	H	1
	F	H	H	1
	Cl	H	H	1
	CF_3	H	H	1
	F	F (o, m, p)[a]	H	1
Cr	H	H	H	2, 3
	F	H	H	1, 4, 5
	Cl	H	H	4, 5
	CF_3	H	H	
	$COOCH_3$	H	H	4, 5
	CH_3	H	H	4, 5
	C_2H_5	H	H	4, 5, 10
	$CH(CH_3)_2$	H	H	10
	F	F (o, m, p)[a]	H	1, 4, 5, 10, 11
	Cl	CF_3 (o, m, p)[a]	H	1
	CF_3	CF_3 (m, p)[a]	H	1
	$CH(CH_3)_2$	$CH(CH_3)_2$	H	10
	OC_6H_5	H	H	36
	$CH(CH_2)_2$ (Cyclopropyl benzene)	H	H	37
		$—CH_2CH_2CH_2CH_2—$[b]		12
		napthalene, methyl napthalenes		14
Mo	H	H	H	6
	F	H	H	6
	Cl	H	H	6
	CH_3	H	H	6
	OCH_3	H	H	6
	$N(CH_3)_2$	H	H	6
	$COOCH_3$	H	H	6
W	H	H	H	6, 13
	F	H	H	6
	CH_3	H	H	6, 13
	OCH_3	H	H	6
	CH_3	CH_3 (o)	H	6
	CH_3	CH_3	CH_3 (1,3,5)	13

[a] X-ray structures completed[38–41].

[b]

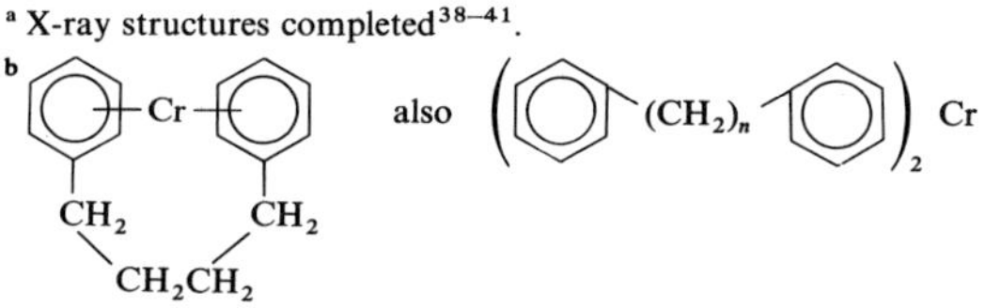

stable, and $(C_6H_6)_2Cr$ is not air stable. Therefore, by warming the matrix and then air exposing the product mixture, the mixed arene complex can be isolated[17]:

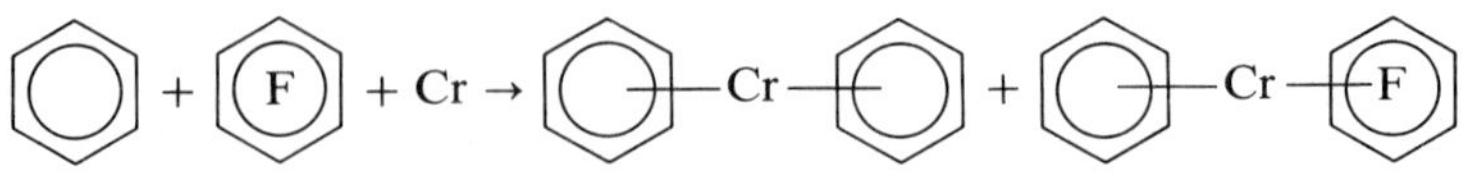

(b)

In a similar way, $C_6H_6CrC_6F_5H$ can be prepared[18]. The H of the complexed C_6F_5H group is readily abstracted by base. Therefore the lithiated compound is readily available, and derivatives [$Sn(CH_3)_3$, $Si(CH_3)_3$, CO_2Li, $C(CH_3)_2OH$ and CH_2OH)][18] prepared from it and studied by ^{19}F NMR and IR[19,20].

Arene–PF_3 and arene–cyclopentadiene mixtures with metal atoms yield mixed π-arene–ligand are stabilized complexes; e.g., C_6F_6–PF_3–Cr yields[10] $C_6F_6Cr(PF_3)_3$ and C_6H_6–C_5H_6–W yields[6] C_6H_6 W(H) C_5H_5. Similarly, Hf and Zr codeposited with arenes and $P(CH_3)_3$ yield the zero-valent Hf and Zr complexes, $(arene)_2Hf[CH_3)_3]$ and $(arene)_2Zr[P(CH_3)_3]$ (arene = benzene or toluene[21], where the arenes are in bent bonding modes[21]. Generally, the arene and ligand are codeposited together with the metal atoms, especially if the ligand is volatile. However, for the later transition metals, pseudo-stable π-arene-metal complexes can be prepared and the second ligand added later, e.g., to yield $(\pi\text{-arene})Fe(L)_2$ complexes[10,22-26]:

(c)

where L = PF_3, BD bipy, disphos, phen, $P(OCH_3)_3$, ChxD.

Table 2 summarizes the mixed arene–arene′ and arene–ligand complexes prepared. π-Arene complexes of polymers are also listed. Iron[24,26] and Ru[27] with benzene or toluene yield π-arene complexes that can be isolated at low temperature. The $(C_6H_6)_2Ru$ species is stable to ~0°C, whereas the $C_6H_5CH_3$–Fe analog is stable to ~ −50°C. Thus, these compounds can be isolated for use as intermediates in other reactions, e.g., the addition of cyclopentadiene to the $(benzene)_nFe$ complex, where on warming hydrogen transfers to benzene to yield (η^5-cyclopentadienyl) (η^5-cyclohexadienyl)Fe[28]:

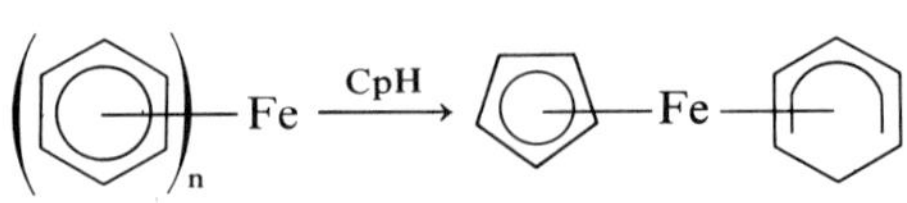

(d)

TABLE 2. MIXED ARENE AND ARENE–LIGAND COMPLEXES OF THE TRANSITION METALS PREPARED BY METAL-ATOM REACTIONS

Metal	Arene	Arene′ or L	Isolated products	Ref.
Hf	C_6H_6	$P(CH_3)_3$	$(C_6H_6)_2Hf[P(CH_3)_3]$	21
Hf	$C_6H_5CH_3$	$P(CH_3)_3$	$(C_6H_5CH_3)_2Hf[P(CH_3)_3]$	21
Zr	C_6H_6	$P(CH_3)_3$	$(C_6H_6)_2Zr[P(CH_3)_3]$	21
Zr	$C_6H_5CH_3$	$P(CH_3)_3$	$(C_6H_5CH_3)_2Zr[P(CH_3)_3]$	21
Cr	C_6H_6	C_6F_6	$(C_6H_6)Cr(C_6F_6)$	10, 17, 20
Cr	C_6H_6	C_6F_5H	$(C_6H_6)Cr(C_6F_5H)$ [a]	18, 19
Cr	C_6H_6	PF_3	$(C_6H_6)Cr(PF_3)_3$	10
Cr	C_6F_6	PF_3	$(C_6F_6)Cr(PF_3)_3$	10
Cr	$C_6H_5CH(CH_3)_2$	PF_3	$[C_6H_5CH(CH_3)_2]Cr(PF_3)_3$	10
Cr	$C_6H_3(CH_3)_3$	PF_3	$[C_6H_3(CH_3)_3]Cr(PF_3)_3$	10
Cr	C_5H_5N	PF_3	$(C_5H_5N)Cr(PF_3)_3$ [b] $(C_5H_5N)Cr(PF_3)_5$ [c]	42
Cr	$C_6H_5CH{=}CH_2$	PF_3	$(C_6H_5CH{=}CH_2)Cr(PF_3)_3$	43
Cr	$2,6\text{-}(CH_3)_2C_5H_3N$	—	$[(CH_3)_2C_5H_3N]_2Cr$ [d]	44, 45
W	C_6H_6	C_5H_6	$(C_6H_6)W(H)(C_5H_5)$	6
Mn	C_6H_6	C_5H_6	$(C_6H_6)Mn(C_5H_5)$	46
Fe	C_6H_6	PF_3	$(C_6H_6)Fe(PF_3)_2$	10, 22
Fe	$C_6H_5CH_3$	PF_3	$(C_6H_5CH_3)Fe(PF_3)_2$	23
Fe	$C_6H_5CH_3$	C_4H_6 (1,3-butadiene)	$(C_6H_5CH_3)Fe(C_4H_6)$	23
Fe	C_6H_6	C_6H_8 (1,3-cyclohexadine)	C_6H_6	25
Fe	$C_6H_5CH_3$	$C_5H_4N\text{-}C_5H_4N$ (2,2-bipyridine)	$(C_6H_5CH_3)Fe(C_5H_4N)_2$	24
Fe	$C_6H_5CH_3$	$Ph_2PCH_2CH_2PPh_2$ (diphos)	$(C_6H_5CH_3)Fe(Ph_2PCH_2CH_2PPh_2)$	24
Ti, V, Cr, Mo		Siloxane polymer	π-Arene-complexed siloxane polymers (colored oils)	47

[a] This compound can be derivatized readily (see text). One of the interesting derivatives prepared is:

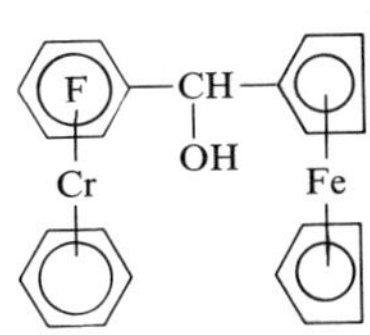

[b] π-C_5H_5N.
[c] σ-C_5H_5N through the N.
[d] Two crystalline forms isolated. X-ray crystal structures completed.

Similarly, Co and Ni form π-arene complexes at low temperatures, but these are not thermally stable enough for isolation. Nevertheless, they have been employed extensively as intermediates for the preparation of highly dispersed Co and Ni catalysts (both unsupported and supported on MgO, SiO_2, Al_2O_3 and C)[29–33]. These solvated metal-atom dispersed (SMAD) catalysts are highly active and often selective for methanation, isomerization, hydrogenation, hydrogenolysis and dehydrogenation reactions. Furthermore, these toluene-solvated metal atoms[29] are useful for the preparation of other organometallic species, such as phosphine or phosphite complexes[34]:

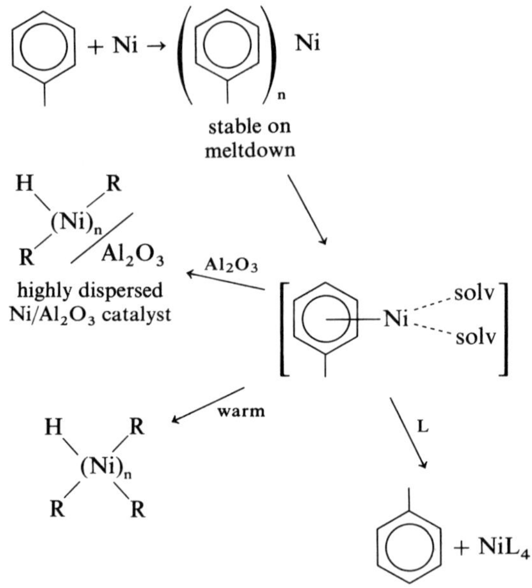

where L = $P(OCH_2CH_3$, $P(CH_2CH_3)_3$, $P(C_6H_5)_3$.

Other substituted arenes have been employed; e.g., C_6F_6 with Ni forms an explosive polymeric complex that is highly reactive[35] with CO, $P(OCH_2CH_3)_3$ and H_2. Further, mesitylene reacts with Fe atoms to form a π-arene complex that is rich in hydrogen, abstracting hydrogen from xs mesitylene so that an 18-electron complex is the final product[22]:

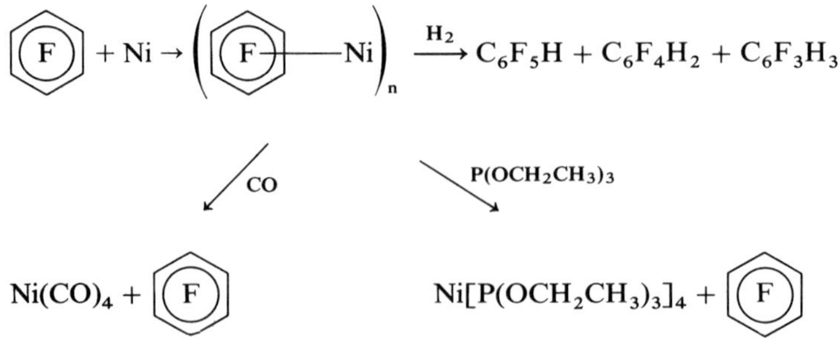

(e)

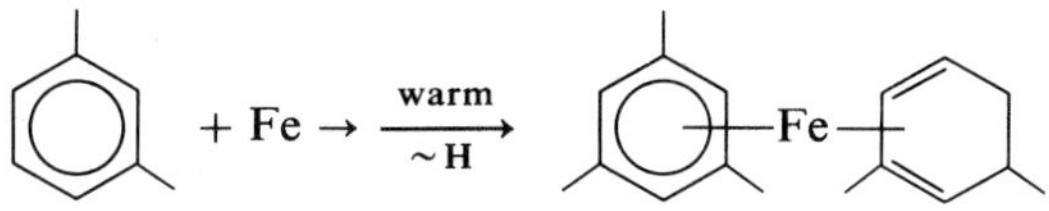

η^6- η^4

(f)

Refer to §5.8.2.9.3 for a discussion of the formation of π-arene complexes subsequent to oxidative addition–disproportionation–metal-atom reactions, e.g.:

$$C_6F_5Br + Ni \rightarrow C_6F_5NiBr$$

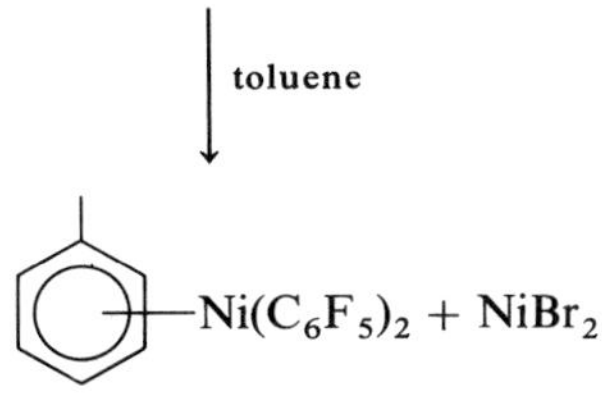

(K. J. KLABUNDE)

1. K. J. Klabunde, H. F. Efner, *Inorg. Chem., 14*, 789 (1975).
2. P. L. Timms, *J. Chem. Soc., Chem. Commun.*, p. 1033 (1969).
3. P. L. Timms, *J. Chem. Educ., 49*, 782 (1972).
4. P. S. Skell, D. L. Williams-Smith, M. J. McGlinchey, *J. Am. Chem. Soc., 95*, 3337 (1973).
5. J. A. Gladysz, J. G. Fulcher, S. J. Lee, A. B. Bocarsley, *Tetrahedron Lett., 39*, 3421 (1977).
6. M. P. Silvon, E. M. Van Dam, P. S. Skell, *J. Am. Chem. Soc., 96*, 1945 (1974).
7. F. W. S. Benfield, M. L. H. Green, J. S. Ogden, D. Young, *J. Chem. Soc., Chem. Commun.*, 866 (1973).
8. M. T. Anthony, M. L. H. Green, D. Young, *J. Chem. Soc., Dalton Trans.*, 1419 (1975).
9. F. G. Cloke, M. L. H. Green, D. H. Price, *J. Chem. Soc., Chem. Commun.*, 431 (1978).
10. R. Middleton, J. R. Hull, S. R. Simpson, C. H. Tomlinson, P. L. Timms, *J. Chem. Soc., Dalton Trans.*, 120 (1973).
11. H. F. Efner, unpublished results from this laboratory.
12. A. N. Neameyanov, N. N. Zaitseva, G. A. Dormrachev, V. D. Zinovev, L. P. Yureva, I. I. Tverdikhlebova, *J. Organomet. Chem., 121* (3), C52 (1976).
13. F. G. Cloke, M. L. H. Green, G. E. Morris, *J. Chem. Soc., Chem. Commun.*, 72 (1978).
14. E. P. Kundig, P. L. Timms, *J. Chem. Soc., Dalton Trans.*, 991 (1980).
15. C. G. Francis, P. L. Timms, *J. Chem. Soc., Dalton Trans.*, 1401 (1980).
16. C. G. Francis, H. Huber, G. A. Ozin, *Inorg. Chem., 19*, 219 (1980).
17. M. J. McGlinchey, T. S. Tan, *J. Am. Chem. Soc., 98*, 2271 (1976).
18. A. Agarwal, M. J. McGlinchey, T. S. Tan, *J. Organomet. Chem., 141*, 85 (1977).
19. M. J. McGlinchey, N. Hao, B. G. Sayer, T. S. Tan, *J. Organomet. Chem., 194*, 325 (1980).
20. J. D. Laposa, N. Hao, B. G. Soyer, M. J. McGlinchey, *J. Organomet. Chem., 195*, 193 (1980).
21. F. G. N. Cloke, M. L. H. Green, *J. Chem. Soc., Chem. Commun.*, 127 (1979).
22. P. L. Timms, *Angew. Chem., Int. Ed. Engl., 14*, 273 (1975).
23. D. L. Williams-Smith, L. R. Wolf, P. S. Skell, *J. Am. Chem. Soc., 94*, 4042 (1972).
24. L. J. Radonovich, M. W. Eyring, T. J. Groshens, K. J. Klabunde, *J. Am. Chem. Soc., 104*, 2816 (1982).
25. S. D. Ittel, C. A. Tolman, *J. Organomet. Chem., 172*, C47 (1979).
26. L. K. Beard, M. P. Silvon, P. S. Skell, *J. Organomet. Chem., 209*, 245 (1981).
27. P. L. Timms, R. B. King, *J. Chem. Soc., Chem. Commun.*, 898 (1978).
28. S. D. Ittel, *J. Organomet. Chem., 195*, 331 (1980).

29. K. J. Klabunde, H. F. Efner, T. O. Murdock, T. Ropple, *J. Am. Chem. Soc.*, *98*, 1021 (1976).
30. S. C. Davis, K. J. Klabunde, *J. Am. Chem. Soc.*, *100*, 5973 (1978).
31. K. J. Klabunde, S. Davis, H. Hattori, Y. Tanaka, *J. Catal.*, *54*, 254 (1978).
32. K. J. Klabunde, D. Ralston, R. Zoellner, H. Hattori, Y. Tanaka, *J. Catal.*, *55*, 213 (1978).
33. K. Matsuo, K. J. Klabunde, *J. Catal.*, *73*, 216 (1982).
34. R. G. Gastinger, K. J. Klabunde, *Transition Met. Chem.*, *4*, 1 (1979).
35. K. J. Klabunde, H. F. Efner, *J. Fluorine Chem.*, *4*, 114 (1974).
36. S. Togashi, J. G. Fulcher, B. R. Cho, M. Hasegawa, J. A. Gladysz, *J. Org. Chem.*, *45*, 3044 (1980).
37. A. J. L. Hanlan, R. C. Ugolick, J. G. Fulcher, S. Togashi, A. B. Bocarsly, J. A. Gladysz, *Inorg. Chem.*, *19*, 1543 (1980).
38. L. Radonovich, C. Zuerner, H. F. Efner, K. J. Klabunde, *Inorg. Chem.*, *15*, 2976 (1976).
39. L. Radonovich, C. Zuerner, unpublished work.
40. V. Graves, J. J. Lagowski, *J. Organomet. Chem.*, *120*, 397 (1976).
41. V. Graves, J. J. Lagowski, *Inorg. Chem.*, *15*, 577 (1976).
42. P. L. Timms, in *Cryochemistry*, M. Moskovits, G. A. Ozin, eds., Wiley (Interscience), New York, 1976, p. 61.
43. J. R. Blackboro, R. Grubbs, A., Miyashita, A. Scrivanti, *J. Organomet. Chem.*, *120* (3), C49 (1976).
44. L. H. Simons, P. E. Riley, R. E. Davis, J. J. Lagowski, *J. Am. Chem. Soc.*, *98*, 1044 (1976).
45. P. E. Riley, R. E. Davis, *Inorg. Chem.*, *15*, 2735 (1976).
46. E. A. Koerner von Gustorf, O. Jaenicke, O. Wolfbeis, C. R. Eady, *Angew. Chem., Int. Ed. Engl.* *14*, 278 (1975).
47. C. G. Francis, P. L. Timms, *J. Chem. Soc., Chem. Commun.*, p. 466 (1977).

5.8.2.7. from Alkynes

5.8.2.7.1. η^2 π-Alkyne Complexes from Reaction of Metal Complexes with Alkynes by Ligand Displacement

η^2-Alkyne complexes are obtained from reactions of transition-metal complexes with alkynes by ligand displacement[1,2]. The relative influence of, e.g., the substituents on the acetylene, the oxidation state of the metal and the additional coordinated ligands can be assessed. Ligands such as CO, PR_3, olefins, dienes and halogens, can be displaced by alkynes under thermal or photolytic conditions.

(i) Titanium. Displacement of CO by diphenylacetylene in $(\eta^5\text{-Cp})_2Ti(CO)_2$ (at RT in heptane) gives rise to $(\eta^5\text{-Cp})_2Ti(CO)(\eta^2\text{-PhC}_2Ph)$, which is catalyst for the hydrogenation of acetylenes, dienes and olefins[3]. This compound is indefinitely stable as a solid under an inert atmosphere but in solution slowly redistributes to $(\eta^5\text{-Cp})_2Ti(CO)_2$ and the titanocyclopentadienyl derivative, $(\eta^5\text{-Cp})_2Ti(PhC_2Ph\text{-}\eta^2)_2$.

(ii) Vanadium, Niobium and Tantalum. Removal of 2 mol of CO from $\eta^5\text{-CpV(CO)}_4$ by HC_2R ($R = n\text{-}C_3H_7$, $n\text{-}C_4H_9$ or $t\text{-}C_4H_9$) under UV irradiation leads to the formation of $\eta^5\text{-CpV(CO)}_2(\eta^2\text{-HC}_2R)$[4]; although unstable products are obtained with PhC_2Ph and PhC_2H, stable $\eta^5\text{-CpV(CO)(PPh}_3)(\eta^2\text{-PhC}_2Ph)$ is synthesized when $\eta^5\text{-CpV(CO)}_3(PPh_3)$ is the starting material.

Further reaction of $\eta^5\text{-CpV(CO)}_2(\eta^2\text{-RC}_2R)$ with another mole of alkyne gives $\eta^5\text{-CpV(CO)}(\eta^2\text{-RC}_2R)_2$; similar behavior occurs also for Ta and Nb compounds, but the nature of the final products may be different[2,5]:

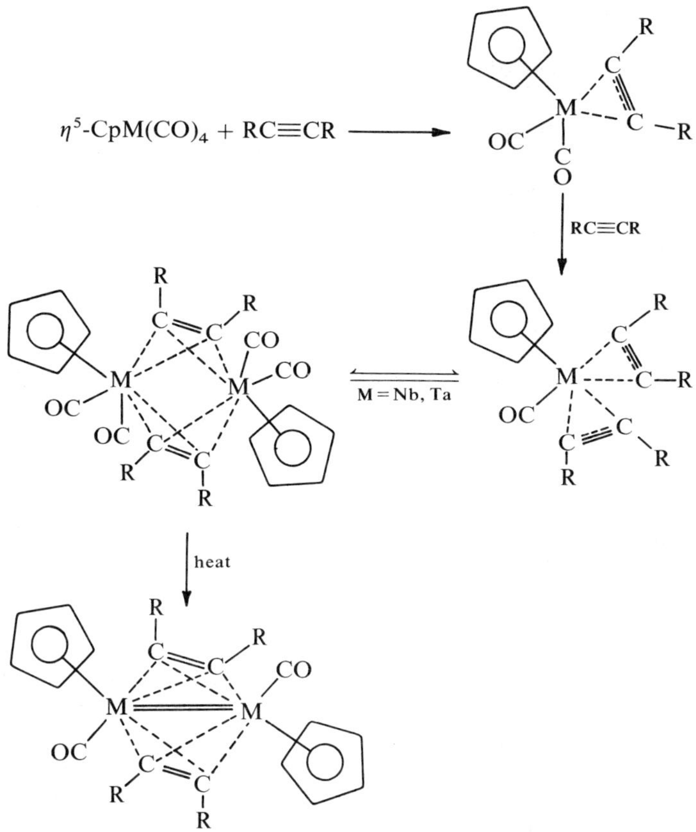

(a)

Iodotantalum (acetylene) derivatives:

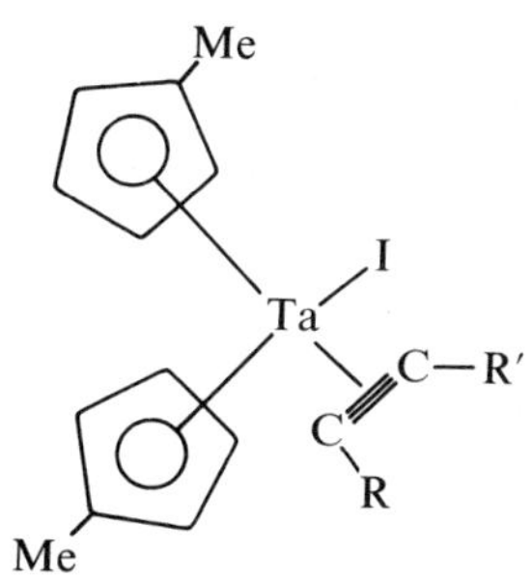

$R = R' = n\text{-}C_3H_7$
$R = CH_3,\ R' = n\text{-}C_4H_9$
$R = CH_3,\ R' = i\text{-}C_3H_7$
$R = CH_3,\ R' = t\text{-}C_4H_9$

are obtained when $(MeCp\text{-}\eta^5)_2TaH_3$ is reacted with RC≡CR′ in the presence of C_6H_5I (refluxing dioxane for 10 min under an inert atmosphere)[6].

(η^2-Alkyne)hydrido complexes are obtained from $(\eta^5\text{-}Cp)_2MH_3$ (M = Ta, Nb)[6,7] and alkynes in contrast to the expected σ-alkenyl derivatives found by M—H insertion (see §5.8.2.7.3). Milder reaction conditions are required for the Nb complex, which can be obtained by refluxing $(\eta^5\text{-Cp})_2NbH_3$ and the dialkylacetylene in benzene for 15 min.

(iii) Chromium, Molybdenum and Tungsten. No simple $(\eta^2$-alkyne)Cr complex can be obtained from $Cr(CO)_6$; but derivatives in which some CO groups are replaced by other ligands are more reactive with alkynes; e.g., the photosensitive $Cr(CO)_2(\eta^2\text{-alkyne})_2$ complex is synthesized at low temperature in the dark from $Cr(CO)_3(MeCN)_3$ and MeSC≡CSMe [8].

The stability of the complexes $(\eta^6\text{-arene})Cr(CO)_2(\eta^2\text{-}RC_2R)$[9] and $(\eta^5\text{-Cp})Cr\text{-}(CO)(NO)(\eta^2\text{-}RC_2R)$[10] depends on the substituents on the arene and acetylenic carbons.

η^2-Alkyne complexes of Mo and W in their low oxidation states do not form oligomers, and compounds containing two or three alkyne molecules not linked together are found.

The compounds $W(CO)(\eta^2\text{-}RC_2R)_3$ (R = Me, Et, Ph)[11] and $M(MeCN)(\eta^2\text{-}CF_3C_2\text{-}CF_3)_3$ (M = Mo, W)[12] are obtained from $M(CO)_3(MeCN)_3$ (M = Mo, W).

Both $MoCl_5$ and WCl_5 react with xs but-2-yne or diphenylacetylene to give $MX_4(\eta^2\text{-}RC_2R)$ products in high yield (M = Mo, X = Cl, R = CH_3; M = Mo, X = Cl, R = C_6H_5; M = W, X = Br, R = Me; M = W, X = Br, R = C_6H_5); similarly, the analogous $WOCl_3(\eta^2\text{-}MeC_2Me)$ may be obtained from $WOCl_4$ and but-2-yne.

The formation of mono- or bis-(acetylene) complexes from the reactions of $\eta^5\text{-CpMX(CO)}_3$ with alkynes depends upon the acetylene substituents and the nature of the E group as it is shown in Table 1.

Complexes $(\eta^5\text{-Cp})_2Mo(\eta^2\text{-}RC_2R)$ (R = H, Me or Ph) are prepared quantitatively by irradiation of a toluene solution of $(\eta^5\text{-Cp})_2MoCO$ in the presence of the alkyne. In the same conditions, the analogous tungstenocene derivative affords a mixture of the alkyne addition product and the arene insertion product, $(\eta^5\text{-Cp})_2WH(C_6H_4CH_3)$. Pure $(\eta^5\text{-Cp})_2W(MeC_2Me\text{-}\eta^2)$ is obtained when the photolysis of the tunstenocene monocarbonyl is carried out in liq but-2-yne at −78°C[19]. Alternatively, $(\eta^5\text{-Cp})_2Mo(HC_2R)$

Table 1

M	E	RC_2R	Product	Ref.
W	Me	HC_2H MeC_2Me	$\eta^5\text{-CpW(CO)}(RC_2R)CH_3$	14
Mo, W	SCF_3 Sc_6F_5	$CF_3C_2CF_3$ PhC_2Ph	$\eta^5\text{-CpMX(CO)}(RC_2R)$	15
Mo, W	Cl	MeC_2Me $CF_3C_2CF_3$ $MeO_2CC_2CO_2Me$ $ClH_2CC_2CH_2Cl$ PhC_2Me	$\eta^5\text{-CpMCl}(RC_2R)_2$	16, 17
Mo, W	Cl	PhC_2Ph	$\eta^5\text{-CpM(CO)Cl}(PhC_2Ph)$	16
Mo	$COCF_3$	PhC_2Ph MeC_2Me	$\eta^5\text{-CpMo}(CF_3)(CO)(RC_2R)$	18
Mo	$COCF_3$	$CF_3C_2CF_3$ $MeO_2CC_2CO_2Me$	$\eta^5\text{-CpMo}(CF_3)(RC_2R)_2$	18

(R = H, Me) can be prepared from (η^5-Cp)$_2$-MoCl$_2$ by reduction with Na-Hg alloy, but the yields are less than 30%[20].

The reaction of [Mo(CO)$_3$(η^5-Cp)]$_2$ with AgBF$_4$ in the presence an excess alkyne (at RT in CH$_2$Cl$_2$) gives [Mo(CO)(η^2-RC$_2$R')$_2$(η^5-Cp)] [BF$_4$] [21], but if the dimer and the alkyne are refluxed in toluene, the compounds [η^5-CpMo(CO)$_2$]$_2$ (μ-η^2-RC$_2$R) are obtained[22].

A facile replacement of a CO molecule by HC$_2$H occurs with Mo(CO)$_2$[S$_2$P(Pr-i)$_2$]$_2$, at RT in CH$_2$Cl$_2$, to form Mo(CO)(HC$_2$H)[S$_2$P(Pr-i)$_2$]$_2$ [23]:

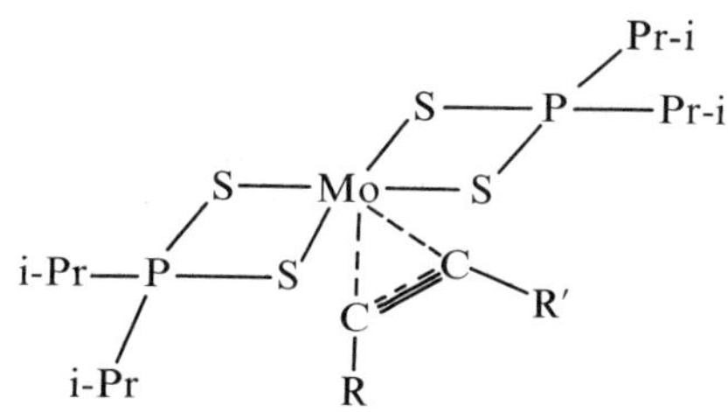

Alkynes give analogous products, and these reactions are claimed to be relevant to the mechanism of nitrogenase catalysis.

(iv) Manganese. The photochemical reaction between η^5-CpMn(CO)$_3$ and alkynes occurs through the replacement of a CO group with the formation of η^5-CpMn(CO)$_2$-(η^2-RC$_2$R), in $\leqslant$20% yields. Similar behavior is observed when (η^5-HMeC$_5$H$_4$)Mn(CO)$_3$ reacts with hexa-2,4-diyne to form the dimer[24]:

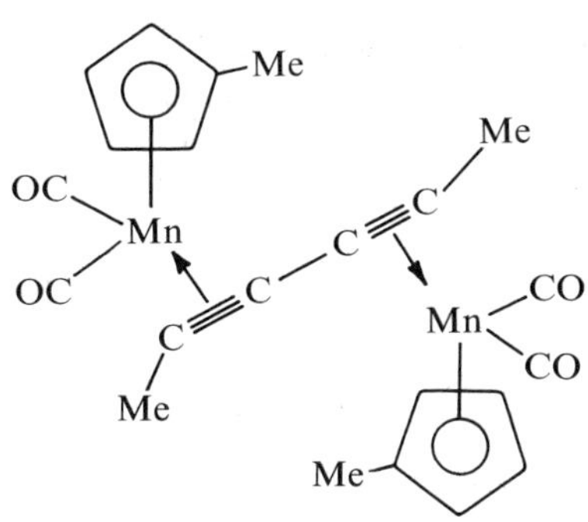

(v) Iron, Ruthenium and Osmium. The reactions of alkynes with the carbonyl derivatives of the iron group depends on both the carbonyl and the acetylene[25]. Since oligomerization of the alkyne eventually coupled with CO insertion occurs easily with iron carbonyls, the mononuclear complexes, Fe(CO)$_4$(η^2-RC$_2$R), are known only when R is bulky such as t-Bu or SiMe$_3$. However, the salts [η^5-CpFe(CO)$_2$(η^2-RC$_2$R)][BF$_4$] (R = Me, Et, Ph) are readily prepared in good yield from [η^5-CpFe(CO)$_2$I] and AgBF$_4$ in the presence of alkyne in CH$_2$Cl$_2$ solution, with [η^5-CpFe(CO)$_2$]$^+$ as an intermediate[26].

Similarly, the cationic acetylene complexes, [Os(CO)(NO)L$_2$(η^2-alkyne)]PF$_6$ (L = PPh$_3$, alkyne = HC$_2$H, PhC$_2$Ph, HC$_2$Ph, MeO$_2$CC$_2$CO$_2$Me L = PChx$_3$, alkyne = HC$_2$H), are obtained when an acetone–CH$_2$Cl$_2$ solution of Os(CO)(NO)L$_2$Cl is reacted with AgPF$_6$ and the alkyne[27].

Ultraviolet irradiation of a hexane solution of trans-$Os(CO)_3[P(OMe)_3]_2$ and xs hexafluorobut-2-yne affords $Os(CO)_2$ $[P(OMe)_3]_2(\eta^2\text{-}CF_3C_2CF_3)$ in 40% yield. No (η^2-alkyne) complex is isolated in the analogous reaction with the corresponding Fe and Ru complexes[28], but an Ru derivative containing a η^2-bonded $CF_3C_2CF_3$:

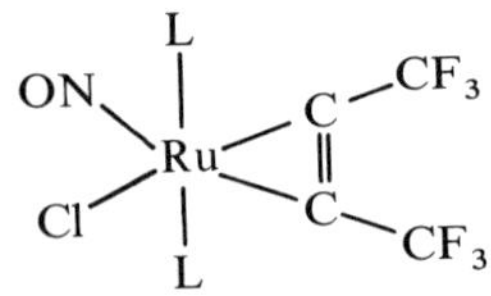

is obtained in good yield[29] when the starting material is $RuCl(NO)L_2$ (L = PPh_3, $PMePh_2$, PMe_2Ph). A compound containing doubly bridge-bonded Fe atoms, **I**, is obtained from the reaction between $Fe_2(CO)_9$ and t-BuC_2Bu-t_2 in hexane at 25°C [30]:

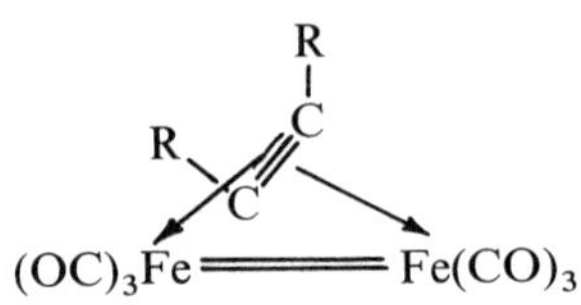

I

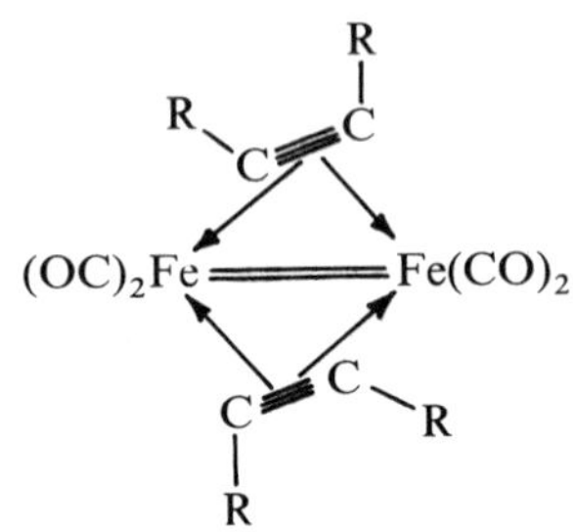

II

The π-related **II** is obtained in 42% when $Fe_3(CO)_{12}$ is refluxed in methylcyclohexane with di-t-butylacetylene[31]. The reactions of alkynes with tri- and tetrametallic carbonyl clusters of Fe, Ru and Os[32] produce few (η^2-alkyne) products, but their formation follows a more complex pathway than the simple CO displacement, e.g., the compound $Ru_4(CO)_{12}(\mu^4\text{-}\eta^2\text{-}RC_2R)$:

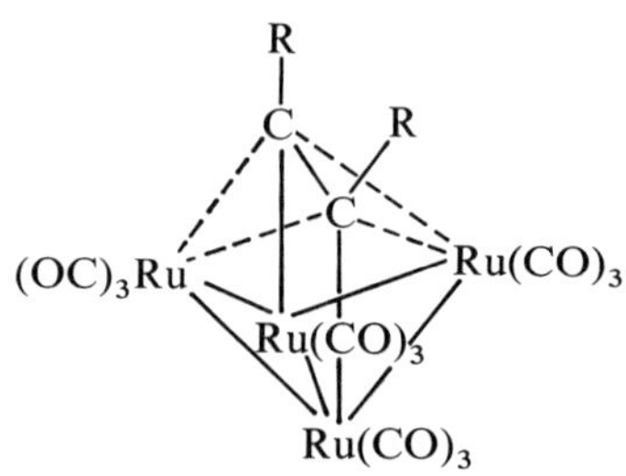

is obtained[33] among other products in the thermal reaction of $Ru_3(CO)_{12}$ with RC≡CR′ (R = R′ = CF_3; R = R′ = CO_2Me; R = Me, R′ = Ph).

(vi) Cobalt, Rhodium and Iridium. η^2-Alkyne mononuclear cobalt complexes are difficult to isolate; e.g., in the photochemical reaction of η^5-$CpCo(CO)_2$ and PhC_2Ph at low temperature the binuclear $(\eta^5\text{-}Cp)_2Co_2(CO)(\mu\text{-}PhC_2Ph)$ and the trinuclear $(\eta^5\text{-}Cp)_3Co_3(\mu\text{-}CO)(\mu\text{-}PhC_2Ph\text{-}\eta^2)$ are obtained among other products: the mononuclear derivative η^5-CpCo(CO) $(\mu^2\text{-}PhC_2Ph)$ is too unstable to be isolated. Similar results are observed when the reaction is carried out by reduction of η^5-$CpCo(CO)I_2$ with sodium amalgam[34]. However, if the CO is replaced by the more basic PPh_3, the compound η^5-$CpCo(PPh_3)$ $(\eta^2\text{-}PhC_2Ph)$ is obtained[35].

Stable bi- and tetranuclear Co complexes are obtained with alkynes by displacement of two CO groups:

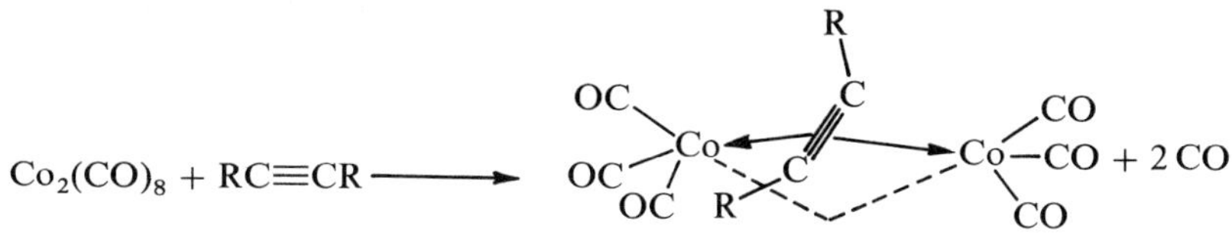
(b)

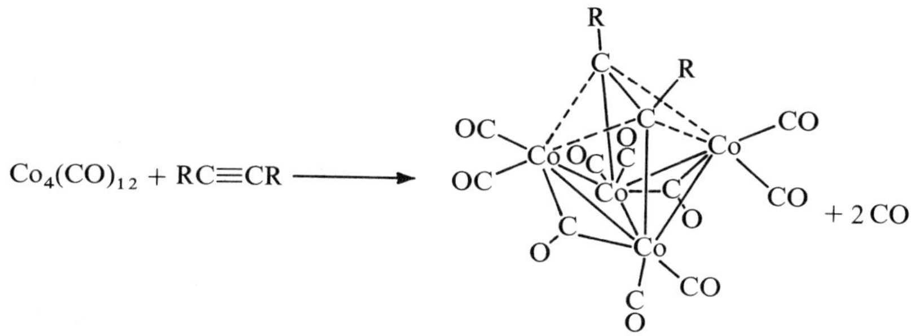
(c)

Reviews of these compounds are available[36]. Similar to the Co analogs, with the Rh complexes it is difficult to isolate simple mononuclear η^2-alkyne derivatives. Stabilization by PPh_3 ligand allows the isolation of $(Ph_3P)_2Rh(RC_2R)Cl$ (R = CF_3 or Ph) by PPh_3 displacement from $(Ph_3P)_3RhCl$ [37,38]. The stable, bridging acetylene complexes, $Rh_2(PF_3)_6$(alkyne), structurally analogous to $Co_2(CO)_6$(alkyne), are obtained in the reaction between $Rh_2(PF_3)_6$ and alkynes[39]. With di-tert-butylacetylene, $Rh_2(PF_3)_5$(t-BuC_2Bu-t) is formed, which is believed to contain a Rh–Rh double bond[40]. When $Rh_4(CO)_{12}$ reacts with PhC_2Ph in the presence of PPh_3 [41] a binuclear derivative is formed in good yield (38%) in which the acetylene moieties are η^2 bonded to each Rh atom:

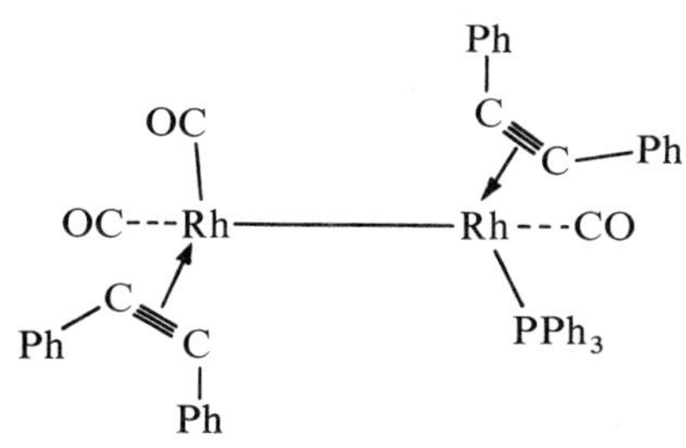

The treatment of trans-$(\eta^5\text{-Cp})_2Rh_2(CO)_2(\mu\text{-}\eta^1\text{-}CF_3C_2CF_3)$ with Me_3NO in acetone at RT affords $(\eta^5\text{-Cp})_2Rh_2(\mu\text{-CO})[\mu\text{-}(\eta^2\text{-}CF_3C_2CF_3)]$, which interconverts to the μ-η^1-alkyne arrangement by reaction with CO[42]:

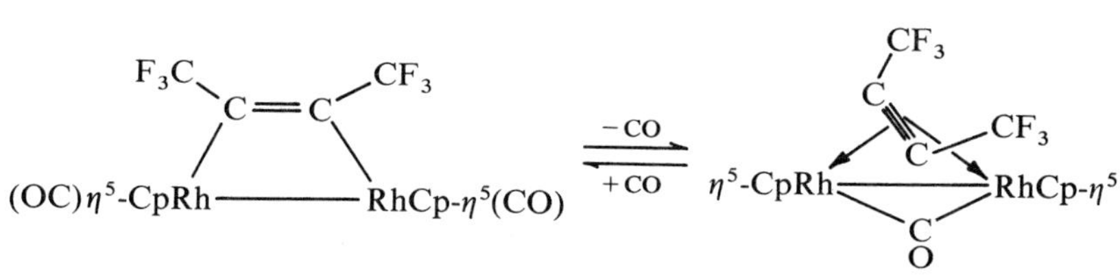
(d)

The reactions of Ir–carbonyl complexes with alkynes require more drastic conditions; e.g., η^5-CpIr(CO)$_2$ reacts slowly at high temperature (120–160°C) with $CF_3C{\equiv}CCF_3$ to form cis and trans isomers of the σ-bridging alkyne complex, (η^5-HCp)$_2$Ir$_2$(CO)$_2$(CF$_3$C$_2$CF$_3$) [43]. Several mononuclear η^2-alkyneiridium complexes are known:

$$IrCl(N_2)(PPh_3)_2 \xrightarrow[-N_2]{+RC\equiv CR} IrCl(PPh_3)_2(\eta^2\text{-}RC_2R) \qquad (e)^{44}$$

where R = CO$_2$Me, COPh, Ph;

$$[Ir(CO)_3(PPh_3)_2]ClO_4 \xrightarrow[-CO]{+RC\equiv CR} Ir(CO)_2(PPh_3)_2(\eta^2\text{-}RC_2R) \qquad (f)^{45}$$

where R = CO$_2$Me;

$$Ir(CO)(PPh_3)_3(\eta^1\text{-}RC{=}CHR) \xrightarrow[-PPh_3]{+RC\equiv CR} Ir(CO)(PPh_3)_2(\eta^1\text{-}RC{=}CHR)(\eta^2\text{-}RC_2R) \qquad (g)^{46}$$

where R = CF$_3$, CN, CO$_2$Et.

(vii) Nickel, Palladium and Platinum. The reactions of alkynes with complexes of this triad form η^2-alkyne ML$_2$ complexes that prove easy to isolate when M = Pt and more difficult when M = Ni. However, the use of bulky phosphines coupled with electron-withdrawing substituents on the acetylenic carbons may help in the isolation of unstable compounds; e.g., Ni(CO)$_4$ reacts with $CF_3C{\equiv}CCF_3$ to form a colorless, volatile liquid complex, Ni(CO)$_2$(η^2-CF$_3$C$_2$CF$_3$), which rapidly decomposes at RT in an open system to give a dark purple tetranuclear species, Ni$_4$(CO)$_4$(CF$_3$C$_2$CF$_3$)$_3$ [47]. More stable analogous products are obtained by[48]:

$$\begin{aligned} &4\ Ni_4[CNC(CH_3)_3]_7 + 15\ C_6H_5C{\equiv}CC_6H_5 \rightarrow \\ &\quad 12\ [(CH_3)_3CNC]_2Ni(\eta^2\text{-}C_6H_5C{\equiv}CC_6H_5) \\ &\qquad + Ni_4[CN(CH_3)_3]_4[\mu_3\text{-}\eta^2C_6H_5C{\equiv}CC_6H_5)]_3 \qquad (h) \end{aligned}$$

A convenient route to NiL$_2$(η^2-RC$_2$R) complexes consists of the replacement of an olefin by an alkyne in NiL$_2$(H$_2$C=CH$_2$) [49].

The bridging acetylene complexes (η^5-Cp)$_2$Ni$_2$(μ-η^2-RC$_2$R) [a close analog of Co$_2$(CO)$_6$(μ-η^2-RC$_2$R)] are obtained in good yield from [η-CpNi(CO)]$_2$ and alkynes[50]. The synthesis of L$_2$Pd(η^2-RC$_2$R) complexes (L = PR$_3$) proceeds by displacement of two phosphines from PdL$_4$ [51] or an olefin from PdL$_2$(olefin) [52].

The reaction (RT in alcohols) of PhC$_2$Ph with Pd(II) acetate leads to the formation of (η^5-C$_5$Ph$_5$)$_2$Pd$_2$(μ-η^2-PhCCPh) [53]:

Pd(OAc)$_2$ + PhC≡CPh ⟶ 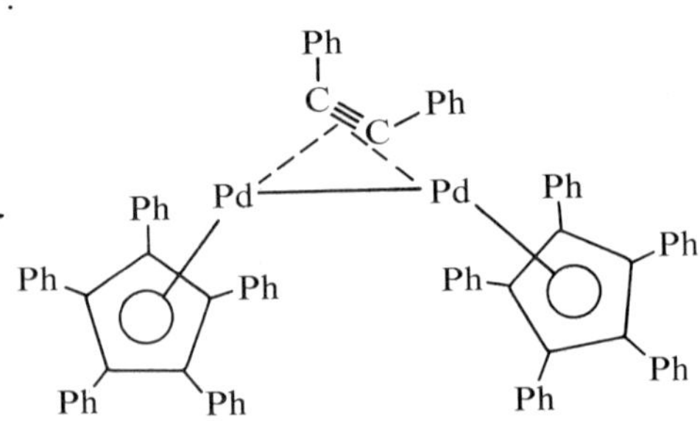 (i)

The formation of the cyclopentadienyl rings from the acetylene involves an unusually facile cleavage of a carbon–carbon triple bond.

The (η^2-alkyne)PtL_2 [54] complexes can be prepared from $Pt(PPh_3)_3$, $Pt(PPh_3)_4$ [55,56] or $PtCl_2(PPh_3)_2$, hydrazine being the reducing agent commonly used[57].

The reaction between $Pt(C_2H_4)(PPh_3)_2$ and $HC_2{\equiv}CC_6F_5$ affords the complex $Pt(\eta^2\text{-}HC_2C_6F_5)(PPh_3)_2$ in contrast to the analogous reaction with $Pt(PR_3)_4$ (R = Ph, Et), which gives the hydride–acetylide complexes trans-$PtH(C_2C_6F_5)(PR_3)_2$ [58].

The displacement of cycloocta-1,5-diene in Pt(1,5-C_8H_{12}) by diphenylacetylene leads to the formation of the bis-acetylene complex:

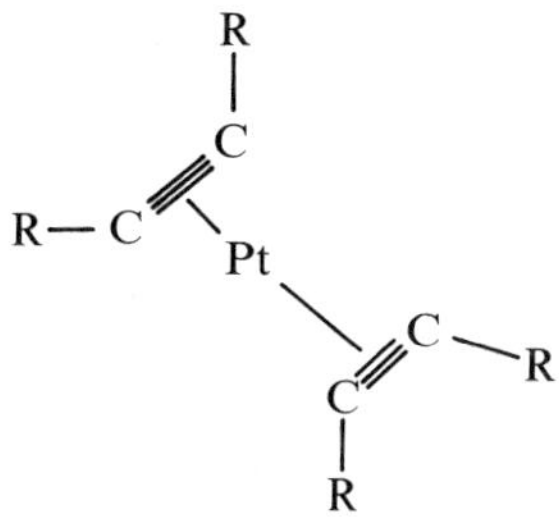

in which the Pt atom is tetrahedral[59].

(S. AIME)

1. F. L. Bowden, A. B. P. Lever, *Organomet. Chem. Rev., 3,* 227 (1968).
2. A. C. Hopkinson, in *The Chemistry of the Triple Bond,* S. Patai, ed. Pergamon Press, Oxford, 1978, Ch. 4, p. 75.
3. G. Fachinetti, C. Floriani, *J. Chem. Soc., Chem. Commun.,* 66 (1974).
4. R. Tsuruma, N. Hagihara, *Bull. Chem. Soc., Jpn., 38,* 1901 (1965).
5. A. N. Nesmeyanov, *Adv. Organomet. Chem., 10,* 1 (1972).
6. J. A. Labinger, J. Schwartz, J. M. Townsend, *J. Am. Chem. Soc., 96,* 4003 (1974).
7. J. A. Labinger, J. Schwartz, *J. Am. Chem. Soc., 97,* 1596 (1975).
8. J. A. Connor, G. A. Hudson, *J. Organomet. Chem., 97,* C43 (1975).
9. W. Strohmeier, H. Helmann, *Chem. Ber., 98,* 1598 (1965).
10. M. Herberhold, H. Alt, *J. Organomet. Chem., 42,* 407 (1973).
11. R. B. King, A. Frouzaglia, *Inorg. Chem., 5,* 1837 (1975).
12. D. P. Tate, J. M. Augl, W. M. Ritchey, B. L. Ross, J. G. Grasselli, *J. Am. Chem. Soc., 86,* 3261 (1964).
13. A. Greco, F. Pirinoli, G. Dall'Asta, *J. Organomet. Chem., 60,* 115 (1973).
14. H. G. Alt, *J. Organomet. Chem., 127,* 349 (1977).
15. P. S. Bratermann, J. L. Davidson, D. W. A. Sharp, *J. Chem. Soc., Dalton Trans.,* 241 (1976).
16. J. L. Davidson, M. Green, D. W. A. Sharp, F. G. A. Stone, A. J. Welch, *J. Chem. Soc., Chem. Commun.,* 706 (1974).
17. J. W. Faller, H. H. Murray, *J. Organomet. Chem., 172,* 171 (1979).
18. J. L. Davidson, M. Green, J. Z. Nyathi, F. G. A. Stone, A. Welch, *J. Chem. Soc, Dalton Trans.,* 2246 (1977).
19. K. L. Tang Wong, J. L. Thomas, H. H. Brintzinger, *J. Am. Chem. Soc., 96,* 3694 (1974).
20. J. L. Thomas, *Inorg. Chem., 17,* 1507 (1978).
21. M. Bottrill, M. Green, *J. Am. Chem. Soc., 99,* 5795 (1977).
22. W. I. Bailey Jr., F. A. Cotton, J. D. Jamerson, J. R. Kolb, *J. Organomet. Chem., 121,* C23 (1976).
23. J. W. McDonald, J. L. Corbin, W. E. Newton, *J. Am. Chem. Soc., 97* 1970 (1975).
24. G. G. Cash, R. C. Petterson, *J. Chem. Soc., Dalton Trans.,* 1630 (1979) and references therein.
25. W. Hubel, in *Organic Synthesis via Metal Carbonyls,* Vol. 1, I. Wender, P. Pino, eds., Interscience, New York, 1967, p. 273.
26. D. L. Reger, C. J. Coleman, P. J. McElligott, *J. Organomet. Chem., 171,* 73 (1979).

27. J. Ashley-Smith, B. F. G. Johnson, J. A. Segal, *J. Organomet. Chem.*, *49*, C38 (1973).
28. R. Burt, M. Cooke, M. Green, *J. Chem. Soc.*, *A*, 2981 (1970).
29. J. Clemens, M. Green, F. G. A. Stone, *J. Chem. Soc.*, *Dalton Trans.*, 375 (1973).
30. F. A. Cotton, J. D. Jamerson, B. R. Stults, *J. Organomet. Chem.*, *94*, C53 (1975).
31. K. Nicholas, L. S. Bray, R. E. Davis, R. Pettit, *J. Chem. Soc.*, *Chem. Commun.*, 608 (1971).
32. A. J. Deeming, in *Transition Metal Clusters*, (B. F. G. Johnson, ed. Wiley, 1980, p. 391. An extensive review of the reactions of alkynes with transition-metal clusters.
33. B. F. G. Johnson, J. Lewis, K. T. Schorpp, *J. Organomet. Chem.*, *91*, C13 (1975).
34. W. S. Lee, H. H. Brintzinger, *J. Organomet. Chem.*, *127*, 93 (1977).
35. H. Yamazaki, Y. Wakatsuki, *J. Organomet. Chem.*, *139*, 157 (1977).
36. R. S. Dickson, P. J. Fraser, *Adv. Organomet. Chem.*, *12*, 323 (1974).
37. M. J. Mays, G. Wilkinson, *J. Chem. Soc.*, 6629 (1965).
38. J. T. Magne, G. Wilkinson, *J. Chem. Soc.*, *A*, 1736 (1966).
39. M. A. Bennett, R. N. Johnson, G. B. Robertson, T. W. Turney, P. O. Whimp, *J. Chem. Soc.*, *94*, 6540 (1972).
40. M. A. Bennett, R. N. Johnson, T. W. Turney, *Inorg. Chem.*, *15*, 111 (1976).
41. Y. Iwashita, F. Tamura, *Bull. Chem. Soc. Jpn.*, *43*, 1517 (1970).
42. R. S. Dickson, G. N. Pain, *J. Chem. Soc.*, *Chem. Commun.*, 277 (1979).
43. P. A. Corrigan, R. S. Dickson, *Aust. J. Chem.*, *32*, 2147 (1979).
44. J. P. Collman, J. W. Kang, *J. Am. Chem. Soc.*, *89*, 844 (1967).
45. M. J. Church, M. J. Mays, R. N. F. Simpson, F. P. Stefanini, *J. Chem. Soc.*, *A*, 2909 (1970).
46. R. M. Kirschner, J. A. Ibers, *J. Am. Chem. Soc.*, *95*, 1095 (1973).
47. J. L. Davidson, M. Green, F. G. A. Stone, *A. J. Welch*, *J. Am. Chem. Soc.*, *97*, 7490 (1975).
48. M. G. Thomas, E. L. Muetterties, R. O. Day, V. W. Day, *J. Am. Chem. Soc.*, *98*, 4645 (1976).
49. E. O. Greaves, C. J. L. Lock, P. M. Maitlis, *Can. J. Chem.*, *46*, 3879 (1968).
50. J. F. Tilney-Bassett, *J. Chem. Soc.*, 577 (1961).
51. E. O. Greaves, P. Maitlis, *J. Organomet. Chem.*, *6*, 104 (1966).
52. G. L. McClure, W. H. Badley, *J. Organomet. Chem.*, *27*, 155 (1971).
53. T. R. Jack, C. J. May, J. Powell, *J. Am. Chem. Soc.*, *99*, 4707 (1977).
54. H. Chatt, G. A. Rowe, A. A. Williams, *Proc. Chem. Soc.*, 208 (1957).
55. J. A. Nelson, H. B. Jonassen, D. M. Rounhill, *Inorg. Chem.*, *8*, 2591 (1969).
56. H. D. Empsall, B. L. Shaw, J. A. Stringer, *J. Chem. Soc.*, *Dalton Trans.*, 185 (1976).
57. S. Jacobson, A. J. Carty, M. Mathew, G. J. Palenik, *J. Am. Chem. Soc.*, 4330 (1974).
58. O. M. Ab-Salah, M. I. Bruce, *Aust. J. Chem.*, *29*, 73 (1976).
59. M. Green, D. M. Grove, J. A. K. Howard, J. L. Spencer, F. G. A. Stone, *J. Chem. Soc.*, *Chem. Commun.*, 759 (1976).

5.8.2.7.2. η^1 σ-Vinyl–Metal Complexes by Addition of HX to η^2 π-Alkyne Complexes

For synthesizing alkenyl–transition-metal complexes (see also §5.8.2.7.3), the addition of HX to η^2-acetylene complexes[1] is applied to η^2-alkynePt–complexes:

$$\eta^2\text{-alkynePtL}_2 + \text{HX} \rightarrow \text{PtX(alkenyl-}\eta^1\text{)L}_2 \qquad \text{(a)}$$

The addition of HX to an (acetylene)Pt(0) complexes results in a cis arrangement of H and Pt on the double bond and takes place at RT, giving high yields of the alkenyl product; e.g., when equimolar amounts of HCl and $Pt[\eta^2\text{-}(C_6H_5)_2C_2](PPh_3)_2$ are reacted at RT in $CHCH_3$ or benzene, the square-planar complex, $Pt[\eta^1\text{-}C_6H_5)C{=}CH(C_6H_5)]\text{-}Cl[P(C_6H_5)_3]_2$ is formed quantitatively.

When terminal alkynes are used, the hydrogen adds to the carbon bearing more hydrogens; i.e., $PtE(CR{=}CH_2)(PPh_3)_2$ complexes are obtained (R = Me, Et, Ph or p-tol; E = Cl, Br, NO_3 or $OCOCF_3$).

Furthermore, the reaction of $Pt[(p\text{-tol-}C{\equiv}CH)](PPh_3)_2$ with DCl forms an alkenyl complex in which the deuterium is exclusively cis to platinum.

Further reaction of the σ-alkenyl complexes with xs HX forms the corresponding olefin together with L_2PtX_2; since both cis- and trans-olefin are obtained the solvolysis of vinyl complexes may not occur with retention of configuration about the double bond[1,3].

The hexafluorobut-2-yne complex, $Pt(CF_3C{\equiv}CCF_3)L_2$ (L = PPh_3, $PMePh_2$, PMe_2Ph, PEt_2Ph; L_2 = $Ph_2PCH_2CH_2PPh_2$) reacts with CF_3COOH (in CH_2Cl_2 at RT) to form the alkenyl complexes, $Pt(OCOCF_3)(CF_3C{=}CHCF_3)L_2$, in which the phosphine ligand is cis. The complex cis-$Pt(OCOCF_3)(CF_3C{=}CHCF_3)(PPh_3)_2$ can be thermally isomerized to its trans isomer, the formation of the cis isomer being kinetically controlled.[4]

The complexes $PtE(CF_3C{=}CHCH_3)(PPh_3)_2$ (E = Cl, Br or OSO_3H) are obtained by addition of gaseous HCl, ethanolic HBr or methanolic H_2SO_4, respectively, to a CH_2Cl_2 solution of the η^2-acetylene complex.

Glacial acetic acid is not sufficiently acidic to effect protonation of this complex, and nitric acid leads only to the dinitrate complex, $Pt(ONO_2)_2(PPh_3)_2$ [4].

The dicyanovinyl complexes $PtX[C(CN){=}CHCN](PR_3)_2$ (X = Cl, R = Ph or Et, X = Br, R = Ph) are obtained when the $(\eta^2\text{-}CNC_2CN)Pt(PR_3)_2$ is reacted with HCl or HBr [5].

Cycloalkynes also complex[6,7]; e.g., the cyclohexyne complexes $Pt(C_6H_8)L_2$ (L = PPh_3, L_2 = dppe) react with an equimolar amount of HCl in benzene at RT, to give $PtCl(C_6H_9)L_2$ in good yield. The corresponding CF_3CO_2 derivative is obtained when $Pt(C_6H_8)(PPh_3)_2$ is similarly treated with CF_3COOH. The cycloheptenyl complex, $PtX(C_7H_{11})(PPh_3)_2$ (X = Cl, CF_3CO_2) is obtained when the corresponding cycloheptyne complex is reacted with HCl or CF_3COOH.

The cyclohexyne complex, $Pt(C_6H_8)$(dppe), can be also protonated by weak protic acid such as nitromethane, acetone, water and methanol, etc.:

$$Pt(C_6H_8)(dppe) + HE \rightarrow PtE(C_6H_9)(dppe) \qquad (b)$$

where E = p-$CH_3C_6H_4O$, CH_2NO_2, CH_2COCH_3, $CH_2COC_6H_5$, $CH(CN)(C_6H_5)$, OH, OCH_3, CH_2CN, $NHCOCH_3$ or $NHCOC_6H_5$.

The products are colorless, air-stáble Pt(II) cyclohexenyls. Nitromethane, acetonitrile or phenylacetonitrile react with $Pt(C_6H_8)$(dppe) under anhydrous conditions, whereas methyl ketones react only in the presence of water via the preformed hydroxo complex.

The σ-cycloalkenylPt(II) complexes are less readily cleaved by xs HX give the corresponding olefin than are the complexes formed from acyclic acetylene–platinum(0) complexes.

σ-Alkenyl complexes are also obtained when $Pd(CF_3C{\equiv}CCF_3)L_2$ (L = PPh_3, L_2 = dppe) are reacted with halocarboxylic acids[8]; e.g., $Pd(CF_3C{\equiv}CCF_3)$(dppe) react with CF_3COOH, CF_2HCOOH and CCl_3COOH and from kinetic data the reaction occurs with the following mechanism:

$$Pd(CF_3C{\equiv}CCF_3)(dppe) + H^+ \xrightarrow{\text{slow}} [PdH(CF_3C{\equiv}CCF_3)(dppe)]^+ \qquad (b)$$

$$\downarrow CO_2CX_3^- \text{ fast}$$

$$Pd(OCOCH_3)[C(CF_3){=}CHCF_3](dppe)$$

These observations support the view that the protonation occurs by initial oxidative addition at the metal atom and subsequent transfer of hydrogen to the coordinated alkyne which is argued on the basis of theory as well as experiment[10]. In fact, the reaction of CH_3COOH with $(Ph_3P)_2Pt(DMA)$ (DMA = dimethylacetylene dicarboxylate) in the presence of PPh_3 proceeds by addition of triphenylphosphonium acetate across the acetylenic triple bond:

$(Pn_3P)_2Pt(DMA) + CH_3COOH + PPh_3 \longrightarrow$ 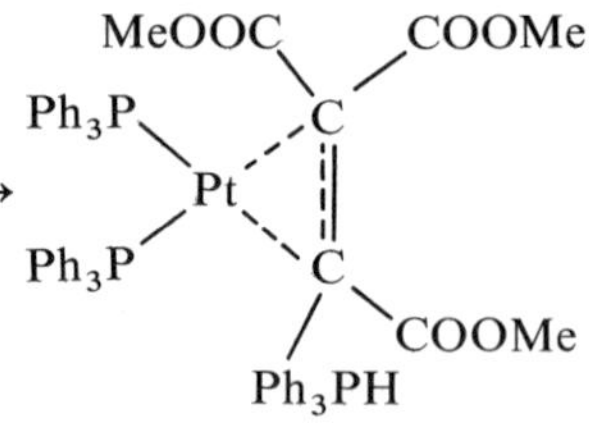 (d)

Thus the formation of alkenyl derivatives by addition of HX to zero-valent acetylene–Pt complexes occurs by the direct addition of HX across the C≡C bond, rather than through oxidative addition to form a 6-coordinated–Pt(IV) intermediate.

Other hexafluorobut-2-yne complexes[4,11] react with HX to form σ-alkenyl complexes:

$$RhCl(CF_3C{\equiv}CCF_3)(PPh_3)_2 + HX \rightarrow RhClX(CF_3C{=}CHCF_3)(PPh_3)_2 \quad (e)$$

$$IrCl(CO)(CF_3C{\equiv}CCF_3)(PPh_3)_2 + HY \rightarrow IrClY(CO)(CF_3C{=}CHCF_3)(PPh_3)_2 \quad (f)$$

$$Os(CF_3C{\equiv}CCF_3)(CO)_2[P(OMe)_3]_2 + HX \rightarrow OsX(CF_3C{=}CHCF_3)(CO)_2[P(OMe)_3]_2 \quad (g)$$

where, X = Cl, Y = CF_3CO_2. On the other hand, the d^6-complex $W(MeCN)(CF_3C{\equiv}CCF_3)_3$ does not react with CF_3COOH.

(S. AIME)

1. D. Max Roundhill, *Adv. Organomet. Chem.*, *13*, 273 (1975).
2. B. E. Mann, B. L. Shaw, N. I. Tucker, *J. Chem. Soc., A*, 2667 (1971).
3. P. B. Tripathy, B. W. Renoe, K. Adzamli, D. M. Roundhill, *J. Am. Chem. Soc.*, *93*, 4406 (1971).
4. R. D. W. Kemmitt, B. Y. Kimura, G. W. Littlecott, *J. Chem. Soc., Dalton Trans.*, 636 (1973).
5. G. L. McClure, W. H. Baddley, *J. Organomet. Chem.*, *25*, 261 (1970).
6. M. A. Benett, G. B. Robertson, P. O. Whimp, T. Yoshida, *J. Am. Chem. Soc.*, *95*, 3028 (1973).
7. M. A. Benett, T. Yoshida, *J. Am. Chem. Soc.*, *100*, 1750 (1978).
8. J. Burgess, R. D. W. Kemmitt, G. W. Littlecott, *J. Organomet. Chem.*, *56*, 405 (1973).
9. K. S. Wheelock, J. H. Nelson, J. D. Kelly, H. B. Jonassen, L. C. Cusachs, *J. Chem. Soc., Dalton Trans.*, 1457 (1973).
10. H. C. Clark, A. B. Goel, *J. Organomet. Chem.*, *190*, C51 (1980).
11. M. Cooke, M. Green, T. A. Kue, *J. Chem. Soc., A*, 1200 (1971).

5.8.2.7.3. η^1 σ-Vinyl Complexes by Addition of Alkynes to Metal Hydrides

Insertion of alkynes into metal–hydrogen bonds to produce σ-vinyl complexes is an important elementary step in catalytic hydrogenation.

The stereochemistry of the acetylene insertion into the M—H bond:

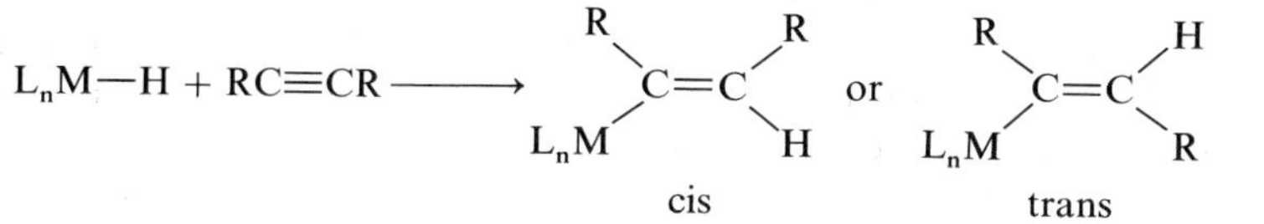

(a)

cannot be straightforwardly related to the mechanism (ionic, radical or concerted)[1].

(i) Zirconium. σ-Vinyl derivatives are obtained by M—H bond insertion in the reactions between $(\eta^5\text{-Cp})_2Zr_2HCl$ and alkynes (in benzene at reflux); trans-alkenyl derivatives are obtained, although in some cases a mixture of isomers is formed. Analogously, $(\eta^5\text{-Cp})_2Zr_2H_2$ reacts with HC≡CR (R = Et, Ph) to form:

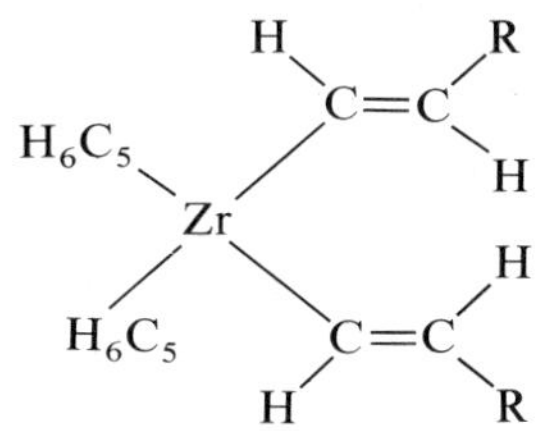

but with diarylacetylenes other products are obtained[2]. On the other hand, $(\eta^5\text{-Cp})_2ZrHR$ (R = CH_3, $CH_2C_6H_5$) with disubstituted alkynes gives rise to an η^2-alkyne derivative by elimination of HR [3].

(ii) Niobium. Alkynes react with $(\eta^5\text{-Cp})_2NbH_3$ to give the η^2-acetylene complex $(\eta^5\text{-Cp})_2NbH(\eta^2\text{-}RC_2R)$ which, only by heating under CO (3.66×10^5 Pa, 75°C, 15 min) gives the expected insertion product, $(\eta^5\text{-Cp})_2Nb(CO)RC = C(H)R$ [4].

(iii) Manganese, Rhenium. The hydrido complexes of these metals react; e.g., hexafluorobut-2-yne reacts with $HMn(CO)_5$ [5] $HRe(CO)_5$ [6] and $(\eta^5\text{-Cp})_2ReH$ [7], forming σ-vinyl derivatives. The insertion product is also the major one from $HRe(CO)_5$ and $CF_3C{\equiv}C{-}H$, but the analogous reaction with $HMn(CO)_5$ affords a mixture of $CH_2 = C(CF_3)Mn(CO)_5$ (as a minor product) and a tetracarbonyl complex $(CO)_4Mn\text{-}CH{:}C(CH_3)C(H)O$ [8].

Other alkynes react with $HMn(CO)_5$ [9]:

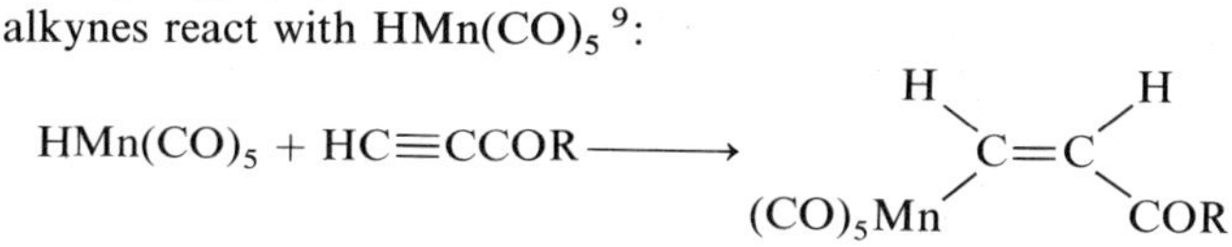

(b)

where R = OM, 5°C, 71%; R = H, −30°C, 14%;

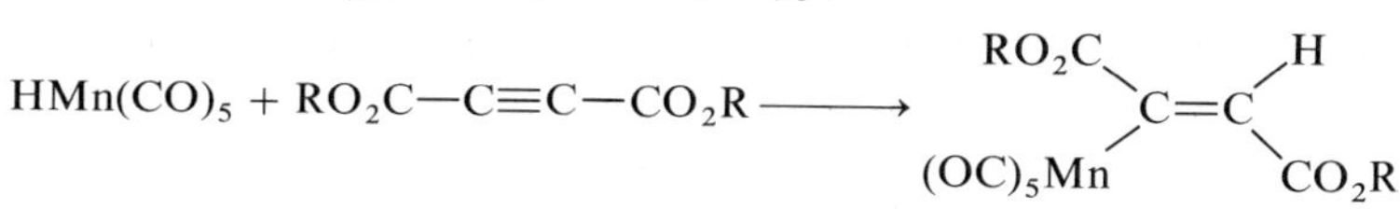

(c)

where R = H, 0°C, 72%; R = Me, 0°C, 86%.

Dimethylacetylenedicarboxylate reacts in ether at RT; the product:

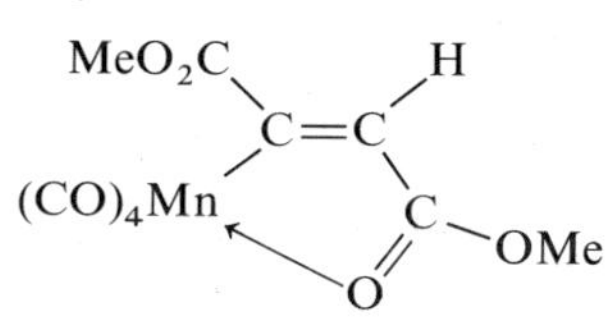

is obtained in 86% yield[8]. Similar behavior is observed in the reactions of $(\eta^5\text{-Cp})_2ReH$ with $HC{\equiv}CCO_2Me$ and $MeO_2C{\cdot}C{\equiv}C{\cdot}CO_2Me$ [10].

(iv) Molybdenum, Tungsten. At low temperature the trans insertion to give $MoH[(CF_3)C{=}CH(CF_3)](\eta^5\text{-Cp})_2$ occurs rapidly on reaction of $MoH_2(Cp\text{-}\eta^2)$ with $CF_3C{\equiv}CF_3$ [11], in contrast to the cis insertion found in the analogous reactions with $CF_3C{\equiv}CH$, $PhC{\equiv}CPh$ and $MeO_2CC{\equiv}CCO_2Me$ [7,11]. The complexes $WH(CO)_3\eta^5\text{-Cp}$ and $MeO_2CC{\equiv}CCO_2Me$ in refluxing THF give $W\{cis[{-}C(CO_2Me){=}CH(CO_2Me)](\eta^5\text{-Cp})_2\}$ and the binuclear complex $\{\eta^5\text{-Cp}(CO)_2W[\mu\text{-}\eta^2\text{-}C_2(CO_2Me)_2]W(CO)_2\eta^5\text{-Cp}\}$ as the major product, but at RT the same reaction gives the trans insertion product[12].

Like the Mo compound, trans insertion to $WH_2(Cp\text{-}\eta^5)_2$ is observed with the reaction with hexafluorobut-2-yne[7].

(v) Iron, Ruthenium, Osmium. Insertion into the Fe—H bond occurs for $\eta^5CpFe(dppe)H$ with $RC{\equiv}CR$ ($R = CO_2Me, CF_3$) to form the σ-alkenyl complexes $\eta^5\text{-}CpFe(dppe)(RC{=}CHR)$ [13]; the reaction between $\eta^5\text{-}CpFe(CO)_2H$ and $CF_3C{\equiv}CH$, at room temperature, affords a mixture of $CpFe(cis\text{-}CH{=}CHCF_3)(CO)_2$ and $\eta^5\text{-}CpFe(CF_3C{=}CH_2)(CO)_2$ [8]. The anion $[FeH(CO)_4]^-$ undergoes trans addition to acetylenes to give a novel η^3-vinylacyl complex[14]:

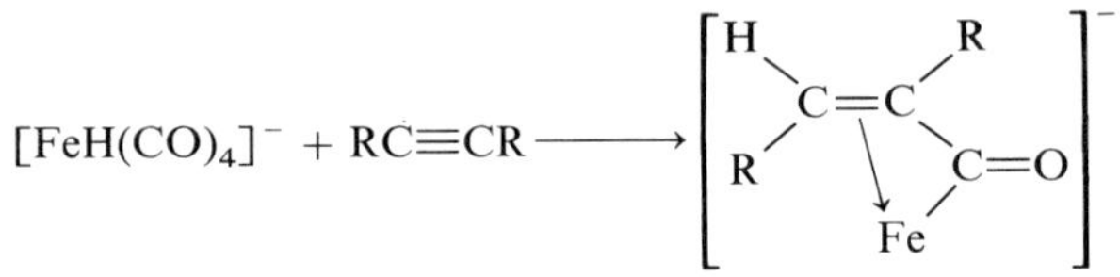

(d)

The complex $\eta^5\text{-}CpRuH(CO)_2$ reacts with $CF_3C{\equiv}CCF_3$ to form $\eta^5\text{-}CpRu\text{-}trans\text{-}(CF_3)C{=}CH(CF_3)(CO)_2$ [15], in contrast to $\eta^5\text{-}CpRuH(PPh_3)_2$, which gives the cis product. Further reaction leads to the insertion of a second molecule of alkyne with retention of the configuration to give, respectively[16]:

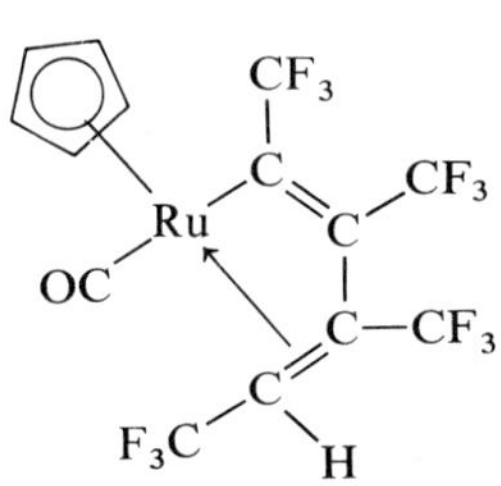

and

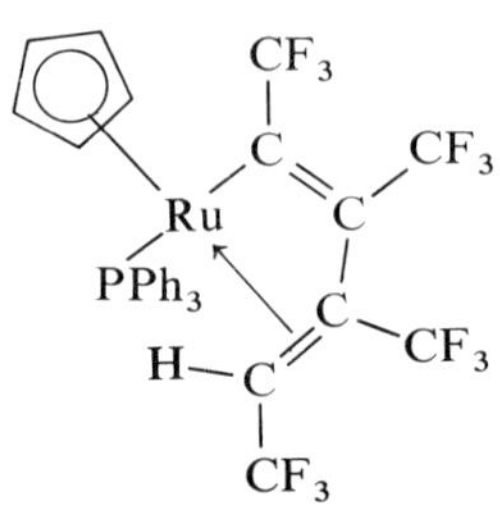

Analogous behavior is observed when $\eta^5\text{-}CpRuH(PPh_3)_2$ is reacted with $MeO_2CC{\equiv}CCO_2Me$, but in this case the 1:1 insertion adduct has the structure:

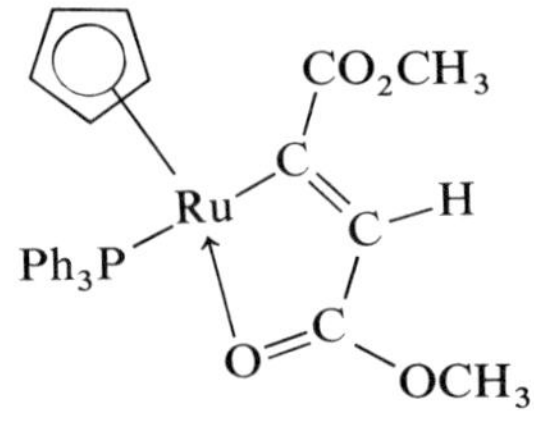

The expected σ-vinyl complex $\eta^5\text{-}CpRu(PPh_3)_2[(MeO_2C)C{=}C(CO_2Me)H]$ is obtained, among other products, in the reaction of $\eta^5\text{-}CpRuH(PPh_3)(CO)$ with

$MeO_2CC{\equiv}CCO_2Me$ [17]. A mixture of 1:1 vinyl adducts is obtained with asymmetrically substituted acetylenes[18]:

$$HM(O_2CCF_3)(CO)(PPh_3)_2 + MeC{\equiv}CPh \rightarrow$$
$$\{M[PhC{=}CHMe](O_2CCF_3(CO)(PPh_3)_2\}$$
$$+ \{M[MeC{=}CHMe](O_2CCF_3)(CO)(PPh_3)_2\} \qquad \text{(e)}$$

where M = Ru, Os. Insertion into the μ-M—H—M bonds of $H_2Os_3(CO)_{10}$ with alkynes gives σ,π-vinyl derivatives[19]:

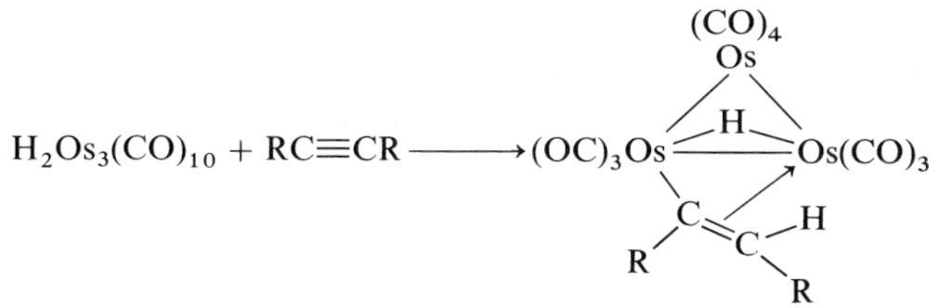

(f)

where the stereochemistry about the double bond is invariably cis.

Similar behavior occurs with $H_4Os_4(CO)_{12}$ [20].

(vi) Rhodium, Iridium. Stable vinyl complexes are obtained from $RhH(CO)(PPh_3)_3$ with the alkynes RC_2R (R = CO_2Me, CO_2Et, CO_2H or C_6H_5)[21]. A trans configuration about the double bond is assigned on the basis of the olefinic products obtained by cleavage of the σ-alkenyl compounds with anhydr HCl. When R = CF_3 the cis insertion product is obtained, implying that a different mechanism is operating here.

Although no Ru compound is isolated in the reaction between $RuH(NO)(PPh_3)_3$ and C_2Ph_2, when the isoelectronic and isostructural complexes $MH(CO)(PPh_3)_3$ (M = Rh, Ir) are reacted with equimolar amounts of this alkyne, high yields of the σ-trans-alkenyl products are obtained[22]. For $IrH(CO)(PPh_3)_3$ with activated alkynes two different paths are observed according to the temperature at which the reaction occurs[23]:

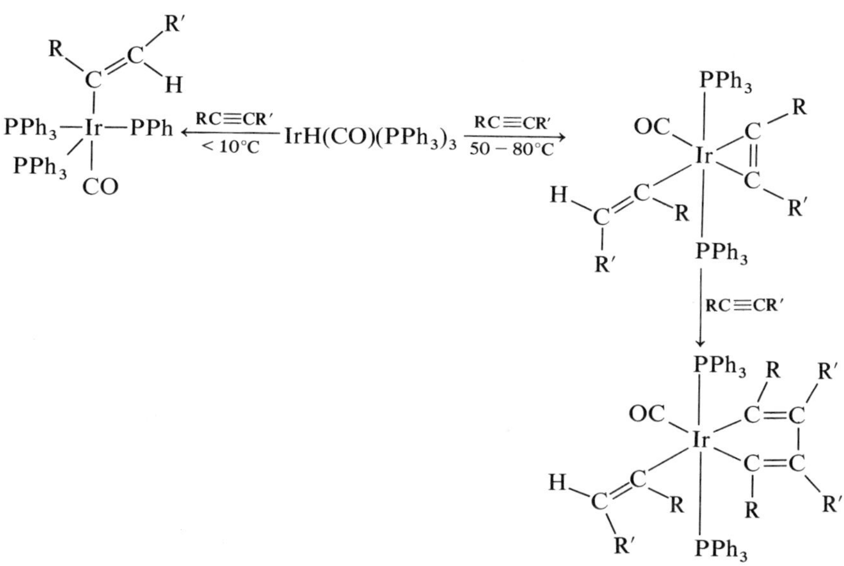

(g)

Insertion into the Ir—H bond is also observed for $IrHCl_2(PEt_3)_3$ reacted with $CF_3C{\equiv}CH$ or $CF_3C_2CF_3$[24], and for $IrHCl_2(Me_2SO)_3$ with PhC_2Ph[25].

(vii) Palladium, Platinum. Cis addition across the triple bond with the hydride adding to the acetylenic carbon bearing the electron-withdrawing group is observed in the insertion of acetylenes into the Pd—H bonds of trans-$PdH(NO_3)(PChx_3)_2$ or trans-[$PdH(MeCN)(PChx)_3$][PF_6], but if the reaction is carried out in the presence of a proton sponge, $Pd(PChx_3)_2(\eta^2\text{-}C_2R_2)$ (R = CF_3, CO_2Me) may be isolated[26].

Insertion of $MeO_2CC{\equiv}CCO_2Me$ also occurs with trans-$MH(C_2Ph)(PEt_3)_2$ to form trans-$M(C_2Ph)$ cis-$(MeO_2C)C{=}CH(CO_2Me)$ $(PEt_3)_2$ (M = Pd, Pt)[27].

Acetylene insertion into the Pt—H bonds of trans-$PtHL_2X$ (X = NO_3, Cl; L = PEt_3 or $PChx_3$) lead to a mechanism based on the initial dissociation of X from the metal and coordination of the alkyne. This gives [cis-$HPtL_2$(alkyne)]$^+$, which collapses to give a 3-coordinated [PtL_2(vinyl)]$^+$ intermediate. Recombination of the metal with X yields the final product[28]. The reaction of trans-$PtHCl(PPh_3)_2$ with C_2Ph_2[29] in the presence of methylfluorosulfonate leads to the formation of $Pt(OSO_2F)(PhC{=}CHPh)(PPh_3)_2$ (65% yield), which by treatment with [Et_4N]Cl gives $PtCl(PhC{=}CHPh)(PPh_3)_2$ quantitatively.

Cis insertion of $RC{\equiv}CR$ into the Pt—H bond of $PtHCl(PEt_3)_2$ is observed when R = CF_3 or CO_2Me is used, but trans insertion is seen with $CF_3C{\equiv}CH$[30].

For dicyanoacetylene the formation of the insertion product with $PHCl(PEt_3)_2$ occurs only when the reaction is carried out in benzene; if THF is used, the $Pt(\eta^2\text{-}C_4N_2)(PEt_3)_2$ complex is formed[31].

(S. AIME)

1. S. Otsuka, A. Nakamura, *Adv. Organomet. Chem., 14*, 245 (1976).
2. P. C. Wailes, H. Weigold, A. P. Bell, *J. Organomet. Chem., 27*, 373 (1971).
3. M. Yoshifuji, K. I. Gell, J. Schwartz, *J. Organomet. Chem., 153*, C15 (1978).
4. J. A. Labinger, J. Schwartz, *J. Am. Chem. Soc., 97*, 1536 (1975).
5. P. M. Treichel, E. Pitcher, F. G. A. Stone, *Inorg. Chem., 1*, 511 (1962).
6. J. B. Wilford, F. G. A. Stone, *Inorg. Chem., 4*, 93 (1965).
7. A. Nakamura, S. Otsuka, *J. Mol. Catal., 1*, 285 (1976).
8. D. A. Harbourne, F. G. A. Stone, *J. Chem. Soc., A*, 1765 (1968).
9. B. L. Booth, G. Hargreaves, *J. Chem. Soc., A*, 2766 (1969).
10. M. Dubeck, R. A. Schell, *Inorg. Chem., 3*, 1757 (1964).
11. A. Nakamura, S. Otsuka, *J. Am. Chem. Soc., 94*, 1886 (1972).
12. R. M. Laine, P. C. Ford, *J. Organomet. Chem., 124*, 29 (1977).
13. P. M. Treichel, D. C. Molzahn, *Inorg. Chim. Acta, 16*, 267 (1979).
14. T. A. Mitsudo, H. Nakanishi, T. Inubishi, I. Morishima, Y. Watanabe, Y. Takegami *J. Chem. Soc., Chem. Commun.*, 416 (1976).
15. J. L. Davidson, M. Green, F. G. A. Stone, A. J. Welch, *J. Chem. Soc., Chem. Commun.*, 286 (1975).
16. T. Blackmore, M. I. Bruce, F. G. A. Stone, *J. Chem. Soc., Dalton Trans.*, 106 (1974).
17. M. I. Bruce, R. C. F. Gardner, F. G. A. Stone, *J. Chem. Soc., Dalton Trans.*, 306 (1979).
18. A. Dobson, D. S. Moore, S. D. Robinson, *J. Organomet. Chem., 177*, C8 (1979).
19. A. J. Deeming, S. Hasso, M. Underhill, *J. Chem. Soc., Dalton Trans.*, 2056 (1975).
20. B. F. G. Johnson, J. W. Kellard, J. Lewis, A. L. Mann, P. R. Raithby, *J. Chem. Soc., Chem. Commun.*, 547 (1980).
21. B. L. Booth, A. D. Lloyd, *J. Organomet. Chem., 35*, 195 (1972).
22. R. A. Saucher-Delgado, G. Wilkinson, *J. Chem. Soc., Dalton Trans.*, 804 (1977).
23. W. H. Baddley, G B. Tupper, *J. Organomet. Chem., 67*, C16 (1974).
24. H. C. Clark, R. K. Mittai, *Can. J. Chem., 51*, 1511 (1973).
25. M. A. Trocha, H. B. Heubest, *J. Chem. Soc., Chem. Commun.* 757 (1968).
26. H. C. Clark, C. R. Milne, *J. Organomet. Chem., 161*, 51 (1978).

27. Y. Tohda, K. Sonogashira, N. Hagihara, *J. Organomet. Chem.*, *110*, C53 (1976).
28. H. C. Clark, C. S. Wong, *J. Organomet. Chem.*, *92*, C31 (1975).
29. C. Eaborn, N. Farrell, A. Pidcock, *J. Chem. Soc., Chem. Commun.*, 766 (1973).
30. D. A. Harbourne, D. T. Rosevear, F. G. A. Stone, *Inorg. Nucl. Chem. Lett.*, *2*, 247 (1966).
31. G. L. McLure, W. H. Baddley, *J. Organomet. Chem.*, *25*, 261 (1970).

5.8.2.7.4. Metallacyclopentadiene and Metallacyclopentene Complexes by Cyclodimerization of Alkynes with Alkynes or Alkenes

Alkynes undergo catalytic and stoichiometric conversions in the coordination sphere of metals proceeding through or leading to metallacyclopentadienes and metallacyclopentenes[1–8]:

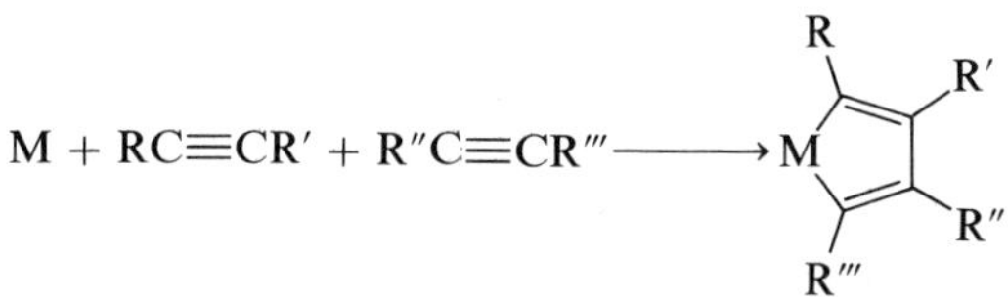

(a)

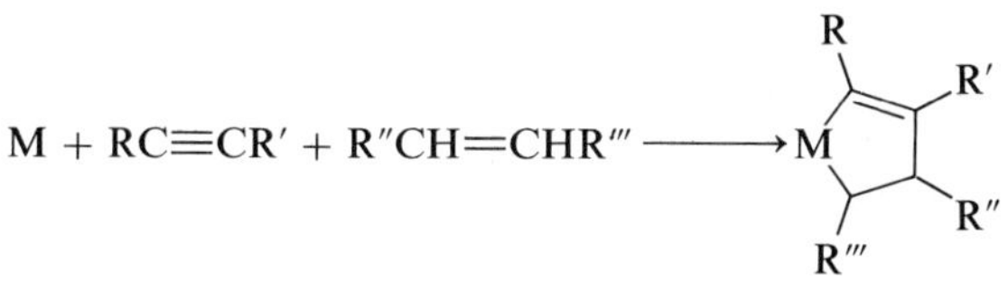

(b)

These transformations usually involve initial coordination of one ligand, followed by the second, and subsequent oxidative coupling to product. Most investigations concern examples of Eq. (a), utilizing either mono- or bimetallic systems, e.g., monometallic complexes of Ti [9–14], Zr [14–21], Hf [14,20,22,23], Nb [24,25], Ta [26,27], U [28,29], Mo [30–32], W [32,33], Fe [34,35], Ru [36,37], Co [38–41], Rh [42–45], Ir [46–48], Ni [49], Pd [50,51] and Pt [52,53]. The precursor metal species M is varied, both with respect to the 'innocent' or reactive ligands it carries and with respect to its oxidation state. Thus, diphenylethyne reacts not only with $(\eta^5\text{-Cp})_2Ti(CO)_2$ to furnish the corresponding bis(cyclopentadienyl)titana cyclopentadiene [via the intermediacy of isolable $(\eta^5\text{-Cp})_2Ti(CO)(C_6H_5C{\equiv}CC_6H_5)$][13], but also with ($\eta^3$-allyl)titanocene (by extrusion of propene)[12], dimethyl)titanocene (on irradiation)[14], and $(\eta^5\text{-Cp})_2TiCl_2$–Mg [14] to the same product. The parent compound, $(\eta^5\text{-Cp})_2Ti(C_4H_4)$, is available through exposure of $(\eta^5\text{-Cp})_2Ti[P(CH_3)_3]_2$ to ethyne [via isolable $(\eta^5\text{-Cp})_2Ti(C_2H_2)P(CH_3)_3$][9]. (Diphenyl)titanocene eliminates benzene on heating and converts to an intermediate benzyne complex that is trapped by alkynes to give benzotitanacycles:

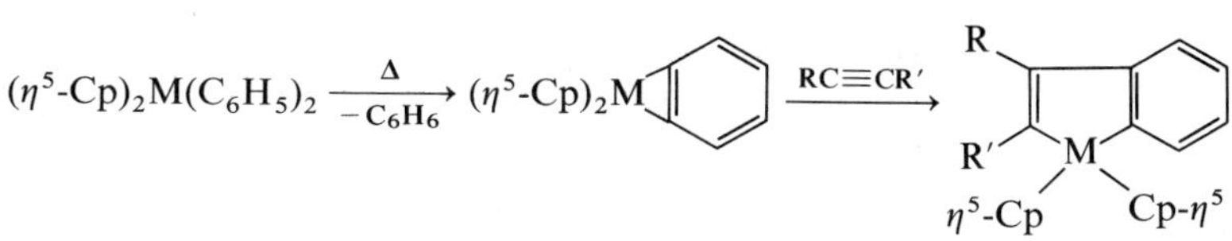

(c)[3]

with occasional regioselectivity [e.g., R = $Si(CH_3)_3$, R = C_6H_5; R = C_6H_5, R′ = C_6F_5]. Zirconocene[15–20] and hafnocene[22,23] show similar behavior, in particular the preferred emergence of the sterically more demanding sustituent [e.g., (a): R, R‴] alpha to the metal, especially when it is a trialkylsilyl group. This behaviour is usual for all metallacycles of this type, although there are exceptions, suggesting a subtle interplay of steric and electronic factors superimposed on kinetic and thermodynamic product control[39,55–58]. Retrocyclization is an important mechanistic option to be considered in this connection, a property that is used in the sequential replacement of the metallacycle ethene units in bis(cyclopentadienyl)hafnacyclopentane by diphenylethyne to isolate the corresponding 2,3-diphenylhafnacyclopenta-2,3-ene and, eventually, the 2,3,4,5-tetraphenylhafnacyclopenta-2,4-diene[23]. Although alkyne-derived metallacycles of the vanadium triad are extremely rare, these metals are only beginning to be explored as centers for organic transformations. For example a bis(cyclopentadienyl)niobatetrakis (trifluoromethyl)cyclopentadiene, is generated by irradiation of $(\eta^5\text{-Cp})_2NbH_3$ and $CF_3C{\equiv}CCF_3$ [25], as is the alkyne $RC{\equiv}CR'$ adducts [$R = R' = C_6H_5$; $R = C(CH_3)_3$, $Si(CH_3)_3$, or C_6H_5, $R' = H$] to tris(2,6-diisopropylphenoxy)tantalum(diphenylethyne), the latter in turn is made from the corresponding metal dichloride by reductive complexation[26,27]. The first actinide representative in this series, $[(CH_3)_5C_5]_2U[C_4(C_6H_5)_4]$, has become accessible in an analogous manner by reduction of various U(IV) chlorides to the hypothetical uranocene intermediate, followed by addition of the alkyne[28]. A transformation according to Eq. (c) [$(CH_3)_5C_5$ instead of η^5-Cp; M = U; $R = R' = C_6H_5$)] has also been observed[29].

All monometallacyclopentadienes whose x-ray structures have been determined contain the 1,3-butadiene-1,4-diyl fragment formulated in Eq. (a), with three exceptions, the 1-metalla-1,3,5-cyclopentadienes **I–III**[30,33,36], formed by the respective alkyne couplings:

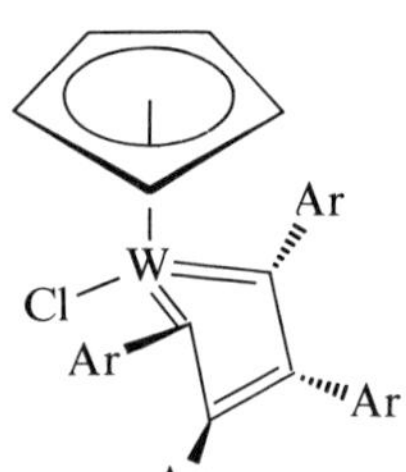

I

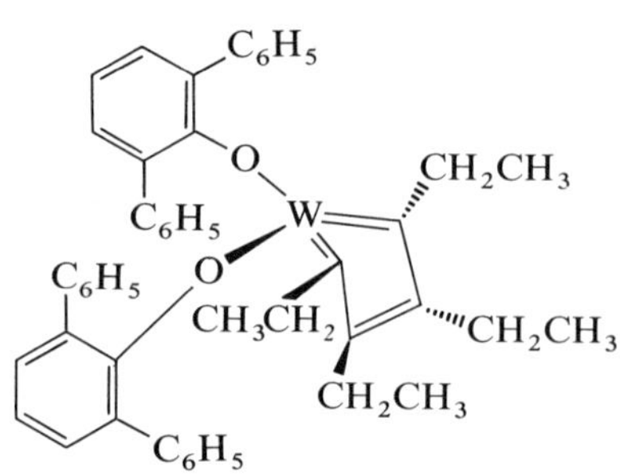

II

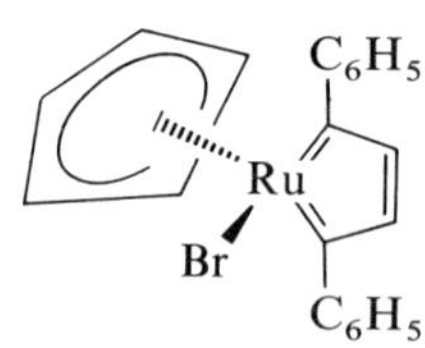

III

Crystallographic analyses confirm the presence of the unusual biscarbene ligands and reveal folded conformations and **1** and **2** to the extent of 60° (as shown), an arrangement that is predicted by calculations[31]. Compound **3** reacts with donor ligands by an associative pathway to give the 'normal' ruthenacyclopentadiene derivative. Isoelectronic Mo and W analogs of the latter, $\eta^5\text{-CpM}[CNC(CH_3)_3]_2(SR)[C_4(CF_3)_4]$, are isolated by inducing the (mechanistically perhaps complicated) coupling of 2-coordinated $CF_3C{\equiv}CCF_3$ ligands in $\eta^5\text{CpM(SR)}L_2$ by the addition of tert-butylisocyanide[32].

Despite the rich catalytic alkyne cyclization chemistry of Fe[6–8], there are few mononuclear ferracyclopentadienes, perhaps because of their high reactivity; e.g., reduction of $FeCl_2[P(CH_3)_3]_2$ with Na-Hg in 2-butyne gives the blue, highly volatile and thermally unstable $[(CH_3)_3P]_4Fe[C_4(CH_3)_4]$ and the cyclotrimer hexamethylbenzene[34]. A similar blue entity **IV** arises when the 4-coordinate complex $(dad)_2Fe$ is exposed to dimethyl butynedioate [dad = ethanedialbis(cyclohexylimine)][35].

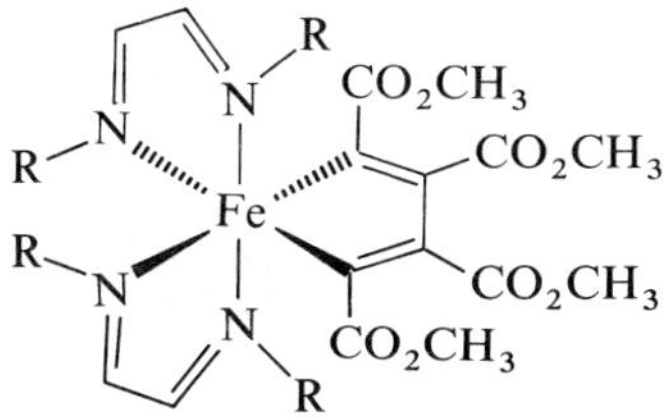

IV: R = C_6H_{11}

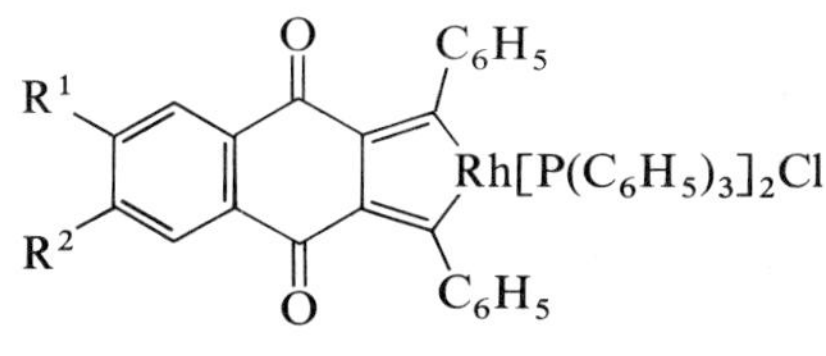

V

The elements of the cobalt triad are most inclined to assemble stable five-membered rings according to Eq. (a)[38–48], e.g., the two-step sequence: in Eq. (d)[38–41]

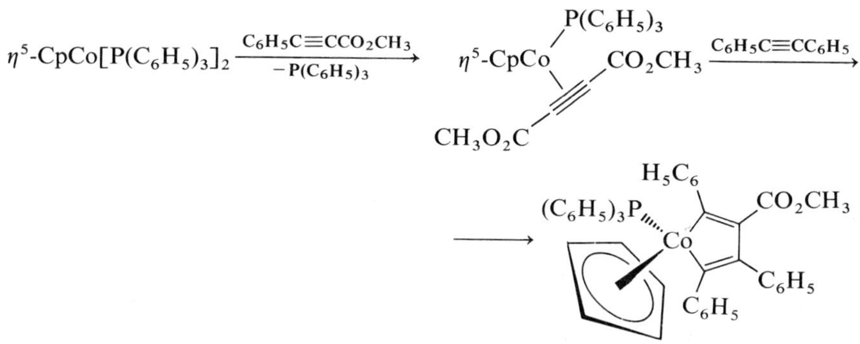

(d)

and complex **V**, derived from 1,2-bis(3-phenylpropynoyl)benzene and $[(C_6H_5)_3P]_3RhCl$[42–45]. The iridium relatives of these systems form by similar chemistry[46–48]. Mechanistic studies[39,40,46,48] implicate either auxiliary ligand predissociation before the alkyne substrates attach themselves to the metal, or intramolecular rearrangement, such as CO insertion [e.g., RM(CO) → RCOM], as prerequisites for metallacyclopentadiene construction. Unfortunately, there is no unambiguous example of the conversion of a transition-metal-bis(alkyne) complex to the corresponding metallacyclopentadiene that is amenable to mechanistic scrutiny. In the (hitherto observed) η^5-CpCo(RC≡CR′)$_2$, calculations imply that oxidative coupling induced by an external ligand[40,41] is energetically highly favorable; in other cases, the electronics and sterics of both the metals and the ligands must be considered[16,39,54–58]. The relatively large opus concerning late transition metals (notably those based on η^5-CpCo) is due to their superior capability of cooligomerizing alkynes with other alkynes or alkenes in stoichoimetric and (better) catalytic transformations; these turn relatively simple acyclic starting materials to the complex, polycyclic frameworks of natural (steroids, alkaloids, terpenes) and unnatural (strained, electronically activated benzenoids) products[1,7,8,15,16,42–45,59–66]. Specifically, cobaltacyclopentadienes function as precursors to pyridines, pyridinoxides, benzenes, cyclohexa- and butadienes, thiophenes, selenophenes, pyrroles, diketones, cyclobutadienes, cyclopentadienones and their imines, and phosphole oxides[67–72].

Mononickelacyclopentadienes are elusive, although they may function as catalysts in alkyne cyclotri- and, particularly, -tetramerizations[49]. In contrast, the equally active, oligomeric palladacyclopentadiene $\{Pd[C_4(CO_2CH_3)_4]\}_n$ has been well studied[50,51]. Such tetraesters are also available for Pt from $Pt_3[CNC(CH_3)_3]_6$ and bis(η^4-1,5-cyclooctadiene)Pt according to Eq. (a), the former by declusterification. The method is general and gives tetraarylmetallacyclopentadienes from diarylalkynes, and a 2:1 mixture of the corresponding methyl-2-butynoate coupling products ($R = R'' = CH_3$, $R' = R''' = CO_2CH_3$ and $R = R''' = CO_2CH_3$, $R' = R'' = CH_3$). via the novel involvement of 1,4-platina-2,5-cyclohexadienes[52]. Finally, there is the highly original, reductive coupling[53]:

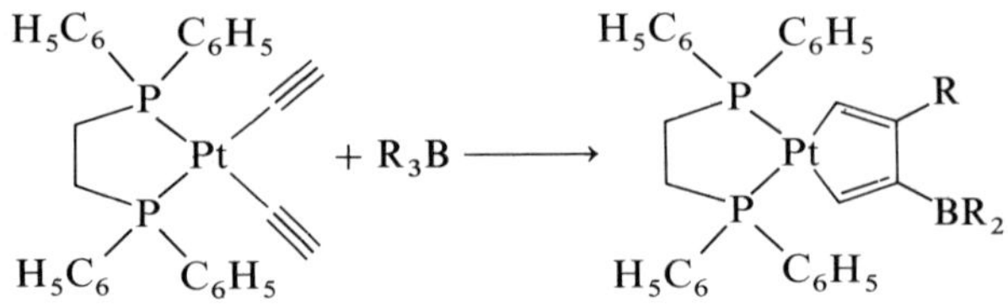

(e)

where $R = CH_3$, CH_3CH_2. $(CH_3)_2CH$. An alternative mode of inter- or intramolecular reagent-induced bisalkyne coupling, leads to unusual 1,3-metallacyclopentadiene derivatives[73–75];

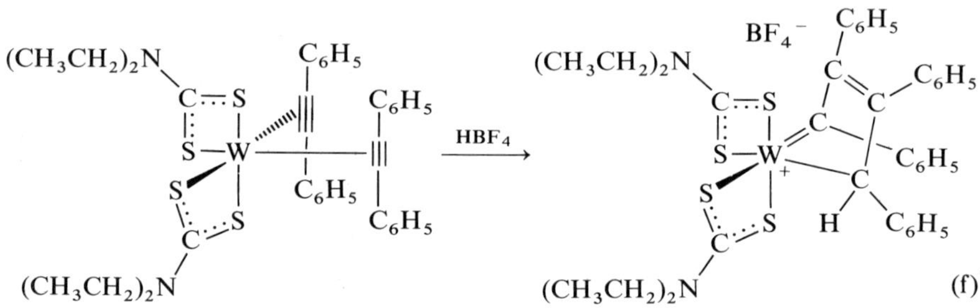

(f)[75]

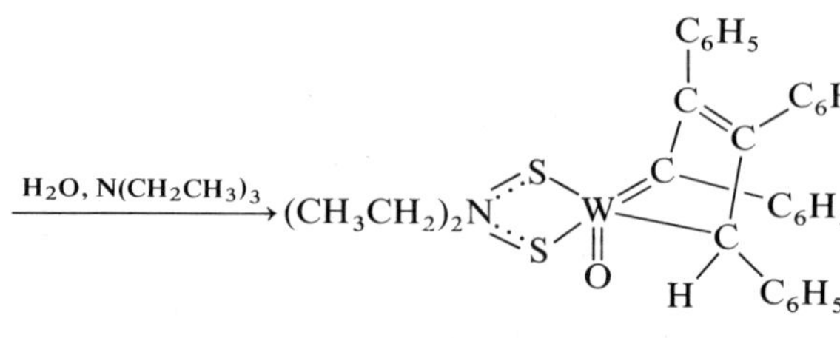

Many of the reactions of transition-metal complexes with alkynes proceed by a dinuclear variant of Eq. (a), either in competition with the latter or to its exclusion[76]:

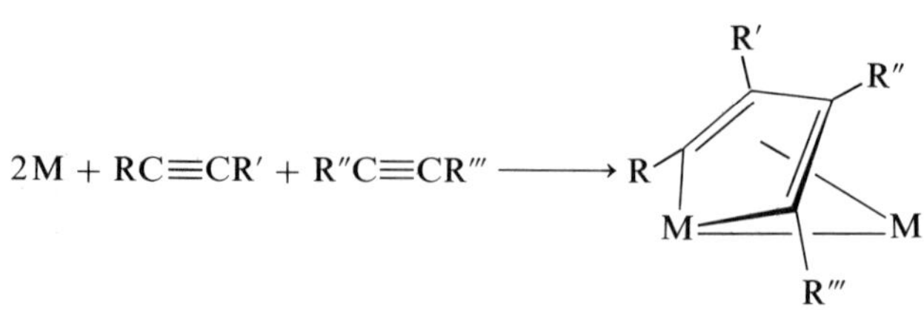

(g)

These systems are very stable thermally, although some are catalytically active. The hydrocarbon fragment adopts various configurations between the metals, occasionally leading to fluxionality equilibrating the environment around the two metal fragments.

The agglomeration of the four pieces making up the dimetallacyclopentadiene core according to Eq. (g) is not restricted to a particular sequence, all permutations (except that of initial alkyne coupling in the absence of M) being possible. The metals that transform in this manner are Ti, the elements of the Cr, Fe and Co triads and Ni, in addition to several examples that include heterobimetallic cores. A review[76] summarizes most of the literature until 1982. The cyclooctatetraene complex $(C_8H_8)TiCl(THF)$ is converted by arylkynes and $(CH_3)_2CHMgCl$ to the fluxional **VI** (for $R = R''' = C_6H_5$, $R' = R'' = CH_3$)[77]. A variation of Eq. (g) starts with the metal-metal triply bonded dinuclear species $(\eta^5\text{-Cp})_2M_2(CO)_4$ (M = Cr, Mo) to furnish the relatives of **VI**, compounds **VII**[78]:

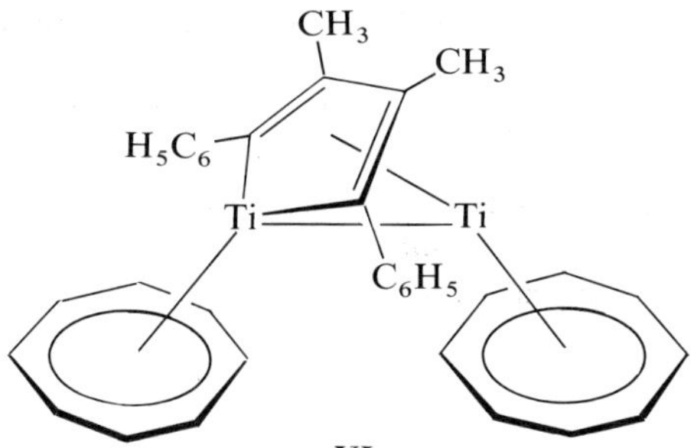

VI

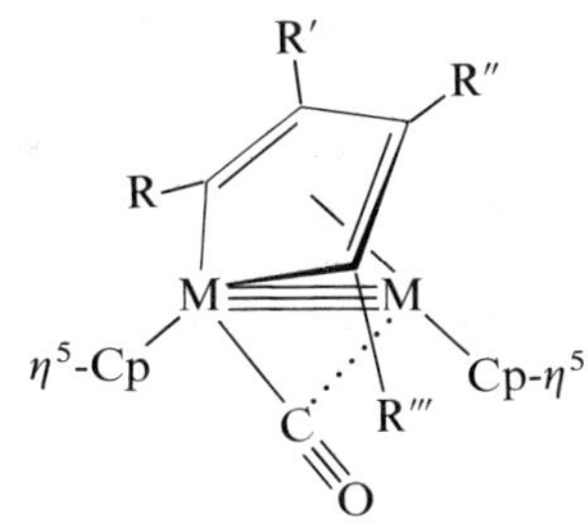

VII

The latter are intermediates in a multiple alkyne coupling sequence[78–81] thought to represent a model for the cyclooctatetraene synthesis from ethyne[82]. Isoelectronic with **VII** is a dimolybdenum by product in the preparation of **1**[30], curiously bearing a bridging $C_4(C_6H_5)_4$ ligand symmetrically disposed between the two metals, its plane perpendicular to the Mo—Mo bond and its terminal carbons adopting bridging carbene-like configurations. When some of the metal alkoxides $M_2(OR)_6$(M = Mo, W) are treated with ethyne, 'normal' bridge formation, as in **VII**, occurs exemplified by the catalytically active $Mo_2[OCH_2C(CH_3)_3]_6$ $(C_5H_5N)(C_4H_4)$ and the inactive $W_2[OCH_2C(CH_3)_3]_6(\eta^2\text{-}C_2H_2)(C_4H_4)$ [83,84] The catalytic inertia of the latter is noteworthy, considering that it is one of the rare examples of a metallacyclopentadiene bearing an alkyne that is expected to undergo ready addition to the neighboring diene fragment en route to benzene. These systems are unique in their ability to cause alkyne cleavage to biscarbyne intermediates[85]. Another variation of Eq. (g) begins with the bisalkyne complexes $\eta^5\text{-CpMCl}(CF_3C{\equiv}CCF_3)_2$ (M = Mo, W). When $Co_2(CO)_8$ is added $\eta^5\text{-Cp(CO)M}(RC{\equiv}CR)_2Co(CO)_2$ are generated and then undergo oxidative alkyne coupling if exposed to carbon monoxide, resulting in the heterodinuclear compounds **VIII**[86]

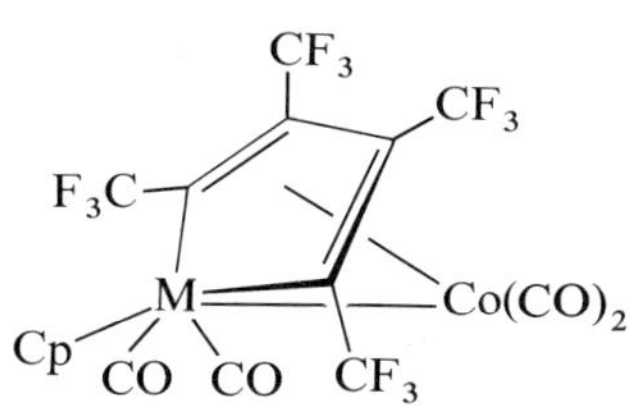

VIII: M = Mo, W

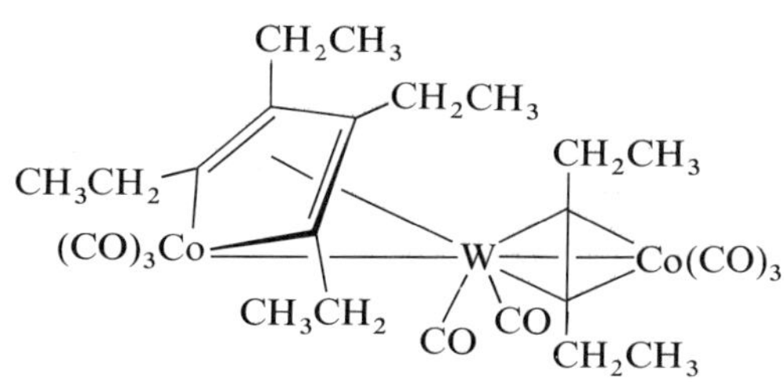

IX

Similar treatment of trisalkyne $W(CO)(CH_3CH_2C{\equiv}CCH_2CH_3)_3$ leads to trinuclear **IX** in one step[87], the (alkyne)$Co(CO)_3$ unit simply satisfying the coordination requirements of W. In contrast, $L_2Ni(C_4H_4)(C_2H_2)NiL_2$, L = $[(CH_3)_2CH]_2PCH_2CH_2P[CH(CH_3)_2]$, is formulated[88] as a monometallacyclopentadiene in which the alkyne links the two metals, although an x-ray structure is lacking. Finally, the symmetrical complexes $(dad)_2Ni_2(CHCRCRCH)$ may be intermediates in the catalytic conversion of RC≡CH to 1,2,5,6-tetrasubstituted cyclooctatetraenes[35,89].

Review articles and the references therein[1–8], examples of metallacyclopentadienes that are either esoteric, owe their origin to obscure pathways that nevertheless fit the topological change associated with Eq. (a), are derived from non-alkyne starting materials, contain additional heteroatoms, or assemble in the coordination sphere of clusters[90,92].

The coupling of alkynes with alkenes according to Eq. (b) is rare, presumably because of the tendency of the alkyne substrates to autocyclize. The resulting metalla-2-cyclopentenes constitute potentially valuable intermediates in enyne cooligomerizations to synthetically versatile dienes[1–3, 59–66], a field that is gaining increasing importance[93]. Early investigations of these systems focused on cobalt and rhodium:

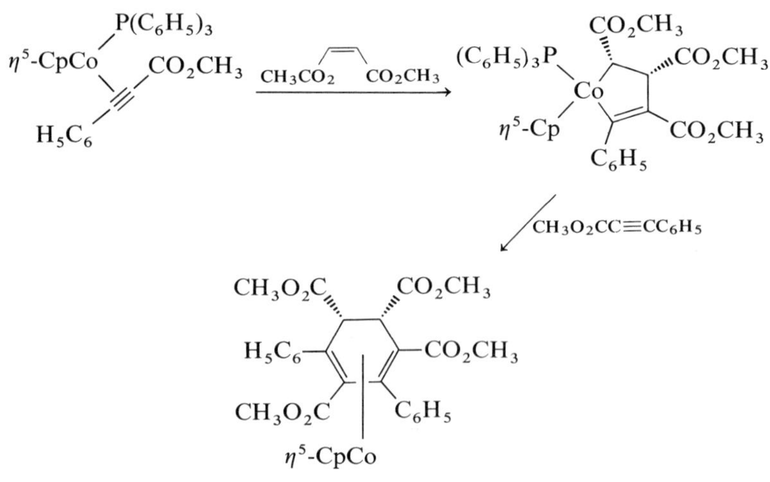

(h)[94]

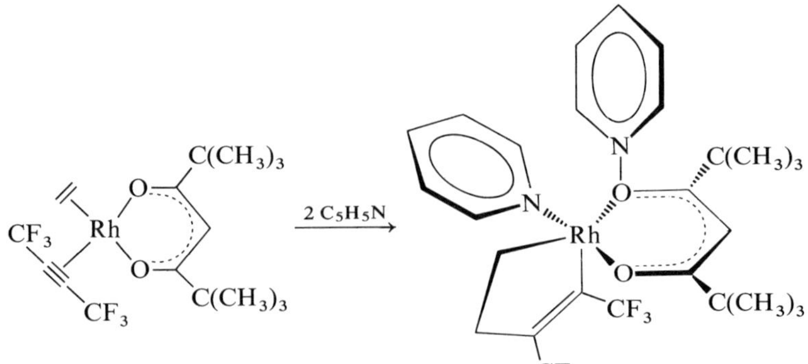

(i)[95]

Synthetic access to in situ generated metallocenes, especially of the titanium triad but also its neighbors, has given a growing number of reactive alkene and alkyne π-complexes that readily transform according to Eq. (b)[15–20,23,96–98]. The chemical behavior of these metallacyclopentenes, such as that indicated in Eq. (h) for cobalt, remains to be fully explored.

(K. P. C. VOLLHARDT)

1. J. P. Collman, L. S. Hegedus, J. R. Norton, R. G. Fincke, *Principles and Applications of Organotransition Metal Chemistry*, University Science Books, Mill Valley, CA, 1987.
2. E. Lindner, *Adv. Heterocyl. Chem., 39* 237 (1986).
3. S. D. Chappell, D. J. Cole-Hamilton, *Polyhedron, 1*, 739 (1982).
4. S. Otsuka, A. Nakamura, *Adv. Organomet. Chem., 14*, 245 (1976).
5. R. S. Dickson, P. J. Fraser, *Adv. Organomet. Chem., 12*, 323 (1974).
6. L. P. Yur'eva, *Russ. Chem. Rev., 43*, 48 (1974).
7. F. L. Bowden, A. B. P. Lever, *Organomet. Chem. Rev., 3*, 227 (1968).
8. I. Wender, P. Pino, *Organic Syntheses via Metal Carbonyls*, Vol. 1, Interscience Publishers, New York, 1968.
9. H. G. Alt, H. E. Engelhardt, M. D. Rausch, L. B. Kool, *J. Organomet. Chem., 329*, 61 (1987).
10. V. B. Shur, E. G. Berkovich, M. E. Vol'pin, B. Lorenz, M. Wahren, *J. Organomet. Chem., 228*, C36 (1982).
11. J. Mattia, D. J. Sikora, D. W. Macomber, M. D. Rausch, J. P. Hickey, G. D. Friesen, L. J. Todd, *J. Organomet. Chem., 213*, 441 (1981).
12. B. Klei, J. H. Teuben, H. J. de Liefde Meijer, *J. Chem. Soc., Chem. Commun.*, 342 (1981).
13. G. Fachinetti, C. Floriani, F. Marchetti, M. Mellini, *J. Chem, Soc., Dalton Trans.*, 1398 (1978).
14. H. Alt, M. D. Rausch, *J. Am. Chem. Soc.*, 96, 5936 (1974).
15. E. Negishi, T. Takahashi, *Synthesis*, 1 (1988) and refs. therein.
16. W. A. Nugent, D. L. Thorn, R. L. Harlow, *J. Am. Chem. Soc., 109*, 2788 (1987).
17. S. L. Buchwald, E. A. Lucas, J. C. Dewan, *J. Am. Chem. Soc., 109*, 4936 (1987).
18. S. L. Buchwald, B. T. Watson, J. C. Huffman, *J. Am. Chem. Soc., 109*, 2544 (1987); *108*, 7411 (1986).
19. S. L. Buchwald, R. T. Lum, J. C. Dewan, *J. Am. Chem. Soc., 108*, 7441 (1986).
20. D. J. Sikora, M. D. Rausch, *J. Organomet. Chem., 276*, 21 (1984).
21. V. Skibbe, G. Erker, *J. Organomet. Chem., 241*, 15 (1983).
22. M. B. Sabade, M. F. Farona, E. A. Zarate, W. J. Youngs, *J. Organomet. Chem., 338*, 347 (1988).
23. G. Erker, U. Dorf, A. L. Rheingold, *J. Organometallics, 7*, 138 (1988) and refs. therein.
24. J. Sala-Pala, J. Amaudrut, J. E. Guerchais, R. Mercier, J. Douglade, J. G. Theobald, *J. Organomet. Chem., 204*, 347 (1981).
25. D. A. Lemenovskii, T. V. Baukova, G. Zysik, V. A. Krizhnikov, V. P. Fedin, E. G. Perevalova, *Koord. Khim., 4*, 1033 (1978); *Sov. J. Coord. Chem.* (*Engl. Transl.*), 4, 560 (1978).
26. M. A. Bruck, A. S. Copenhaver, D. E. Wigley, *J. Am. Chem. Soc., 109*, 6525 (1987).
27. J. R. Strickler, P. A. Wexler, D. E. Wigley, *Organometallics 7*, 2067 (1988).
28. J. M. Manriquez, P. J. Fagan, T. G. Marks, S. H. Vollmer, C. S. Day, V. W. Day, *J. Am. Chem. Soc., 101*, 5075 (1979).
29. P. J. Fagan, J. M. Manriquez, E. A. Maatta, A. M. Seyam, T. J. Marks, *J. Am. Chem. Soc., 103*, 6650 (1981).
30. W. Hirpo, M. D. Curtis, *J. Am. Chem. Soc., 110*, 5218 (1988).
31. M. D. Curtis, J. Real, *J. Am. Chem. Soc., 108*, 4668 (1986).
32. J. L. Davidson, W. F. Wilson, *J. Chem. Soc., Dalton Trans.*, 27 (1988).
33. J. L. Kerschner, P. E. Fanwick, I. P. Rothwell, *J. Am. Chem. Soc., 110*, 8235 (1988).
34. J. W. Rathke, E. L. Muetterties, *J. Am. Chem. Soc., 97*, 3272 (1975).
35. H. T. Dieck, L. Stamp, R. Diercks, C. Müller, *Nouv. J. Chim., 9*, 289 (1985).
36. M. O. Albers, D. J. A. de Waal, D. C. Liles, D. J. Robinson, E. Singleton, M. B. Wiege, *J. Chem. Soc., Chem. Commun.*, 1680 (1986).
37. R. Burt, M. Cooke, M. Green, *J. Chem. Soc. A*, 2981 (1970)
38. H. Yamazaki, Y. Wakatsuki, *J. Organomet. Chem., 272*, 251 (1984) and refs. therein.

39. Y. Wakatsuki, O. Nomura, K. Kitaura, K. Morokuma, H. Yamazaki, *J. Am. Chem. Soc.*, *105*, 1907 (1983).
40. D. R. McAlister, J. E. Bercaw, R. G. Bergman, *J. Am. Chem. Soc.*, *89*, 1666 (1977).
41. P. Hofmann, Technical University of Munich, private communication.
42. J. T. Mague, M. O. Nutt, E. H. Gause, *J. Chem. Soc., Dalton Trans.*, 2578 (1978).
43. J. Hambrecht, E. Müller, *Z. Naturforsch, Teil B*, *32*, 68 (1977).
44. R. G. Gastinger, M. D. Rausch, D. A. Sullivan, G. J. Palenik, *J. Organomet. Chem.*, *117*, 355 (1976).
45. E. Müller, *Synthesis*, 1 (1974).
46. B. J. Rappoli, M. R. Churchill, T. S. Janik, W. M. Rees, J. D. Atwood, *J. Am. Chem. Soc.*, *109*, 5145 (1987) and refs. therein.
47. W. H. Baddley, G. B. Tupper, *J. Organomet. Chem.*, *67*, C16 (1974).
48. J. P. Collman, J. W. Kang, W. F. Little, M. F. Sullivan, *Inorg. Chem.*, *7*, 1298 (1968).
49. J. J. Eisch, J. G. Galle, A. A. Aradi, M. P. Boleslawski, *J. Organomet. Chem.*, *312*, 399 (1986).
50. L. D. Brown, K. Itoh, H. Suzuki, K. Hirai, J. Ibers, *J. Am. Chem. Soc.*, *100*, 8232 (1978).
51. P. M. Maitlis, *Acc. Chem. Res.*, *9*, 93 (1976).
52. N. M. Boag, G. H. M. Dias, M. Green, J. L. Spencer, F. G. A. Stone, J. Vincente, *J. Chem. Soc., Dalton Trans.*, 1981 (1981).
53. A. Sebald, B. Wrackmeyer, *J. Chem. Soc., Chem Commun.*, 1293 (1983).
54. K. Tatsumi, A. Nakamura, *J. Am. Chem. Soc.*, *109*, 3195 (1987).
55. R. J. McKinney, D. L. Thorn, R. Hoffmann, A. Stokis, *J. Am. Chem. Soc.*, *103*, 2595 (1981).
56. A. Stokis, R. Hoffmann, *J. Am. Chem. Soc.*, *102*, 2952 (1980).
57. Y. Wakatsuki, O. Nomura, H. Tone, *J. Chem. Soc., Perkin Trans. II*, 1344 (1980).
58. D. L. Thorn, R. Hoffmann, *Nouv. J. Chim.*, *3*, 39 (1979); *Inorg. Chem.*, 17, 126 (1978).
59. P. J. Fagan, W. A. Nugent, *J. Am. Chem. Soc.*, *110*, 2310 (1988).
60. N. E. Schore, *Chem. Rev.*, *88*, 1081 (1988).
61. B. M. Trost, D. C. Lee, *J. Am. Chem. Soc.*, *110*, 7255 (1988).
62. R. Grigg, R. Scott, P. Stevenson, *J. Chem. Soc., Perkin Trans. I*, 1357 (1988).
63. G. P. Chiusoli, *J. Mol. Catal.*, *41*, 75 (1987).
64. K. P. C. Vollhardt, *Heterocycles*, *24*, S59 (1987); *Angew. Chem., Int. Ed. Engl.*, *23*, 539 (1984).
65. R. Boese, H.-J. Knölker, K. P. C. Vollhardt, *Angew. Chem., Int. Ed. Engl.*, *26*, 1035 (1987).
66. L. Blanco, H. E. Helson, M. Hirthammer, H. Mestdagh, S. Spyroudis, K. P. C. Vollhardt, *Angew. Chem., Int. Ed. Engl.*, *26*, 1246 (1987).
67. Y. Wakatsuki, S. Miya, H. Yamazaki, S. Ikuta, *J. Chem. Soc., Dalton Trans.*, 1201 (1986).
68. Y. Wakatsuki, S. Miya, H. Yamazaki, *J. Chem. Soc., Dalton Trans.*, 1207 (1986).
69. Y. Wakatsuki, H. Yamazaki, *J. Chem. Soc., Dalton Trans.*, 1278 (1978); *Bull. Chem. Soc. Jpn.*, *52*, 1239 (1979); *J. Organomet. Chem.*, *139*, 169 (1977); *149*, 377 (1978); *149*, 385 (1978).
70. P. Hong, Y. Yamazaki, *Synthesis*, 50 (1977).
71. F.-W. Grevels, Y. Wakatsuki, Y. Yamazaki, *J. Organomet. Chem.*, *141*, 331 (1977).
72. Y. Wakatsuki, T. Kuramitsu, H. Yamazaki, *Tetrahedron Lett.*, 4549 (1974).
73. J. L. Davidson, *J. Chem. Soc., Dalton Trans.*, 2715 (1987).
74. L. Carlton, J. L. Davidson, P. Ewing, L. Manojlovic-Muir, K. W. Muir, *J. Chem. Soc., Chem Commun.*, 1474 (1985).
75. J. R. Morrow, T. L. Tonker, J. L. Templeton, *J. Am. Chem. Soc.*, *107*, 5004 (1985).
76. J. Holton, M. F. Lappert, R. Pearce, P. I. W. Yarrow, *Chem. Rev.*, *83*, 135 (1983).
77. M. E. E. Veldman, H. R. van der Wal, S. J. Veenstra, H. J. de Liefde Meijer, *J. Organomet. Chem.*, *197*, 59 (1980).
78. S. A. R. Knox, R. F. D. Stansfield, F. G. A. Stone, M. J. Winter, P. Woodward, *J. Chem. Soc., Dalton Trans.*, 173 (1982).
79. M. Green, P. A. Kale, R. J. Mercer, *J. Chem. Soc., Chem. Commun.*, 375 (1987).
80. L. Brammer, M. Green, A. G. Orpen, K. E. Paddick, D. R. Saunders, *J. Chem. Soc., Dalton Trans.*, 657, (1986).
81. D. B. Pourreau, R. R. Whittle, G. L. Geoffroy, *J. Organomet. Chem.*, *273*, 333 (1984).
82. W. Peppe, O. Schlichting, K. Kloger, T. Toebel, *Justis Liebigz Ann. Chem.*, *560*, 1 (1948).
83. M. H. Chisholm, K. Folting, J. C. Huffman, I. P. Rothwell, *J. Am. Chem. Soc.*, *104*, 4389 (1982).
84. M. H. Chisholm, D. M. Hoffman, J. C. Huffman, *J. Am. Chem. Soc.*, *106*, 6806 (1984).
85. M. H. Chisholm, B. K. Conroy, J. C. Huffman, *Organometallics*, *5*, 2384 (1986).
86. J. L. Davidson, *J. Chem. Soc., Dalton Trans.*, 1667 (1983).

87. L. J. Farrugia, M. J. Went, *J. Chem. Soc., Chem. Commun.*, 973 (1987).
88. K.-R. Pörschke, *Angew. Chem., Int. Ed. Engl., 26*, 1288 (1987).
89. R. Dierks, L. Stamp, J. Kopf, H. tom Dieck, *Angew. Chem., Int., Ed. Engl., 23*, 893 (1984).
90. B. C. Gates, L. Guczi, H. Knözinger, (eds.), *Metal Clusters in Catalysis*, Elsevier, Amsterdam, Netherlands 1986.
91. M. Moskovits, (ed.), *Metal Clusters*, John Wiley and Sons, New York, 1986.
92. E. Sappa, A. Tiripicchio, P. Braunstein, *Coord. Chem. Rev., 65*, 219 (1985).
93. For recent reports, see: T. V. RajanBabu, W. A. Nugent, D. F. Taber, P. J. Fagan, *J. Am. Chem. Soc., 110*, 7128 (1988); K. Tamkao, K. Kobayashi, Y. Ito, *J. Am. Chem. Soc., 110*, 1287 (1988); B. M. Trost, J. M. Tour, *J. Am. Chem. Soc., 109*. 5268 (1987); B. M. Trost, *G. J. Tanoury, J. Am. Chem. Soc., 109*. 4753 (1987); B. M. Trost, F. Rise, *J. Am. Chem. Soc., 109*, 3161 (1987).
94. Y. Wakatsuki, K. Aoki, H. Yamazaki, *J. Am. Chem. Soc., 101*, 1123 (1979); see also M. Scozzafava, A. M. Stolzenberg, *Organometallics, 7*, 1073 (1988).
95. M. E. Howden, R. D. W. Kemmitt, M. D. Schilling, *J. Chem. Soc., Dalton Trans.*, 2459 (1983).
96. S. A. Cohen, J. E. Bercaw, *Organometallics, 4*, 1006 (1985).
97. C. McDade, J. E. Bercaw, *J. Organomet. Chem., 279*, 281 (1985).
98. G. Erker, K. Kropp, *J. Am. Chem. Soc., 101*, 3659 (1979).

5.8.2.7.5. η^6 π-Arene Complexes by Cyclooligomerization of Alkynes

Transition-metal compounds catalyze the cyclooligomerization of alkynes, giving rise to benzene and cyclooctatetraene derivatives[1].

Insertion of the carbonyl group is observed in the reaction of alkynes with metal carbonyls with the consequent formation of cyclopentadienones, tropones, quinones and their η^6 complexes[2].

In the cyclotrimerization of alkynes to give substituted benzenes few intermediates are isolated owing to their thermal lability or their high reactivity. Metallocyclopentadiene derivatives are involved in these reactions[3]:

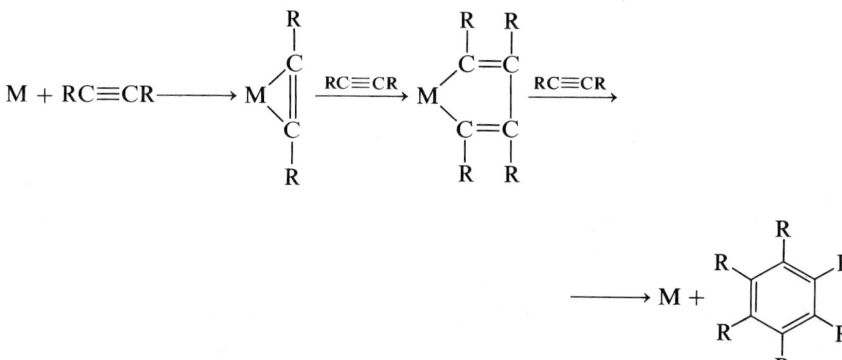

(a)

Furthermore, addition of an alkyne to a metallocyclopentadiene derivative can proceed through two distinct mechanisms, according to the nature of the R groups: one path involves entry of the alkyne into the coordination sphere of the metal (by replacement of another ligand), and the other occurs by direct complexation of the alkyne with the metallocycle[4].

But-2-yne undergoes cyclic condensation when reacted with triphenylchromium(III) tetrahydrofuranate to yield hexamethylbenzene and bis-hexamethylbenzene–chromium complex[5,6]; in the same reaction 1,2,3,4-tetramethylnaphthalene and its Cr η^6 complex is also recovered. Also dimesitylchromium and but-2-yne (in THF) give rise to hexamethylbenzene and the bis-hexamethylbenzeneCr π complex, but here the formation of the naphthalenic derivatives is avoided by the presence of the σ-methyl groups of the mesitylmetal organics.

Good yields of $(\eta^6\text{-}C_6R_6)Cr(I)(C_5R_5\text{-}\eta^5)$ complexes (R = Me, Et) are obtained[7] when $CrCl_3$ is allowed to react with but-2-yne or hex-3-yne (in benzene or toluene) in the presence of AlR_3:

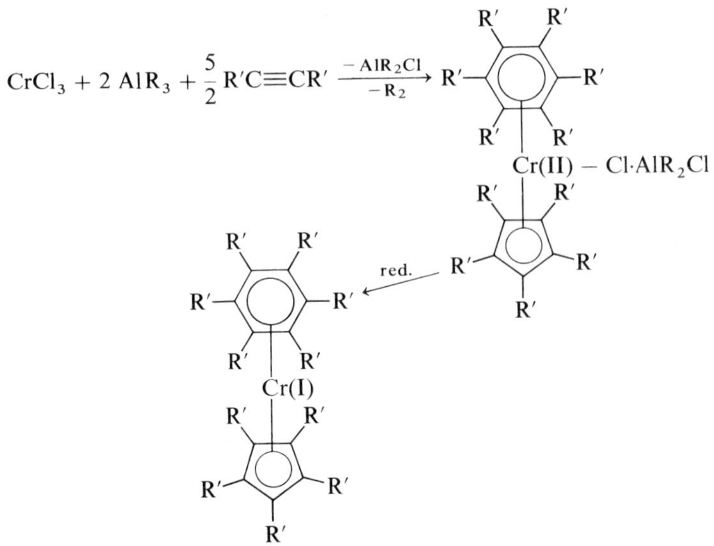

(b)

The latter stage of reduction can be affected by treatment with $LiAlH_4$ or by hydrolysis with dithionite.

The formation of the five-membered ring is related to the cleavage of the triple bond of the alkyne with the subsequent reaction of each half with 2 mols alkyne.

Diphenylmanganese, prepared from phenylmagnesium bromide and Mn(II) chloride, with but-2-yne in diethyl (31%) and the corresponding η^6 complex (10%), isolated as the tetraphenylboron salt[8].

Diphenylmanganese (or its mesityl analog) fails to react with but-2-yne in THF, but the less basic diethyl ether allows incorporation of acetylene in the complex and the subsequent reaction. No naphthalenic product is detected, suggesting that they form only when arylchromium(II or III) is used as the condensation reagent.

Cyclotrimerization of acetylenedicarboxylic acid and hexafluorobut-2-yne occurs with $Fe_2(CO)_9$ [9] and trans-$Ru(CO)_3L_2$ [10] (L = tertiary phosphine or phosphite) to give $[\eta^6\text{-}C_6(CO_2Me)_6]Fe(CO)_3$ and $[\eta^6\text{-}C_6(CF_3)_6]Ru(CO)_2L$, respectively.

Ultraviolet irradiation of $\eta^6\text{-}C_6H_6(C_6H_8)Ru$ with RC≡CR, in n-hexane, at 40°C for 1-2 d offers a convenient route to a new bis-arene derivatives, $\eta^6\text{-}C_6H_6(\eta^6\text{-}C_6R_6)Ru$ ($R = C_6H_5$, η^6 $COOCH_3$) and $(\eta^6\text{-}C_6R_6)_2Ru$ ($R = CH_3$)[11].

Dimesitylcobalt(II) (prepared from mesitylmagnesium bromide and $CoCl_2$) reacts at −30°C in THF with but-2-yne[8]. The bis-hexamethylbenzenecobalt(I) π complex is isolated as halide, picrate and tetraphenylboron salts. This complex is catalytic in the cyclization of the alkyne, since its amount remains constant while the yields of hexamethylbenzene increase as the mole ratio between the dimesithylCo(II) and but-2-yne is increased.

Facile replacement of a NBD unit is observed when $Rh(NBD)_2$ is reacted with but-2-yne (in excess at 25°C in acetone) to form $[Rh(NBD)(\eta^6$-hexamethylbenzene)$]^+$ (yield 15%, in a few hours)[12]. The η-cyclopentadienyl-14-η-hexakis(trifluoro-
methyl)benzene complex is obtained in the reactions of hexafluorobut-2-yne with η^5-$CpM(CO)_2$ (M = Co and Rh): its formation is favored for M = Rh provided forcing reaction conditions (e.g., 120°C, excess of alkyne) are used[13].

Cyclotrimerization of $(CF_3)_2C_2$ also occurs in the reactions with Rh(I)-β-keto-enolate rings and related systems[14]:

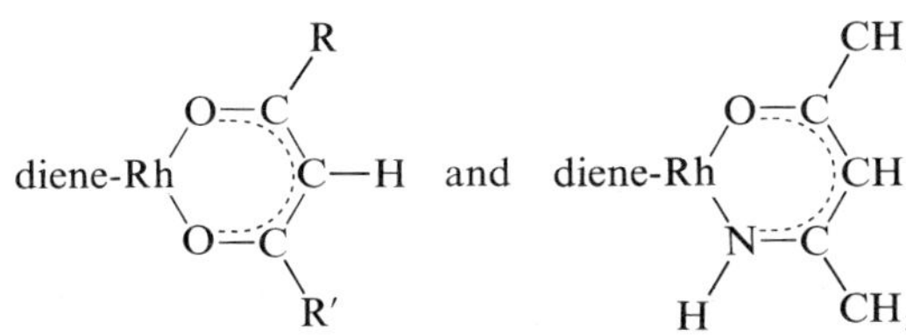

where if R = R′ = Ph, diene = COD; diene = COD, NBD, R = Me, R′ = NHPh diene- = COD, NBD.

The reaction occurs in diethylether at RT and gives rise to the yellow crystalline adducts:

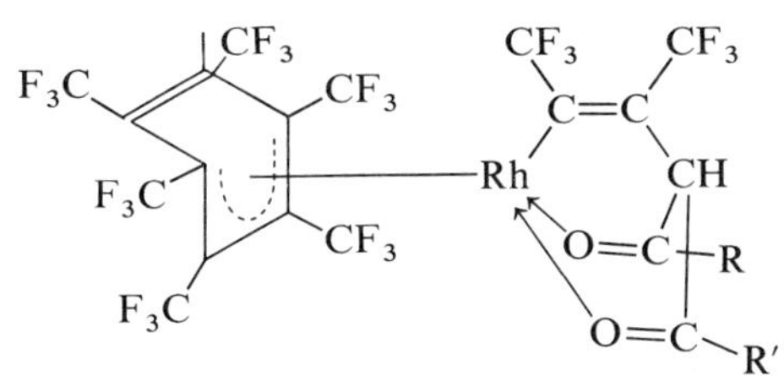

and

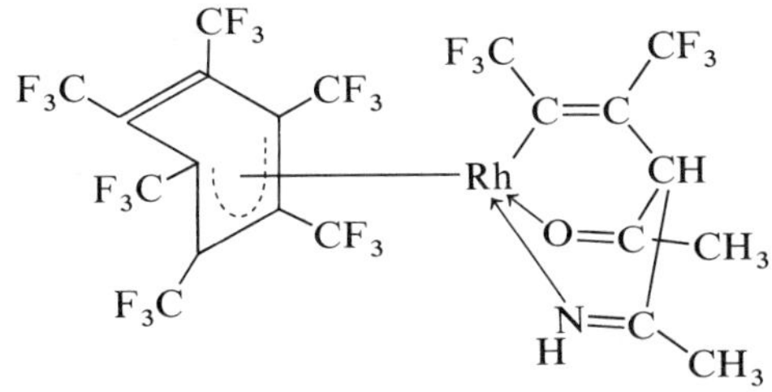

where if R = R[1] = Ph; if R = Me, R′ = NHPh. When the reaction is carried out in the presence of a calculated amount of air (ca. 2 mol of O_2 per 1 mol of complex) η^5-$CpRh(PPh_3)_2$ gives η^5-$CpRh(RC{\equiv}CR')_3$ (R = CO_2Me; R′ = Me or Ph)[15]; under a N_2 atmosphere the complex η^5-$CpRh(PPh_3)L$ (L = MeO_2-$CC{\equiv}CCO_2Me$) is formed.

A route to an η^6-arene complex is represented by the use of o-phenylethynyltriphenylphosphane with $RhCl(PPh_3)_3$[16]. The octahedral rhodacyclopentadiene chelate complex is easily formed, which further reacts with tolan in boiling xylene:

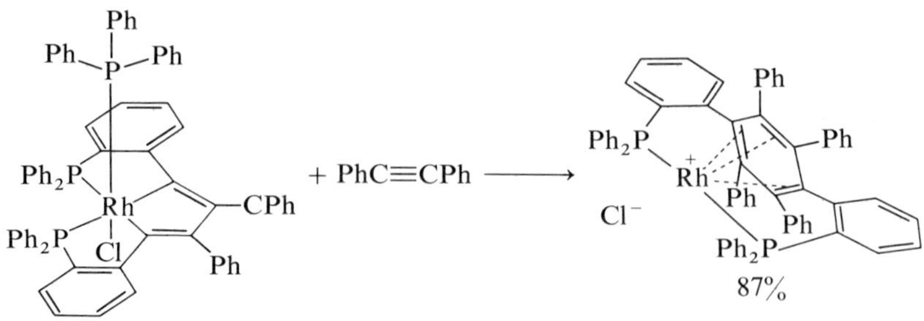

(c)

The chelate function of the tertiary bis-phosphane then allows then the isolation of this η^6-arene complex, which represents a good model for the well-known Rh-catalyzed trimerization of alkynes.

The (C∩D)$_2$Ni complex reacts in benzene in a sealed tube with excess hexafluorobut-2-yne for 7 d at RT to give (COD)Ni[$C_6(CF_3)_6$] (29% yield) and $(COD)_2Ni_2[C_6(CF_3)_6]$ (63% yield)[17], the latter compound having the structure:

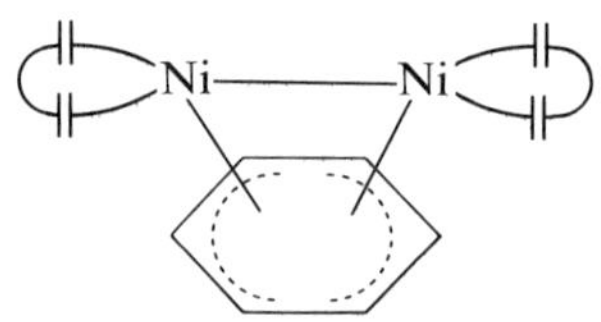

The derivative $(AsMe_2Ph)_2Ni[C_6(CF_3)_6$-$\eta^6]$ can be obtained in 69% yield from $Ni(AsMe_2Ph)_4$ and $C_2(CF_3)_2$.

(S. AIME)

1. L. P. Yur'eva, *Russ. Chem. Rev. (Engl. Transl.), 43*, 48 (1974).
2. R. F. Heck, *Organotransition Metal Chemistry*, Academic Press, New York, 1967.
3. S. Otsuka, A. Nakamura, *Organomet. Chem. Rev., 14*, 245 (1976).
4. D. R. McAlister, J. E. Bercaw, R. G. Bergman, *J. Am. Chem. Soc., 99*, 1666 (1977).
5. H. Zeiss, W. Herwig, *J. Am. Chem. Soc., 80*, 291, (1958).
6. W. Herwig, W. Metlesies, H. Zeiss, *J. Am. Chem. Soc., 81*, 6203 (1959).
7. H. Benn, G. Wilke, D. Henneberg, *Angew. Chem., Int. Ed. Engl., 12*, 1001 (1973).
8. M. Tsutsui, H. Zeiss, *J. Am. Chem. Soc., 83*, 825 (1961).
9. M. Avram, I. G. Dinnlescu, Gh. D. Mateescu, C. D. Nenitzescu, *Rev. Roum. Chim., 14*, 1191 (1969).
10. R. Burt, M. Cook, M. Green, *J. Chem. Soc., A*, 1981 (1970).
11. A. Lucherini, L. Porri, *J. Organomet. Chem., 155*, C45 (1978).
12. R. R. Schrock, J. A. Osborn, *J. Am. Chem. Soc., 93*, 3098 (1971).
13. R. S. Diekson, H. P. Kirsch, *Aust. J. Chem., 27*, 61 (1974).
14. A. C. Jarvis, R. D. W. Kemmett, *J. Organomet. Chem., 136*, 121 (1977).
15. Y. Wakatsuki, H. Yamazaki, *J. Organomet. Chem., 64*, 393 (1974).
16. W. Winter, *Angew. Chem., Int. Ed. Engl., 15*, 241 (1976).
17. J. Browning, C. S. Cundy, M. Green, F. G. A. Stone, *J. Chem. Soc., A*, 448 (1971).

5.8.2.7.6. η^1 σ-Alkenyl Complexes by Nucleophilic Attack on π-Alkyne– and Allene–Metal Complexes.

As outlined in §5.8.2.3.4, coordination of an alkene to a neutral or cationic transition metal activates the alkene toward nucleophilic attack, leading to an alkylmetal product. Development of the analogous synthesis of alkenylmetal complexes starting with η^2-alkyne–metal complexes is more recent. An early reaction of this type is[1,2]:

$$\text{trans-PtCl(Me)(PMe}_3)_2 + \text{AgPF}_6 + \text{MeC}{\equiv}\text{CMe} \xrightarrow{\text{MeOH}} \text{trans-[PtMe(MeC}{\equiv}\text{CMe)(PMe}_3)_2]\text{PF}_6 + \text{AgCl}$$

$$\xrightarrow[\text{12 h, 30°C}]{\text{MeOH}} [\text{Pt[C(Me)=C(OMe)Me](PMe}_3)_2]\text{PF}_6 + \text{CH}_4$$

$$\xrightarrow{\text{NaCl}} \text{trans-PtCl[C(Me)=C(OMe)Me](PMe}_3)_2 + \text{NaPF}_6 \qquad \text{(a)}$$

The alkenyl group is assigned an E configuration based on NMR coupling constant arguments. A second example, in which the stereo- and regiochemistry of the alkenyl group is established, is[3]:

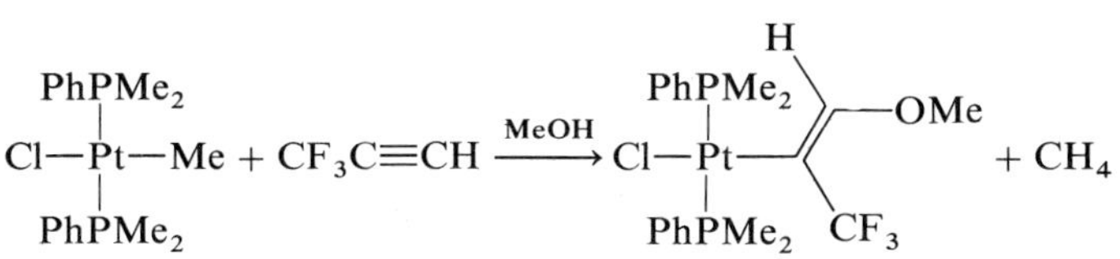

(b)

From these data the reactions take place by overall trans addition to the π-alkyne complex.

Hydride addition can occur to a π-alkyne complex[4]:

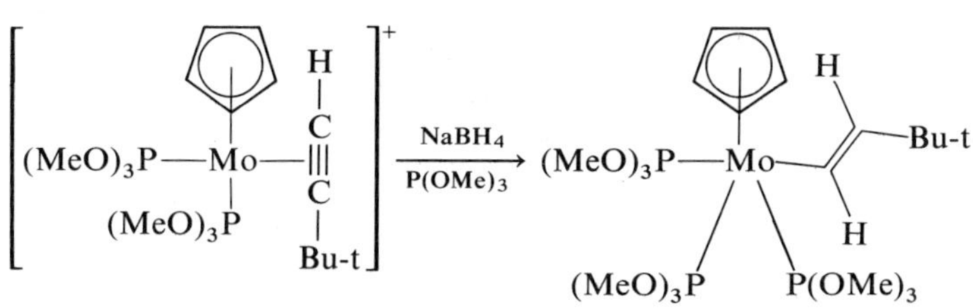

(c)

The E configuration is verified by crystallographic techniques[5]. This reaction is suggested to take place by hydride addition to the metal, followed by alkyne insertion[5].

Similar reactions in the absence of added phosphite with hydride and organocuprate reagents lead to the η^2-alkenyl complexes[7–9]:

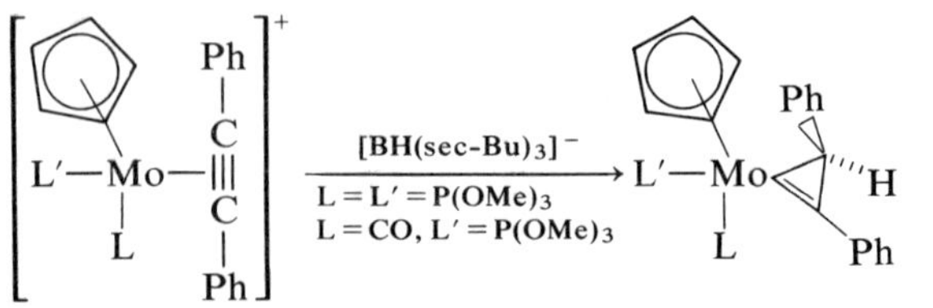

(d)

The formation of η^2-alkenyl products is reasonable for these complexes because NMR studies[10] indicate that the alkyne is a four-electron donor in these starting materials (making them 18-electron species). Many other examples of reactions of this type that lead to η^2-alkenyl complexes are reported[11–13]. An interesting example is:

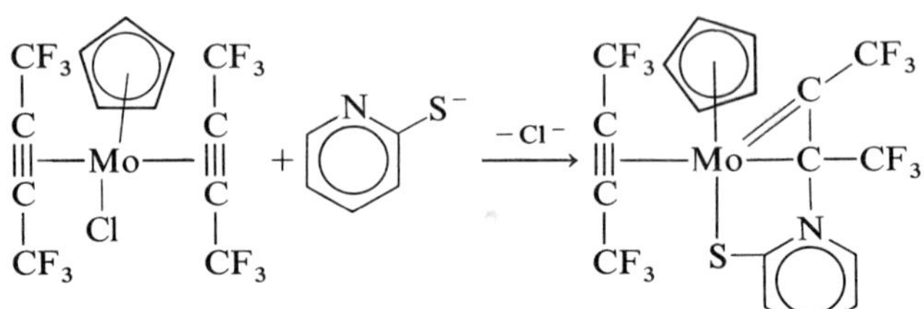

(e)

the product of which is confirmed crystallographically[11]. Phosphines and isocyanides also add to the alkynes similarly[14]:

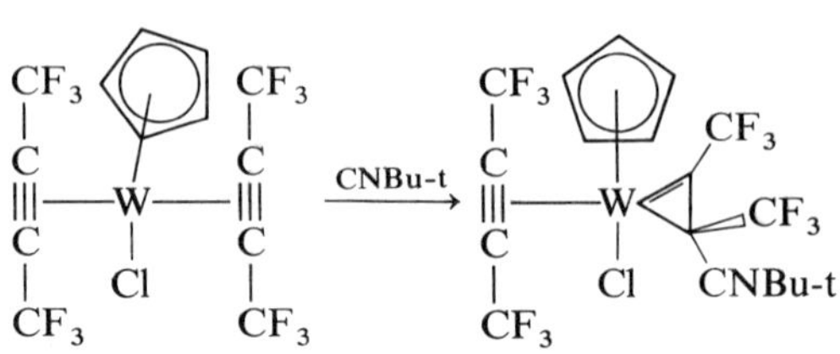

(f)

Nucleophiles; including H^- [15,16] from $NaBH_3CN$ or $KBH(s\text{-}Bu)_3$, Me^-, Ph^-, $CH{=}CH_2^-$, $C{\equiv}CMe^-$, from LiR_2Cu or $R_2Cu(CN)Li_2$, $CH(CO_2Et^-$, CN^- and RS^-, add to $[\eta^5\text{-}CpFe(CO)(L)(\eta^2\text{-alkyne})]BF_4$ [L=CO,PPh_3, $P(OPh)_3$] complexes to form a large family of η^1-alkenyl complexes[17–19]. A review of this chemistry is published[20]. The reactions are both regio- and stereo-selective. The products of:

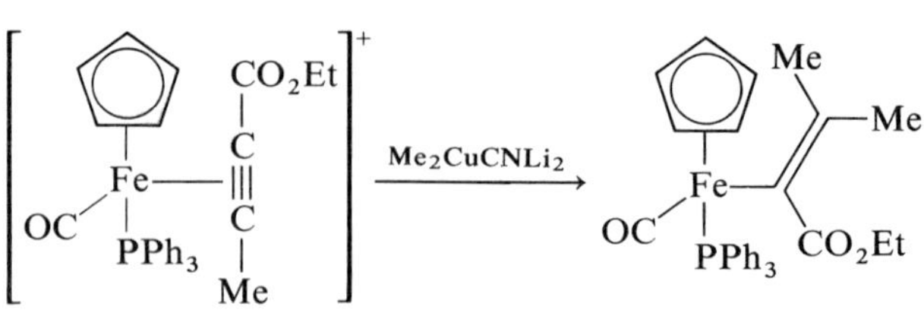

(g)

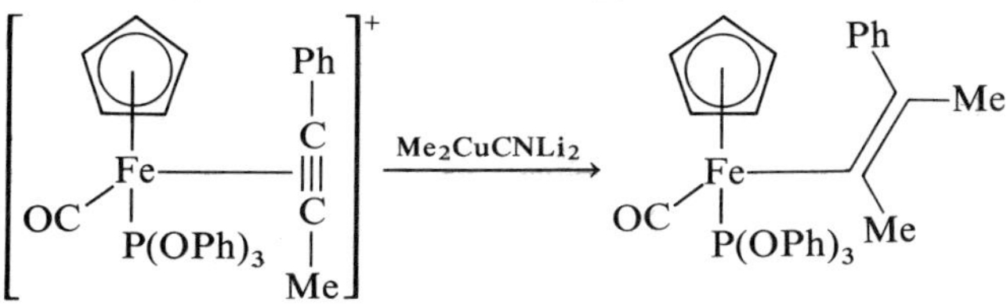
(h)

are established crystallographically[21,22]. Addition thus takes place away from an alkyne carbon atom substituted with an electron-withdrawing ester group [note also Eq. (b)] and at an alkyne carbon atom substituted with a group capable of electron donation, a result that fits with a theory of the addition of nucleophiles to π-alkene complexes[23]. Reaction (h) shows that the stereochemistry of the addition is trans. The possibility of cis addition followed by a rapid cis–trans isomerization reaction about the double bond (a process known for Ni-alkenyl complexes[24]) can be ruled out by the synthesis of the E isomer:

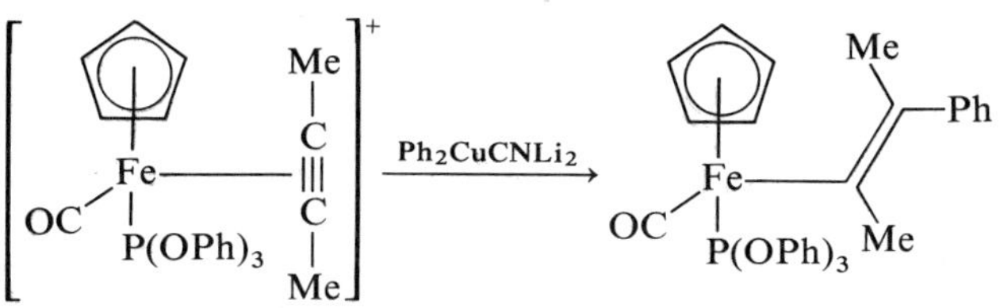
(i)

The Z and E isomers do not interconvert under the reaction conditions or isolation procedures[22].

Trans addition also takes place with most other nucleophiles[17–19], with the main exception of hydride, which adds cis:

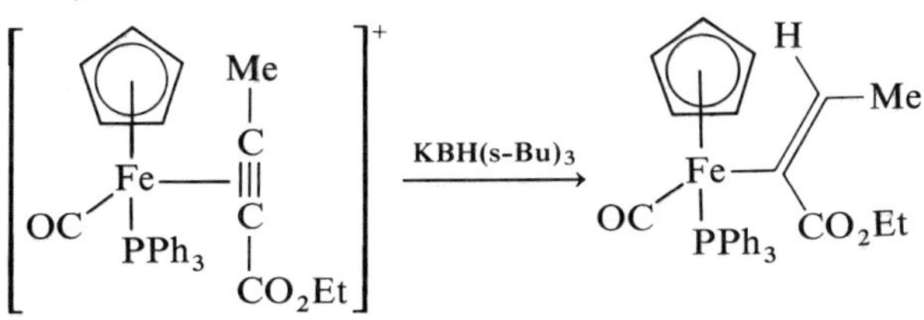
(j)

The stereochemistry is proved crystallographically[25]. Deuterium-labeling studies show that the cis-addition product arises from addition of hydride (deuteride) to the Cp ring followed by intramolecular cis addition to the alkyne of the endo hydrogen of the Cp ring formed in the first step[25]. Additional examples of this ring-substitution chemistry are observed with amides and MeLi as the nucleophile, although these reactions are indiscriminate, with the nucleophile also adding to or deprotonating the alkyne, depending on the nucleophile and the alkyne complex[26].

Alcohols also react with these Fe–π-alkyne complexes. The σ-alkenyl complex is not isolated but reacts with the proton released from the alcohol to form a π-alkene product [Fp = η^5-CpFe(CO)$_2$][27]:

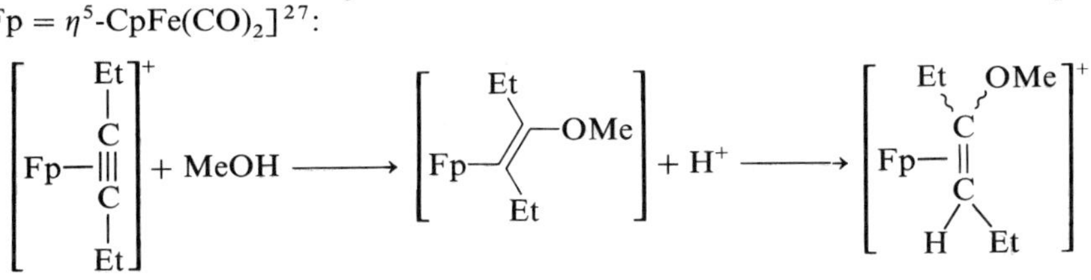
(k)

Acetylinic alcohols react similarly to form cyclic products[28,29]. Olefins also add to $[Fp\text{-}(\eta^2\text{-}RC{\equiv}CCO_2Me)]^+$ (R = H, Me) complexes to yield cyclic alkenyl complexes in certain cases[30].

Metalalkenyl complexes also form in nucleophilic addition of η^2-allene complexes. Both anionic and neutral nucleophiles react with $[Fp\text{-allene}]^+$:

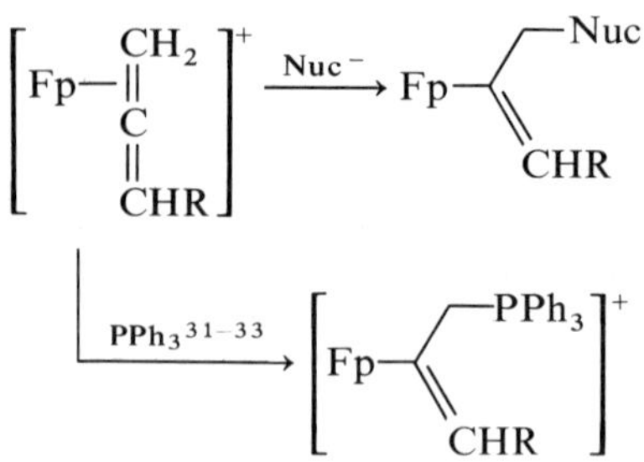

(l)

where Nuc = H, NEt_2 [31,32], $CH(CO_2Me)_2$ [34], Me (from $LiMe_2Cu$)[34], OMe[33]; R = H, Me, Ph. Starting with either the syn or anti isomer for Re = Me, a specific alkenyl product forms as expected for trans addition of the nucleophile[33,34]. Addition takes place at a terminal allene carbon and away from the R substituent. Addition at the internal allene carbon is observed in a hydration, but alkenyl products are not produced[35].

A wide range of nucleophiles add to the allene ligand in the $P(OPh)_3$-substituted systems:

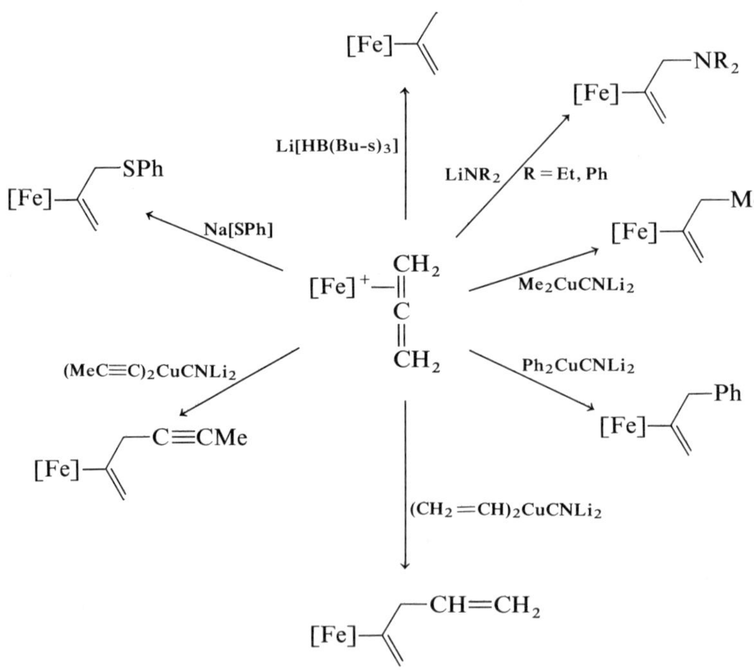

(m)

where [Fe] = η^5-CpFeCO[P(OPh)$_3$]. In this case, both hydride and amides add directly to the allene ligand. Many of these terminal alkenyl complexes undergo isomerization to yield the internal E-alkenyl isomers in the presence of alumina or trace amounts of acid[36].

σ-Yne-ol complexes, M = η^5-CpMo(CO)$_3$, rearrange to π-allene complexes and cyclize[37]:

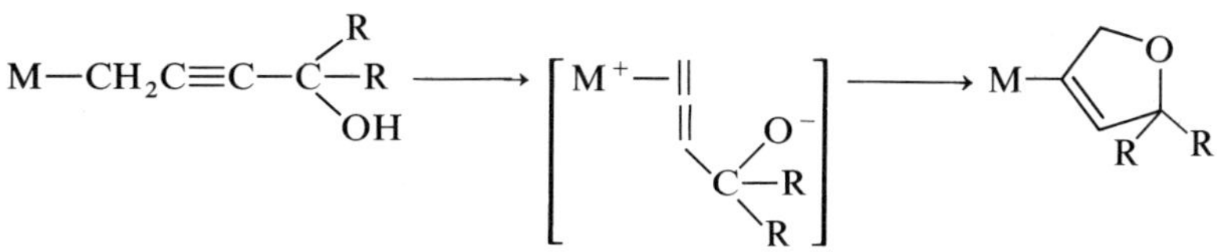

(n)

Neutral Pt-allene complexes add amines to form zwitterionic-alkenyl complexes[38]:

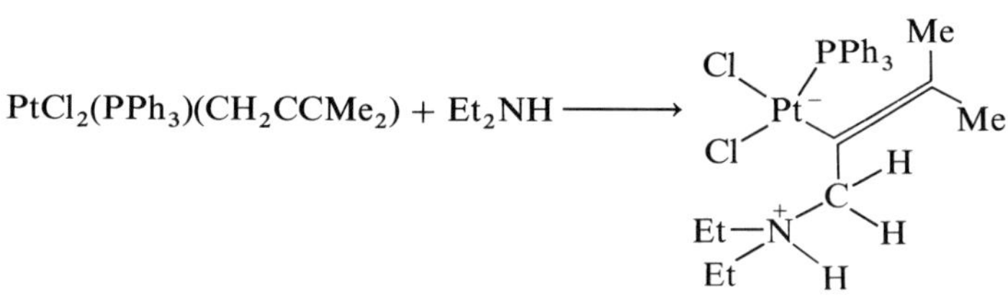

(o)

Similarly using excess primary amine, two cyclic products form, depending on the amine[39]:

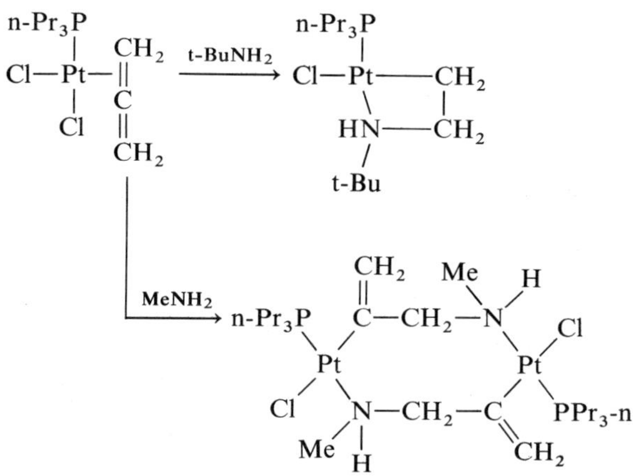

(p)

The structures of these Pt-alkenyl products are known. Other nucleophiles add to these types of Pt-alkene complexes[40–42].

(D. L. REGER)

1. M. H. Chisholm, H. C. Clark, *Inorg. Chem.*, *10*, 2557 (1971).
2. M. H. Chisholm, H. C. Clark, *J. Am. Chem. Soc.*, *94*, 1532 (1972).
3. M. H. Chisholm, H. C. Clark, L. E. Manzer, *Inorg. Chem.*, *11*, 1269 (1972).

4. M. Bottrill, M. Green, *J. Am. Chem. Soc.*, *99*, 5795 (1977).
5. S. R. Allen, P. K. Baker, S. G. Barnes, M. Botrill, M. Green, A. J. Welch, *J. Chem. Soc., Dalton Trans.*, 927 (1983).
6. M. Green, N. C. Normal, A. G. Orpen, *J. Am. Chem. Soc.*, *103*, 1267 (1981).
7. S. R. Allen, P. K. Baker, S. G. Barnes, M. Botrill, M. Green, A. G. Orpen, *J. Chem. Soc., Dalton Trans.*, 927 (1983).
8. S. R. Allen, R. G. Beevor, M. Green, N. C. Norman, A. G. Orpen, I. D. Williams, *J. Chem. Soc., Dalton Trans.*, 435 (1985).
9. R. G. Beevor, M. Green, A. G. Orpen, I. D. Williams, *J. Chem. Soc., Dalton Trans.*,
10. J. L. Templeton, B. C. Ward, *J. Am. Chem. Soc.*, *102*, 3288 (1980).
11. J. L. Davidson, I. E. P. Murray, P. N. Preston, M. V. Russo, L. Manojlovic-Muir, K. W. Muir, *J. Chem. Soc., Chem. Commun.*, 1059 (1981).
12. J. L. Davidson, I. E. P. Murray, P. N. Preston, M. V. Russo, *J. Chem. Soc., Dalton Trans.*, 1783 (1983).
13. J. R. Morrow, T. L. Tonker, J. L. Templeton, *J. Am. Chem. Soc.*, *107*, 6956 (1985).
14. J. L. Davidson, G. Vasapollo, L. Manoklovic-Muir, K. W. Muir, *J. Chem. Soc., Chem. Commun.*, 1025 (1982).
15. D. L. Reger, C. J. Coleman, P. J. McElligott, *J. Organomet. Chem.*, *171*, 73 (1979).
16. D. L. Reger, C. J. Coleman, *Inorg. Chem.*, *18*, 3155 (1979).
17. D. L. Reger, P. J. McElligott, *J. Am. Chem. Soc.*, *102*, 5923 (1980).
18. D. L. Reger, K. A. Belmore, E. Mintz, P. J. McElligott, *Organometallics*, *3*, 134 (1984).
19. D. L. Reger, S. A. Klaeren, J. E. Babin, R. D. Adams, *Organometallics*, *7*, 181 (1988).
20. D. L. Reger, *Acc. Chem. Res.*, *21*, 229 (1988).
21. D. L. Reger, P. J. McElligott, N. G. Charles, E. A. H. Griffith, E. L. Amma, *Organometallics*, *1*, 443 (1982).
22. D. L. Reger, K. A. Belmore, E. Mintz, N. G. Charles, E. A. H. Griffith, E. L. Amma, *Organometallics*, *2*, 101 (1983).
23. O. Eisenstein, R. Hoffman, *J. Am. Chem. Soc.*, *103*, 4308 (1981).
24. J. M. Huggins, R. G. Bergman, *J. Am. Chem. Soc.*, *103*, 3002 (1981).
25. D. L. Reger, K. A. Belmore, J. A. Atwood, W. E. Hunter, *J. Am. Chem. Soc.*, *105*, 5710 (1983).
26. D. L. Reger, S. A. Klaeren, L. Leboida, *Organometallics*, *5*, 1072 (1986).
27. D. J. Bates, M. Rosenblum, S. B. Samuels, *J. Organomet. Chem.*, *209*, C55 (1981).
28. D. F. Marten, *J. Chem. Soc., Chem. Commun.*, 341 (1980).
29. J. Benaim, A. L'Honore, *J. Organomet. Chem.*, *202*, C53 (1980).
30. M. Rosenblum, D. Scheck, *Organometallics*, *1*, 397 (1982).
31. D. W. Lichtenberg, A. Wojcicki, *J. Am. Chem. Soc.*, *94*, 8271 (1972).
32. D. W. Lichtenberg, A. Wojcicki, *J. Organomet. Chem.*, *94*, 311 (1975).
33. P. Lennon, M. Madhavarao, A. Rosan, M. Rosenblum, *J. Organomet. Chem.*, *108*, 93 (1976).
34. P. Lennon, A. Rosan, M. Rosenblum, *J. Am. Chem. Soc.*, *99*, 8426 (1977).
35. P. Klemarczyk, M. Rosenblum, *J. Org. Chem.*, *43*, 3488 (1978).
36. D. L. Regers, K. A. Belmore, *Organometallics*, *4*, 305 (1985).
37. I. Benaim, F. Giulieri, *J. Organomet. Chem.*, *202*, C9 (1980).
38. A.-DeRenzi, B. Di Blasio, A. Panunzi, C. Pedono, A. Vitagliana, *J. Chem. Soc., Dalton Trans.*, 1392 (1978).
39. J. R. Briggs, C. Crocker, W. S. McDonald, B. L. Shaw, *J. Chem. Soc., Dalton Trans.*, 575 (1981).
40. A.-DeRenzi, A. Panunzi, A. Saporito, A. Vitagliano, *J. Chem. Soc., Perkin Trans. II*, 993 (1983).
41. A.-DeRenzi, A. Panunzi, A. Saporito, A. Vitagliano, *J. Chem. Soc., Perkin Trans. I*, 1095 (1985).
42. A. T. Hutton, D. M. McEwan, B. L. Shaw, S. W. Wilkinson, *J. Chem. Soc., Dalton Trans.*, 2011 (1983).

5.8.2.7.7. by Metal-Atom and Related Reactions.

Although alkynes are quite reactive toward transition-metal atoms, in only few cases do stable metal–carbon bonds result from their interaction. Instead, carbon–carbon bonds form through trimerization of the alkyne to an arene. For example, Cr atoms serve as templates for the trimerization of 1-butyne, 2-butyne and 1-pentyne[1]. Cyclic diynes can be trimerized by Cr atoms to yield a unique arene–triyne derivative[2]:

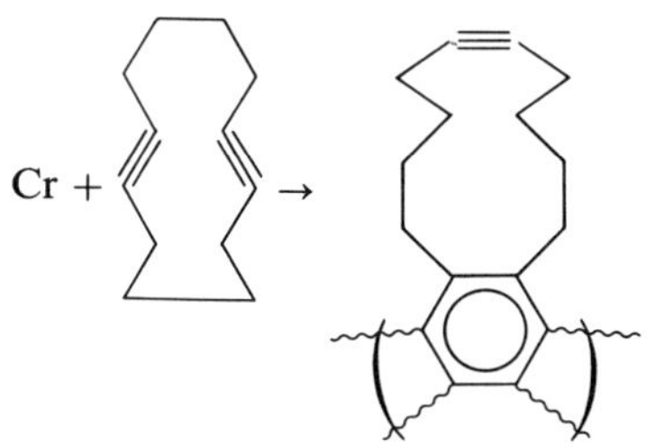

(a)

Iron atoms codeposited with alkynes sometimes yield stable organometallics of the formula[3] Fe(alkyne)$_5$ which may possess a ferrocene-like structure. Nickel atoms with normal alkynes yield an uncharacterized organometallic mixture that serves as an effective homogeneous catalyst for oligomerization of terminal alkynes to novel organic materials[4].

Hexafluoro-2-butyne shows unique complexing abilities with the later transition-metal atoms, Co, Ni, Pd and Pt. Each metal–$CF_3C{\equiv}CCF_3$ pair yields an organometallic that is stable but tends toward decomposition to butyne trimer and metal particles[5]. With Ni and Pd, 1:1 metal:butyne complexes are isolated as solvent adducts. The structures are unknown, but with Pd, a dimer is formed[5].

Addition of CO to the $CF_3C{\equiv}CCF_3$–Ni or –Pd complexes (no solvent) before warmup of the matrices forms $(CF_3C{\equiv}CCF_3)Ni(CO)_2$ and $(CF_3C{\equiv}CCF_3)Pd(CO)_2$. These are volatile, unstable complexes that upon workup spontaneously convert to cluster complexes[5]:

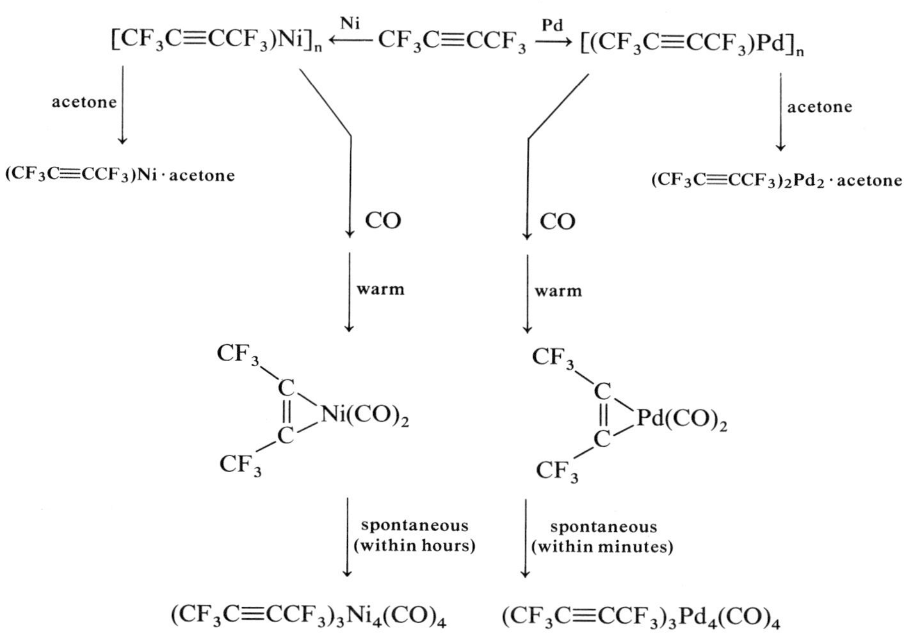

TABLE 1. MAJOR PRODUCTS FORMED IN Co ATOM–ALKYNE–CYCLOPENTADIENE REACTIONS

Alkyne	Diene	Products	Yield (%)
$CF_3C{\equiv}CCF_3$	C_5H_6	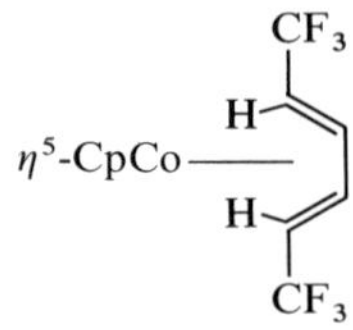	13
		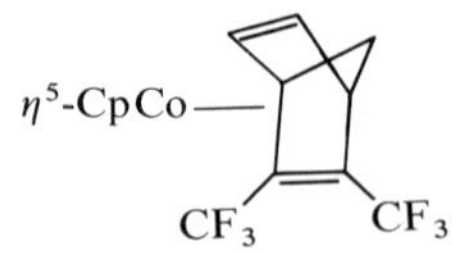	4
		$(\eta^5\text{-Cp})_3Co_3(CF_3C{\equiv}CCF_3)$	5
$CF_3C{\equiv}CCF_3$	$(CH_3)_5C_5H$	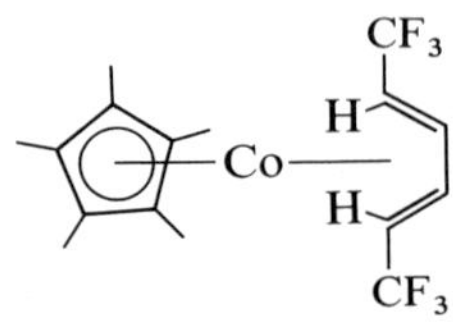	17
		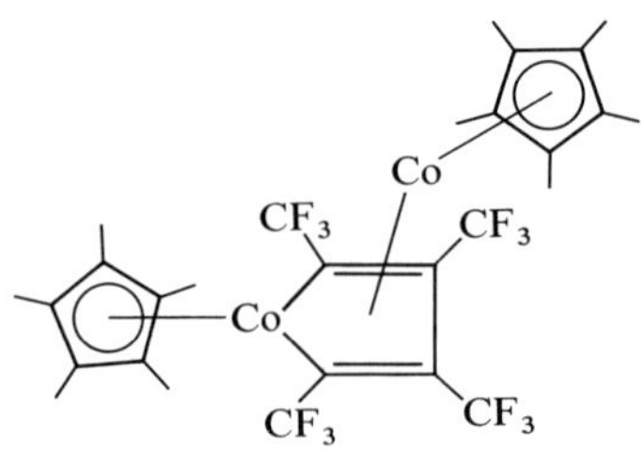	13
$C_6H_5C{\equiv}CC_6H_5$	C_5H_6	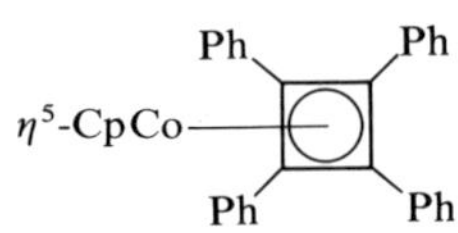	6
		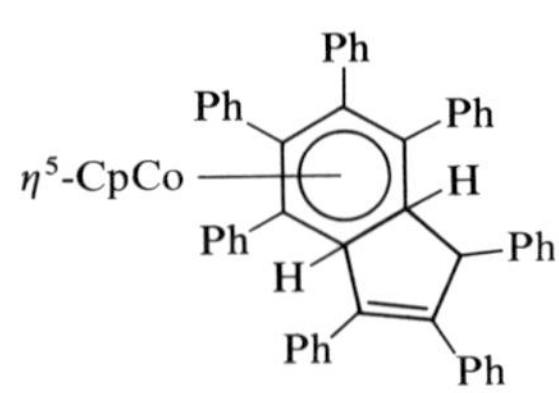	7

Mixtures of alkynes and cyclopentadiene with Co atoms yield a variety of air-stable organometallics[6]. These result from hydrogen transfer from cyclopentadiene, dimerization or trimerization of alkyne, clustering of Co atoms and simple addition. The nature of the products implies that a variety of reactive intermediates must be formed en route. Table 1 shows the major products that are formed.

Mixtures of alkynes, cyclopentadiene and boranes with Co atoms yield metallocarborane clusters in small amounts[7,8]; e.g., with 2-butyne and B_5H_9:

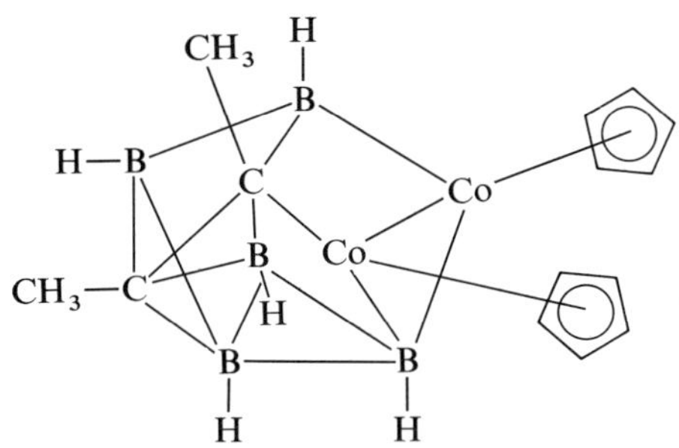

Several similar examples are reported including metalloborane clusters[8].

A review of metal-atom–alkyne reactions has appeared[9].

(K. J. KLABUNDE)

1. P. S. Skell, D. L. Williams-Smith, M. J. McGlinchey, *J. Am. Chem. Soc.*, *95*, 3337 (1973).
2. J. A. Gladysz, J. G. Fulcher, S. J. Lee, A. B. Bocarsley, *Tetrahedron Lett.*, 3421 (1977).
3. L. H. Simons, J. J. Lagowski, *Proc. Int. Conf. Organomet. Chem.*, *8th, 1977* Pap. 5BOY (1977).
4. L. H. Simons, J. J. Lagowski, *J. Org. Chem.*, *43*, 3247 (1978).
5. K. J. Klabunde, T. Groshens, M. Brezinski, W. Kennelly, *J. Am. Chem. Soc.*, *100*, 4437 (1978).
6. M. B. Freeman, L. W. Hall, L. G. Sneddon, *Inorg. Chem.*, *19*, 1132 (1980).
7. L. W. Hall, G. J. Zimmerman, L. G. Sneddon, *J. Chem. Soc., Chem. Commun.*, 45 (1977).
8. G. J. Zimmerman, L. W. Hall, L. G. Sneddon, *Inorg. Chem.*, *19*, 3642 (1980).
9. R. W. Zoellner, K. J. Klabunde, *Chem. Rev.*, *84*, 545 (1984).

Abbreviations

abs	absolute
a.c.	alternating current
Ac	acetyl, CH_3CO
acac	acetylacetonate anion
acacH	acetylacetone, $CH_3C(O)CH_2C(O)CH_3$
AcO	acetate anion, $CH_3C(O)O$
Ad	adamantyl
ads	adsorbed
AIBN	2,2′-azobis(isobutyronitrile), 2,2′-$[(CH_3)_2CCN]_2N_2$
Alk	alkyl
am	amine
amt	amount
Am	amyl, C_5H_{11}
amu	atomic mass unit
anhyd	anhydrous
aq	aqueous
Ar	aryl
asym	asymmetrical, asymmetric
at	atom (not atomic, except in atomic weight)
atm	atmosphere (not atmospheric)
av	average
BBN	9-Borabicyclo[3.3.1]nonane
bcc	body-centered cubic
BD	butadiene
bipy	2,2′-bipyridyl
bipyH	protonated 2,2′-bipyridyl
bp	boiling point
Bu	butyl, C_4H_9
Bz	benzyl, $C_6H_5CH_2$
c-	cyclo (used in formulas)
ca.	circa, about, approximately
catal	catalyst (not catalyzing, catalysis, catalyzed, etc.)
CDT	cyclododecatriene
cf.	compare
Ch.	chapter
CHD	1,3-cycloheptadiene
Chx	cyclohexyl
ChxD	1,3-cyclohexadiene
CI	configuration interaction
Cob	cobalamine
COD	cyclooctadiene
COE	cyclooctene
conc	concentrated (not concentration)
const.	constant
COT	cyclooctatriene
COTe	cyclooctatetraene
Cp	cyclopentadienyl, C_5H_5
CPE	controlled-potential electrolysis

cpm	counts per minute
CT	charge-transfer
CV	cyclic voltammetry
CVD	chemical vapor deposition
CW	continuous wave
d	day, days
DABIP	N,N′-diisopropyl-1,4-diazabutadiene
DBA	dibenzylideneacetone
d.c.	direct current
DCM	dicyclopentadienylmethane
DCME	$Cl_2CHC(O)CH_3$
DCP	1,3-dicyclopentadienylpropane
DDT	dichlorodiphenyltrichloroethane, 1,1,1,′-trichloro-2,2-bis-(4-chlorophenyl)ethane
dec	decomposed
DED	1,1-bis(ethoxycarbonyl)ethene-2,2-dithiolate, $[[(H_5C_2OC(O)]_2C{=}CS_2]^2$
depe	1,2-bis(diphenylphosphino)ethene, $(C_6H_5)_2PCH{=}CHP(C_6H_5)_2$
DIAD	diindenylanthracenyl
diars	1,2-bis(dimethylarsino)benzene, o-phenylenebis(dimethylarsine), 1,2-$(CH_3)_2AsC_6H_4As(CH_3)_2$
dien	diethylenetriamine, $[H_2N(CH_2)_2]_2NH$
diglyme	diethyleneglycol dimethylether, $CH_3O(CH_2CH_2O)CH_2$
dil	dilute
diop	2,3-O-isopropylidene-2,3-dihydroxy-1,4-bis(diphenylphosphino)butane, $(C_6H_5)_2PCH_2CH[OCH(CH_3){=}CH_2]CH$ $[OCH(CH_3){=}CH_2]CH_2P(C_6H_5)_2$
dipda	p-i-$PrC_6H_4CH{=}CHC_6H_4$-c-p
diphos	1,2-bis(diphenylphosphino)benzene, 1,2-$(C_6H_5)_2PC_6H_4P(C_6H_5)_2$
Div.	division
DMA	dimethylacetamide
dme	dropping mercury electrode
DME	1,2-dimethoxyethane, glyme, $CH_3O(CH_2)_2OCH_3$
DMF	N,N-dimethylformamide, $HC(O)N(CH_3)_2$
DMG	dimethylglyoxime, $CH_3C({=}NOH)C({=}NOH)CH_3$
DMP	1,2-dimethoxybenzene, 1,2-$(CH_3O)_2C_6H_4$
dmpe	1,2-bis(dimethylphosphino)ethane, $(CH_3)_2P(CH_2)_2P(CH_3)_2$
DMSO	dimethylsulfoxide, $(CH_3)_2SO$
dpam	bis(diphenylarsino)methane, $[(C_6H_5)_2As]_2CH_2$
dpic	dipicolinate ion
DPP	differential pulse polarography
dppb	1,4-bis(diphenylphosphino)butane, 1,4-$(C_6H_5)_2P(CH_2)_4P(C_6H_5)_2$
dppe	1,2-bis(diphenylphosphino)ethane, 1,2-$(C_6H_5)_2P(CH_2)_2P(C_6H_5)_2$
dppm	bis(diphenylphosphino)methane, $[(C_6H_5)_2P]_2CH_2$
dppoe	bis(diphenylphosphoryl)ethane
dppp	1,3-bis(diphenylphosphino)propane, 1,3-$(C_6H_5)_2P(CH_2)_3P(C_6H_5)_2$

dptpe	1,2-bis(di-p-tolylphosphino)ethane, 1,2-(4-$CH_3C_6H_4)_2P(CH_2)_2P$ $(C_6H_4CH_3$-4$)_2$
DTA	differential thermal analysis
DTBQ	3,5-di-t-butyl-o-benzoquinone
DTH	1,6-dithiahexane, butane-1,4-dithiol, 1,4-$HS(CH_2)_4SH$
DTS	dithiosquarate
ed.	edition, editor
eds.	editors
EDTA	ethylenediaminetetraacetic acid, $[HOC(O)]_2N(CH_2)_2N[C(O)OH]_2$
e.g.	exempli gratia, for example
EHMO	extended Hückel molecular orbital
emf	electromotive force
en	ethylenediamine, $H_2N(CH_2)_2NH_2$
enH	protonated ethylenediamine
EPR	electron paramagnetic resonance
equimol	equimolar
equiv	equivalent
EPR	electron paramagnetic resonance
Eq.	equation
ERF	effective reduction factor
ES	excited state
ESR	electron-spin resonance
esu	electrostatic unit
Et	ethyl, CH_3CH_2
etc.	et cetera, and so forth
Et_2O	diethyl ether, $(C_2H_5)_2O$
EtOH	ethanol, C_2H_5OH
et seq.	et sequentes, and the following
eu	entropy unit
fac	facial
Fc	ferrocenyl
fcc	face-centered cubic
ff.	following
Fig	figure
Fl	fluorenyl
Fp	η^5-$C_5H_5Fe(CO)_2$
fp	freezing point
g	gas
g-at	gram-atom
GLC	gas–liquid chromatography
glyme	1,2-dimethoxyethane, $CH_3O(CH_2)_2OCH_3$
graph	graphite
GS	ground state
h	hour, hours
H-Cob	cobalamine
HD	1,5-hexadiene
hept	heptyl
Hex	hexyl
HMDB	hexamethyl(Dewar benzene)
hmde	hanging mercury drop electrode
MHI	heptamethylindenyl

HMPA	hexamethylphosphoramide, $[(CH_3)_2N]_3PO$
HOMO	highest occupied molecular orbital
HPLC	high-pressure liquid chromatography
i.e.	id est, that is
Im	imidazole
inter alia	among other things
IPC	isopinocamphylborane
IR	infrared
irrev	irreversible
ISC	intersystem crossing
isn	isonicotinamide
l	liquid
L	ligand
LC	ligand centered
LF	ligand field
LFER	linear free-energy relationship
liq	liquid
LMCT	ligand-to-metal charge transfer
Ln	lanthanides, rare earths
LSV	linear-scan voltammetry
LUMO	lowest unoccupied molecular orbital
m	meta
max	maximum
M	metal
MC	metal centered
Me	methyl, CH_3
Men	menthyl
mes	mesitylene, 1,3,5-trimethylbenzene derivative
MeOH	methanol, CH_3OH
mer	meridional; the repeating unit of an oligomer or polymer
mhp	2-hydroxy-6-methylpyridine, 2-HO, 6-$CH_3C_5H_3N$
min	minimum, minute, minutes
MLCT	metal-to-ligand charge transfer
MO	molecular orbital
mol	molar
mp	melting point
MV	methyl viologen, 1,1′-dimethyl-4,4′-bipyridinium dichloride
n.a.	not available
napy	naphthyridine
NBD	norbornadiene, [2.2.1]bicyclohepta-2,5-diene
neg	negative
nhe	normal hydrogen electrode
NMR	nuclear magnetic resonance
No.	number
np	tris-[2-(diphenylphosphino)ethyl]amine, $N[CH_2CH_2P(C_6H_5)_2]_3$
Np	naphthyl
Nuc	nucleophile
NPP	normal pulse polarography
NQR	nuclear quadrupole resonance
NTA	nitrilotriacetate
o	ortho
obs	observed

Oct	octyl
ocp	octaethylporphyrin
O_F	oxidation factor
O_h	octahedral
Oq	oxyquinolate
ox.	oxidation
p	para
p.	page
P	pressure
Pat.	patent
pet.	petroleum
Ph	phenyl, C_6H_5
phen	1,10-phenanthroline
Ph_2PPy	2-(diphenylphosphino)pyridine, 2-$(C_6H_5)_2PC_5H_4N$
pip	piperidine, $C_5H_{10}N$
PMDT	pentamethyldiethylenetriamine, $(CH_3)_2N(CH_2)_2N(CH_3)(CH_2)_2N(CH_3)_2$
PMR	proton magnetic resonance
pn	propylene-1,3-diamine, 1,3-$H_2NCH_2CH_2CH_2NH_2$
pos	positive
Po-tol_3	tri-o-tolylphosphine
pp.	pages
ppb	parts per billion
ppm	parts per million
ppn	bis(diphenylphosphino)amine, $[(C_6H_5)_2P]_2NH$
ppt	precipitate
Pr	propyl, C_3H_2
PSS	photostationary state
PVC	poly(vinyl chloride)
py	pyridine, C_5H_5N
pyr	pyrazine
pz	pyrazolyl
PZE	potential of zero charge
rac	racemic mixture, racemate
R	organic group; universal gas constant
RDE	rotated disk electrode
RE	rare earths, lanthanides
red.	reduction
Redox	reduction-oxidation reactions
ref.	reference
rev	reversible
rf	radiofrequency
RF	reduction factor
R_F	R group with substituted F
rh	rhombohedral
rms	root mean square
rpm	revolutions per minute
RT	room temperature
s	second, seconds; solid
sce	saturated calomel electrode
SCE	standard calomel electrode
sec	secondary
Sep	sepulcrate, 1,3,6,8,10,13,16,19-octaazabicyclo[6.6.6]eicosane

Sia	Diisamyl
SMAD	solvated metal-atom dispersed
soln	solution
solv	solvated
sp	specific
STP	standard temperature and pressure
subl	sublimes
Suppl.	supplement
sym	symmetrical, symmetric
t	time; tertiary
T	temperature
T_d	tetrahedral
TCNE	tetracyanoethylene
TEA	tetraethylammonium ion, $[(C_2H_5)_4N]^+$
terpy	2,2′2″-terpyridyl
tetraphos	$Ph_2PCH_2CH_2PPhCH_2CH_2PPhCH_2CH_2PPh_2$
TGA	thermogravimetric analysis
TGL	triethyleneglycol dimethylether
THF	tetrahydrofuran
THP	tetrahydropyran
THT	tetrahydrothiophene
Thx	thexyl
TLC	thin-layer chromatography
TMED	N,N,N′,N′-tetramethylethylenediamine, $(CH_3)_2N(CH_2)_2N(CH_3)_2$
tmen	N,N,N′,N′-tetramethylethylenediamine
TMP	2,2,6,6-tetramethylpiperidyl
TMPH	2,2,6,6-tetramethylpiperidine, 2,2,6,6-$(CH_3)_4C_5H_7N$
Tol	tolyl, $C_6H_4CH_3$, p-tolyl
Tos	tosyl, tolylsulfonyl, 4-$CH_3C_6H_4SO_2$
TPA	tetraphenylarsonium ion, $[(C_6H_5)_4As]^+$
TPPO	triphenylphosphineoxide
triars	bis-[-(dimethylarsino)phenyl]methylarsine, $[2\text{-}(CH_3)_2AsC_6H_4]_2AsCH_3$
triphos	1,1,1-tris(diphenylphosphinomethyl)ethane, $[(C_6H_5)_2PCH_2]_3CCH_3$
trien	triethylenetetraamine, $H_2N(CH_2)_2NH(CH_2)_2NH(CH_2)_2NH_2$
UV	ultraviolet
v	vicinal
Vi	(E)-$[2\text{-}(CH_3)_2NCH_2C_6H_4]C{=}C(CH_3)C_6H_4CH_3$-4
viz.	videlical, that is to say, namely
vol., Vol.	volume
VPE	vapor-phase epitaxy
vs.	versus
wk.	week
wt	weight
X	halogen or pseudohalogen
xs	excess
Y	often used for S, Se
yr.	year
§	section
η	hapto designator

Author Index

The entries of this index were derived directly by computer program from the lists of references. The accuracy of the references was the sole responsibility of the authors. No editorial check, except for format and journal-title abbreviation,was applied. Consequently, errors occurring in authors' names in the references will recur in this index.
Each entry in the index refers to the appropriate section number. Sections 5.8 - 5.8.2.7.7 are in volume 12A and Sections 5.8.2.8 - 5.10.4.2 are in volume 12B.

B

D

F

G

H

I

J

L

M

N

P

Q

W

X

Y

Compound Index

This index lists individual, fully specified compositions of matter that are mentioned in the text. It is an index of empirical formulas, ordered according to the following system: the elements within a given formula occur in alphabetical sequence except for C, or C and H if present, which always come first. The formulas are ordered alphanumerically without exception.

The index is augmented by successively permuted versions of all empirical formulas. As an example, $C_3H_3AlO_9$ will appear as such and, at the appropriate positions in the alphanumeric sequence, as $H_3AlO_9*C_3$, $AlO_9*C_3H_3$ and $O_9*C_3H_3Al$. The asterisk identifies a permuted formula and allows the original formula to be reconstructed by shifting to the front the elements that follow the asterisk.

Whenever an empirical formula does not show how the elements are combined in groups, it is followed by a linearized structural formula, which reveals the connectivity of the compound(s) underlying the empirical formula and serves to distinguish substances which are identical in composition but differ in the arrangement of elements.

The nonpermuted empirical formulas are followed by keywords. They describe the context in which the compounds represented by the empirical formulas are discussed. Section numbers direct the reader to relevant positions in the book.

Subject Index

This index supplements the compound index and the table of contents by providing access to the text by way of methods, techniques, reaction conditions, properties, effects and other phenomena. Reactions of specific bonds and compound classes are noted when they are not accessed by the heading of the section in which they appear.

For multiple entries, additional keywords indicate contexts and thereby avoid the retrieval of information that is irrelevant to the user's need.

Section numbers are used to direct the reader to those positions in the volume where substantial information is to be found. Sections 5.8 - 5.8.2.7.7 are in volume 12A and Sections 5.8.2.8 - 5.10.4.2 are in volume 12B.

G

H

I

K

L

M

N

O

Periodic Tab

Period	Group IA	Group IIA	Group IIIA	Group IVA	Group VA	Group VIA	Group VIIA	
1 1*s*	1 **H**							
2 2*s*2*p*	3 **Li**	4 **Be**						
3 3*s*3*p*	11 **Na**	12 **Mg**						
4 4*s*3*d* 4*p*	19 **K**	20 **Ca**	21 **Sc**	22 **Ti**	23 **V**	24 **Cr**	25 **Mn**	26 **Fe**
5 5*s*4*d* 5*p*	37 **Rb**	38 **Sr**	39 **Y**	40 **Zr**	41 **Nb**	42 **Mo**	43 **Tc**	44 **Ru**
6 6*s* (4*f*) 5*d* 6*p*	55 **Cs**	56 **Ba**	57* **La**	72 **Hf**	73 **Ta**	74 **W**	75 **Re**	76 **Os**
7 7*s* (5*f*) 6*d*	87 **Fr**	88 **Ra**	89** **Ac**	104	105	106	107	108

*Lanthanide series 4*f*	58 **Ce**	59 **Pr**	60 **Nd**	61 **Pm**	62 **Sm**	63 **Eu**
**Actinide series 5*f*	90 **Th**	91 **Pa**	92 **U**	93 **Np**	94 **Pu**	95 **Am**

Adapted from F.A. Cotton, G. Wilkinson, *Advanced Inorganic Che*